RNA-Protein Interaction Protocols

METHODS IN MOLECULAR BIOLOGY™

For other titles published in this series, go to
www.springer.com
select the subdiscipline
search for your title

METHODS IN MOLECULAR BIOLOGY™

RNA-Protein Interaction Protocols

Edited by

Ren-Jang Lin

Division of Molecular Biology, Beckman Research Institute of the City of Hope, Duarte, California

Editor
Ren-Jang Lin
Division of Molecular Biology
Beckman Research Institute of the City of Hope
Duarte, CA
USA

Series Editor
John M. Walker
School of Life Sciences
University of Hertfordshire
Hatfield, Hertfordshire Al10 9 AB
UK

ISBN: 978-1-61737-555-2 e-ISBN: 978-1-60327-475-3
DOI: 10.1007/978-1-60327-475-3

Printed on acid-free paper

9 8 7 6 5 4 3 2 1

springer.com

Preface

"Real answers need to be found in dialogue and interaction"

– *Malcolm Boyd*

RNA and protein are the two main workhorses in the cell and, quite often, they function together to accomplish a biological task. For example, the maintenance of chromosome ends by the telomerase (Chapter 2), the splicing of messenger RNA precursors by the spliceosome (Chapters 1, 4, 5, 12, 13, 14, 16), the processing of tRNA transcripts by RNase P (Chapters 3, 19), the protein synthesis by the ribosome (Chapters 7, 8, 9, 17), and the RNA interference by the RISC complex (Chapters 20, 24) all involve RNA-protein complexes or ribonucleoproteins (RNPs). Thus, a protocol book that describes the detail of experimental procedure used to study the components and the interacting partners within an RNA-protein complex should find a home in many laboratories. The protocols collected in the two editions of *RNA-Protein Interaction Protocols* serve such a purpose.

The first edition, nicely put together by Susan Haynes, who is now at the National Institute of General Medical Sciences, describes many techniques commonly used for studying RNA and RNA-protein interaction. Those protocols are still extremely helpful when purifying RNPs, identifying interacting partners, analyzing interaction details between partners, or assaying the function of RNPs.

This second edition updates a number of techniques described in the first edition but is mainly focused on techniques established or refined in the past few years. The book is generally organized in the following order: the isolation and identification of RNP components, the analysis and measurement of RNA-protein interaction, and related novel techniques and strategies. The two editions are not redundant but complement each other.

Chapter 1 describes the affinity purification of proteins that bind to specific RNA that is tagged with biotin and bound to streptavidin beads. Chapter 2 describes the affinity purification of an entire RNP using 2'-O-methyl antisense RNA followed by elution with a displacement oligo. Chapter 3 describes the affinity purification of proteins that bind *in vivo* to specific RNA fused to an aptamer that can bind to streptavidin or Sephadex. Chapter 4 describes the reconstitution of RNA-protein complexes *in vitro* using synthetic RNA followed by isolation of the RNP through gradient sedimentation. Chapter 5 describes tagging a protein component *in vivo* with an epitope and the isolation of the entire RNP *in vitro* using the antibody; the intact RNP is then eluted from the antibody resin with the epitope peptide under mild conditions.

Selection of RNA from a pool that binds to a specific protein is described in two chapters. Jensen and Darnell describe the isolation of all cellular RNAs that physically interact *in vivo* with a protein of interest (Chapter 6). The protocol combines UV-induced crosslinking *in vivo*, the isolation of RNA-protein crosslinks, and the amplification and identification of the RNA. Stelzl and Nierhaus describe the generation of a pool of many different RNA segments from a large, known RNA and the selection from that pool of RNA sequences that bind to the interacting protein partner (Chapter 17).

Chapters 7 and 8 describe the measuring of specificity and strength of the interaction between RNA and protein using gel mobility shift and isothermal titration calorimetry methods, respectively.

Nucleotide analogue footprinting or interference is described in two chapters: A. Özlem Tastan Bishop et al. describe the use of phosphorothioate analogues to map the RNA residues that contribute to the interaction with the protein *in vitro* (Chapter 9), while Weinstein Szewczak uses a similar technique to study the interaction *in vivo* (Chapter 10).

Crosslinking induced by photoactivation is an effective way to study the physical interaction between RNA and protein. Jensen and Darnell use crosslinking to isolate RNAs that interact with a known protein (Chapter 6). Here are chapters that use crosslinking to map the protein domains that interact with a given RNA. Banerjee and Singh describe a procedure to use crosslinking and chemical cleavage of the protein to map the protein domain that contacts the RNA (Chapter 12). Turner et al. describe an elegant method that randomly inserts TEV protease cleavage sites into the protein for fine mapping purpose (Chapter 14). Gaur describes a protocol to efficiently incorporate azido-adenine to RNA to expand and aid in crosslinking studies (Chapter 11).

Chemical or protein nucleases are used to map the contact between RNA and protein. Ken and MacMillan tether iron-EDTA to the amino-termial cysteine residue in a protein of interest; hydrogen peroxide is then added for cleavage of the RNA in contact as a way to map the interacting nucleotides (Chapter 13). Trang and Liu describe the use of ribonucleases and iron-EDTA to cleave a naked or a protein-bound RNA (Chapter 19). Chen et al. describe a protocol utilizing terminal transferase and PCR to analyze *in vivo* RNAs that are resistant to ribonuclease due to structural constrain or protein binding (Chapter 21).

Mass spectrometry is used to analyze protein contacts in RNPs. Kvaratskhelia and Le Grice compare the accessibility to biotinylation of lysine residues in a protein with or without the bound RNA; the biotinylated and unbiotinylated lysine-containing peptides are distinguished by mass spec (Chapter 15). Urlaub et al. crosslink the RNA and the protein and then determine the crosslinking sites by measure the mass of the oligonucleotide-peptide crosslinks (Chapter 16).

A protocol to study the export of RNA from nucleus to the cytoplasm and the association of the RNA with specific proteins in vivo is described in Chapter 18. Accurate quantification of highly homologous or isomeric RNAs is difficult. Eis and Garcia-Blanco describe a method to overcome the problem, which can also be used to measure microRNAs (Chapter 20). A protocol to measure the disruption of RNA duplex or RNA-protein interaction by RNA helicases is described in Chapter 22. Preparation and comparison of extracts from nuclei, cytoplasm, or whole cells for pre-mRNA splicing are described in Chapter 23. Protocols to design effective siRNAs to down regulate RNA-binding proteins are described in Chapter 24. A protocol to isolate proteins that are involved in a specific RNA metabolism, without going through biochemical purification, is described in Chapter 25. The method takes advantage of screening a library that contains essentially the entire protein set from the yeast genome.

I am indebted to all the authors, who contributed their refined and detailed protocols to this book, as well as their patience with me during the editing stages. I am grateful to John M. Walker, the series editor, who gave me this opportunity and guided me patiently through the entire process. I thank Faith Osep who compiled the manuscripts and corresponded with the authors.

Ren-Jang Lin
March 2008

Contents

Contributors

HIREN BANERJEE • *Department of Molecular, Cellular, and Developmental Biology, University of Colorado at Boulder, Boulder, Colorado, USA*
MARK BEHLKE • *Integrated DNA Technologies (IDT), Coralville, Iowa, USA*
A. ÖZLEM TASTAN BISHOP • *Bioinformatics and Computational Biology Unit, Department of Biochemistry, University of Pretoria, Pretoria, South Africa*
HSIU-HUA CHEN • *Biology Division, Beckman Research Institute of the City of Hope, Duarte, California, USA*
MARK J. CHURCHER • *MRC Laboratory of Molecular Biology, Hills Road, Cambridge, United Kingdom*
GAËL CRISTOFARI • *INSERM, Ecole Normale Supérieure de Lyon, Lyon, France*
ROBERT B. DARNELL • *Howard Hughes Medical Institute and the Laboratory of Molecular Neuro-Oncology, The Rockefeller University, New York, New York, USA*
KENNETH J. DERY • *Molecular Biology Division, Beckman Research Institute of the City of Hope, Duarte, California, USA*
GIDEON DREYFUSS • *Howard Hughes Medical Institute and Department of Biochemistry and Biophysics, University of Pennsylvania School of Medicine, Philadelphia, Pennsylvania, USA*
PEGGY S. EIS • *Roche NimbleGen, Inc., Madison, Wisconsin, USA*
DAVID R. ENGELKE • *Department of Biological Chemistry, University of Michigan, Ann Arbor, Michigan, USA*
MARGARET E. FAIRMAN • *Department of Biochemistry and Center for RNA Molecular Biology School of Medicine, Case Western Reserve University, Cleveland, Ohio, USA*
MARIANO A. GARCIA-BLANCO • *Departments of Molecular Genetics and Microbiology and of Medicine, Duke University Medical Center, Durham, North Carolina, USA*
RAJESH K. GAUR • *Molecular Biology Division, Beckman Research Institute of the City of Hope, Duarte, California, USA*
ELIZABETH J. GRAYHACK • *Department of Biochemistry and Biophysics, University of Rochester School of Medicine, Rochester, New York, USA*
JANE E. JACKMAN • *Department of Biochemistry and Biophysics, University of Rochester School of Medicine, Rochester, New York, USA*
ECKHARD JANKOWSKY • *Department of Biochemistry and Center for RNA Molecular Biology School of Medicine, Case Western Reserve University, Cleveland, Ohio, USA*

KIRK B. JENSEN • *School of Molecular and Biomedical Science and the Centre for the Molecular Genetics of Development, University of Adelaide, Adelaide, Australia*
NAOYUKI KATAOKA • *Tokyo Medical and Dental University, Medical Research Institute, Tokyo, Japan*
OLIVER A. KENT • *Department of Biochemistry, University of Alberta, Alberta, Canada*
DONG-HO KIM • *Molecular Biology Division, Beckman Research Institute of the City of Hope, Duarte, California, USA*
EVA KÜHN-HÖLSKEN • *Bioanalytical Mass Spectrometry Group, Max-Planck-Institute for Biophysical Biochemistry, Göttingen, Germany*
ISABEL KURTH • *The O'Donnell Laboratory, Rockefeller University, New York, New York, USA*
MAMUKA KVARATSKHELIA • *Center for Retrovirus Research and Comprehensive Cancer Center, College of Pharmacy, The Ohio State University, Columbus, Ohio, USA*
JEANNE LEBON • *Biology Division, Beckman Research Institute of the City of Hope, Duarte, California, USA*
STUART F.J. LE GRICE • *RT Biochemistry Section, HIV Drug Resistance Program, National Cancer Institute at Frederick, Frederick, Maryland, USA*
REN-JANG LIN • *Molecular Biology Division, Beckman Research Institute of the City of Hope, Duarte, California, USA*
JOACHIM LINGNER • *Swiss Institute for Experimental Cancer Research (ISREC), Epalinges s/Lausanne, Switzerland*
FENYONG LIU • *Division of Infectious Diseases, School of Public Health, University of California at Berkeley, Berkeley, California, USA*
REINHARD LÜHRMANN • *Department of Cellular Biochemistry, Max-Planck-Institute for Biophysical Chemistry, Göttingen, Germany*
NAOTO MABUCHI • *Institute for Virus Research, Kyoto University, Japan*
ANDREW M. MACMILLAN • *Department of Biochemistry, University of Alberta, Alberta, Canada*
KAORU MASUYAMA • *Institute for Virus Research, Kyoto University, Japan*
ANDREW J. NEWMAN • *MRC Laboratory of Molecular Biology, Cambridge, United Kingdom*
KNUD H. NIERHAUS • *Max-Planck-Institut für Molekulare Genetik, AG Ribosomen, Berlin, Germany*
CHRIS M. NORMAN • *MRC Laboratory of Molecular Biology, Cambridge, United Kingdom*
MUTSUHITO OHNO • *Institute for Virus Research, Kyoto University, Japan*

Markus Pech • *Max-Planck-Institut für Molekulare Genetik, AG Ribosomen, Berlin, Germany*
Eric M. Phizicky • *Department of Biochemistry and Biophysics, University of Rochester School of Medicine, Rochester, New York, USA*
Michael I. Recht • *Palo Alto Research Center, Palo Alto, California, USA*
Arthur D. Riggs • *Biology Division, Beckman Research Institute of the City of Hope, Duarte, California, USA*
John J. Rossi • *Molecular Biology Division, Beckman Research Institute of the City of Hope, Duarte, California, USA*
Sean P. Ryder • *Department of Biochemistry and Molecular Pharmacology, UMass Medical School, Worcester, Massachusetts, USA*
Felicia H. Scott • *Department of Biological Chemistry, University of Michigan, Ann Arbor, Michigan, USA*
Shalini Sharma • *Howard Hughes Medical Institute, University of California at Los Angeles, Los Angeles, California, USA*
Ravinder Singh • *Department of Molecular, Cellular, and Developmental Biology, University of Colorado at Boulder, Boulder, Colorado, USA*
Chatchawan Srisawat • *Department of Biochemistry, Mahidol University, Bangkok, Thailand*
Ulrich Stelzl • *Max-Planck-Institut für Molekulare Genetik, AG Ribosomen, Berlin, Germany*
Scott W. Stevens • *Section in Molecular Genetics and Microbiology, Institute for Cellular and Molecular Biology, University of Texas at Austin, Austin, Texas, USA*
Lara B. Weinstein Szewczak • *Department of Molecular Biophysics and Biochemistry, Yale University, New Haven, Connecticut, USA*
Phong Trang • *Division of Infectious Diseases and Immunity, Program of Comparative Biochemistry, School of Public Health, University of California at Berkeley, Berkeley, California, USA*
Ian A. Turner • *MRC Laboratory of Molecular Biology, Cambridge, United Kingdom*
Henning Urlaub • *Bioanalytical Mass Spectrometry Group, Max-Planck-Institute for Biophysical Chemistry, Göttingen, Germany*
Scott C. Walker • *Department of Biological Chemistry, University of Michigan, Ann Arbor, Michigan, USA*
James R. Williamson • *Department of Molecular Biology, Department of Chemistry, and the Skaggs Institute for Chemical Biology, The Scripps Research Institute, La Jolla, California, USA*
Shyue-Lee Yean • *Molecular Biology Division, Beckman Research Institute of the City of Hope, Duarte, California, USA*

1

Isolation of a Sequence-Specific RNA Binding Protein, Polypyrimidine Tract Binding Protein, Using RNA Affinity Chromatography

Shalini Sharma

Summary

Many important cellular processes are mediated by sequence-specific RNA binding proteins, and it is often necessary to purify these proteins. When the RNA binding site is known, it is convenient to use this RNA as a matrix for affinity purification. The intronic splicing silencer (ISS) element present upstream of the N1 exon of the c-src pre-mRNA is a high-affinity binding site for the polypyrimidine tract binding protein (PTB). Using a 5′-biotinylated ISS RNA and PTB as an example, I describe a one-step method for affinity chromatography of RNA binding proteins from nuclear extracts.

Key Words: Affinity chromatography; c-src N1 exon; intronic splicing silencer; polypyrimidine tract binding protein; RNA binding proteins.

1. Introduction

The RNA binding proteins regulate gene expression and RNA processing at several levels. They are known to affect transcription, pre-mRNA (messenger RNA) processing, mRNA stability, RNA transport, localization, and translation *(1–3)*. These proteins generally interact with their target RNAs via specific sequences, to which they bind with high affinity *(4–6)*. The polypyrimidine tract binding protein (PTB) is involved in several posttranscriptional events, including pre-mRNA splicing and polyadenylation and mRNA localization, translation, and degradation *(7–10)*. The 3′-splice site of the c-src N1 exon has a pyrimidine-rich intronic splicing silencer (ISS) element that binds PTB *(11, 12)*. Electrophoretic mobility shift assays (EMSAs) and filter-binding assays show that PTB binds with high affinity ($K_d \sim 10^{-9}$ *M*) to this repressor element and also to pyrimidine-rich sequences in other target RNAs *(13–15)*. Mutations

From: *Methods in Molecular Biology, vol. 488: RNA-Protein Interaction Protocols*
Edited by: Ren-Jang Lin © Humana Press Inc, Totowa, NJ

in the ISS sequence cause a significant reduction in binding *(16)*. In nuclear extracts from HeLa cells, this RNA forms a single discrete complex that can be completely supershifted by a PTB-specific antibody (S. Sharma and D.L. Black, unpublished observation; **ref.** ***12***). This shows a stoichiometric binding of PTB to the ISS RNA, indicating that each RNA molecule has at least one PTB protein bound to it. These properties make the ISS RNA a potential ligand for affinity purification of PTB.

We describe here a method for isolating PTB from cell extracts by RNA affinity chromatography. This method can potentially be used for purification of any RNA binding protein if its specific target sequence is known. Unknown factors can also be purified from cell extracts and identified if they are known to form stable complexes with specific RNAs.

The coupling of RNA to a chromatography matrix can be carried out in several ways. The substrate RNA can be covalently coupled to cyanogen bromide-activated Sepharose or to adipic acid hydrazide agarose *(17, 18)*. Noncovalent attachment can also be used if one can insert additional sequences into the RNA. These include (1) poly-A for attachment to poly-U Sepharose *(18, 19)*; (2) RNA aptamers for small molecules, such as tobramycin and streptomycin *(20, 21)*, or for proteins such as streptavidin (see Chapter 3); or (3) binding sequences for the MS2 phage coat protein *(22)*. Finally, biotinylated nucleotides can be incorporated into the RNA to act as high-affinity ligands for avidin or streptavidin *(18, 23)*. We took advantage of the high-affinity biotin–streptavidin interaction ($K_d \sim 10^{-15}$ *M*) and used a 5′-biotinylated RNA. The availability of synthetic RNA oligonucleotides allows the homogeneous coupling of the RNA at just one end. This is preferred over RNA carrying biotin substitutions along its length, as is produced by in vitro transcription. Internally tagged RNA may not show the same homogeneous binding by the target protein. This coupling method is simple and does not require any harsh treatment of the RNA. Another advantage of this method is that the isolated proteins are devoid of the substrate RNA and can be used directly for other applications, such as in vitro splicing assays or mass spectrometric (MS) analysis.

A general scheme for purification is outlined in **Fig. 1**. First, the biotinylated RNA is bound to the chromatography matrix, streptavidin Sepharose. Then, the unbound RNA is removed, and the nuclear extract is passed through the RNA column. The unbound proteins are first removed by washing the column with the low-salt binding buffer. Many RNA binding proteins exhibit a low-affinity, nonspecific binding to RNA. A wash with buffer containing an intermediate salt concentration is used to separate these nonspecifically bound proteins from the specific ones. Then, a high-salt buffer is used to elute the specifically bound proteins. RNA elements can be monospecific or may bind more than one factor. Cofractionating proteins can be separated using an additional

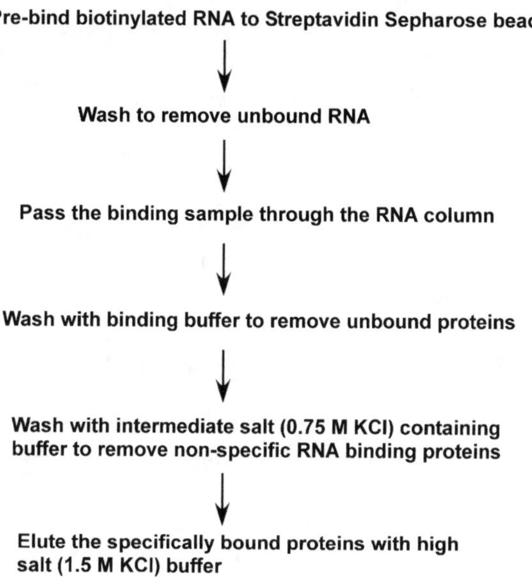

Fig. 1. Schematic outline of the protocol for RNA affinity purification.

chromatography step. For example, PTB can be separated from other proteins by passing it through a Blue Sepharose column ***(12)***.

2. Materials

1. 5′-Biotinylated RNA with the sequence 5′-BiAGCCUCUCCUUCUCUCUGCUU CUCUCUCGCUGGCCCUU (Dharmacon Research).
2. Streptavidin Sepharose (Amersham Biosciences).
3. Colloidal blue staining solution (Invitrogen).
4. HeLa nuclear extract.
5. Yeast transfer RNA (tRNA).
6. PMSF (phenylmethylsulfonyl fluoride).
7. Buffer DG: 20 m*M* HEPES-KOH, pH 7.9, 80 m*M* potassium glutamate, 0.1 m*M* EDTA (ethylenediaminetetraacetic acid), 0.1 m*M* PMSF, 1 m*M* DTT (dithiothreitol), and 20% glycerol.
8. Buffer D: 20 m*M* HEPES-KOH, pH 7.9, 0.1 m*M* EDTA, 0.1 m*M* PMSF, 1 m*M* DTT, 20% glycerol, and KCl at the indicated concentration.
9. Disposable Poly-Prep columns (Bio-Rad).
10. SDS-PAGE (sodium dodecyl sulfate polyacrylamide gel electrophoresis) apparatus.
11. TBST: 20 m*M* Tris-HCl, pH 7.5, 150 m*M* NaCl, and 0.1% Tween-20.
12. Blocking solution: 5% Blotto: nonfat dry milk (Bio-Rad) in TBST.
13. Novex immunoblotting apparatus.
14. Monoclonal anti-PTB antibody, BB7 ***(11)***.

15. Monoclonal anti-KSRP (KH-type splicing regulatory protein), AB5 *(24)*.
16. Monoclonal anti-hnRNP (heterogeneous nuclear ribonucleoprotein) A1 antibody.
17. Goat antimouse immunoglobulin G (IgG)–horseradish peroxidase conjugate.
18. SuperSignal West Pico Chemiluminescent Substrate (Pierce Biotechnology).

3. Methods

The steps described below (1) binding of RNA to chromatography matrix, (2) affinity purification of PTB from HeLa nuclear extract, and (3) identification of the isolated proteins.

3.1. Binding of Biotinylated RNA to Streptavidin Sepharose

All steps are carried out at 4 °C.

1. In a 1.5-mL polypropylene microcentrifuge tube, equilibrate 250 μL of streptavidin Sepharose beads by washing them four times with 1 mL (4 volumes) of buffer DG.
2. Add 25 nmol of biotinylated RNA (stock 1 nmol/μL) to 225 μL of buffer DG. Save about 2 μL of this solution to check A_{260} for the determination of binding efficiency. Add the RNA solution to the beads. Incubate at 4 °C for 1 h with end-over-end rotation.
3. Centrifuge the beads at 5000 rpm for 5 min at 4 °C and transfer the supernatant to another tube. Check A_{260} of the supernatant and calculate binding efficiency (*see* **Note 1**).
4. Wash the beads with buffer DG (four washes, 1 mL each) to remove any unbound RNA. The beads can be packed into a column at this stage. Make a 50% slurry of the beads in buffer DG, transfer it to a disposable Poly-Prep column using a pipet, and allow the buffer to drain. Leave a small volume of buffer so that the column surface does not dry.

3.2. Affinity Purification of RNA Binding Proteins

Described next are steps involved in isolation of the proteins that bind the c-src ISS RNA in HeLa nuclear extracts (*see* **Note 2**). The extract was prepared as described previously *(25–27)* and stored at −80 °C in buffer DG at a protein concentration of 6–8 mg/mL. Isolation of PTB was followed by SDS-PAGE and immunoblotting using an anti-PTB antibody.

3.2.1. Binding of Proteins to RNA-Streptavidin Sepharose

1. Make up 30 mL of loading sample, comprising 24 mL of HeLa nuclear extract (60%), 2.2 m*M* $MgCl_2$, 0.1 mg/mL yeast tRNA, 1 m*M* DTT, and 0.1 m*M* PMSF and add water to 30 mL.
2. Mix the solution gently; clarify by centrifugation (*see* **Note 3**). Cap the column containing the RNA beads at the bottom and then add the loading sample.
3. Cap the column at the top and incubate at 4 °C for 30 min with end-over-end rotation.

3.2.2. Removal of Nonspecifically Bound Proteins

1. After incubation, set the column to stand for 15 min to allow the beads to settle.
2. Remove the top and bottom caps and collect the flowthrough. To remove the unbound proteins, wash the column with 5 mL of buffer DG, which should be added as soon as the binding solution flows through to prevent the column from drying.
3. Wash the column with 2.0 mL of buffer D containing 0.75 *M* KCl to remove nonspecific RNA binding proteins that might still be bound to the beads (**Fig. 2A**, lane 3).

3.2.3. Elution of the Specifically Bound Proteins

1. Elute the bound proteins with four successive additions of 0.5 mL of buffer D containing 1.5 *M* KCl (*see* **Note 4**).
2. Separate an aliquot (15 µL) from each wash and each high-salt fraction on a 10% SDS-PAGE gel and stain with colloidal blue stain (*see* **Fig. 2A**).
3. In addition to the characteristic about 57-kDa doublet of PTB, the fractions eluted with 1.5 *M* KCl contain two other proteins with apparent molecular weights of about 100 and 45 kDa (**Fig. 2A**, lanes 1–4; *see* **Note 5**).

A Commassie Staining

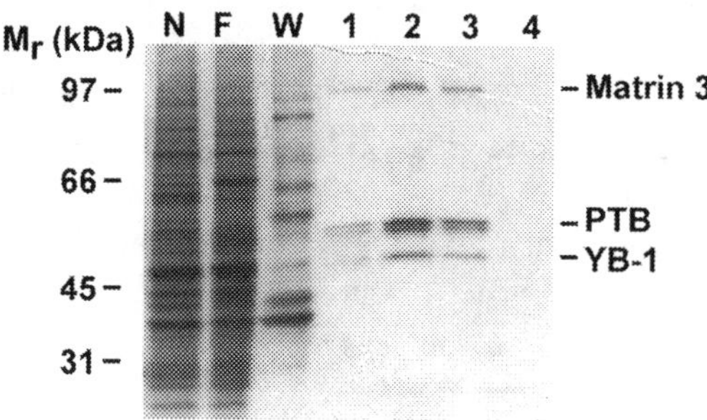

B Immunoblotting

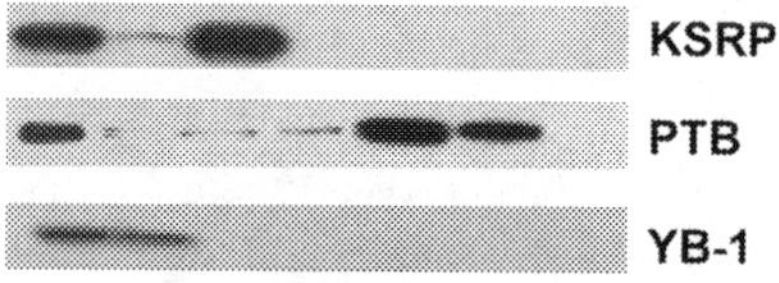

Fig. 2. Affinity isolation of PTB (polypyrimidine tract binding protein) from HeLa nuclear extract. (**A**) Proteins from 5 µL of HeLa nuclear extract (N), 15 µL of flowthrough from the RNA column (F), 15 µL of 0.75 *M* KCl wash (W), and 15 µL of 1.5 *M* KCl eluted protein fractions (1–4) were separated on a 4–20% SDS-polyacrylamide gel and stained with colloidal blue stain (Invitrogen). (**B**) Proteins separated on SDS-polyacrylamide gel were blotted onto a nitrocellulose membrane and probed with antibodies against KSRP (KH-type splicing regulatory protein), PTB, and hnRNP (heterogeneous nuclear ribonucleoprotein) A1.

3.3. Immunoblot Analysis

Verify the presence of PTB protein by probing a Western blot with anti-PTB antibody (*see* **Fig. 2B**, lanes 4–7).

1. Separate the flowthrough and eluted fractions on an SDS-PAGE gel as described.
2. For immunoblotting, transfer the proteins from the gel to a nitrocellulose membrane using the Novex transfer apparatus.
3. Add blocking solution (1 mL/cm^2) and incubate at room temperature for 30 min.
4. Add primary antibodies against hnRNP A1, KSRP, and PTB at a dilution of 1:1000 and incubate at room temperature for 1 h.
5. Discard the primary antibody solution and wash the nitrocellulose membrane twice with 10 mL of TBST.
6. Dilute the secondary antibody in blocking solution at 1:5000, add it to the nitrocellulose membrane, and incubate at room temperature for 1 h.
7. Wash the membrane three times with 20 mL of TBST with 10-min incubation during each wash.
8. Develop the blot with SuperSignal West Pico Chemiluminescent Substrate according to manufacturer's instructions.

The hnRNP A1 does not bind to this RNA column and flows through (**Fig. 2B**, lane 2). The KSRP binds with high affinity to poly(U) RNA but has lower affinity for the c-src ISS element. The 0.75 *M* KCl wash removes KSRP and several other proteins from the column (**Fig. 2B**, lane 3). Probing the blot with PTB antibody confirms the identity of the 57-kDa doublet as PTB in the high-salt eluate. PTB can be further purified using a Blue Sepharose column as described previously ***(12)***. The purified protein was dialyzed against buffer DG and stored in aliquots at −80 °C.

The RNA column, as prepared here, has a very high binding capacity (*see* **Note 6**). However, the abundance of the RNA binding proteins in nuclear extracts differ, and this will determine the yield of purification (*see* **Note 7**). Using this method, from 24 mL of HeLa nuclear extract, about 70 µg of PTB can be obtained.

4. Notes

1. The binding efficiency can be calculated as a ratio of the A_{260} of the RNA solution after and before binding to the streptavidin Sepharose beads, Binding efficiency = $(A_{260(before)} - A_{260(after)})/A_{260(before)}$. This value depends on the extent of biotinylation of the RNA and is usually about 80–90% for the RNA obtained from Dharmacon.
2. Although the method described here uses nuclear extract from a cell line as a source of protein, it can be used for isolation of RNA binding proteins from tissue extracts as well.
3. Due to high concentration and repeated freezing and thawing, proteins in nuclear extracts often precipitate out of solution. These precipitated proteins do not bind to the column but clog it, and this can drastically reduce the flow rate of the column. If the loading solution appears cloudy, centrifuge it at 10,000*g* for 20 min at 4 °C.

4. Elution conditions required for separating other RNA binding proteins from their specific ligand RNAs will vary and must be optimized for each protein by testing the elution at different salt concentrations. For PTB, the binding is stable below 1 *M* salt. Thus, we wash the column with 0.75 *M* salt and elute the bound PTB with 1.5 *M*.
5. Unknown proteins can be identified by MS. Bands from stained SDS-PAGE gels can be cut out and subjected to in-gel trypsin digestion, extraction, and peptide mass fingerprinting using matrix-assisted laser desorption ionization MS (MALDI-MS). The 100- and 45-kDa bands in **Fig. 2** were found to be matrin 3 (Accession no. AAH15031) and the Y box-1 protein (Accession no. NP_004550).
6. The high RNA/protein ratio used here helps maximize the yield. The high-affinity ISS RNA–PTB interaction combined with the high-salt elution keeps nonspecific binding low.
7. The presence of ribonucleases in some crude extracts may reduce the column capacity and hence the yield. This can sometimes be overcome by carrying out a preliminary purification using heparin-Sepharose ***(18)***.

Acknowledgments

I am grateful to Douglas Black, in whose laboratory this work was performed, for advice and support during the course of this work. I also thank the members of the Black lab for helpful suggestions.

References

1. Black, D. L. (2003) Mechanisms of alternative pre-messenger RNA splicing. *Annu. Rev. Biochem.* **72**, 291–336.
2. Reed, R., and Hurt, E. (2002) A conserved mRNA export machinery coupled to pre-mRNA splicing. *Cell* **108**, 523–531.
3. Maniatis, T., and Reed, R. (2002) An extensive network of coupling among gene expression machines. *Nature* **416**, 499–506.
4. Graveley, B. R. (2000) Sorting out the complexity of SR protein functions. *RNA* **6**, 1197–211.
5. Krecic, A. M., and Swanson, M. S. (1999) hnRNP complexes: composition, structure, and function. *Curr. Opin. Cell Biol.* **11**, 363–371.
6. Dreyfuss, G., Kim, V. N., and Kataoka, N. (2002) Messenger-RNA-binding proteins and the messages they carry. *Nat. Rev. Mol. Cell Biol.* **3**, 195–205.
7. Belsham, G. J., and Sonenberg, N. (1996) RNA–protein interactions in regulation of picornavirus RNA translation. *Microbiol. Rev.* **60**, 499–511.
8. Wagner, E. J., and Garcia-Blanco, M. A. (2001) Polypyrimidine tract binding protein antagonizes exon definition. *Mol. Cell. Biol.* **21**, 3281–3288.
9. Valcarcel, J., and Gebauer, F. (1997) Post-transcriptional regulation: the dawn of PTB. *Curr. Biol.* **7**, R705–R708.
10. Castelo-Branco, P., Furger, A., Wollerton, M., et al. (2004) Polypyrimidine tract binding protein modulates efficiency of polyadenylation. *Mol. Cell. Biol.* **24**, 4174–4183.

11. Chou, M. Y., Underwood, J. G., Nikolic, J., Luu, M. H., and Black, D.L. (2000) Multisite RNA binding and release of polypyrimidine tract binding protein during the regulation of c-src neural-specific splicing. *Mol. Cell* **5**, 949–957.
12. Chan, R. C., and Black, D. L. (1997) The polypyrimidine tract binding protein binds upstream of neural cell-specific c-src exon N1 to repress the splicing of the intron downstream. *Mol. Cell. Biol.* **17**, 4667–4676.
13. Perez, I., McAfee, J. G., and Patton, J. G. (1997) Multiple RRMs contribute to RNA binding specificity and affinity for polypyrimidine tract binding protein. *Biochemistry* **36**, 11881–11890.
14. Liu, H., Zhang, W., Reed, R. B., Liu, W., and Grabowski, P. J. (2002) Mutations in RRM4 uncouple the splicing repression and RNA-binding activities of polypyrimidine tract binding protein. *RNA* **8**, 137–149.
15. Yuan, X., Davydova, N., Conte, M. R., Curry, S., and Matthews, S. (2002) Chemical shift mapping of RNA interactions with the polypyrimidine tract binding protein. *Nucleic Acids Res.* **30**, 456–462.
16. Amir-Ahmady, B., Boutz, P. L., Markovtsov, V., Phillips, M. L., and Black, D. L. (2005) Exon repression by polypyrimidine tract binding protein. *RNA* **11**, 699–716.
17. Caputi, M., and Zahler, A. M. (2001) Determination of the RNA binding specificity of the heterogeneous nuclear ribonucleoprotein (hnRNP) H/H′/F/2H9 family. *J. Biol. Chem.* **276**, 43850–43859.
18. Kaminski, A., Ostareck, D. H., Standart, N. M., and Jackson, R. J. (1998) Affinity methods for isolating RNA binding proteins. In: *RNA: Protein Interactions* (Smith, C. W. J., ed.), Oxford University Press, New York, pp. 137–160.
19. Siebel, C. W., Kanaar, R., and Rio, D. C. (1994) Regulation of tissue-specific P-element pre-mRNA splicing requires the RNA-binding protein PSI. *Genes Dev.* **8**, 1713–1725.
20. Hartmuth, K., Vornlocher, H. P., and Luhrmann, R. (2004) Tobramycin affinity tag purification of spliceosomes. *Methods Mol. Biol.* **257**, 47–64.
21. Bachler, M., Schroeder, R., and von Ahsen, U. (1999) StreptoTag: a novel method for the isolation of RNA-binding proteins. RNA **5**, 1509–1516.
22. Zhou, Z., Sim, J., Griffith, J., and Reed, R. (2002) Purification and electron microscopic visualization of functional human spliceosomes. *Proc. Natl. Acad. Sci. U. S. A.* **99**, 12203–12207.
23. Rodgers, J. T., Patel, P., Hennes, J. L., Bolognia, S. L., and Mascotti, D. P. (2000) Use of biotin-labeled nucleic acids for protein purification and agarose-based chemiluminescent electromobility shift assays. *Anal. Biochem.* **277**, 254–259.
24. Hall, M. P., Huang, S., and Black, D. L. (2004) Differentiation-induced colocalization of the KH-type splicing regulatory protein with polypyrimidine tract binding protein and the c-src pre-mRNA. *Mol. Biol. Cell* **15**, 774–786.
25. Dignam, J. D. (1990) Preparation of extracts from higher eukaryotes. *Methods Enzymol.* **182**, 194–203.
26. Black, D. L., Chan, R., Min, H., Wang, J., and Bell, L. (1998) The electrophoretic mobility shift assay for RNA binding proteins. In: *RNA:Protein Interactions* (Smith, C. W. J., ed.), Oxford University Press, New York, pp. 109–136.
27. Rooke, N., and Underwood, J. (2003) In vitro RNA splicing in mammalian cell extracts. *Curr. Protoc. Cell Biol.* **2**, 11.17.1–11.17.20.

2

An Affinity Oligonucleotide Displacement Strategy to Purify Ribonucleoprotein Complexes Applied to Human Telomerase

Isabel Kurth, Gaël Cristofari, and Joachim Lingner

Summary

Antisense oligonucleotides have been used to study the structure and function of small nuclear ribonucleoprotein (snRNP) complexes and were adapted and modified for the purification of a variety of RNPs. We describe methods for recombinant expression and reconstitution of catalytically active human telomerase and the purification of native and recombinant telomerase using antisense affinity oligonucleotides. The purification procedure involves binding of the RNP complex to NeutrAvidin beads via a biotinylated 2´-O-methyl (2´-OMe) RNA oligonucleotide complementary to the RNA subunit. The complex is eluted from the beads through competition with a displacement oligonucleotide. Thus, the purified RNP is eluted under mild conditions, retaining its catalytic activity.

Key Words: Biotinylated 2´OMe RNA oligonucleotides; oligonucleotide affinity purification; recombinant telomerase; reverse transcriptase; ribonucleoprotein.

1. Introduction

Antisense DNA oligonucleotides have been applied to inhibit the function of ribonucleoproteins (RNPs) by cleaving the RNA moiety in DNA-RNA hybrids with ribonuclease (RNase) H. The introduction of 2´-O-alkyl-RNA oligonucleotides to disrupt RNA structures or RNA–protein interactions within RNPs took advantage of the exceptionally stable binding of these modified RNA oligonucleotides to their target RNA sequences while preventing target RNA cleavage by RNase H *(1)*.

These oligonucleotides were adapted for the purification of RNP complexes as an alternative strategy to conventional chromatographic methods or immunoaffinity chromatography. A biotinylated 2´-O-methyl (2´-OMe)-RNA

From: *Methods in Molecular Biology, vol. 488: RNA-Protein Interaction Protocols*
Edited by: Ren-Jang Lin © Humana Press Inc, Totowa, NJ

oligonucleotide hybridizing to the U2 small nuclear RNA (snRNA) was used to bind U2 small nuclear ribonucleoprotein (snRNP) to streptavidin agarose beads, and the bound components were analyzed on release under denaturing conditions *(2)*.

A refinement of this method has been described for the purification of telomerase from *Euplotes aediculatus* **(3)**, for which the active enzyme was released from the beads using a displacement oligonucleotide, which binds to the affinity oligonucleotide with higher affinity. This method was adapted for the purification of native and recombinant human telomerase ***(4,5)*** and is schematically depicted in **Fig. 1**. Human telomerase is a ribonucleoprotein with a core that contains a protein catalytic subunit (human telomerase reverse transcriptase,

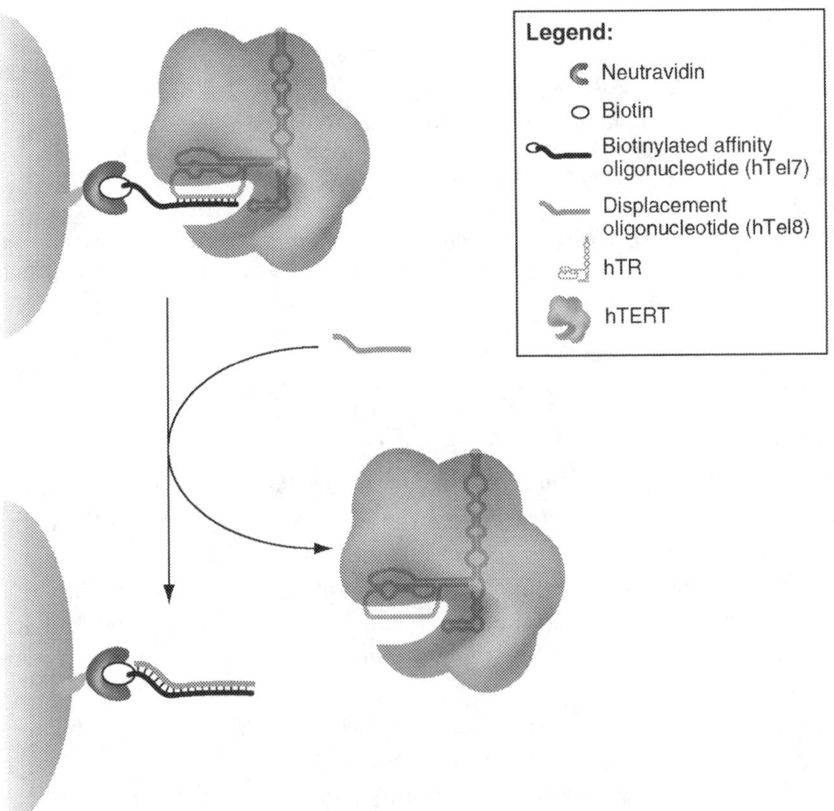

Fig. 1. Principle of affinity purification of telomerase. A biotinylated affinity oligonucleotide hybridizing to the RNA subunit is used to bind telomerase to NeutrAvidin beads. After washing off nonbound proteins, the telomerase complex is eluted from the beads with a displacement oligonucleotide.

hTERT) related to viral reverse transcriptases and a stably associated RNA moiety (human telomerase RNA, hTR). A small region in hTR contains a template sequence that is used for reverse transcription by the protein moiety to add telomeric repeats at the end of chromosomes.

We describe the expression and reconstitution of recombinant human telomerase and its purification using affinity and displacement oligonucleotides. This purification procedure is also compared to RNA tag-mediated affinity chromatography approaches.

2. Materials

1. Bac-n-Blue Transfection kit manual (Invitrogen).
2. TNM-FH (Pharmingen).
3. InsectXpress (Cambrex, BioWhittaker).
4. Insect cells: sf9, High five (Invitrogen).
5. T7 Ribomax kit (Promega).
6. Neutravidin beads (Pierce).
7. Phosphate-buffered saline (PBS): 137 m*M* NaCl, 2.7 m*M* KCl, 10 m*M* Na_2HPO_4, 2 m*M* KH_2PO_4.
8. Lysis buffer: 20 m*M* HEPES-KOH, pH 7.9, 25 m*M* KCl, 10 m*M* NaCl, 1 m*M* $MgCl_2$, 0.1 m*M* EDTA (ethylenediaminetetraacetic acid), 10% v/v glycerol, 5 m*M* β-mercaptoethanol, 0.1 m*M* phenylmethylsulfonyl fluoride (PMSF), protease inhibitor cocktail (complete mini-EDTA free, Roche).
9. Buffer A: 10 m*M* HEPES-KOH, pH 7.9, 10 m*M* KCl, 0.1 m*M* EDTA, 0.1 m*M* EGTA (ethylene glycol tetraacetic acid), 1 m*M* DTT (dithiothreitol), 0.5 m*M* PMSF.
10. Buffer C: 20 m*M* HEPES-KOH, pH 7.9, 0.4 *M* KCl, 1 m*M* EDTA, 1 m*M* EGTA, 1 m*M* DTT, 1 m*M* PMSF.
11. Wash buffer: 300 m*M* (WB300): 20 m*M* HEPES-KOH, pH 7.9, 300 m*M* KCl, 1 m*M* EDTA, 10% v/v glycerol, 0.5 m*M* DTT, 0.5 m*M* PMSF.
12. Bovine serum albumin (BSA): RNase-free (acetylated) BSA (Fluka or NEB).
13. Yeast tRNA (transfer RNA) (Sigma).
14. Glycogen solution, 10 mg/mL (Roche).
15. TRAP (telomeric repeat amplification protocol) buffer: 20 m*M* Tris-HCl, pH 8.3, 1.5 m*M* $MgCl_2$, 63 m*M* KCl, 1 m*M* EGTA, 0.1 m*M* BSA, 0.005% Tween-20.
16. LightCycler (Roche) or conventional thermocycler.
17. FastStart SybrGreen II kit (Roche) or Taq polymerase.
18. TS primer: 5´AATCCGTCGAGCAGAGTT3´.
19. AXC primer: 5´GCGCGGCTTACCCTTACCCTTACCCTAACC3´.
20. Tris-borate-EDTA (TBE) electrophoresis buffer: 90 m*M* Tris-borate, 4 m*M* EDTA.
21. 15% nondenaturing polyacrylamide gel electrophoresis (PAGE) in 0.6X TBE for TRAP.
22. Formamide sample buffer: 98% deionized formamide, 10 m*M* EDTA.
23. Church buffer: 1% (w/v) BSA, 1 m*M* EDTA, 0.5 *M* phosphate buffer, 7% sodium dodecyl sulfate (SDS).

3. Methods

The methods described outline (1) the preparation of cell lysates containing native human telomerase, (2) the expression and in vitro reconstitution of recombinant human telomerase, (3) the affinity purification protocol, (4) methods for additional purification steps, and (5) characterization of the purified RNP complex.

3.1. Preparation of Cell Lysate From Cell Lines

The preparation of native human telomerase is described in **Subheadings 3.1.1.** and **3.1.2.** and includes (a) recommended cell lines and (b) a protocol for the preparation of nuclear extracts.

3.1.1. Human Telomerase-Positive Cell Lines

Human cell lines that express active telomerase include HeLa, 293T, HT1080, and HL60.

3.1.2. Extract Preparation

HeLa nuclear extracts are prepared following an adapted procedure described by Schreiber et al. *(6)*. The method involves hypotonic swelling of the cells, disruption of the cellular membrane to release the nuclei, and incubation with high concentration of salts to extract nuclear proteins and RNPs. Other protocols for extract preparation have also been used successively (*see* **Note 1**).

1. Harvest cells from 50 confluent 135-cm^2 plates and wash them three times in PBS.
2. Resuspend the cell pellet in 20 mL of buffer A and transfer the cell suspension into precooled centrifuge tubes. Allow the cells to swell for 15 min on ice before adding NP-40 to a final concentration of 0.6%.
3. Vortex gently 10 s, avoiding the formation of foam.
4. Centrifuge at 4 °C at 10,000 *g* for 30 s.
5. Remove the supernatant (cytoplasmic fraction) and resuspend the nuclei containing pellet in 5 mL of buffer C.
6. Rock the resuspended nuclei vigorously for 30 min at 4 °C.
7. Centrifuge at 4 °C at 10,000 *g* for 10 min.
8. Collect the supernatant, which corresponds to the nuclear extract.
9. Add glycerol to a final concentration of 20% and quick freeze aliquots in liquid nitrogen before storing them at –80 °C.

3.2. Preparation of Recombinant Human Telomerase

This section describes (1) the expression of recombinant hTERT in insect cells, (2) the in vitro transcription of hTR, and (3) the in vitro reconstitution of active telomerase RNP.

3.2.1. Expression of hTERT

The open reading frame of the hTERT cDNA (complementary DNA) was fused with an N-terminal $(His)_6$ tag and cloned in between the SmaI and XbaI sites of the pVL1393 baculovirus transfer vector (BD Biosciences, Pharmingen). To generate recombinant baculoviruses, sf9 cells that were grown in TNM-FH medium were transfected with the recombinant transfer plasmid and linearized Bac-N-Blue DNA according to the instructions of the supplier (Invitrogen). After 4 d at 27 °C, the virus-containing supernatant was harvested, and single recombinant viruses were isolated by plaque assay and amplified using High five cells that were cultivated in Insect XPRESS medium. The virus suspensions carrying the hTERT transgene were titrated by endpoint dilution. The following is the protocol for hTERT production:

1. Seed High five cells at a density of 1.5×10^7 cells per 135-cm^2 dish (*see* **Note 2**) and let cells attach for 30 min.
2. Remove the medium and add recombinant virus in a total volume of 6 mL of medium at a multiplicity of infection (MOI) of 10.
3. Rock dishes gently (2 min^{-1}) for 1 h at room temperature (RT).
4. Add 14 mL of fresh medium and incubate the cells for 40 h at 27 °C.
5. Harvest the cells and wash them twice in PBS.
6. Resuspend the cell pellet in 0.5 mL of cold lysis buffer per 135-cm^2 dish.
7. Lyse the cells by sonication (three pulses of 15 s) on ice.
8. Remove the nonsoluble fraction by centrifugation (30 min at 10,000 *g*).
9. Determine the protein concentration by Bradford assay (Bio-Rad).
10. Quick freeze aliquots on dry ice and store them at –80 °C. (*see* **Note 5**).

3.2.2. In Vitro Transcription of hTR

For the construction of plasmid pT7hTER, the hTR gene was fused at the 5′ end with the T7 promoter sequence and cloned in between the EcoRI and BamHI sites of pUC18. For in vitro transcription, pT7hTER was linearized with BamHI and phenol chloroform extracted. In vitro transcription was performed with T7 RNA polymerase according to the supplier (Ribomax kit, Promega).

1. Incubate 5 μg of linearized template DNA in 40 m*M* Tris-HCl, pH 8.0, 6 m*M* $MgCl_2$, 10 m*M* DTT, 2 m*M* spermidine, 4 m*M* rNTPs, 1.2 U/μL RNasin (Promega), and 150 U of T7 RNA polymerase (Promega) in a total volume of 75 μL for 3 h at 37 °C.
2. Digest the template DNA by addition of RNase-free DNase (Promega) to a concentration of 1 U/μg of template DNA for 15 min at 37 °C.
3. Phenol-chloroform extract the RNA.
4. Purify the RNA-containing aqueous phase by gel filtration using mini Quick Spin columns (Roche).
5. Determine the RNA quantity by measuring the optical density (OD) at 260 nm and verify the intactness of the RNA by electrophoresis on agarose or polyacrylamide gels stained with ethidium bromide.

3.2.3. In Vitro Reconstitution of Telomerase

In a standard reconstitution reaction, 1 mL of insect cell lysate containing 7.8×10^{-12} to 3.9×10^{-11} mol of recombinant hTERT (see **Notes 3** and **5**) was adjusted to 150 m*M* KCl, 25 m*M* $MgCl_2$, and 20 m*M* ATP (adenosine triphosphate). Sixfold molar excess of hTR (*see* **Note 4**) was heat denatured for 30 s at 95 °C in 20 m*M* HEPES-KOH, pH 7.9, 2 m*M* $MgCl_2$, 50 m*M* KCl and allowed to cool slowly to RT for 20 min prior to the addition of cell lysate. RNA concentrations were kept high enough to ensure that the final dilution of the cell lysate did not exceed 20%. The reconstitution mixture was incubated at 30 °C for 90 min. For storage, reconstituted telomerase was quick frozen on dry ice and kept at −70 °C (*see* **Notes 6** and **10**).

3.3. Affinity Purification

The following steps describe the procedure for the purification using the affinity oligonucleotide displacement strategy. A biotinylated affinity oligonucleotide is used, which hybridizes to the RNA subunit and allows the binding of the RNP complex to NeutrAvidin beads. The complex is released on competition with a displacement oligonucleotide (**Fig. 1**). This method allows purification under mild conditions, maintaining higher-order structures while retaining the enzymatic activity of the complex. **Subheadings 3.3.1.** and **3.3.2.** describe the design of appropriate oligonucleotides and the protocol for the purification.

3.3.1. Design of Affinity and Displacement Oligonucleotides

The affinity oligonucleotide (26 nucleotides) used for the purification of human telomerase was designed as follows (**Fig. 2**): The 14 most 3′ nucleotides are complementary to the template sequence in the RNA subunit accessible in the telomerase RNP. The 11 most 3′ terminal nucleotides are 2′-O-methylated ribonucleotides (*see* **Note 7**), whereas the others are deoxyribonucleotides. The 12 most 5′ terminal nucleotides contain a sequence that is not related to the RNA to be purified. The oligonucleotide is biotinylated at the 5′ end and contains for the purification of human telomerase the following sequence: 5′-CTAGACCTG TCATCAGUUAGGGUUAG-3′ (hTel7). The 26-nucleotide displacement DNA oligonucleotide (hTel8) contains a sequence that is fully complementary to the affinity oligonucleotide: 5′-CTAACCCTAACTGATGACAGGTCTAG-3′.

All centrifugation steps of bead-containing suspensions are done at 4 °C at 700 *g* for 2 min.

Preparation of beads:

1. Use a volume of beads that corresponds to 10% of the reconstitution mixture or nuclear extract volume.
2. Wash the beads once with 10 bead volumes (bv) of WB300.

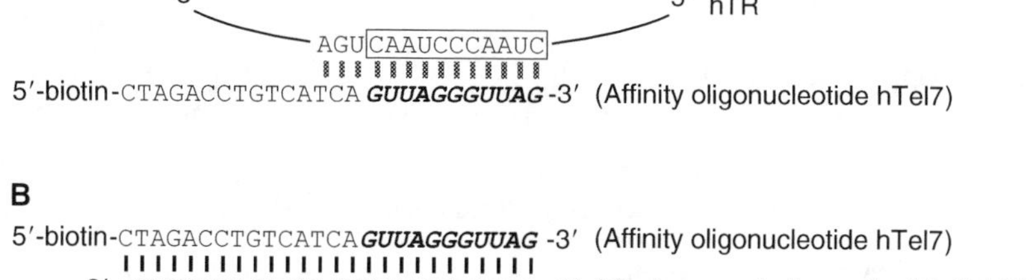

B

5′-biotin-CTAGACCTGTCATCA*GUUAGGGUUAG* -3′ (Affinity oligonucleotide hTel7)

3′-GATCTGGACAGTAGTCAATCCCAATC -5′ (Displacement oligonucleotide hTel8)

Fig. 2. Sequences of oligonucleotides used for the purification of human telomerase. (**A**) The 5′ biotinylated affinity oligonucleotide hybridizes to the template region of hTR (human telomerase RNA). The RNA template is shown in the box. Italic nucleotides are 2′-O-methylribonucleotides. (**B**) The sequence of the DNA displacement oligonucleotide is complementary to the affinity oligonucleotide.

3. Preblock the beads in WB300 supplemented with 0.5 mg/mL BSA for 10 min on a rotating wheel at RT.
4. Wash the beads three times with 10 bv of WB300.

For the purification protocol,

5. In the meantime, adjust the samples to be purified to WB300 buffer salt concentration and remove an aliquot (IN) for later analysis. Add Triton X-100 to 0.5%, 50 µg/µL yeast tRNA, and 1 µ*M* of hTel7 (for nuclear extracts) or a threefold molar excess of hTel7 over the RNA (for reconstituted telomerase).
6. Incubate the mixture at 30 °C for 10 min (*see* **Note 8**).
7. Add the sample to the beads and incubate for 10 min at 30 °C followed by 2 h at 4 °C on a rotating wheel.
8. Centrifuge and save the supernatant (SN).
9. Wash the beads with 10 bv each time with the following buffers:
    a. Once with WB300 containing 0.5% Triton-X100 at 4 °C.
    b. Twice with WB300 containing 0.5% Triton-X100 for 15 min at 4 °C on a rotating wheel.
    c. Once in WB300 adjusted to 600 mM KCl for 5 min at 30 °C.
    d. Once in WB300 for 5 min at RT on a rotating wheel.
    e. Rinse once with WB300.
10. Elute the RNP complex three times in 1 bv of WB300 containing 0.15% Triton-X100 and 3 µ*M* hTel8 (for nuclear extracts) or three times molar excess of hTel8 over the affinity oligonucleotide (for reconstituted telomerase) for 30 min at RT.
11. Measure protein concentration by Bradford (*see* **Note 9**).
12. Quick freeze the eluates on dry ice and store them at –80 °C.

3.4. Glycerol Gradient Ultracentrifugation

Affinity purification of telomerase yields typically a 70- to 150-fold enrichment of recombinant telomerase or a 200- to 300-fold enrichment of endogenous telomerase from nuclear extracts. To achieve further purification and to separate telomerase from the displacement oligonucleotide, glycerol gradient ultracentrifugation can be performed as follows (*see* **Note 10**):

1. Pool the eluates from the affinity purification and concentrate them 10 times by centrifugation in Ultrafree centrifugal filters (50 K cut off, Millipore) at 4 °C and 2000 *g*.
2. Pour 5 mL 15–40% glycerol gradients in WB300 buffer in Ultra-Clear Centrifuge tubes (13 × 51 mm, Beckman).
3. Overlay 100 μL of the concentrate on top of the gradient.
4. Centrifuge at 4 °C for 7 h at 55,000 rpm (382,800 *g*) using a TST 55.5 rotor and a Centrikon T-1080 ultracentrifuge.
5. Collect 250-μL fractions from the top and check for the presence of the RNP complex (*see* **Note 11**).

3.5. Analysis of the Purified Telomerase Complex

The quantitative analysis of recovery and enrichment is essential for a successful reconstitution and purification. **Subheadings 3.5.1.** and **3.5.2.** describe methods to determine the enzymatic activity of telomerase and hTR RNA levels (**Table 1**).

3.5.1. Telomeric Repeat Amplification Protocol Assay

The following protocols describe methods to detect telomerase activity. Briefly, a telomeric DNA oligonucleotide substrate is incubated in the presence of dNTPs (deoxyribonucleotide 5′-triphosphates) with telomerase, which extends the substrate-adding telomeric repeats. The elongated products are amplified subsequently by polymerase chain reaction (PCR). In the classical TRAP assay, a radioactive-labeled dNTP is added to the reaction mixture to allow the visualization of the reaction products by PAGE *(7)* (**Fig. 3**). Recently, the TRAP protocol has been combined with quantitative real-time PCR (RQ-TRAP), allowing a more accurate quantification of the activity *(8)*.

For the TRAP protocol:

1. Add to a thermocycler tube 50 μL of TRAP reaction in 1X TRAP buffer containing 25 μ*M* of dNTPs, 0.1 μg of TS primer, 2 μCi of [α-^{32}P] dGTP (deoxyguanosine 5′-triphosphate; 3000 Ci/mmol, Amersham), and 1 μL of telomerase fraction (<0.5 μg protein).
2. Place the tube in a thermocycler and incubate at 30 °C for 30 min.
3. Heat to 94 °C and add 2 U of Taq DNA polymerase and 0.1 μg of ACX primer in 1 μL TRAP buffer.
4. Cycle the reaction 27 times at 94 °C for 30 s and at 60 °C for 30 s.

Table 1
Purification of Reconstituted Recombinant Human Telomerase[a]

	Protein	hTR				Telomerase activity			
Fraction	Total amount (mg)	Total amount (μg)	Relative amount (μg/mg protein)	Yield (%)	Purification factor	Total activity (AU)	Specific activity (AU/mg protein)	Yield (%)	Purification factor
Load	10	310	31	100	1	100	10	100	1
Supernatant	8.4	86	10	28	0.3	0	0	0	NA
Eluate	0.024	58	2417	19	78	17	725	17	73

hTR, human telomerase RNA; NA, not applicable.

[a]Protein concentration was measured by Bradford, hTR concentration by Northern blot, and telomerase activity by TRAP (telomeric repeat amplification protocol).

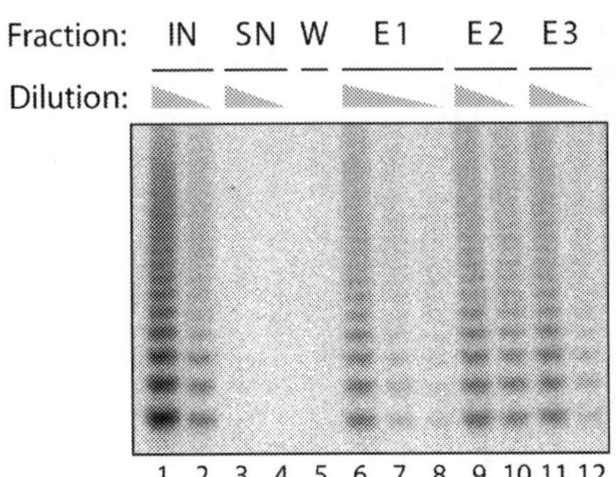

Fig. 3. Activity of telomerase fractions during affinity purification as determined by TRAP. The indicated fractions are incubated with a telomeric substrate in the presence of radiolabeled dNTPs (deoxyribonucleotide 5′-triphosphates). Telomerase adds telomeric repeats, elongated products are amplified by polymerase chain reaction (PCR) and resolved by a 15% native polyacrylamide gel electrophoresis (PAGE). TRAP (telomeric repeat amplification protocol) uses 1 µL of the indicated fractions in the following respective dilutions: lanes 1, 2: input (IN) 1:20 and 1:40; lanes 3, 4: supernatant (SN) 1:10 and 1:20; lane 5: wash (W) undiluted; lanes 6 to 8: elution 1 (E1) 1:10, 1:20 and 1:40; lanes 9, 10: elution 2 (E2) 1:5 and 1:10; lanes 11, 12: elution 3 (E3) 1:5 and 1:10.

5. Separate the PCR products on a 15% nondenaturing polyacrylamide gel in 0.6X TBE. Analyze dried gels using a PhosphorImager (Fuji) and the AIDA software.

For the RQ-TRAP protocol, the protocol has been adapted for the LightCycler (Roche) PCR machine.

1. Mix 20-µL reactions containing 100 ng of TS primer, 50 ng of ACX primer, 2 µL of 10X LC-FastStart SYBR Green Mix (Roche), and 1 µL of telomerase fraction in glass capillaries (Roche) and adjust $MgCl_2$ to a final concentration of 2 m*M*.
2. Place the capillaries in the LightCycler and incubate them for 30 min at 30 °C followed by 10 min at 95 °C to heat activate Taq polymerase.
3. Cycle the reaction 40 times at 94 °C for 30 s and at 60 °C for 30 s.

Controls should include serial dilution of extract from telomerase-positive cells and RNase A-treated samples (*see* **Note 12**).

3.5.2. Northern Blot Analysis (**Fig. 4**)

1. Digest fractions with 1 mg/mL proteinase K (Roche) in 10 m*M* Tris-HCl, pH 8.0, 0.2% SDS for 15 min at 37 °C.
2. Phenol-chloroform extract the reaction mixture and precipitate the RNA-containing aqueous phase with 0.1 volume of 3 *M* sodium acetate (pH 5.2) and 2.5 volumes of 95% ethanol in the presence of 5 µg of glycogen. Mix and precipitate overnight at –20 °C.

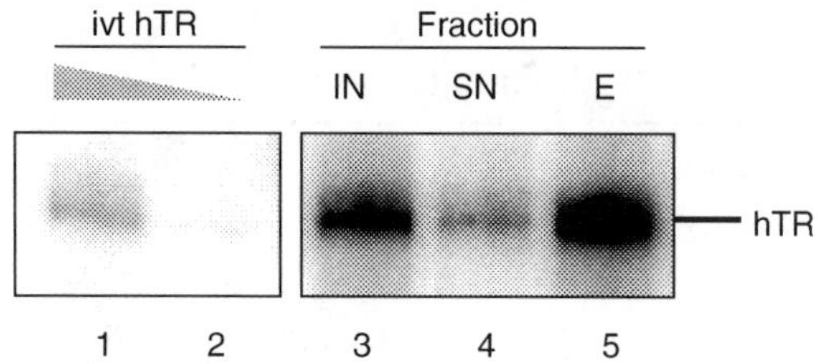

Fig. 4. Northern blot analysis of human telomerase RNA (hTR). RNA from 1 µL of the indicated fractions is extracted and re-solved by 6% denaturing polyacrylamide gel electrophoresis (PAGE). Transferred membranes are revealed with a probe specific for hTR. Lanes 1 and 2: 0.5 and 0.05 ng in vitro-transcribed hTR, respectively; lane 3: input (I); lane 4: supernatant (SN); lane 5: eluate (E).

3. Spin at top speed in a microcentrifuge for 30 min, carefully remove the supernatant, and wash the pellet with 1 mL of 70% ethanol. Resuspend the pellet in 10 µL formamide sample buffer.
4. Heat denature the samples for 1 min at 95 °C, quick cool on ice, and re-solve the fractions by 6% PAGE in 7 *M* urea and 1X TBE.
5. Transfer the gel onto a Hybond N^+ membrane (Amersham) and UV-crosslink the RNA to the membrane using a Stratalinker 2000 (Stratagene).
6. Block the membranes in Church buffer for 2 h at 65 °C and incubate overnight at 65 °C with a random-primer-labeled DNA probe corresponding to nucleotides 91 to 374 of hTR.
7. Wash the blots in 0.2X SSC and 0.5% SDS at 65 °C and analyze them using a PhosphorImager (Fuji) and the AIDA software.

3.6. Alternative Elution Methods for Affinity Oligonucleotide Purified RNPs

Several variations for the elution of oligonucleotide affinity-purified complexes have been described. Boiling the beads in SDS-PAGE sample buffer efficiently elutes bound proteins and RNAs under denaturing conditions. Apart from destroying the complexes, this method has the disadvantage of also efficiently eluting contaminant proteins that would not elute with mild RNP-specific elution methods. Also, successful elution with free biotin has been reported for affinity-purified hepatitis C virus RNA replication complexes ***(9)***. In this article, the biotinylated affinity oligonucleotide was also 3′ end labeled with digoxigenin, allowing a second step of purification by immunopurification with antidigoxigenin antibody. Since in this method the affinity oligonucleotide remains bound to the target RNA, it may interfere with its function.

Table 2
Comparison of Purification via Affinity Oligonucleotides and RNA Tags

Oligonucleotide affinity purification	RNA-tag-mediated affinity purification
Cloning the target RNA is not required.	The target RNA is cloned into an expression vector, and an aptamer-encoding sequence is added.
Endogenous complexes are purified.	Expression of the tagged RNA from vectors may lead to abnormal expression levels. The endogenous RNA may compete for RNP assembly with the RNA expressed from the transgene.
An accessible region must be identified in the RNA to allow annealing of the affinity oligonucleotide.	A position in the RNA needs to be identified where the RNA tag is accessible but does not interfere with function.
Salt conditions allowing the specific annealing of the affinity and displacement oligonucleotides may affect the intactness of the RNP.	Most RNA tags can be bound and eluted in a broad range of salt conditions, allowing recovery of intact RNPs under mild conditions.

RNP, ribonucleoprotein.

3.7. Comparison With RNA-Tag-Mediated Affinity Purification of RNPs

A number of natural or artificial RNA sequences referred to as aptamers bind with high affinity to specific ligands or proteins. These sequences can be used to tag the RNA subunit of an RNP, allowing its purification by affinity chromatography. Therefore, this method, which is described in chapter 3, offers an alternative approach to purify RNPs. Advantages and limitations of the two methods are listed in **Table 2**. They are reminiscent of the differences that exist between specific antibodies and epitope tags to purify complexes via protein subunits.

4. Notes

1. Nuclear extracts are recommended as starting material for the affinity purification of endogenous telomerase. As an alternative to the nuclear extract protocol described here, cells can be lysed on hypotonic swelling with a Dounce homogenizer, on which the pelleted nuclei are extracted with high salt ***(10)***.
2. Optimal expression conditions vary with different proteins. Viral titers, infection times, and cell lines should be optimized. Good results for hTERT were also obtained in sf21, a cell line that is adapted to grow in suspension, facilitating expression at a larger scale.

3. hTERT levels were estimated by quantitative Western blotting comparing the signal intensity to a known concentration of an hTERT C-terminal peptide fragment. Molecular weight differences were adjusted in the quantification.
4. In vitro transcribed and purified hTERT was quantified by measuring the OD at 260 nm and applying Beer's law: $A = \varepsilon Cl$. where A is the measured absorption at 260 nm, ε is the RNA extinction coefficient (25 µL/µg/cm), C is the RNA concentration, and l is the pathlength (1 cm).
5. The yield of hTERT was 0.5–1 µg/mg total protein. Approximately 50% of hTERT was soluble. Preparation of nuclear extracts resulted in an enrichment of hTERT, but the reconstitution efficiency was very low, probably due to the lack of cytoplasmic chaperones involved in RNP assembly. Telomerase interacts with the chaperones p23 and Hsp90, and blocking this interaction inhibits assembly of active telomerase ***(11,12)***.
6. The in vitro reconstitution is very inefficient, and only a minor fraction of hTERT is assembled with hTR correctly to yield an enzymatically active complex.
7. The 2´-O-alkyl-substituted oligoribonucleotides are more resistant to chemical and nucleolytic degradation. In addition, they form very stable interactions with their target RNA sequence with significantly higher melting temperature values than the corresponding RNA-RNA hybrids. Thus, they have also been used to disrupt natural RNA-RNA helices. Good results were obtained with oligonucleotides 11–16 nucleotides long. If the structure of the RNA target is unknown, accessible sequences can be identified by hybridization of complementary DNA oligonucleotides and RNase H cleavage.
8. Binding of 2´-OMe-ribonucleotides to snRNA was stimulated at higher temperatures ***(6)***, and initial incubation at 30 °C has been found to be critical for efficient binding of the affinity oligonucleotide to telomerases from *Euplotes aediculatus* and humans ***(3)***.
9. Triton X-100 interferes with the Bradford reagent and can lead to inaccurate measurement of the protein concentration.
10. The excess of RNA used during in vitro reconstitution leads to a high yield of protein-free RNA in the purified samples. Chromatography on a DEAE column prior to affinity purification partially separated the free RNA from the assembled RNP. Free hTR could not be well separated from hTERT-hTR complexes by glycerol gradient ultracentrifugation, gel filtration, or chromatography using Heparin, SP, or MonoQ columns.
11. In parallel glycerol gradients, protein markers can be run, allowing determination of the sedimentation coefficient of the purified complex.
12. The amount of product obtained by TRAP is proportional to the amount of telomerase contained in the fraction as long as the total protein used per reaction does not exceed 0.5 µg. At higher protein concentrations, the reaction reaches a plateau or is inhibited. To ensure linearity of the assay, it is recommended to test several dilutions of each sample.

Acknowledgments

We thank Christian Wenz for his contributions to the described protocols. This work was supported by an ISREC PhD fellowship to I.K., an EMBO Long-Term Postdoctoral Fellowship to G.C., and a Swiss National Science Foundation grant to J.L.

References

1. Lamond, A. I., Sproat, B., Ryder, U., and Hamm, J. (1989) Probing the structure and function of U2 snRNP with antisense oligonucleotides made of 2′-OMe RNA. *Cell* **58**, 383–390.
2. Lamond, A. I., and Sproat, B. S. (1994) Isolation and characterisation of ribonucleoprotein complexes. In: *RNA Processing: A Practical Approach* (Hames, B. D., and Higgins, S. J., eds.), IRL Press at Oxford University Press, Oxford, U.K., pp. 103–140.
3. Lingner, J., and Cech, T. R. (1996) Purification of telomerase from Euplotes aediculatus: requirement of a primer 3′ overhang. *Proc. Natl. Acad. Sci. USA* **93**, 10712–10717.
4. Schnapp, G., Rodi, H. P., Rettig, W. J., Schnapp, A., and Damm, K. (1998) One-step affinity purification protocol for human telomerase. *Nucleic Acids Res.* **26**, 3311–3313.
5. Wenz, C., Enenkel, B., Amacker, M., Kelleher, C., Damm, K., and Lingner, J. (2001) Human telomerase contains two cooperating telomerase RNA molecules. *EMBO J.* **20**, 3526–3534.
6. Schreiber, E., Matthias, P., Muller, M. M., and Schaffner, W. (1989) Rapid detection of octamer binding proteins with "mini-extracts," prepared from a small number of cells. *Nucleic Acids Res.* **17**, 6419.
7. Kim, N. W., and Wu, F. (1997) Advances in quantification and characterization of telomerase activity by the telomeric repeat amplification protocol (TRAP). *Nucleic Acids Res.* **25**, 2595–2597.
8. Wege, H., Chui, M. S., Le, H. T., Tran, J. M., and Zern, M. A. (2003) SYBR Green real-time telomeric repeat amplification protocol for the rapid quantification of telomerase activity. *Nucleic Acids Res.* **31**, E3–3.
9. Waris, G., Sarker, S., and Siddiqui, A. (2004) Two-step affinity purification of the hepatitis C virus ribonucleoprotein complex. *RNA* **10**, 321–329.
10. Dignam, J. D., Lebovitz, R. M., and Roeder, R. G. (1983) Accurate transcription initiation by RNA polymerase II in a soluble extract from isolated mammalian nuclei. *Nucleic Acids Res.* **11**, 1475–1489.
11. Holt, S. E., Aisner, D. L., Baur, J., et al. (1999) Functional requirement of p23 and Hsp90 in telomerase complexes. *Genes Dev.* **13**, 817–826.
12. Forsythe, H. L., Jarvis, J. L., Turner, J. W., Elmore, L. W., and Holt, S. E. (2001) Stable association of hsp90 and p23, but Not hsp70, with active human telomerase. *J. Biol. Chem.* **276**, 15571–15574.

3

RNA Affinity Tags for the Rapid Purification and Investigation of RNAs and RNA–Protein Complexes

Scott C. Walker, Felicia H. Scott, Chatchawan Srisawat, and David R. Engelke

Summary

Isolation of ribonucleoprotein particles from living cells and cell lysates has allowed the identification of both simple bimolecular interactions and the members of large, extended complexes. A number of different strategies have been devised to isolate these complexes by using affinity purification methods that are specific for the RNA rather than the protein components of these complexes. We describe the use of two such RNA affinity tags: small RNAs that bind with high affinity and specificity to either Sephadex beads or streptavidin affinity resins and can be eluted under mild, native conditions that retain intact complexes. The tags can be inserted into appropriate locations in genes encoding the RNA components, and ribonucleoproteins can be assembled either in vivo or in vitro before affinity isolation. Strategies toward the design and production of these tagged RNA sequences are discussed, and the purification procedure is outlined.

Key Words: Aptamer; ribonucleoprotein; RNA; RNP isolation; SELEX; Sephadex; streptavidin.

1. Introduction

The use of small, genetically introduced affinity tags for the recombinant production and purification of proteins as well as for the isolation of defined protein complexes has greatly facilitated the study of individual protein function and chemistry in a variety of research fields *(1)*. These protein affinity tags include polyhistidine *(2)*, the hemagglutinin epitope *(3)*, myc epitope *(4)*, the tandem affinity protein (TAP) tag *(5)*, protein A *(6)*, glutathione S-transferase *(7)*, Strep-tag *(8)*, and the FLAG epitope *(9)*. These affinity tags all bind with high affinity to a ligand that can either be immobilized on a chromatography resin or be conjugated to an antibody detection system. In most cases, the isolated complexes can either be released from the resin by competitive elution or

From: *Methods in Molecular Biology, vol. 488: RNA-Protein Interaction Protocols*
Edited by: Ren-Jang Lin © Humana Press Inc, Totowa, NJ

cleaved off by a protease with a recognition site that is incorporated along with the affinity tag as part of the fusion protein. The widespread use of protein affinity tags has led to the development of similar tags for nucleic acids. Particular emphasis has been placed on developing affinity tags for RNA molecules since this versatile nucleic acid is involved in a wide variety of cellular processes.

There are four known methods in use for tagging an RNA: (1) chemical tagging during in vitro transcription, (2) incorporation of a well-characterized protein-binding RNA sequence during in vitro or in vivo transcription, (3) hybridization of affinity-tagged oligonucleotides (biotinylated), and (4) incorporation of an artificially selected RNA motif during in vivo or in vitro transcription. Each of these methods for the tagging and affinity isolation of RNA molecules has particular advantages and disadvantages, discussed here.

There are several ways to chemically tag an in vitro transcribed RNA; these include the incorporation of modified ribonucleotide triphosphates (rNTPs) containing biotin, fluorescent dyes, digoxeginin, or other compounds. RNAs chemically modified with biotin and purified using streptavidin beads have been used to isolate RNA–protein complexes in which the chemically modified RNA is incubated in vitro with cellular extracts or recombinant proteins ***(10)***. However, there are two major drawbacks to this technique. In some cases, the chemical modifications can lead to structural perturbations that can inhibit complex formation. Furthermore, this method can only be used for in vitro studies, and the complexes formed may not reflect the true nature of the complexes formed in vivo.

The second method of tagging is the incorporation of an RNA sequence from a well-characterized RNA–protein interaction. The most widely used RNA–protein interaction is that of the MS2 coat protein and its cognate RNA ***(11)***. The U1A protein and its interaction with its cognate RNA have also been used to a lesser extent. Generally, the limiting step when using a known RNA–protein interaction as part of the affinity purification is the inability to efficiently elute or release the purified complex under native conditions since the binding affinity of the known RNA–protein interaction is usually high. One way to circumvent this challenge is to fuse an additional peptide such as the maltose-binding peptide (MBP) to the known protein-binding partner. In this case, there is no requirement for the MS2 coat protein to be immobilized, and the isolation can be performed using the MBP and not the MS2 coat protein. The complex can be released from the affinity resin by elution with maltose ***(12)***. Alternatively, a protease cleavage site can be engineered between the MS2 coat protein site and the protein that binds directly to the affinity resin. In this case, purified complexes are released from the resin on specific protease cleavage ***(13)***. A bipartite affinity tag has been used to purify RNAs under nondenaturing conditions from in vitro transcription reactions ***(14)***. The tag consists of a variant of the hepatitis delta virus (HδV) ribozyme that is activated by imidazole and the tandem stem-loop motifs from the *Thermotoga maritima* signal recognition particle (SRP) RNA that specifically binds the SRP protein Ffh. On the induced ribozyme cleavage, the target RNA is released from the column, while the ribozyme remains bound through its fused SRP stem-loop motifs.

The use of biotinylated oligonucleotides that are complementary to accessible single-stranded regions of a particular ribonucleoprotein (RNP) complex has also been a successful purification strategy for the isolation of these complexes. Elution of the complex can be achieved under denaturing conditions or through the use of a competitor oligonucleotide to release the RNP under native conditions. This technique has been used to isolate the U4/U6 small nuclear ribonucleoprotein (snRNP) and telomerase ***(15,16)***. In both cases, an unstructured and accessible region of RNA sequence was available to facilitate the technique. While this method can be difficult for highly structured RNAs and RNPs, it has several advantages over methods that incorporate a foreign RNA sequence into the target. (1) No misfolding of the RNA occurs in vivo due to the foreign sequence. (2) It can be used when genetic manipulations of the organism to introduce a foreign tag are not possible. (3) Several different sites in the RNA target can be quickly tested for accessibility to synthetic oligonucleotide probes.

The development of in vitro selection or SELEX (systematic evolution of ligands by exponential enrichment) technology has led to the discovery of several RNA or DNA sequences (aptamers) with the ability to bind specifically and with high affinity to small molecules as well as more complex macromolecules (reviewed in **ref.** ***17***). One application of the RNA aptamer sequences is their use as tags for RNA affinity purification. RNA aptamers to antibiotics have been used to study RNA–protein interactions in vitro with crude extracts or recombinant proteins ***(18,19)***.

To further extend the application of aptamers as RNA affinity tags, our lab has identified two artificially selected RNA aptamers that have compact, defined structures and are particularly useful for the study of RNA–protein complexes formed in vivo and isolated under native conditions ***(20,21)***. Their applications to the study of RNA–protein complexes as well as the strengths and weaknesses of each aptamer are described here. Several factors influenced the selection of the target molecule: (1) availability and price of the potential affinity resins, (2) low background affinity (i.e., the resin did not have a high affinity for nonspecific RNAs), and (3) the complexes could be eluted under native conditions. The two target molecules that met these criteria were streptavidin and dextran B512 (in the insoluble form of Sephadex beads), and aptamers to each of these resins were successfully produced ***(20,21)***. The consensus sequence for each of these aptamers is shown in **Fig. 1**. The distinguishing property that makes these aptamers particularly useful for the isolation of active RNP complexes is that both can be eluted from their affinity resins under native conditions.

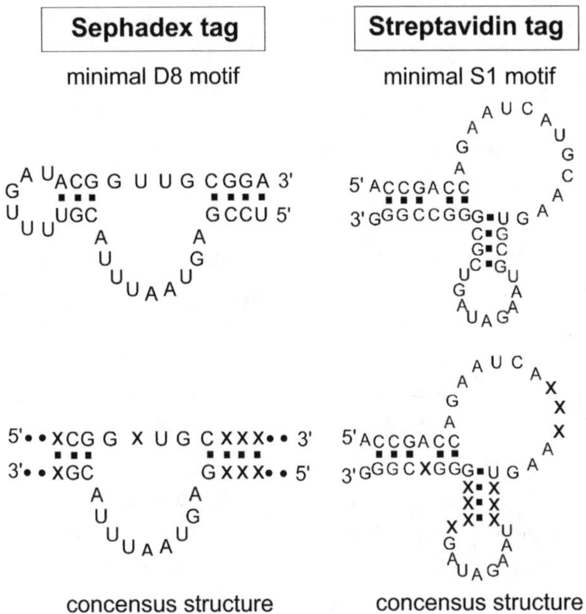

Fig. 1. The minimal binding motifs of both the Sephadex and streptavidin-binding RNA affinity tags *(25)*. These sequences were originally identified using a SELEX (systematic evolution of ligands by exponential enrichment) procedure designed to isolate sequences that displayed binding to the affinity matrix under approximately physiological conditions (50 m*M* HEPES, pH 7.4, 10 m*M* $MgCl_2$, 100 m*M* NaCl). The consensus structures of the two families of aptamers isolated in SELEX are also shown, where **X** indicates a nonconserved nucleotide identity and ■ indicates basepairing.

Further advantages and disadvantages of using each aptamer are summarized in **Table 1**.

The streptavidin aptamer has been used successfully in the affinity purification of ribonuclease P (RNase P) from *Saccharomyces cerevisiae* ***(20)*** and human S3 cells ***(22)*** as well as complexes as large as the ribosome ***(13)***. The Sephadex aptamer has been used in an extensive study of *S. cerevisiae* RNase P in which the differential protein composition of the precursor and mature forms of the RNP complex were analyzed ***(21)***.

In this chapter, we describe the use of the streptavidin and Sephadex aptamers. We have employed these within the yeast system and have previously isolated active eukaryotic RNase P holoenzyme from *S. cerevisiae* using either of the two aptamer tags ***(20,21,23)***; these are used as illustrative examples within this chapter. Specific examples of the use of the streptavidin aptamer to facilitate isolations from *Escherichia coli* and *Homo sapiens* can be found elsewhere ***(13,22)***.

Table 1
Advantage-Disadvantage Comparison Between Two RNA Affinity Tags

	D8 Sephadex RNA motif	S1 Streptavidin RNA motif
Advantages	Sephadex (G-200 is best choice) is cheap, and the concentration of ligand on the beads is nearly infinite; purification from large starting quantities of cell extract is practical; elution can be either with denaturants (such as urea) or by competition with soluble dextran B512 (average MW 6,000 or 10,000)	High affinity for streptavidin (K_d ~ 70 n*M*), but not for egg white avidin (allows blocking of cellular biotin and biotinylated proteins with avidin); avidin and streptavidin reagents for affinity purification and detection are readily available from multiple commercial sources; elutes cleanly and quickly with biotin under native conditions; binding stable to high salt (400 m*M* NaCl)
Disadvantages	Affinity of RNA for the ligand is not as high as with the streptavidin tag, so extensive washing of the resin after binding leads to slow loss of bound RNA; native elution by competition with dextran leaves dextran in the eluate, which is harder to remove than biotin	Resin more expensive; number of binding sites per bead much lower than with Sephadex; egg white avidin is usually needed to block biotin in crude cellular lysates

2. Materials

2.1. Design of Hybrid RNAs

Computer with a secondary structure prediction program or access to Web-based servers for RNA-folding prediction. Downloadable programs can be found at the following addresses, links to Web-based RNA folding servers can also be found there: for the PC, RNAstructure at http://rna.urmc.rochester.edu/rnastructure.html and RNAdraw at http://www.rnadraw.com; for the Mac, Mulfold at http://iubio.bio.indiana.edu/soft/molbio/mac/.

2.2. Construction of the Hybrid RNA Sequence

2.2.1. Recursive Polymerase Chain Reaction

1. Template DNA for polymerase chain reaction (PCR), either genomic DNA or plasmid-borne sequence.
2. Oligonucleotides.

3. Taq DNA polymerase, supplied with commercial buffer (New England Biolabs).
4. Agarose gel electrophoresis equipment.
5. Gel purification miniprep kit (Qiagen).
6. Restriction enzymes, supplied with commercial buffer (New England Biolabs).
7. T4 DNA ligase, supplied with commercial buffer (New England Biolabs).
8. Competent *E. coli* cells DH5α or XL1-Blue.
9. Access to an automated DNA sequencing facility.

2.2.2. Site-Directed Mutagenesis

1. Suitable vector bearing the target RNA sequence.
2. QuikChange® mutagenesis kit (Stratagene™).
3. Oligonucleotides.
4. Restriction enzymes, supplied with commercial buffer (New England Biolabs).
5. T4 DNA ligase, supplied with commercial buffer (New England Biolabs).
6. Competent *E. coli* cells DH5α or XL1-Blue.
7. Access to an automated DNA sequencing facility.

2.3. Preparing Extracts for Binding

1. Transformed or transfected cell line suitable for biological system under study.
2. Appropriate growth media and incubator system for above.
3. 1X lysis buffer: 50 m*M* HEPES, pH 7.4, 10 m*M* $MgCl_2$, 100 m*M* NaCl, 1 m*M* dithiothreitol (DTT), 0.1% Triton X-100, 10% glycerol, Complete® protease inhibitors (Roche).
4. Acid-washed glass beads 425–600 μm (Sigma).
5. Protein content assay, Micro Bicinchoninic acid assay (Pierce).

2.4. Affinity Purification Using Streptavidin

1. Streptavidin agarose (Sigma).
2. Avidin from egg white (Sigma).
3. 1X lysis buffer, as in **Subheading 2.3., step 3**, but without Complete.
4. Ultrafree-MC centrifugal filter device (Millipore).
5. D-Biotin (Sigma).

2.5. Affinity Purification Using Sephadex G-200

1. Sephadex G-200 (bead size 40–120 μm) (Pharmacia).
2. 1X lysis buffer, as in **Subheading 2.3., step 3**, but without Complete.
3. Ultrafree-MC centrifugal filter device (Millipore).
4. Dextran produced by *Leuconostoc mesenteroides* strain B-512 (approx molecular weight 6,000 or 10,000 Da) (Fluka) (Sigma).

3. Methods

The choice of aptamer to use can be guided by their individual properties, which are summarized in the **Heading 1**. and **Table 1**. The methods described next outline (1) the design of the hybrid aptamer-target RNA molecule, (2) the construction of the expression plasmid and strains (in yeast), and (3) the affinity purification of the tagged RNA/complex from cell extracts.

3.1. Guidelines for Designing Aptamer-Tagged RNAs

The most significant step in the use of aptamer tags for the isolation of RNAs and their complexes is in the successful introduction of an aptamer into the target RNA; there are three main considerations. First, the affinity tag must be inserted such that both the aptamer and the target RNA are able to maintain their correct folds (**Subheadings 3.1.1.–3.1.4.**). Second, it is important to consider the steric effects of using such tags and the potential for interference with the normal function or interactions of the target particle under study (**Subheading 3.1.5.**). Finally, the aptamer tag may introduce a sensitive target for nuclease degradation, directly affecting the purification and potentially the overall integrity of the particle under study (**Subheading 3.1.6.**).

3.1.1. Maintaining the Correct Fold of the RNA

Aptamer insertion is most successful where some structural information regarding the target RNA is known or can be obtained. A predicted, and preferably experimentally verified, secondary structure provides an essential starting point for design. It can be very difficult to predict the best position for the aptamer tag, and it is best to simultaneously test several hybrid RNA designs. The aptamer sequence can be inserted internally within the RNA structure at an appropriate position or at either the 5′ or 3′ end.

Not only the position of the aptamer tag should be varied, but also various linkers and spacers that may affect the folding of the RNA or how the tag is presented to the affinity matrix. In addition to testing whether a particular tagged RNA can be used to facilitate isolation, wherever possible, the experiment should be designed to test whether the tagged RNA functions normally in the cell (*see* **Note 1**).

3.1.2. Internal Tagging of the RNA

The use of internal tags has been quite successful, and it is recommended that the aptamer be positioned in the terminal loop of a long, nonconserved stem that is known to protrude into solution. When such structural information is available, the design of tagged RNAs is easier, but it is still prudent to design and test several tagged RNAs to increase the chance of success. In the case of the *RPR1* RNA from the yeast RNase P, chemical probing has shown that the terminal loops of several helices protrude into solution ***(24)***, and aptamer tags at some of these positions have proven to be useful ***(20,25)*** (**Fig. 2**).

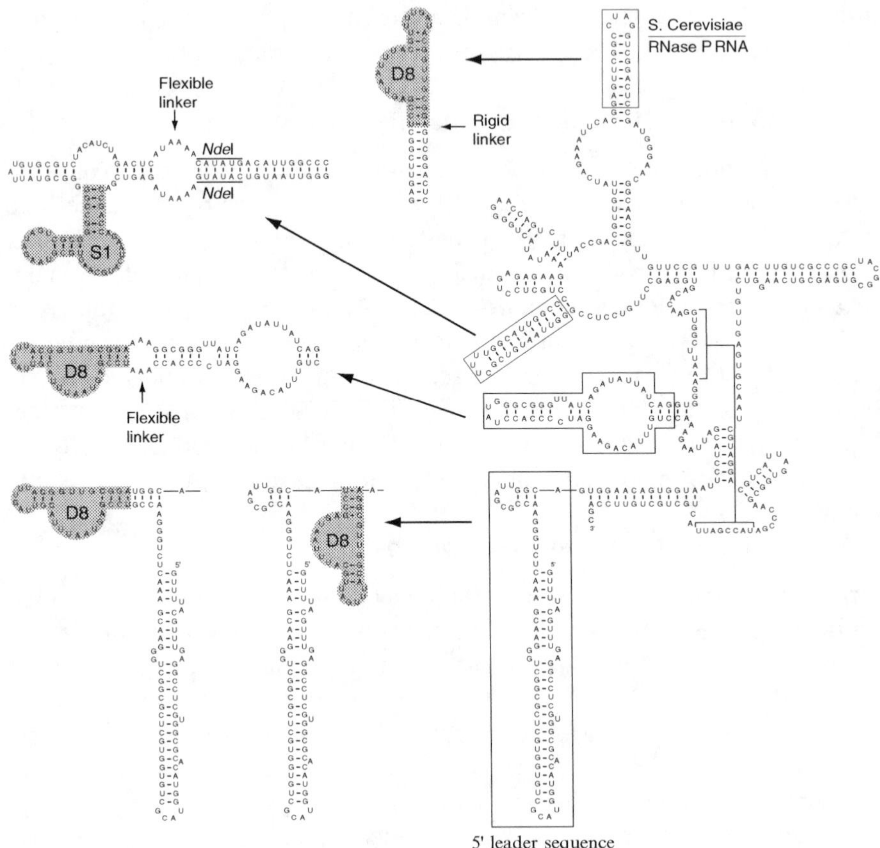

Fig. 2. Affinity tag insertions into both the precursor and the mature *RPR1* RNA subunit of RNase P. Individual insertions of both the streptavidin (S1) and Sephadex (D8) aptamers into the *RPR1* and pre-*RPR1* RNA are shown. The minimal tag sequences are shown with a gray background. The streptavidin (S1) aptamer insertion was carried out using the full, originally identified aptamer sequence, prior to the identification of the minimal sequence. The minimal S1 aptamer sequence has also been used within the RPR1 RNA with equivalent results. The use of both flexible and rigid linking regions of various lengths is illustrated here (refer to **Subheading 3.1.** for discussion). Constructs were generated by either of the techniques discussed in **Subheading 3.2.**, and an *Nde*I restriction site used to insert the streptavidin (S1) aptamer is indicated. Each position shown has been used to isolate either the mature RNase P or pre-RNase P from cellular extracts using a singly tagged RNA (*see* **Note 8**).

3.1.3. Tagging the Regions Flanking the Target RNA

An alternative strategy is to insert the tag at either the 5′ or 3′ end of a target RNA. This approach has been successful for the isolation of the pre-RNase P holoenzyme through the tagging of its known 5′ leader sequence *(23)* (**Fig. 2**). However, it should be noted that the use of these aptamers for the isolation of messenger RNA (mRNA)–protein complexes has not been as successful (*see* **Note 2**).

3.1.4. Testing the Design of Hybrid RNAs

After designing the hybrid RNA, it is useful to predict the folding of these sequences *in silico*. Secondary structure prediction programs based on the Mfold algorithm *(26)*, such as RNAstructure/Mulfold, or the Vienna algorithm *(27)*, such as RNAdraw, are freely available, and there are several Web-based RNA-folding servers available (**Subheading 2.1.**). The predictions generated by such programs are not always reliable and should be used as a guide only (*see* **Note 3**). Ultimately, the best way to test the designed hybrid sequences is to attempt to isolate an RNP from a strain that displays no detectable phenotypic defects. If such attempts fail, it may be due to factors such as steric effects or nuclease degradation, discussed in the next two subheadings. To investigate this possibility, attempting to bind the sequence to the affinity matrix as an in vitro transcribed RNA will confirm whether the tag itself is able to fold correctly within the context of the target RNA sequence.

3.1.5. The Steric Effects of Tagging an RNA

It is difficult to predict whether an aptamer will sterically hinder RNP formation or whether RNP formation will hinder the aptamer function. We reiterate that designing and testing several tagged RNAs is recommended. It can be useful to insert a spacer region between the tag and the RNA, whether the tag is positioned at the end of the RNA or internally. By varying the length of a spacer region or by introducing either a rigid or flexible linker, it is possible to affect how the aptamer tag interacts with both the target complex and the affinity matrix. **Figure 2** shows examples of tags in the *RPR1* RNA sequence that have been used to successfully isolate RNase P or its precursor complexes from *S. cerevisiae*.

3.1.6. Maintaining the Tag on the RNA: Considering Nucleases

In general, RNA-based tags are less stable than protein-based tags as they can be susceptible to nuclease degradation. The potential for nuclease degradation is specific to the target RNA, the positioning of the tag, and the system from which purification is being attempted (*see* **Note 4**). Here, we discuss some general strategies toward reducing potential nuclease problems. In our experience, we have found that internal tags appear to be less susceptible to nucleases than flanking tags. Degradation by 3′ to 5′ exonucleases can be especially rapid, and a tag on the 3′ end of the RNA may be prone to this. However, flanking tags can be protected from degradation through the use of a strong "closing" stem. Placing a GC-rich stem, with a tetraloop, immediately downstream from the tag will increase the resistance of the RNA to single-strand-specific 3′ to 5′ exonucleases. In addition, any spacers and flexible linkers could potentially introduce targets for endonucleases, and this should be considered.

3.2. Constructing the Hybrid Aptamer Target RNA Sequence

Once a set of hybrid RNAs has been designed, the specific sequences must be prepared and cloned into an appropriate expression vector for transcription within a chosen system. We have used the pRS315 (*LEU2*-marked) vector for our own work in yeast. The recombinant clones contain the *RPR1* RNA sequence along with its natural upstream promoter regions and downstream terminator regions. There are at least two methods available for the construction of a particular hybrid RNA sequence. The first involves the construction of a synthetic gene via recursive, or nested, PCR. The second method involves the introduction of a restriction site via site-directed mutagenesis; the site can be used to allow the direct insertion of any designed aptamer and spacer/linker at this position (*see* **Note 5**). The methods are outlined in **Subheadings 3.2.1.** and **3.2.2.** and shown schematically in **Fig. 3**. DNA manipulations such as performing a PCR reaction, gel isolation of DNA, restriction digests, ligations, transformation of plasmid DNA, or the isolation of plasmid DNA can be performed by any standard protocols.

3.2.1. Generation of a Hybrid RNA Sequence Using Recursive PCR

Once a particular site for the introduction of the aptamer sequence has been designed, it is useful to split the sequence into upstream and downstream sections between which the aptamer sequence is to be inserted (refer to **Fig. 3**). To design primers that amplify the downstream fragment: For primer 1, there must be sufficient complementarity to the 5′ end of the fragment sequence (20–25 nt), and the aptamer sequence, which does not anneal to the template, is added to the 5′ end of this oligonucleotide. Primer 2 must have sufficient complementarity to the 3′ end of the fragment sequence (20–25 nt), and a suitable restriction site can be added at the 5′ end of the oligonucleotide to facilitate cloning. For the upstream fragment, the aptamer is added at the 3′ end of the fragment (primer 4) and the restriction site, or other required sequence (*see* **Note 6**), at the 5′ end of the fragment (primer 3). The entire construction can be carried out in three PCR reactions using these four designed primers as outlined next.

1. The downstream fragment (**Fig. 3A(i)**) is produced by PCR using primers 1 and 2 using a suitable template DNA encoding the target RNA sequence.
2. The upstream fragment [**Fig. 3A(ii)**] is produced by PCR using primers 3 and 4 using a suitable template DNA encoding the target RNA sequence.

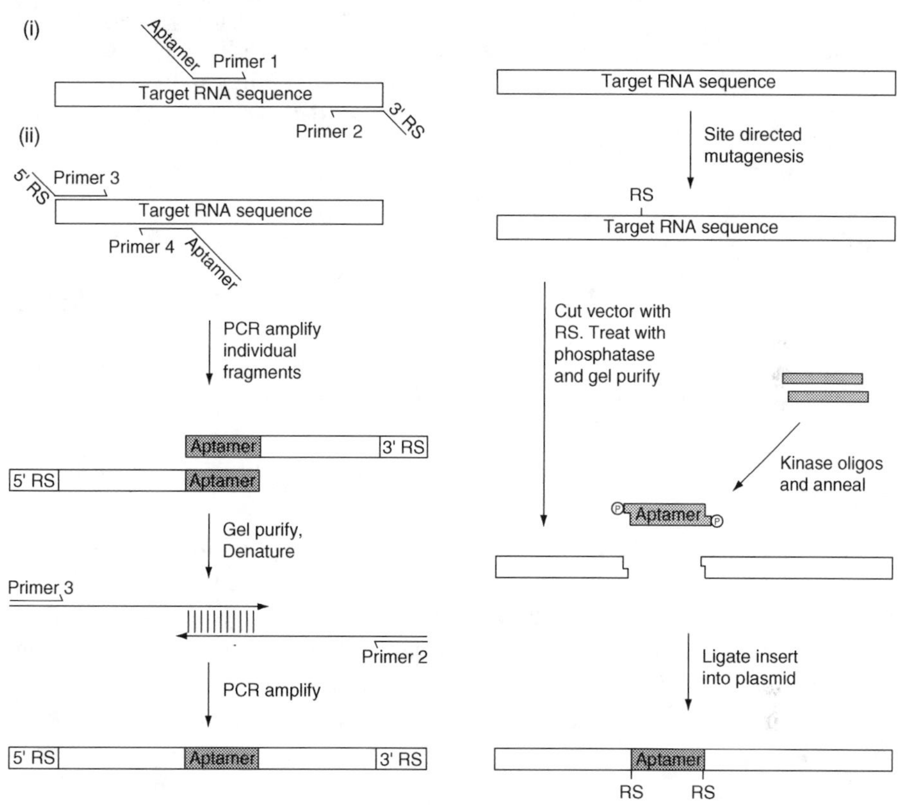

Fig. 3. Strategies for the construction of aptamer-tagged RNA sequences. (**A**) The insertion of the aptamer at the required position using a strategy based on the polymerase chain reaction (PCR). Individual PCR fragments are produced, (**i**) and (**ii**), where the aptamer sequence or a required restriction site (RS) is introduced at the desired location by incorporation into the 5′ regions of the primers. After the two aptamer-tagged sequence fragments have been produced and isolated, they are used in trace amounts as template for a further round of PCR. The outlying primers (2 and 3) can be used to amplify the entire region and produce the final product bearing an accurately inserted aptamer within the target RNA sequence. (**B**) The insertion of an aptamer at a required position using a cloning-based strategy. First, site-directed mutagenesis is used to introduce a suitable restriction site into the target RNA sequence (plasmid borne). The restriction site is used to facilitate the introduction of a synthetic insert bearing the required aptamer sequence. The presence of a correctly oriented insert can be confirmed by PCR screening colonies prior to sequencing.

3. The PCR inserts are checked on an agarose gel next to size markers and are gel purified.
4. The two PCR fragments are used as a template (1–5 ng each fragment) in PCR with primers 2 and 3 to amplify the entire fragment bearing the tagged RNA sequence and introduced 5′ and 3′ restriction sites. The first round of the PCR creates a full-length gene fragment due to the overlapping half-gene fragments, which will be extended by the DNA polymerase.
5. The correct size of the PCR fragment is checked on an agarose gel next to size markers, and the DNA is gel isolated.
6. The fragment is cut with the appropriate restriction enzymes at the termini and ligated into a suitable plasmid at the correct restriction sites.
7. The ligation reaction is transformed to competent DH5α or XL1-Blue cells and plated onto selective media.
8. Plasmid is isolated from individual colonies, and the PCR insert is verified by appropriate restriction digests and automated DNA sequencing.

3.2.2. Site-Directed Mutagenesis and Aptamer Cloning at the Introduced Restriction Sites

For the specific introduction of restriction sites into a particular sequence, we routinely use the QuikChange kit (Stratagene). This kit uses two user-designed oligonucleotides bearing the mutated sequence to generate two mutant strands from a plasmid template. The parental (methylated) DNA is digested, using DpnI, and the resulting annealed double-stranded, nicked complementary DNA (cDNA) molecules are transformed into *E. coli*, in which the nicked cDNA is repaired to generate a plasmid bearing the required mutations. For more exact details of the technique, including oligonucleotide design, refer to the product manual. For insertion of an aptamer sequence, within a stem internal to the target RNA, it is possible to mutate the terminal loop of the stem to a single-hexameric restriction site. The site can then be used to introduce the tag by the insertion of synthetic DNA into this restriction site. The result is that the introduced tag is flanked by two copies of the initial restriction site, which could potentially be used for future insertion of altered tags. The overall cloning strategy is represented in **Fig. 3B**. When using a single restriction site, the direction of the insert must be screened by PCR or by sequencing. In some cases, the sequential introduction of two restrictions sites, to facilitate the insertion of the aptamer sequence, may be necessary.

1. Restriction site(s) are introduced into the plasmid sequence by site-directed mutagenesis using the QuikChange kit (Stratagene).
2. The plasmid is cut with the appropriate restriction enzyme. After digestion, the cut vector is treated with alkaline phosphatase to remove terminal phosphates and prevent religation to itself. The DNA is then purified from an agarose gel in preparation for ligation with the insert.

3. The designed aptamer for insertion is purchased as two individual DNA oligonucleotides bearing the correct overhangs for ligation. To facilitate ligation, these oligonucleotides are individually phosphorylated with polynucleotide kinase. The enzyme is removed by phenol/chloroform extraction, and the two oligonucleotides are annealed in equal concentration by heating to 65 °C and rapid cooling on ice to create a double-stranded insert.
4. The insert is ligated into the vector at the restriction site.
5. The ligation reaction is transformed to competent DH5α or XL1-Blue cells and plated onto selective media.
6. Plasmid is isolated from individual colonies, and the presence and correct orientation of the inserted aptamer are verified by appropriate restriction digests and automated DNA sequencing.

3.3. Preparing Extracts for Binding

Recombinant DNA clones that express the aptamer-tagged RNA sequence should be transformed or transfected into a suitable cell line, dependant on the biological system under study. We have used a yeast strain bearing a chromosomal knockout of the essential *RPR1* gene, allowing us to introduce the pRS315-*RPR1* plasmid bearing the modified RNA sequence ***(23)***. This has allowed us to directly assess the impact of the tagging modifications on the phenotype of the transformed strains. In other systems in which such genetic manipulation may not be possible, the tagged RNA can be introduced into cells containing the chromosomal wild-type copy and will result in the isolation of only the introduced, aptamer-tagged, RNA and its complexes. In these cases, efforts to assess the integrity of the isolated complexes through an appropriate activity assay are required. An example is the successful processing of pre-transfer RNA (tRNA) substrates that was shown by the human RNase P holoenzyme, isolated using the streptavidin aptamer ***(22)***. In contrast, cotranscription of the tagged RNA alongside the chromosomal wild-type sequence has been exploited to specifically isolate known lethal mutants of the 23S ribosomal RNA sequence for further study ***(13)***.

The particular method used to generate crude extracts will vary depending on the system in which the tagged RNA is being expressed. In general, any protocol for the preparation of crude extracts should focus on first maintaining the temperature at 4 °C throughout and second producing the extract and binding it to beads as quickly as possible to minimize ribonuclease (and protease) degradation of the RNP complexes. For the preparation of extracts from yeast cultures, we use the following protocol:

1. A single colony of the appropriate yeast strain is grown to saturation in YPD media at 30 °C and used to inoculate 1 L YPD media.
2. The culture is grown at 30 °C until an OD_{600} (optical density at 600 nm) of around 1–2, and the cells are harvested by centrifugation at 4000 *g* for 15 min at 4 °C.

3. The pellet is washed and resuspended in 10 pellet volumes of ice-cold sterile water, then repelleted by centrifugation at 4000 *g* for 5 min at 4 °C.
4. The pellet is resuspended in 3 mL of lysis buffer containing Complete protease inhibitors (Roche).
5. The cells are lysed by vortexing with one-third volume (~1 mL) of acid-washed beads, 425–600 µm (Sigma) for 20–30 min. Efforts to keep the mixture cool during the lysis procedure are strongly recommended. Periodic cooling on crushed dry ice can be used to keep the sample cool, taking care to avoid freezing the sample. For larger-scale preparations (>20 L culture), we have carried out lysis by passing the resuspended cells (**step 4**) through a microfluidizer three to four times (model 110Y, Microfluidics). Again, efforts to keep the mixture well cooled throughout the lysis process are strongly recommended.
6. The lysate is cleared by centrifugation at 14,000 *g* for 20 min at 4 °C. A second ultracentrifugation step is optional.
7. The protein content in the cleared lysate is determined using a Micro BCA assay (Pierce).

3.4. Isolation of Aptamer-Tagged RNA Complexes From Lysates Using Streptavidin

In the case of the streptavidin aptamer, a background of free biotin and biotinylated cellular material can be blocked by the addition of egg white avidin (Sigma) to the lysate prior to binding to the streptavidin affinity resin (the RNA affinity tag does not bind to egg white avidin). For optimal yields from different lysates, it can be useful to experimentally determine the correct amount of avidin required to block biotin in the lysate. In our hands, lysates produced from yeast cultures can be sufficiently blocked by the addition of 10 µg avidin per milligram protein in the lysate. The streptavidin aptamer can stably bind to the affinity resin at up to 400 m*M* NaCl, allowing purification under different salt conditions to be attempted if desired.

1. Block the lysate with 5–20 µg of avidin per milligram of protein in the extract. Incubate at 4 °C for 10 min prior to binding to the affinity resin.
2. Incubate the lysate with 10–20 µL of streptavidin beads per milligram of protein in the extract. Binding should be carried out at 4 °C for 1 h in 1X lysis buffer at an appropriate volume (5–10 times the bed volume). The beads are separated by centrifugation at 4000 *g* for 5 min at 4 °C.
3. Wash the beads with 20 bed volumes of 1X lysis buffer five times for 5 min each at 4 °C.
4. Transfer the beads to an Ultrafree-MC centrifugal filter unit (0.45-µm pore size Millipore) and wash twice with 5 bed volumes of 1X lysis buffer at 4 °C.
5. Elute the beads by incubating for 0.5–1 h at 4 °C with 2 bed volumes of 1X lysis buffer containing 5 m*M* D-biotin.

3.5. Isolation of Aptamer-Tagged RNA From Lysates Using Sephadex G-200

1. The Sephadex G-200 beads are prepared by swelling 1 g in 50–100 mL of lysis buffer (without Complete) overnight at room temperature, resulting in approx 20 mL of resin. The beads are then washed several times with lysis buffer, and a 50% suspension is prepared.
2. Incubate the lysate with 10 µL of Sephadex beads per milligram of protein in the extract. Binding should be carried out at 4 °C for 1 h in 1X lysis buffer at an appropriate volume (5–10 times the bed volume). The beads are separated by centrifugation at 4,000 *g* for 5 min at 4 °C.
3. Wash the beads with 5 volumes of 1X lysis buffer five times for 3 min each at 4 °C.
4. Transfer the beads to an Ultrafree-MC centrifugal filter unit (0.45-µm pore size Millipore) and wash twice with 2 bed volumes of 1X lysis buffer at 4 °C.
5. Elute the beads by incubating for 0.5–1 h at 4 °C with 2 bed volumes of 1X lysis buffer containing 50 mg/mL dextran B512 (molecular weight 6,000 or 10,000 Da) (*see* **Note 7**).

4. Notes

1. If an essential RNA is the target of aptamer tagging, then if an appropriate genetic system is available, a knockout strain can be used such that the only copy of the RNA present is the tagged version. This will enable a thorough assessment of the growth phenotype to determine the impact of the tagged RNA on normal functions. In the case of the *RPR1* RNA from the *S. cerevisiae* RNase P, the growth phenotype has been assessed at various temperatures. Northern blot analysis has been carried out to determine that the relative levels of precursor and product *RPR1* RNA are normal, and the pre-tRNA processing profile was also shown to be the same as wild type ***(20,25)***.
2. Attempts to purify mRNA and pre-mRNA complexes have not worked well, even though the purified tagged mRNAs themselves were able to bind the affinity resins. However, their complexes did not isolate well from the cellular extracts. It is not clear whether this is because the tags are obscured by the protein, the aptamer structure is antagonized by RNP formation, or both. When the RNA structure is not constrained, we recommend that the investigator consider of the use of biotinylated oligonucleotide tagging (see **Heading 1**.) as an alternative approach should the use of aptamers fail.
3. When testing a hybrid RNA design by *in silico* folding, it can be useful to analyze both the full sequence and also smaller fragments, such as the aptamer and 50 nt of upstream and downstream sequence. Smaller sequence fragments of RNA and their interactions with local sequence elements are often more reliably predicted by such programs.
4. In the yeast system, it is possible to reduce the overall nuclease activity by constructing a *PEP4* deletion strain. The *PEP4* gene encodes a protease that is responsible for the maturation of many cellular nucleases; when *PEP4* is deleted, these nucleases are not processed and no longer function.

5. The use of restriction sites to introduce aptamer sequences can place some constraints on the sequence design in these areas. However, these can also be used for future manipulations, allowing alternative tags and spacer/linker designs to be introduced via these sites. Although not necessary for the construction of a gene via PCR, it can be useful to include such sites in the design for this future purpose.
6. For the rapid production of templates to be used for in vitro transcription of the target RNA, the T7 RNA polymerase promoter sequence (5′-TAATACGACTCACTATAGG-3′) can be encoded by the outlying primer and thus incorporated into the final PCR product. In vitro RNA transcripts produced from such templates can be used to test the binding of the tagged RNA to the affinity matrix in the absence of cellular components.
7. It has also been possible to elute the aptamer using lower molecular weight dextran, such as enzymatically synthesized dextran ($M_r \sim 1500$). However, higher concentrations are required (100 mg/mL) as the minimal binding unit for the D8 aptamer is estimated as dextran with 9–10 glucose subunits (molecular weight ~1600–1800). These elution conditions can be used to help facilitate concentration or dialysis.
8. We have successfully performed dual-affinity steps on a doubly tagged *RPR1* RNA construct (both Sephadex and streptavidin aptamers). The doubly tagged strains show no phenotypic defects, and the isolation of active RNase P can be successfully carried out from either of the two inserted tags. The second affinity tag can suffer nuclease degradation during the first affinity step (when it is not being used). We have been able to overcome this by taking great care to keep our extracts cold and working quickly. The use of the Sephadex tag as the initial step is recommended since the low cost allows a larger volume of resin to be used. Employing the streptavidin aptamer second allows the eluted sample to be more concentrated and avoids potential problems associated with removing dextran from the sample. Although successful, the use of dual aptamer tags to purify yeast nuclear RNase P resulted in significantly lower yields when compared to a dual protein affinity tag (TAP-tag).

Acknowledgments

This work was supported by a National Institute of Health grant (R01 6M34869) to D.R.E. We thank both Shaohua Xiao and Rebecca Haeusler for their helpful comments and review of the text.

References

1. Nygren, P. A., Stahl, S., and Uhlen, M. (1994) Engineering proteins to facilitate bioprocessing. *Trends Biotechnol.* **12**, 184–188.
2. Porath, J. (1992) Immobilized metal ion affinity chromatography. *Protein Expr. Purif.* **3**, 263–281.
3. Field, J., Nikawa, J., Broek, D., et al. (1988) Purification of a RAS-responsive adenylyl cyclase complex from *Saccharomyces cerevisiae* by use of an epitope addition method. *Mol. Cell. Biol.* **8**, 2159–2165.

4. Evan, G. I., Lewis, G. K., Ramsay, G., Bishop, J. M. (1985) Isolation of monoclonal antibodies specific for human c-myc proto-oncogene product. *Mol. Cell. Biol.* **5**, 3610–3616.
5. Rigaut, G., Shevchenko, A., Rutz, B., Wilm, M., Mann, M., Séraphin, B. (1999) A generic protein purification method for protein complex characterization and proteome exploration. *Nat. Biotechnol.* **17**, 1030–1032.
6. Uhlen, M., et al. (1983) Gene fusion vectors based on the gene for staphylococcal protein A. *Gene* **23**, 369–378.
7. Simons, P. C., and Vander Jagt, D.L. (1981) Purification of glutathione S-transferases by glutathione-affinity chromatography. *Methods Enzymol.* **77**, 235–237.
8. Schmidt, T. G., Koepke, J., Frank, R., Skerra, A. (1996) Molecular interaction between the Strep-tag affinity peptide and its cognate target, streptavidin. *J. Mol. Biol.* **255**, 753–766.
9. Prickett, K. S., Amberg, D. C., and Hopp, T. P. (1989) A calcium-dependent antibody for identification and purification of recombinant proteins. *Biotechniques* **7**, 580–589.
10. Rouault, T. A., Hentze, M. W., Haile, D. J., Harford, J. B., Klausner, R. D. (1989) The iron-responsive element binding protein: a method for the affinity purification of a regulatory RNA-binding protein. *Proc. Natl. Acad. Sci. U. S. A.* **86**, 5768–5772.
11. Bardwell, V. J., and Wickens, M. (1990) Purification of RNA and RNA-protein complexes by an R17 coat protein affinity method. *Nucleic Acids Res.* **18**, 6587–6594.
12. Das, R., Zhou, Z., and Reed, R. (2000) Functional association of U2 snRNP with the ATP-independent spliceosomal complex E. *Mol. Cell.* **5**, 779–787.
13. Leonov, A. A., Sergiev, P. V., Bogdanov, A. A., Brimacombe, R., Dontsova, O. A. (2003) Affinity purification of ribosomes with a lethal G2655C mutation in 23 S rRNA affects the translocation. *J. Biol. Chem.* **278**, 25664–25670.
14. Kieft, J. S., and Batey, R. T. (2004) A general method for rapid and nondenaturing purification of RNAs. *RNA* **10**, 988–995.
15. Blencowe, B. J., Sproat, B. S., Ryder, U., Barabino, S., Lamond, A. I. (1989) Antisense probing of the human U4/U6 snRNP with biotinylated 2′-OMe RNA oligonucleotides. *Cell* **59**, 531–539.
16. Lingner, J., and Cech, T. R. (1996) Purification of telomerase from *Euplotes aediculatus*: requirement of a primer 3′ overhang. *Proc. Natl. Acad. Sci. U. S. A.* **93**, 10712–10717.
17. Wilson, D. S., and Szostak, J. W. (1999) In vitro selection of functional nucleic acids. *Annu. Rev. Biochem.* **68**, 611–647.
18. Bachler, M., Schroeder, R., and von Ahsen, U. (1999) StreptoTag: a novel method for the isolation of RNA-binding proteins. *RNA* **5**, 1509–1516.
19. Hartmurth, K., Vornlocher, H. P., and Luhrmann, R. (2004) Tobramycin affinity tag purification of spliceosomes. *Methods Mol. Biol.* **257**, 47–64.
20. Srisawat, C., and Engelke, D. R. (2001) Streptavidin aptamers: affinity tags for the study of RNAs and ribonucleoproteins. *RNA* **7**, 632–641.
21. Srisawat, C., Goldstein, I. J., and Engelke, D. R. (2001) Sephadex-binding RNA ligands: rapid affinity purification of RNA from complex RNA mixtures. *Nucleic Acids Res.* **29**, e4.

22. Li, Y., and Altman, S. (2002) Partial reconstitution of human RNase P in HeLa cells between its RNA subunit with an affinity tag and the intact protein components. *Nucleic Acids Res.* **30**, 3706–3711.
23. Srisawat, C., Houser-Scott, F., Bertrand, E., Xiao, S., Singer, R. H., Engelke, D. R. (2002) An active precursor in assembly of yeast nuclear ribonuclease P. *RNA* **8**, 1348–1360.
24. Tranguch, A. J., Kindelberger, D. W., Rohlman, C. E., Lee, J. Y., Engelke, D. R. (1994) Structure-sensitive RNA footprinting of yeast nuclear ribonuclease P. *Biochemistry* **33**, 1778–1787.
25. Srisawat, C., and Engelke, D. R. (2002) RNA affinity tags for purification of RNAs and ribonucleoprotein complexes. *Methods* **26**, 156–161.
26. Zuker, M. (2003) Mfold Web server for nucleic acid folding and hybridization prediction. *Nucleic Acids Res.* **31**, 3406–3415.
27. Hofacker, I. L. (2003) Vienna RNA secondary structure server. *Nucleic Acids Res.* **31**, 3429–3431.

4

Assembly and Glycerol Gradient Isolation of Yeast Spliceosomes Containing Transcribed or Synthetic U6 snRNA

Kenneth J. Dery, Shyue-Lee Yean, and Ren-Jang Lin

Summary

Studies of RNA–protein interactions often require assembly of the RNA–protein complex using in vitro synthesized RNA or recombinant protein. Here, we describe a protocol to assemble a functional spliceosome in yeast extracts using transcribed or synthetic RNAs. The in vitro assembled spliceosome is stable and can be isolated by sedimentation through glycerol gradients for subsequent analysis. The protocols describe two procedures to prepare RNA: using bacteriophage RNA polymerases or ligation of RNA oligos using T4 DNA ligase. We also describe the preparation of splicing competent yeast extracts, the assembly of the spliceosome, and the isolation of the spliceosome by glycerol gradient sedimentation. To allow exogenously added U6 RNA to be incorporated into the spliceosome, the endogenous U6 small nuclear RNA (snRNA) in the extract is eliminated by an antisense U6 DNA oligo and ribonuclease H; a "neutralizing" U6 DNA oligo was then added to protect the incoming U6 RNA. This protocol allows study of the role individual bases or the phosphate backbone of U6 plays in splicing and of the interaction between U6 snRNA and the spliceosomal proteins.

Key Words: glycerol gradient sedimentation; pre-mRNA splicing; RNA ligation; *Saccharomyces cerevisiae*; SP6 transcription; spliceosome; splicing extracts; T7 transcription; U6 reconstitution.

1. Introduction

Ribonucleoproteins (RNPs) are often composed of an RNA molecule complexed to multiple proteins. RNPs serve many functions, ranging from general protection and stabilization of RNA, regulation of RNA processing, to assisting in RNA transport *(1)*. RNPs can vary in size from the large ribosome and spliceosome to the small nuclear RNPs (snRNPs). A critical step to understanding the molecular basis of cellular processes performed by RNP complexes

From: *Methods in Molecular Biology, vol. 488: RNA-Protein Interaction Protocols*
Edited by: Ren-Jang Lin © Humana Press Inc, Totowa, NJ

involves the successful assembly, isolation, and characterization of constituent components in these RNA–protein complexes.

The yeast spliceosome forms through a series of RNA–protein associations involving small nuclear ribonucleoprotein particles (U1, U2, U4/U6, U5 snRNPs) and non-snRNP proteins with the pre-mRNA (pre-messenger RNA). Through a highly structured and ordered process, the mature spliceosome attains its catalytic conformation necessary to perform the two transesterification reactions characteristic of pre-mRNA splicing (reviewed in **refs.** ***2*** and ***3***). The study of how spliceosome complexes assemble in vitro has revealed important insights into the overall mechanisms of splicing. Originally identified by glycerol gradient isolation as a 40S particle in yeast ***(4,5)*** or 60S particle in mammalian extracts ***(6)***, the spliceosome is responsible for recognizing sites of chemistry that delineate the exon (expressed sequence) and intron (intervening sequence) boundaries.

To investigate the snRNP contribution to splicing in yeast or human cells, it is necessary to deplete endogenous small nuclear RNA (snRNA) from whole-cell or nuclear extracts and reconstitute the RNP with synthetic RNA. Many studies, for example, have reported the successful depletion of endogenous U6 RNA by nuclease digestion by ribonuclease (RNase) H in yeast ***(7)***, nematode ***(8)***, and human extracts ***(9)***. Other snRNPs have been depleted in nuclear extract using anti-m_3G affinity chromatography ***(10,11)***.

Purification of RNPs can be achieved by immunoprecipitating U1, U2, U4/U6, and U5 snRNPs using anti-m_3G antibodies ***(12)***, density gradient sedimentation ***(4,13)***, nondenaturing gel electrophoresis ***(14,15)***, and MS2 affinity purification ***(16,17)***. We describe our protocol for the glycerol gradient isolation of yeast spliceosomes.

We present details to assemble a functional spliceosome in yeast extracts using transcribed or synthetic RNAs. Whole-cell extract prepared from a temperature-sensitive *prp2-1* yeast strain is used to assemble precatalytic spliceosomes ***(18)***. We show how complexes are visualized and analyzed for their ability to retain biological activity as well as how to deplete endogenous U6 and add back synthetic U6 to be incorporated into the assembling spliceosome. The protocols described here are derived from our work characterizing the *S. cerevisiae* spliceosome ***(18–20)*** but should be applicable to study protein or RNA constituents of other RNP complexes.

2. Materials

All buffers and solutions were made using sterile, filtered, deionized (by a Barnstead MP-3A Megapure system), or glass double-distilled water.

2.1. RNA Preparation

1. α-[^{32}P] uridine triphosphate (UTP) (800 Ci/mmol, 10 μCi/μL MP Biomedical, Solon, OH).

2. γ-[^{32}P] adenosine triphosphate (ATP) (7000 Ci/mmol, 10 µCi/µL MP Biomedical).
3. pSP6ACT: A low copy number plasmid containing a fragment of the *ACT1* gene ***(21)***.
4. pKDACT: A high copy number plasmid derived from pSP6ACT and containing an ACT1 DNA fragment.
5. pT7U6: the *SNR6* gene encoding U6 snRNA cloned into pUC18 *(7)*.
6. RNasin (Promega, Madison, WI). Use 10 U per 10-µL reaction.
7. SP6 RNA polymerase (New England Biolabs, Ipswich, MA). Use 10 U per 10-µL reaction.
8. T7 RNA polymerase (New England Biolabs). Use 10 U per 10-µL reaction.
9. Taq DNA polymerase, 5000 U/mL (New England Biolabs).
10. T4 polynucleotide kinase, 10,000 U/mL (New England Biolabs).
11. T4 DNA Ligase, 2000 U/µL (New England Biolabs).
12. Dithiothreitol (DTT). To prepare a 0.1 *M* stock, add 0.154 g of DTT to 10 mL of water.
13. TE-saturated phenol, chloroform, and isoamyl alcohol (25:24:1), pH 5.2 (Fisher Scientific, Fair Lawn, NJ).
14. 3 *M* sodium acetate, pH 5.3. Dissolve 408.1 g of sodium acetate · $3H_20$ in 800 mL of water. Adjust to pH 5.3 with glacial acetic acid, and the volume is then adjusted to 1 L with water.
15. 10X T4 polynucleotide kinase (T4 PNK) buffer: 700 m*M* Tris-HCl, pH 7.6, 100 m*M* $MgCl_2$, 50 m*M* DTT.
16. 10X T4 DNA ligase buffer: 500 m*M* Tris-HCl, 100 m*M* $MgCl_2$, 10 m*M* ATP, 100 m*M* DTT, 250 µg/mL bovine serum albumin (BSA).
17. 10X RNA polymerase reaction buffer: 400 m*M* (pH 7.9), 60 m*M* $MgCl_2$, 20 m*M* spermadine, 100 m*M* DTT.
18. 10X ThermoPol reaction buffer: 200 m*M* Tris-HCl, 100 m*M* ammonium sulfate, 100 m*M* KCl, 20 m*M* $MgSO_4$, 1% Triton X-100, pH 8.8 (New England Biolabs).
19. 10X transcription master mix. Add 2.5 µL of 20 m*M* guanosine triphosphate (GTP), 2.5 µL of 20 m*M* ATP, 2.5 µL of 20 m*M* cytosine triphosphate (CTP), 2.5 µL of 4 m*M* UTP, 10 µL of 10X RNA polymerase reaction buffer, and 10 µL of 0.1 *M* DTT. The mixture is gently agitated and briefly microcentrifuged. Dispense 3 µL per 10-µL reaction.
20. T7/U6 forward primer (5′TAATACGACTCACTTAGGGGTTCGCGAAGTAACC CTTCGTGG).
21. U6-specific reverse primer (5′AAAACGAAATAAATCTCTTTGTAAAACGG). DNA oligos are stored at 100 µ*M* concentration.
22. DNA ligation bridge ($cdU6_{19-100}$) or U6 DNA ligation template (DLT) oligo (5'ATCTCTTTGTAAAACGGTTCATCCTTATGCAGGGGAACTGCTGATCAT CTCTGTATTGTTTCAAATTGACCAAATGTCCACG).
23. U6 RNA oligos: Oligo 1, U6/1–37 (5′GUUCGCGAAGUAACCCUUCGUGGA CAUUUGGUCAAUU); oligo 2, U6/38–69 (5'UGAAACAAUACAGAGAUGA UCAGCAGUUCCCC); oligo 3, U6/70–83 (5′UGCAUAAGGAUGAA); oligo 4, U6/84–112 (5′CCGUUUUACAAAGAGAUUUAUUUCGUUUU).

2.2. Gel Isolation

1. ANE buffer: 10 m*M* sodium acetate, pH 5.2, 1 m*M* ethylenediaminetetraacetic acid (EDTA), 10 m*M* NaCl.
2. Chloripane (chloroform/isoamyl alcohol/phenol = 50:1:50). Mix 467 mL phenol (density 1.07), 467 mL chloroform, 9 mL isoamyl alcohol, 0.47 g 8-hydroxyquinolin in a total volume of 943 mL. This chloripane solution is saturated with 1X volume of ANE buffer by stirring for 1 h at room temperature.
3. 10X TBE buffer: Add 108 g Tris-HCl, 55 g boric acid and 9.8 g EDTA (pH 8.0). Add water to 1 L volume.
4. Polyacrylamide 7.5 *M* urea solution (29:1; 20%): Add 96.67 g acrylamide, 3.33 g bis-acrylamide, 225 g ultrapure urea (MP Biomedicals), and water to a final volume of 500 mL. To the solution, add 4 g of Mixed Bed Resin (Sigma, cat. no. M8032) and stir at room temperature for 1 h. Add 50 mL of 10X TBE buffer and filter using a Buckner funnel (*see* **Note 1**). Store at room temperature.
5. 7.5 *M* urea TBE buffer: Add 450 g of ultrapure urea to 400 mL of water. Add 8 g of Mixed Bed Resin and stir at room temperature for 1 h. Add 100 mL of 10X TBE buffer and filter using a Buckner funnel. Store at room temperature.
6. 5% denaturing polyacrylamide 7.5 *M* urea gel. Mix 15 mL of 7.5 *M* urea TBE buffer and 5 mL of polyacrylamide 7.5 *M* urea solution, then add 200 µL of 10% ammonium persulfate and 20 µL of TEMED.
7. Splicing load buffer: Mix 40 µL of 0.5 *M* EDTA, 2 mg of xylene cyanol, and 2 mg of bromophenol blue in 10 mL of deionized formamide.
8. RNA elution buffer: 0.5 *M* ammonium acetate, pH 5.3, 1 m*M* EDTA, and 0.1% sodium dodecyl sulfate (SDS).

2.3. Extract Preparation

1. 0.5 *M* phenylmethanesulfonyl fluoride (PMSF). Dissolve 0.871 g of PMSF into 10 mL of dimethyl sulfoxide.
2. 3 *M* potassium chloride (KCl). Dissolve 223.68 g of KCl into 1 L of water.
3. Buffer A: 10 m*M* HEPES, pH 7.8, 1.5 m*M* $MgCl_2$, 10 m*M* KCl, 0.5 m*M* DTT, 1 µ*M* leupeptin, 1 µ*M* pepstatin A, 1 µ*M* benzamidine, 1 µ*M* µ*M* PMSF.
4. Buffer D: 20 m*M* HEPES, pH 7.8, 20% glycerol, 50 m*M* KCl, 0.2 m*M* EDTA, 0.5 m*M* DTT, 0.5 m*M* PMSF.
5. SB30 buffer: 1 *M* sorbitol, 50 m*M* Tris-HCl, pH 7.8, 10 m*M* $MgCl_2$, 30 m*M* DTT.
6. SB3 buffer: 1 *M* sorbitol, 50 m*M* Tris-HCl, pH 7.8, 10 m*M* $MgCl_2$, 3 m*M* DTT.
7. TE buffer: 50 m*M* Tris-HCl, pH 8.0, 0.2 m*M* EDTA.
8. Yeast strain CRL2101 ***(18)***: (α, *prp2-1, ade2, his3, lys2-801, ura3*).
9. Yeast strain EJ101 ***(22)***: (α, *his1, prb1-1122, prc1-126*).
10. YPD: Dissolve 20 g peptone, 20 g dextrose, and 10 g yeast extract into 1 L of water; aliquot if desired and autoclave.
11. Zymolyase-100T, 100,000 U/g (MP Biomedicals).
12. Zymolyase buffer: 20 m*M* potassium phosphate, pH 7.8, 5% glucose.
13. Dialysis tubing with a molecular weight cutoff between 6000 and 8000 Da (Spectrum Medical Industries, CA).

2.4. In Vitro Splicing

1. d_1 (U6/28–54) oligonucleotide (5' ATCTCTGTATTGTTTCAAATTGACCAA).
2. αd_1 oligo (5' TTGGTCAATTTGAAACAATACAGAGAT).
3. 8% denaturing polyacrylamide gel: Mix 12 mL of 7.5 *M* urea TBE buffer and 8 mL of polyacrylamide 7.5 *M* urea solution and then add 200 µL of 10% ammonium persulfate and 20 µL of TEMED.
4. Splicing stop buffer: 50 m*M* sodium acetate, pH 5.3, 1 m*M* EDTA, and 0.1% SDS.

2.5. Glycerol Gradient Isolation

1. Polyallomer ultracentrifuge tube (13 × 51 mm, Beckman, CA).
2. 10X gradient buffer containing triton (GFT) solution: 400 m*M* potassium phosphate, pH 7.4, 15 m*M* $MgCl_2$, 2 m*M* EDTA, 0.5% Triton X-100.
3. Glycerol gradient solutions (27%, 23%, 19%, 15%). Add 5 mL of 10X GFT, *X* mL of 50% glycerol, and (45 – *X*) mL of water, where *X* equals the percentage of glycerol. For example, to make 27% glycerol gradient solution, mix 5 mL of 10X GFT, 27 mL of 50% glycerol, and 18 mL of water.
4. Gradient splicing buffer: 2 m*M* ATP, 2.5 m*M* $MgCl_2$, 7 µL of gradient fraction, 1 µL of complementing factor.
5. 40% (w/v) acrylamide. Dissolve 40 g acrylamide into 100 mL of water. Filter sterilize and store in a Pyrex bottle wrapped in aluminum foil.
6. 40% (w/v) acrylamide:bis-acrylamide (59:1) solution. Mix equal parts of 40% (w/v) acrylamide with ultrapure AccuGel (40% acrylamide, 29:1, sequencing grade, National Diagnostics, cat. no. EC-852). Store in a Pyrex bottle wrapped in aluminum foil.
7. Low-melting agarose.
8. 50% glycerol.
9. SurfaSil (Pierce, Rockford, IL, cat. no. 42800.) *Use with caution.*
10. 3MM chromatography paper (Whatman International, Maidstone, England, cat. No. 3030-347).

3. Methods

3.1. RNA Preparation

3.1.1. SP6 Reaction

Plasmid pKDACT harbors an ACT1 DNA fragment used to prepare ACT1 pre-mRNA (*see* **Note 2**). Transcription reactions are typically prepared in 10-µL volumes composed of 1X RNA polymerase reaction buffer, 0.5 m*M* (ATP, CTP, GTP), 0.1 m*M* UTP, 10 m*M* DTT, 50 ng of HindIII linearized pKDACT1 (*see* **Note 3**), 20 µCi of α-[^{32}P] UTP (800 Ci/mmol), 0.5 U RNasin, and 0.5 U SP6 RNA polymerase (*see* **Note 4**). The assembly of this reaction is greatly simplified by preparing a 10X transcription master mix (*see* **Subheading 2.1.**). To assemble a 10-µL reaction, the following are added to a prechilled microcentrifuge tube:

1. 3 µL of RNase-free water.
2. 3 µL of 10X transcription master mix.
3. 1 µL of 0.5 µg pKDACT1 DNA template.
4. 0.5 µL of both SP6 RNA polymerase and RNasin (10 U each).
5. 2 µL of α-[^{32}P] UTP.
6. Incubate at 37 °C for 60 to 120 min (*see* **Note 5**)
7. Reactions are terminated by adding an equal volume of splicing load buffer, heated and resolved by denaturing polyacrylamide gel electrophoresis (PAGE; *see* **Subheading 3.1.4.** for further details).

3.1.2. T7 Reaction

The plasmid pT7U6 was used as a DNA template for the generation of the U6 RNA transcript. To prepare a linear template for runoff transcription, a 50-µL polymerase chain reaction (PCR) is performed:

1. 40.5 µL of RNase-free water.
2. 1 µL of pT7U6 plasmid (25 ng).
3. 5 µL of 10X ThermoPol Reaction buffer.
4. 2 µL of 2.5 m*M* dNTPs.
5. 0.5 µL of the gene-specific T7/U6 forward and U6-specific reverse primer.
6. 0.5 µL of Taq DNA polymerase (5 U).
7. Follow the PCR cycle conditions: 1 cycle (94 °C, 30 s), 25 cycles (94 °C, 30 s; 55 °C, 30 s; 72 °C, 30 s), 1 cycle (72 °C, 10 min).
8. After verifying by agarose gel electrophoresis, the PCR product is isolated by adding an equal volume of TE-saturated phenol, pH 5.2, to the reaction mixture and vortexed for 15–30 s. The sample is centrifuged for 5 min at room temperature to separate the phases. Next, 90% of the upper, aqueous layer is transferred to a clean tube, and an equal volume of ice-cold chloroform is added, vortexed briefly, and centrifuged for 3 min at room temperature. The lower chloroform layer is discarded, and the DNA is precipitated by adding 5 µL of 3 *M* sodium acetate, pH 5.3, and 2.5 to 3 volumes of 100% ethanol. The pellet is dried and resuspended in RNase-free water, and the concentration is adjusted to 100 ng/µL.

As with the SP6 transcription reaction, T7 transcription reactions are prepared in 10-µL volumes containing 1X RNA polymerase reaction buffer, 0.5 m*M* (ATP, CTP, GTP), 0.1 m*M* UTP, 10 m*M* DTT, 200 ng of PCR-generated U6 DNA template, 20 µCi of α-[^{32}P] UTP (800 Ci/mmol), 0.5 U of RNasin, and 0.5 U of T7 RNA polymerase (final concentrations listed). The assembly of this reaction is simplified by preparing a 10X transcription master mix (*see* **Subheading 2.1.**). To a prechilled Eppendorf tube, the following are added in order:

1. 2 µL of RNase-free water.
2. 3 µL of 10X transcription master mix.
3. 2 µL of PCR-generated U6 DNA template (100 ng/µL).
4. 0.5 µL of both T7 RNA polymerase and RNasin (10 U each).

5. 2 μL of α-[^{32}P] UTP.
6. Incubate at 37 °C for 60–120 min.
7. Reactions are terminated by adding an equal volume of splicing load buffer, heated and re-solved by denaturing PAGE (*see* **Subheading 3.1.4.** for further details).

3.1.3. Ligation of RNA Oligonucleotides

Joining of two or more RNA fragments by T4 DNA ligase-mediated, DNA oligo-splinted ligation ***(23,24)*** can be used to introduce modified nucleotides within the RNA molecule. Previously, our lab and others have investigated how individual bases ***(25)*** and the phosphate backbone ***(26)*** of U6 snRNA regulate pre-mRNA splicing. U6 RNA oligonucleotides (U6/1–37, U6/38–69, U6/70–83, and U6/84–112, hereafter described as oligo 1, 2, 3, or 4, respectively), obtained from Dharmacon Research (*see* **Note 6**), were used to ligate and generate U6 snRNA ***(26)***. To set up a 20-μL ligation reaction, add

1. 0.5 μL of 1X T4 PNK buffer.
2. 1 μL each of the second, third, and fourth oligos (10 pmol/μL stock).
3. 0.5 μL of both T4 PNK and RNasin.
4. 0.5 μL (5 μCi) of γ-[^{32}P] ATP.
5. Incubate at 37 °C for 30 min.

To ensure all 5′ ends are phosphorylated, add 1.2 μL of 0.5 m*M* ATP and incubate the reaction at 37 °C for another 30 min (*see* **Note 7**). The T4 PNK is heat inactivated at 65 °C for 20 min before continuing to the following ligation step:

1. Add 1.8 μL of RNase-free water.
2. Add 1 μL of the RNA oligo 1 (10 pmol/μL).
3. Add 1 μL of the DLT oligo (8 pmol/μL).
4. The mixture is incubated at 90 °C for 1 min and then left at room temperature for 20 min to promote slow annealing.
5. Add
   a. 2 μL of 10X T4 DNA ligase buffer.
   b. 3.92 μL of 24% polyethylene glycol (PEG).
   c. 0.5 μL RNasin.
   d. 2 μL of T4 DNA ligase (2000 U/μL, NEB).
   e. Add RNase-free water to a final volume of 20 μL.
6. The tube is incubated for 4 h at 30 °C.
7. Reactions are terminated by adding an equal volume of splicing load buffer, heated and re-solved by denaturing PAGE (*see* **Subheading 3.1.4.**).

3.1.4. Gel Isolation

The RNA generated by transcription or ligation can be purified away from the DNA and other reaction components on a denaturing polyacrylamide gel.

The ^{32}P-labeled RNA band can be visualized through the aid of an X-ray film and can be cut out using a razor blade. The RNA can then be eluted with an overnight incubation in RNA elution buffer.

1. For all RNA species, reactions are terminated by adding an equal volume of splicing load buffer, heated at 95 °C for 1 min, and then chilled on ice.
2. The samples are separated on a 5% polyacrylamide 7.5 *M* urea denaturing gel (195 × 160 × 0.4 mm) at 500 V for 1–2 h in 1X TBE running buffer. The ACT1 sample is electrophoresed until the xylene cyanol dye reaches the bottom of the gel. The expected size of this modified pre-mRNA is 517 nucleotides, with both exon 1 and 2 containing 73 and 135 nucleotides, respectively, while the intervening sequence contains 309 nucleotides. The U6 sample is electrophoresed for 1 h at 500 V in 1X TBE buffer. The full-length U6 produced by transcription and ligation migrates near xylene cyanol. The expected size of this product is 112 nucleotides.
3. The ^{32}P-labeled RNAs are visualized by exposing X-ray film to the gel for 1 min followed by a very short pulse of light (*see* **Note 8**).
4. The RNA band corresponding to the full-length product is excised, transferred to a microcentrifuge tube, and 420 µL of RNA elution buffer is added and eluted overnight in a shaking incubator (at 250 rpm) at 37 °C.
5. The next day, 390 µL of the elution buffer containing the RNA (use caution to avoid small gel fragments) is recovered and divided into 195-µL aliquots.
6. 19.5 µL of 3 *M* sodium acetate, pH 5.3, and 536 µL of 100% ethanol are added to each aliquot, and the tubes are mixed and placed at –80 °C for at least 30 min (*see* **Note 9**).
7. Precipitated RNA is recovered by microcentrifugation at 4 °C for 30 min. The pellet is washed with ice-cold 70% ethanol, dried, resuspended in 20 µL of RNase-free water, and the radioactivity is counted by Cerenkov counting (*see* **Note 10**).

3.1.5. RNA Concentration Determination

The amount of RNA in a mole produced from a transcription reaction using ^{32}P-labeled nucleoside triphosphates can be estimated by using the Cerenkov counts per minute (cpm) with the assumption that 1 µCi equals 1 million Cerenkov cpm. Three factors are needed for the conversion (for example, if the radioisotope used is ^{32}P-UTP): the amount of "hot" UTP (*h*, in pmol) and the amount of "cold" UTP (*c*, in pmol), the specific activity of the radioisotope (*s*, in µCi/pmol), and the number of uridines in the RNA product (*n*).

$$1\,\text{pmol of RNA} = s \times [h/(h+c)] \times n \times 10^{6}\ \text{cpm}$$

For example, in a 10-µL transcription reaction, 2 µL of 10 µCi/µL of α-[^{32}P] UTP (800 Ci/mmol) is used to make an RNA that contains 150 uridine residues in a reaction containing 100 µ*M* of unlabeled UTP.

1. *s* equals 800 Ci/mmol or 0.8 µCi/pmol.

2. h equals 25 pmol (20 µCi divided by 0.8 µCi/pmol).
3. c equals 1000 pmol (100 mM × 10 µL).
4. n equals 150.
5. Assuming 1 µCi equals 1×10^6 cpm.
6. 1 pmol of RNA = $0.8 \times 25/(25 + 1000) \times 150 \times 10^6 = 0.8 \times 1/40 \times 150 \times 10^6 = 3 \times 10^6$ cpm. Thus, 3000 cpm equals 1 fmol of RNA.

3.2. Preparation of Whole-Cell Yeast-Splicing Extracts

1. *Saccharomyces cerevisiae prp2* mutant strain is cultured in YPD at 26 °C until mid-to-late logarithmic phase (A_{600} of 2–4). Then, 2 mL of 0.5 M PMSF/L culture are added with vigorous mixing, and the cultures are incubated for another 30 min at 26 °C. For protease-deficient strains like EJ101 or BJ2168, the PMSF is often omitted.
2. Cells are collected by centrifugation at room temperature for 10 min at 7277 *g* (5000 rpm in a Sorvall H-6000A rotor). The wet weight of the cells is measured (*see* **Note 11**), and this value is referenced for the rest of the extract preparation procedure.
3. Cell pellets are resuspended in TE buffer at 50 mL per 10 g of cells. Cells are collected by centrifugation at room temperature for 6 min at 6084 *g* (6000 rpm in a Sorvall GS3 rotor). Pellets are then resuspended in SB30 buffer at 30 mL per 10 g of cells (*see* **Note 12**).
4. Cells are again collected by centrifugation at room temperature. Pellets are next resuspended in SB3 buffer at 20 mL per 10 g of cells.
5. To follow the digestion of the cell wall by zymolyase in the next step, it is important to aliquot 10 µL of the suspended cells to a new Eppendorf tube. To this, add 1 mL of 10% SDS and measure the optical density at 800 nm (OD_{800}; typically between 0.6 and 0.8 absorbance). This reading will serve as the baseline comparison for later digestion readings.
6. The suspended cells are treated with 3 mg of zymolyase in 200 µL of zymolyase buffer per 10 g of cells (*see* **Note 13**). The cells are shaken at 30 °C for 45 min at 120 rpm in an incubator shaker (for example, Innova 4080, New Brunswick Scientific) and then remeasure the OD_{800} by aliquoting 10 µL of the zymolyase-treated cell suspension plus 1 mL of 10% SDS. The absorbance reading must decrease by at least 80% of the original untreated absorbance reading before continuing (*see* **Note 14**).
7. Zymolyase-treated cells, now called *spheroplasts*, are collected by centrifugation at room temperature for 5 min at 1463 *g* (3000 rpm in a Sorvall GSA rotor). The pellets are gently resuspended in SB3 containing 0.5 mM PMSF and centrifuged under the same conditions. All steps from this point onward are carried out in a cold room.
8. Spheroplast pellets are chilled on ice for 30 min and then resuspended in buffer A at 10 mL per 10 g of cells.
9. Resuspended pellets are poured into a prechilled dounce glass homogenizer type B and lysed by 14 hard strokes (7 up and 7 down) with 2 min between each pair

of strokes to allow heat to dissipate (*see* **Note 15**). The homogenate volume is measured and transferred to a beaker that has been prechilled with a stir bar.

10. With stirring, 1/10 volume of 3 *M* KCl is added dropwise to a final concentration of 0.2 M. The solution is stirred on ice for 30 min and then centrifuged at 4 °C for 30 min at 34,541 *g* (17,000 rpm in a Sorvall SS-34 rotor).
11. The supernatant is collected and centrifuged at 4 °C for 60 min at 137,472 *g* (37,000 rpm in a Beckman 60Ti rotor).
12. The clear middle layer, which contains the splicing active extract, is collected by a Pasteur pipet and transferred to a new 50-mL polypropylene conical tube. The top and bottom layers, which contain lipids and cellular debris, are discarded. The top and bottom layers, though containing some splicing activity, should be avoided (*see* **Note 16**).
13. The whole-cell extracts are placed in dialysis tubing with a molecular weight cutoff between 6000 and 8000 Da and dialyzed against 1 L of buffer D for 3 h at 4 °C with two changes. Excessive dialysis decreases overall splicing efficiency *(27)*. The dialyzed mixture is centrifuged at 4 °C for 30 min at 34,541 *g* (17,000 rpm in a Sorvall SS-34 rotor). The supernatant is carefully aliquoted into 1.5-mL microcentrifuge tubes (0.5- to 1-mL aliquots), flash frozen in liquid nitrogen, and stored at –80 °C.
14. Whole-cell extracts prepared in this manner retain activity over multiple freeze-thaw cycles and are active for at least 2 yr at –80 °C.

3.3. In Vitro Pre-mRNA Splicing

Typically, 10-µL splicing reactions are done containing 2 m*M* ATP, 60 m*M* KPO_4, pH 7.4, 2.5 m*M* $MgCl_2$, 2.4% PEG-8000, 3.3 fmol ACT1 RNA, and 40%(v/v) yeast-splicing extract.

1. Mix in a microcentrifuge tube:

   a. 2.5 µL RNase-free water.
   b. 0.5 µL of 40 m*M* ATP.
   c. 0.5 µL of 1.2 *M* KPO_4, pH 7.4.
   d. 0.5 µL of 50 m*M* $MgCl_2$.
   e. 1 µL of 24% PEG-8000.
   f. 1 µL of ACT1 RNA (10,000 cpm).
   g. 4 µL of yeast-splicing extract.

2. The reaction is performed at 23 °C for 30 min.
3. After incubation, the reaction is stopped by the addition of 200 µL of splicing stop buffer (*see* **Note 17**) and 200 µL of chloripane solution (*see* **Note 18**).
4. The solution is vortexed for 2 min and centrifuged for 1 min, and 180 µL of the supernatant are transferred to a new tube.
5. Add 20 µL of 3 *M* sodium acetate, pH 5.3, and 500 µL of 100% ethanol. The mixture is incubated at –80 °C for 30 min followed by centrifugation at 4 °C for 15 min.
6. The pellet is washed with 70% ethanol, dried, and resuspended in 3 µL of RNase-free water and 7 µL of splicing load buffer.

7. The samples are heated at 95 °C for 2 min.
8. Load onto an 8% denaturing polyacrylamide 7.5 *M* urea gel.
9. Run in 1X TBE buffer at 500 V for about 100 min or until the xylene cyanol tracking dye reaches the bottom.
10. The products of this reaction are visualized by autoradiography or phosphoimager.

To heat-inactive mutant Prp2 protein, extracts derived from temperature-sensitive *prp2-1* strain are incubated at 37 °C for 20 min in the presence of 20 m*M* KPO_4, pH 7.4, and 2 m*M* $MgCl_2$ *(28)* before use in splicing assay. To complement or rescue the heat-inactivated *prp2-1* extracts, 1 μL of purified wild-type Prp2 protein (10 ng/μL) is added after **step 2**, and the reaction is incubated further at 23 °C for 20 min.

Figure 1 shows the activity levels of three extracts made from yeast strain CRL2101 by dounce homogenization. Extracts 1 and 3 show high activity, as

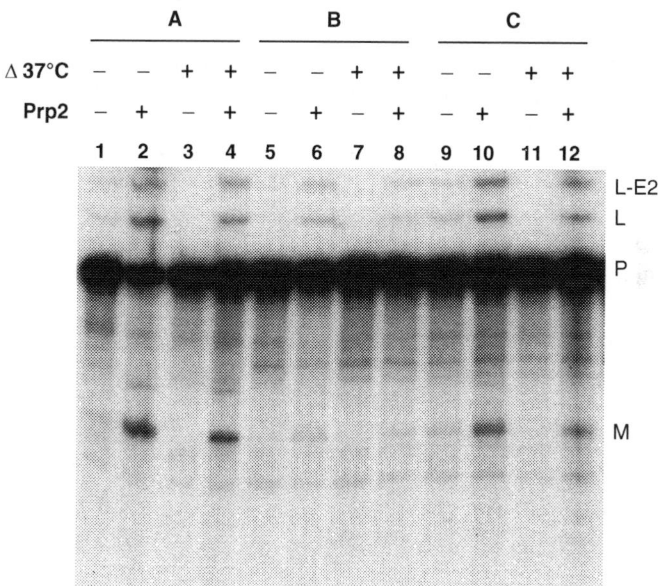

Fig. 1. Yeast in vitro pre-mRNA (messenger RNA) splicing of the ACT1 RNA. Three *prp2-1* extracts (A, B, C) were isolated and tested for their ability to convert pre-mRNA to mRNA. ^{32}P-labeled ACT1 pre-mRNA was incubated in extracts with (+) or without (–) a 10-min heat inactivation step at 37 °C (Δ37 °C). Purified Prp2 protein was added to reactions labeled with + above the lanes. Extracts B and C were isolated using separate type B dounce homogenizers. After the reaction, RNA was extracted and separated on a denaturing 8% polyacrylamide-urea gel. The RNA species from the reaction are marked on the right-hand side: L-E2 is lariat intron-exon 2, L is lariat intron, P is pre-mRNA, and M is mRNA. Exon 1 is not shown.

evidenced by the amount of mRNA and splicing intermediate produced as compared to extract 2. Each extract was heat inactivated (lanes 3, 4, 7, 8, 11, and 12) to demonstrate extract dependence on Prp2 for activity. This can be seen by the restoration of splicing on complementation with purified Prp2 (lanes 4, 8, and 12). Extracts 1 and 3 show weak splicing activity without the heat inactivation step (lanes 1 and 9), indicating some activity of the mutant Prp2 protein.

3.4. U6 Reconstitution

RNase H is a unique endoribonuclease that recognizes RNA–DNA duplexes and cleaves the phosphodiester backbone of the RNA molecule ***(29)***. Previous experiments have established DNA oligo d_1, which targets nucleotides 28–54 of U6 and can direct degradation of U6 snRNA in the extract in the presence of ATP *(7)*. Other studies have also demonstrated that oligonucleotide-directed RNase H digestion can result in extracts becoming splicing incompetent ***(30,31)***. Reconstitution studies have identified mutations ***(25,32,33)*** or phosphorothioates ***(26,34)*** that are important in snRNA function in splicing. We further modified the protocol to include a neutralizing oligo that minimizes the degradation of added synthetic U6 RNA (*see* **Note 19**); our protocol was recently used by others *(35)*.

1. To deplete endogenous U6, 6 pmol of the d_1 oligo and 2 m*M* ATP (final concentration) are added to 4 µL wild-type or *prp2* mutant extract for a total volume of 4.7 µL.
2. Gently mix and incubate at 33 °C for 10 min.
3. Chill on ice and add 15 pmol (0.3 µL of 50 µ*M* stock) of the αd_1 oligo to neutralize the remaining d_1 oligo for 5 min at 4 °C.
4. Add 2.5 µL of the following splicing cocktail and 1 µL of 120 n*M* of U6 RNA (made by transcription or by ligation):
    a. 0.5 µL of 40 m*M* ATP.
    b. 0.5 µL of 1.2 *M* KPO_4, pH 7.4.
    c. 0.5 µL of 50 m*M* $MgCl_2$.
    d. 1 µL of 24% PEG-8000.
5. Incubate at 23 °C for 5 min.
6. Add 1 µL of the Prp2 protein (10 ng/µL stock) or buffer D if Prp2 is not needed and 0.5 µL of ACT1 pre-mRNA (20,000 cpm/µL stock) to bring the total volume to 10 µL.
7. Incubate at 23 °C for 20 min.

Figure 2 shows a typical reconstitution experiment using wild-type and mutant extracts. Each extract was treated with oligos d_1 and αd_1 (lanes 2, 3, 5, and 6) or mock treated (lanes 1 and 4) and then incubated with T7-transcribed U6 (lanes 2, 3, 5, and 6) or water (lanes 1 and 4). We use the C48A allele (lanes 2 and 5) of U6 as a negative control. Previously, it has been shown this allele does

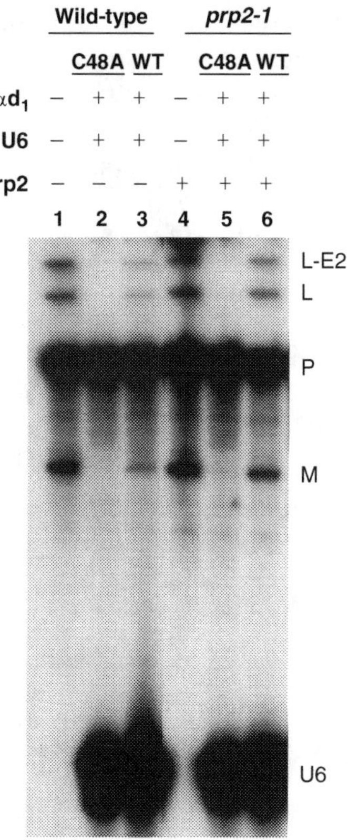

Fig. 2. U6 reconstitution assay using wild-type and *prp2* mutant extracts. Extracts were treated with the oligo d_1 (+) or mock treated (–) and incubated at 33 °C for 20 min. The same samples were treated with the antisense oligo αd_1 or water and incubated at 4 °C for 5 min. Other splicing ingredients were added with (+) or without (–) T7-transcribed U6 RNA and incubated at 23 °C for 5 min. Then, 3.3 fmol of ACT1 pre-mRNA (messenger RNA) was added with (+) or without (–) the Prp2 protein, and the samples were incubated at 23 °C for 30 min. RNA was extracted and separated on a denaturing 8% polyacrylamide-urea gel. See **Fig. 1** for the RNA species marked. The added ^{32}P-labeled U6 is shown near the bottom of the gel.

not associate with the spliceosome and leads to a failed in vitro splicing event under reconstitution conditions ***(32)***. Note the levels of U6 reconstitution differ between the two extracts (*see* **Note 20**).

3.5. Glycerol Gradient Isolation

Discontinuous density gradients composed of sucrose or glycerol solution can resolve various types of macromolecules depending on the size and shape of the sedimenting complex relative to the centrifugal force. The resolution of spliceosomes ***(4,5)***, prespliceosomes ***(36)***, and heterogeneous nuclear ribonucleoprotein (hnRNP) complexes ***(37)*** have been demonstrated previously. Consideration of salt concentration is important as KCl concentrations ranging from 25 to 400 m*M* retard the migration of the spliceosome from a 40S particle to 25S ***(38)***. Purifying a high concentration of spliceosome particles remains a challenge. It is estimated that only 0.1 fmol of spliceosome is produced from yeast-splicing reactions ***(38)***. However, the gradient purification method is gentle, so spliceosomes remain intact after fractionation and can be chased in the presence of splicing factors.

A discontinuous glycerol gradient appropriate for isolating the yeast spliceosome is prepared by layering successively decreasing glycerol density solutions (27%, 23%, 19%, and 15% glycerol in 1X gradient buffer) on one another in an ultracentrifuge tube. This method is simple, quick, and reproducible.

1. Set the ultracentrifuge tube firmly on a test tube rack.
2. Use a 2-mL pipet to draw 1.2 mL of 27% glycerol gradient solution and dispense into the ultracentrifuge tube. The same pipet can be used for another gradient to be made at the same time. A P-1000 Pipetman can also be used to make gradient, but the result is less satisfactory.
3. Use another 2-mL pipet to draw 1.2 mL of 23% glycerol gradient solution. The pipet is then inserted into the ultracentrifuge tube at an angle so that the tip touches the wall and the top of the solution that has been laid. The solution is slowly dispensed from the pipet with little mixing of the two solutions. Do not empty the last drop of the solution from the pipet. If layered properly, a distinct line separating the two layers can be seen.
4. Lay the third layer (19% glycerol gradient solution) and the fourth layer (15%) with the same caution.
5. The glycerol gradient tubes are kept at 4 °C for 6 h to overnight to produce a linear gradient before loading the samples (*see* **Note 21**).
6. A 50-μL splicing reaction is assembled as described in **Subheading 3.3.** or **3.4.**
7. The reaction is then diluted with 150 μL of ice-cold 1X GFT buffer. Usually, 190 μL of this mixture is layered on top of the 15–27% gradient made from **steps 1–5**, and the remaining 10 μL is kept for reference.
8. The tubes are inserted into the buckets and centrifuged at 302,986 *g* (37,000 rpm in a Beckman SW 55-Ti rotor) for 2 h at 4 °C.
9. After centrifugation, the buckets are removed and kept at 4 °C. To fractionate the gradients, the tip of a P-200 micropipettor is placed just below the surface of the gradient, and 0.2 mL is drawn out into an ice-cold microcentrifuge tube. A total of 24 fractions are taken. This method of hand fractionation takes less than 5 min to perform and is reproducible.
10. Each fraction is then counted by Cerenkov counting in a scintillation counter and stored at −80 °C until further analysis.

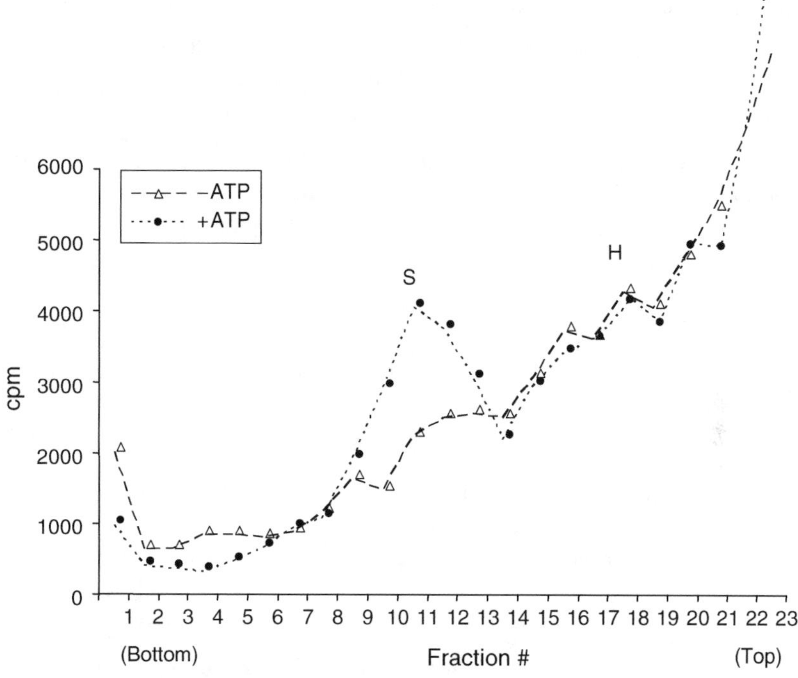

Fig. 3. Glycerol gradient analysis of splicing reactions. ^{32}P-labeled ACT1 RNA was incubated in heat-inactivated *prp2* extracts under splicing conditions with or without adenosine triphosphate (ATP) for 30 min to allow assembly of the spliceosome. In the −ATP reaction, endogenous ATP was depleted by incubating the extract at 23 °C for 5 min with 5 m*M* glucose. The 50-µL splicing reaction was diluted with 1X GFT buffer, loaded on a 15–27% glycerol gradient, and sedimented for 2 h. Fractions of 200 µL were collected, and the amount of the ^{32}P-labeled RNA present in the fractions was determined as Cerenkov counts per minute (cpm) using a scintillation counter. S, spliceosomes; H, heterogeneous nuclear ribonucleoprotein (hnRNP) complexes.

Figure 3 shows two gradients, one from a splicing reaction containing *prp2* mutant extract with ATP and the other from the same reaction but without ATP. The spliceosome (S) sediments to a peak at fraction 11, while the hnRNP or H complexes sediment to around fraction 18. Fractions at the top contain degraded RNA.

3.6. Evaluating Isolated Yeast Spliceosomes by Native Gel Electrophoresis

The prespliceosome, precatalytic, and mature spliceosome in a splicing reaction mixture or in a gradient fraction derived from a splicing reaction can be separated on a vertical, nondenaturing, polyacrylamide gel ***(14,15,25,39,40)***. Spliceosome assembly intermediates have also been separated on native gels and characterized biochemically ***(40,41)***. Here is a protocol for analyzing spliceosomes after gradient isolation on a nondenaturing gel (20 × 20 × 1.5 mm).

1. Wearing gloves, coat one glass plate with a Kimwipe tissue and two drops of SurfaSil (a siliconizing reagent) in a fume hood. Let it dry for 10 min and then assemble one coated plate and one uncoated plate together with spacers.
2. Prepare the gel mix (52 mL):
   a. Melt 0.26 g of low-melting agarose in 40.3 mL of water plus 2.6 mL 10X TBE with low heat.
   b. Add 5.2 mL of 50% glycerol and 3.9 mL of 40% (w/v) acrylamide:bis-acrylamide (59:1) and mix by swirling gently. The temperature of the solution should be around 50 °C to keep the agarose melted and to prevent acrylamide from polymerizing.
   c. Add 40 mg of ammonium persulfate and 11 µL of TEMED; pour the gel solution immediately between two glass plates.
   d. The gel is allowed to sit for 2–4 h at room temperature and then for 1 h at 4 °C.
   e. The gel is prerun without samples for 1 h at 4 °C in 0.5X TBE.
3. Pipet 10 µL of the gradient fraction of interest into a prechilled microcentrifuge tube.
4. Add a trace amount (0.25%) of xylene cyanol and bromophenol blue to the tube and load onto the gel.
5. Run the gel at 50 V for 18–24 h at 4 °C or until the xylene cyanol migrates to 15 cm into the gel.
6. After running, the glass plates are carefully separated, and the gel is transferred to a 3 MM Whatman paper, wrapped in plastic wrap, and exposed to an X-ray film or in a phosphoimager cassette for the desired length of time.

Figure 4 shows the nondenaturing gel results of analyzing peak fractions (fractions 10 and 12) from the gradients of splicing reactions carried out with ATP or without ATP (**Fig. 3**). Spliceosomes are detected in the +ATP reaction but not in the –ATP reaction. These spliceosomes are precatalytic since no Prp2 activity is present in the *prp2* mutant extracts.

The protocols have been used to successfully isolate the precatalytic and activated spliceosomes reconstituted with synthetic U6 snRNA ***(26)***, and they can be easily adapted to isolate RNA–protein complexes assembled in other in vitro reactions.

4. Notes

1. A membrane-filtering unit is assembled by first inserting a sterile Buckner funnel into the neck of a 1-L side-arm flask. Ensure a tight seal is created. A rubber hose is preconnected to a vacuum source. Filter paper is added to the funnel, and

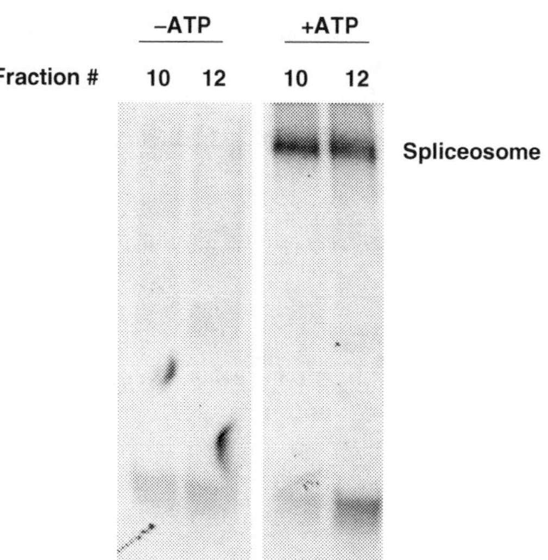

Fig. 4. Native gel electrophoresis of gradient fractions. Fractions from the gradient fractions isolated from the –ATP (adenosine triphosphate) and +ATP reactions shown in **Fig. 3** were analyzed by nondenaturing gel electrophoresis. Then, 10-µL aliquots of fractions 10 and 12 from each gradient were loaded onto a nondenaturing 3% polyacrylamide, 0.5% agarose composite gel.

the solution is poured slowly into the Buckner funnel. Once all the solution has passed through, the pump is turned off, and the solution is poured into a bottle for storage.

2. pKDACT was generated to yield large amounts of plasmid DNA for ACT1 pre-mRNA production. This plasmid was constructed by using pSP6 vector-primed oligonucleotide 5′GACACTATAGGTCGACGGATCCCC and ACT1 internal gene-primed oligonucleotide 5′CAAGAATTAATTCCGCGGAATTCCC to amplify a fragment of pSP6ACT (where 27 nucleotides were deleted from the 3' end of the SP6-ACT1 construct), and the PCR fragment was cloned into pCR2.1 (Invitrogen, CA). No detrimental effect of deleting this portion of the downstream exon was observed in our in vitro splicing reaction.
3. Although the 10X RNA polymerase reaction buffer contains DTT, we often added additional DTT. A linear DNA template can be a cut plasmid, PCR product, or oligonucleotide duplex. There must be a double-stranded promoter region to which the phage polymerase can bind to initiate RNA synthesis. The quality and concentration of the DNA can be determined by UV (ultraviolet) spectrophotometry or estimated from a stained agarose gel.
4. SP6 RNA polymerase recognizes its promoter and starts transcription at the final G in the promoter sequence, resulting in a product that begins with a G. SP6 polymerase lacks proofreading exonuclease activity, so the last nucleotide in the transcript can include an extra base.

5. When using SP6 RNA polymerase, incubation at 39 °C increases product formation by up to 40% over the more routine 37 °C incubation (New England BioLabs product notes).
6. RNA oligonucleotides are provided as dried pellets and shipped at room temperature. On receipt, RNA oligos should be resuspended in an RNase-free solution buffered to pH 7.4 to 7.6 at a convenient stock concentration and stored in smaller volumes in a −80 °C freezer. RNA oligos should not be allowed to freeze-thaw more than five cycles.
7. T4 DNA ligase uses ATP as a cofactor in catalyzing the formation of a phosphodiester bond between juxtaposed 5'-phosphate and 3'-hydroxyl termini in duplex DNA or RNA. The successful ligation of RNA molecules is then dependent on the efficiency of 5'-end phosphorylation. It is therefore useful to consider doing the first incubation in the presence of hot ATP and a second incubation in the presence of cold ATP. It is also necessary to optimize the ratio of DNA splint (DLT) to RNA oligos. Too much DLT can cause ligation of two or three oligos and result in less full-length product.
8. After gel electrophoresis, the gel is let set on one of the glass plates and covered with plastic wrap. In a dark room, the plastic-wrapped gel is placed face down on a piece of X-ray film and exposed for the desired length of time (e.g., 1 min). Ambient light is then allowed to flush the film (e.g., the door can be opened quickly for 5 s), which will create a shadow of the gel and comb wells on the film. Keep track of the orientation of the film in respect to the gel. The film is developed and place on a light box. The gel sitting on the glass plate and facing up is placed on the film and aligned to the shadow created. The RNA band corresponding to the full-length product is cut out with a sharp razor blade.
9. To avoid further denaturing the RNA, the eluted sample is not treated with phenol or chloroform but simply precipitated in ethanol. If inadvertent nucleases might be present, addition of an equal volume of chloripane followed by a quick vortex and spin at this step is performed. Denaturation can lead to energetically stable alternative states that do not resemble the native biological state ***(42)***. In these cases, consider renaturation by heating and cooling.
10. Cerenkov counting is a measurement technique for samples that deliver high-energy β-particles, such as ^{32}P, and has the advantage of not requiring the addition of scintillation cocktail. The tube containing the entire prep or an aliquot transferred to another tube is placed in vial and counted with the full window of detection in a scintillation counter.
11. The cell biomass concentration can be quantified as grams of dry or wet weight per liter of culture. We typically harvest the cells from the broth and weigh them while they are wet. It is suggested an empty centrifuge bottle be weighed before and after the centrifugation step to calculate this value.
12. An incubation step of the cells for 15 min at 25 °C with no shaking at this point *(27)* is optional.
13. Digestion of yeast and fungal cell walls is necessary for many experimental procedures, including spheroplasting, immunofluorescence, transformation, protein

purification, and others. Lytic enzymes like zymolyase are routinely used for digestion. The two main enzymatic factors are β-1,3-glucanase and β-1,3-glucan laminaripentao-hydrolase, which hydrolyze glucose polymers at the β-1,3-glucan linkages, releasing laminaripentaose *(43)*. The optimized concentration for lytic activity is 5 U/μL at the optimal temperature between 30 °C and 37 °C. Lytic activity ceases at higher temperatures. Although the lyophilized form is free of DNase/RNase activity, it does contain trace amount of proteases. Zymolyase is partially soluble in zymolyase buffer. It is recommended that, after briefly mixing the enzyme and buffer, the tube is centrifuged briefly, and the pellet is discarded. Add the supernatant to the treated cells. The OD_{800} of the suspension must decrease by more than 80% from the initial absorbance reading before proceeding to the dounce homogenizer step. Another way of estimating spheroplast formation is to put some cell suspension on a standard microscope slide, cover with a coverslip, and add to the edge 10% (w/v) SDS solution. Estimate the fraction of cells that have been converted to spheroplast, which will "pop" on contacting SDS.

14. It is not recommended to lyse the cells for longer than 90 min. If the digestion is not complete by this time, verify whether the amount of zymolyase per volume of sorbital-treated cells is correct. Alternatively, this may be an indication the zymolyase is no longer as active as it should be.
15. Both Kontes and Wheaton glass dounce homogenizers can be used for preparing active extracts. The dounce pestle must be tight-fitting to be effective, so type B, but not type A, is recommended. Different type B dounce homogenizers can yield different splicing activities (Fig. 1), so it may be necessary to test a number of mortar and pestles. Pulling the dounce pestle too fast can break the mortar. The exact speed and number of strokes can be adjusted empirically since too little or too much lysis may lead to inactive extracts. Although it is difficult to monitor cell lysis during the dounce step, an overall clear supernatant with a tight pellet after the high-speed spin suggests overlysis. This dounce procedure also works for another yeast mutant strain, SS304 *(44)*.
16. Although the top and bottom layers can be active for splicing, RNAs from these reactions aggregate abnormally in the wells during denaturing gel electrophoresis. To obtain splicing yeast extracts that give clean gel results requires being conservative in collecting extracts at this step.
17. The addition of proteinase K at a final concentration of 10 mg/mL to the splicing stop buffer followed by heat treatment at 37 °C for 10 min does not increase the quality of processed RNA samples as determined by denaturing electrophoresis. This is in contrast to human extracts, which require this protease as most samples will degrade during RNA processing (not shown).
18. RNA species produced from a splicing reaction are extracted under denaturing conditions, deproteinized, and analyzed by denaturing electrophoresis. A comparison of the production of intermediates and products with respect to the input pre-mRNA should be determined. A yeast extract with good splicing activity will never splice more than approximately 50% of ACT1 pre-mRNA to mRNA in 30 min at 23 °C (not shown). When the nuclease activity of the extract is low, it

is possible to quantitate the RNA species ***(45)***. Alternatively, analyses at different time points (e.g., 0, 5, 30 min) can reveal the kinetics of pre-mRNA turnover. An active extract can produce products in as little as 2 min when assayed on a denaturing acrylamide gel (data not shown).

19. Significant challenges to this in vitro reconstitution system exist for the user. First, the concentration of the knock-out oligonucleotide and temperature of incubation must be optimized for maximum depletion of endogenous U6 snRNA. Second, the degradation of the oligo afterward by endogenous nucleases also needs to be efficient. Third, the amount of the added synthetic U6 RNA needs to be at a sufficient quantity for reconstitution. These processes are all extract dependent, and a failure in any of them will result in inefficient or no reconstitution.
20. While the in vitro reconstitution of the U6 snRNP can be efficient, the yield at which it is incorporated into the yeast spliceosome is often less than 50% of a mock-treated sample (*see* Fig. 2; compare lane 3 with 1 and lane 6 with 4). The degree to which this varies is extract dependent (compare lanes 3 and 6).
21. After a period of time, discontinuous gradient layers will diffuse to form a linear gradient. Caution should be taken not to use gradients prepared after several days. The layers will have diffused greatly, overhomogenizing, and leading to a flattening of the spliceosome peak.

Acknowledgments

We thank members of the Lin laboratory for sharing reagents and protocols. K.J.D. was supported in part by Ruth L. Kirschstein Predoctoral Fellowship F31 GM67579 from the National Institutes of Health (NIH). This work was supported by NIH R01 grant GM40639 and funds from City of Hope Beckman Research Institute to R.J.L.

References

1. Haynes, S. R. (1992) The RNP motif protein family. *New Biol.* **4**, 421–429.
2. Brow, D. A. (2002) Allosteric cascade of spliceosome activation. *Annu. Rev. Genet.* **36**, 333–360.
3. Staley, J. P., and Guthrie, C. (1998) Mechanical devices of the spliceosome: motors, clocks, springs, and things. *Cell* **92**, 315–326.
4. Brody, E., and Abelson, J. (1985) The "spliceosome": yeast pre-messenger RNA associates with a 40S complex in a splicing-dependent reaction. *Science* **228**, 963–967.
5. Lin, R. J., Lustig, A. J., and Abelson, J. (1987) Splicing of yeast nuclear pre-mRNA in vitro requires a functional 40S spliceosome and several extrinsic factors. *Genes Dev.* **1**, 7–18.
6. Grabowski, P. J., Seiler, S. R., and Sharp, P. A. (1985) A multicomponent complex is involved in the splicing of messenger RNA precursors. *Cell* **42**, 345–353.
7. Fabrizio, P., McPheeters, D. S., and Abelson, J. (1989) In vitro assembly of yeast U6 snRNP: a functional assay. *Genes Dev.* **3**, 2137–2150.

8. Yu, Y. T., Maroney, P. A., and Nilsen, T. W. (1993) Functional reconstitution of U6 snRNA in nematode *cis*- and *trans*-splicing: U6 can serve as both a branch acceptor and a 5' exon. *Cell* **75**, 1049–1059.
9. Black, D. L., and Steitz, J. A. (1986) Pre-mRNA splicing in vitro requires intact U4/U6 small nuclear ribonucleoprotein. *Cell* **46**, 697–704.
10. Kramer, A., Keller, W., Appel, B., and Luhrmann, R. (1984) The 5' terminus of the RNA moiety of U1 small nuclear ribonucleoprotein particles is required for the splicing of messenger RNA precursors. *Cell* **38**, 299–307.
11. Winkelmann, G., Bach, M., and Luhrmann, R. (1989) Evidence from complementation assays in vitro that U5 snRNP is required for both steps of mRNA splicing. *EMBO J.* **8**, 3105–3112.
12. Bringmann, P., Rinke, J., Appel, B., Reuter, R., and Luhrmann, R. (1983) Purification of snRNPs U1, U2, U4, U5 and U6 with 2,2,7-trimethylguanosine-specific antibody and definition of their constituent proteins reacting with anti-Sm and anti-(U1)RNP antisera. *EMBO J.* **2**, 1129–1135.
13. Frendewey, D., and Keller, W. (1985) Stepwise assembly of a pre-mRNA splicing complex requires U-snRNPs and specific intron sequences. *Cell* **42**, 355–367.
14. Konarska, M. M., and Sharp, P. A. (1986) Electrophoretic separation of complexes involved in the splicing of precursors to mRNAs. *Cell* **46**, 845–855.
15. Das, R., and Reed, R. (1999) Resolution of the mammalian E complex and the ATP-dependent spliceosomal complexes on native agarose mini-gels. *RNA* **5**, 1504–1508.
16. Zhou, Z., Sim, J., Griffith, J., and Reed, R. (2002) Purification and electron microscopic visualization of functional human spliceosomes. *Proc. Natl. Acad. Sci. U. S. A.* **99**, 12203–12207.
17. Jurica, M. S., Licklider, L. J., Gygi, S. R., Grigorieff, N., and Moore, M. J. (2002) Purification and characterization of native spliceosomes suitable for three-dimensional structural analysis. *RNA* **8**, 426–439.
18. Yean, S. L., and Lin, R. J. (1991) U4 small nuclear RNA dissociates from a yeast spliceosome and does not participate in the subsequent splicing reaction. *Mol. Cell. Biol.* **11**, 5571–5577.
19. Kim, S. H., and Lin, R. J. (1996) Spliceosome activation by PRP2 ATPase prior to the first transesterification reaction of pre-mRNA splicing. *Mol. Cell. Biol.* **16**, 6810–6819.
20. Yean, S. L., and Lin, R. J. (1996) Analysis of small nuclear RNAs in a precatalytic spliceosome. *Gene Expr.* **5**, 301–313.
21. Lin, R. J., Newman, A. J., Cheng, S. C., and Abelson, J. (1985) Yeast mRNA splicing in vitro. *J. Biol. Chem.* **260**, 14780–14792.
22. Aebi, M., Clark, M. W., Vijayraghavan, U., and Abelson, J. (1990) A yeast mutant, PRP20, altered in mRNA metabolism and maintenance of the nuclear structure, is defective in a gene homologous to the human gene RCC1 which is involved in the control of chromosome condensation. *Mol. Gen. Genet.* **224**, 72–80.
23. Kurschat, W. C., Muller, J., Wombacher, R., and Helm, M. (2005) Optimizing splinted ligation of highly structured small RNAs. *RNA* **11**, 1909–1914.

24. Moore, M. J., and Query, C. C. (2000) Joining of RNAs by splinted ligation. *Methods Enzymol.* **317**, 109–123.
25. Ryan, D. E., and Abelson, J. (2002) The conserved central domain of yeast U6 snRNA: importance of U2-U6 helix Ia in spliceosome assembly. *RNA* **8**, 997–1010.
26. Yean, S. L., Wuenschell, G., Termini, J., and Lin, R. J. (2000) Metal-ion coordination by U6 small nuclear RNA contributes to catalysis in the spliceosome. *Nature* **408**, 881–884.
27. Cheng, S. C., Newman, A. N., Lin, R. J., McFarland, G. D., and Abelson, J. N. (1990) Preparation and fractionation of yeast splicing extract. *Methods Enzymol.* **181**, 89–96.
28. Lustig, A. J., Lin, R. J., and Abelson, J. (1986) The yeast RNA gene products are essential for mRNA splicing in vitro. *Cell* **47**, 953–963.
29. Nakamura, H., Oda, Y., Iwai, S., et al. (1991) How does RNase H recognize a DNA.RNA hybrid? *Proc. Natl. Acad. Sci. U. S. A.* **88**, 11535–11539.
30. Kretzner, L., Rymond, B. C., and Rosbash, M. (1987) *S. cerevisiae* U1 RNA is large and has limited primary sequence homology to metazoan U1 snRNA. *Cell* **50**, 593–602.
31. McPheeters, D. S., Fabrizio, P., and Abelson, J. (1989) In vitro reconstitution of functional yeast U2 snRNPs. *Genes Dev.* **3**, 2124–2136.
32. Fabrizio, P., and Abelson, J. (1990) Two domains of yeast U6 small nuclear RNA required for both steps of nuclear precursor messenger RNA splicing. *Science* **250**, 404–409.
33. McGrail, J. C., Tatum, E. M., and O'Keefe, R. T. (2006) Mutation in the U2 snRNA influences exon interactions of U5 snRNA loop 1 during pre-mRNA splicing. *EMBO J.* **25**, 3813–3822.
34. Fabrizio, P., and Abelson, J. (1992) Thiophosphates in yeast U6 snRNA specifically affect pre-mRNA splicing in vitro. *Nucleic Acids Res.* **20**, 3659–3664.
35. McGrail, J. C., Tatum, E. M., and O'Keefe, R. T. (2006) Mutation in the U2 snRNA influences exon interactions of U5 snRNA loop 1 during pre-mRNA splicing. *EMBO J.* **25**, 3813–3822.
36. Roscigno, R. F., and Garcia-Blanco, M. A. (1995) SR proteins escort the U4/U6.U5 tri-snRNP to the spliceosome. *RNA* **1**, 692–706.
37. Schenkel, J. (1991) Isolation of HnRNP particles from *Drosophila melanogaster* embryos. *Biochem. Int.* **24**, 423–428.
38. Clark, M. W., Goelz, S., and Abelson, J. (1988) Electron microscopic identification of the yeast spliceosome. *EMBO J.* **7**, 3829–3836.
39. Konarska, M. M., Grabowski, P. J., Padgett, R. A., and Sharp, P. A. (1985) Characterization of the branch site in lariat RNAs produced by splicing of mRNA precursors. *Nature* **313**, 552–557.
40. Cheng, S. C., and Abelson, J. (1987) Spliceosome assembly in yeast. *Genes Dev.* **1**, 1014–1027.
41. Moore, M. J., and Sharp, P. A. (1993) Evidence for two active sites in the spliceosome provided by stereochemistry of pre-mRNA splicing. *Nature* **365**, 364–368.

42. Uhlenbeck, O. C. (1995) Keeping RNA happy. *RNA* **1**, 4–6.
43. Kitamura, K., and Yamamoto, Y. (1972) Purification and properties of an enzyme, zymolyase, which lyses viable yeast cells. *Arch. Biochem. Biophys.* **153**, 403–406.
44. Vijayraghavan, U., Company, M., and Abelson, J. (1989) Isolation and characterization of pre-mRNA splicing mutants of Saccharomyces cerevisiae. *Genes Dev.* **3**, 1206–1216.
45. Rymond, B. C., Pikielny, C., Seraphin, B., Legrain, P., and Rosbash, M. (1990) Measurement and analysis of yeast pre-mRNA sequence contribution to splicing efficiency. *Methods Enzymol.* **181**, 122–147.

5

Purification of Ribonucleoproteins Using Peptide-Elutable Antibodies and Other Affinity Techniques

Scott W. Stevens

Summary

Recently developed affinity purification methods have revolutionized our understanding of the higher-ordered structures of multisubunit, often low-abundance macromolecular complexes, including ribonucleoproteins (RNPs). Often, purification by classical, non-affinity-based techniques subjects salt-labile complexes to an ionic strength incompatible with the integrity of the RNP, leading to a misrepresentation of the true higher-ordered structure of these complexes. A family of plasmids has been generated that can be used to introduce a number of different epitope tags, including peptide-elutable affinity tags, into the genome of the yeast *Saccharomyces cerevisiae*. Alternatively, these plasmids may be used for plasmid-borne expression of epitope-tagged proteins in either yeast or *Escherichia coli*. The gentle elution of the complex from the antibody affinity matrix can be performed at 4 °C and is compatible with a range of salt and pH conditions. RNPs purified by this method are active and suitable for downstream analyses such as RNA sequencing, structural analysis, or mass spectrometry peptide identification.

Key Words: Affinity purification; peptide antibody; ribonucleoprotein.

1. Introduction

Ribonucleoproteins (RNPs) are composed of at least one molecule of RNA and one or more proteins and are ubiquitously found in all organisms. The most abundant RNP, the ribosome, has been studied extensively for many decades due to its high abundance and ease of preparation. Prior to the advent of genetically engineered epitope tags, low-to-moderate abundance RNPs had been difficult to purify from eukaryotic cells, especially yeast. Recently, high-efficiency, high-affinity epitope tags have been developed that can be engineered into the native chromosomal context of the gene of interest to reduce nonspecific effects sometimes seen with expression of proteins at nonnative levels ***(1–3)***. A system has been developed for using peptide-specific monoclonal antibodies to affinity

From: *Methods in Molecular Biology, vol. 488: RNA-Protein Interaction Protocols*
Edited by: Ren-Jang Lin © Humana Press Inc, Totowa, NJ

purify peptide-tagged proteins from complex extracts alone or in conjunction with other affinity techniques to produce highly pure RNP complexes. This method involves the use of free peptide to disrupt the antibody–antigen interaction. In this chapter, details are provided for (1) epitope tagging genes in the chromosome of yeast, (2) preparation of mid- to large-scale whole-cell extracts, (3) affinity purification of the species of interest, and (4) downstream analysis of the purified material.

2. Materials

1. Appropriate tagging vector.
2. Monoclonal antibody of interest, protein G agarose, Ni-NTA agarose, calmodulin agarose.
3. Oligonucleotide primers for polymerase chain reaction (PCR) amplification, dNTPs (deoxynucleotide 5'-triphosphate), PCR enzymes.
4. Yeast strain of interest with desired genotype.
5. YPD liquid: 1% yeast extract (Difco), 2% Bacto Peptone (Difco), 2% dextrose.
6. Selective dropout media: 0.67% yeast nitrogen base without amino acids (Difco), required amino acid/nucleotide supplements (Bio101), 2% dextrose. (For plates, add 2% agar.)
7. Solutions used in yeast transformations: 50% PEG3350, 1 *M* LiOAc, 2 mg/mL salmon sperm DNA.
8. Lyticase solution: 1 *M* sorbitol, 10 m*M* Tris-HCl, pH 7.5, 10 mg/mL lyticase (Sigma cat. no. L4025).
9. Spheroplast wash buffer: 1 *M* sorbitol, 10 m*M* Tris-HCl, pH 7.5.
10. Lysis buffer A solution: 2 *N* NaOH, 8% β-mercaptoethanol.
11. Lysis buffer B solution: 50% trichloroacetic acid.
12. Sodium dodecyl sulfate polyacrylamide gel electrophoresis (SDS-PAGE) sample buffer: 100 m*M* Tris, pH 6.8, 2% SDS, 100 m*M* dithiothreitol (DTT), 15% glycerol, 0.05% bromophenol blue.
13. SDS-PAGE gels and electrophoresis apparatus, electroblot apparatus, nitrocellulose membrane.
14. Horseradish peroxidase (HRP)-conjugated antimouse immunoglobulin G (IgG) secondary antibody or rabbit PAP (Sigma P1291).
15. Protein G agarose (GE Biosciences).
16. IPP100: 10 m*M* Tris-HCl, pH 8.0, 100 m*M* NaCl (*note*: IPP150 contains 150 m*M* NaCl), 0.1% NP-40.
17. 0.2 *M* sodium borate, pH 9.0.
18. 0.2 *M* ethanolamine, pH 8.0.
19. Dimethylpimelimidate (DMP).
20. Phosphate-buffered saline (PBS), 1X: 10 m*M* Na_2HPO_4, 2 m*M* KH_2PO_4, 2.7 m*M* KCl, 137 m*M* NaCl.
21. Buffer A: 10 m*M* HEPES, pH 8.0, 10 m*M* KCl.
22. Chrontrol XT timer (Chrontrol, San Diego, CA).

23. Bead-beater apparatus (BioSpec), 0.5-mm glass beads.
24. Ti-45 or Ti-60 (or equivalents) ultracentrifugation rotors and ultracentrifuge
25. Chromatography columns (0.7 × 5 cm, Bio-Rad)
26. Protease inhibitors: Phenylmethylsulfonyl fluoride (PMSF), leupeptin, pepstatin, ethylenediaminetetraacetic acid (EDTA).
27. Synthetic peptide.
28. Antibody column-stripping solution: 1 *M* urea, 50 m*M* Tris, pH 8.0.
29. Phenol/chloroform (1:1), pH > 6.7.
30. 3 *M* NaOAc, pH 5.3.
31. 100% ethanol (–20 °C).
32. 80% ethanol (–20 °C).
33. 100% acetone (–20 °C).
34. RNA sample buffer: 8 *M* urea, 10 m*M* Tris, pH 7.0, 0.05% bromophenol blue, 0.05% xylene cyanol.
35. Optional: SW41 or equivalent rotor and tubes for velocity sedimentation gradients.

3. Methods

3.1. Creation of a Yeast Strain Containing a Genomically Encoded Epitope Tag

We have created a family of vectors with which one can amplify any of several tags with two gene-specific oligonucleotides. In this section, methods for inserting a tag in the genomic locus of interest are outlined.

3.1.1. Targeting Fragment PCR Amplification and Purification

In **Table 1**, the various tags available in the family of vectors are shown. In addition to the peptide-antibody affinity tags, others are available that can be used alone or in conjunction with the peptide epitopes on other polypeptides in multiprotein complexes (*see* **Note 1**). Oligonucleotides are designed using the format presented in **Fig. 1**. Essentially, the 3' ends of the oligos are fixed (cyan and red in **Fig. 1**) and will anneal to the appropriate locations in all of the plasmids described in **Table 1**. The 5'-most 50 nt of the oligos are designed to be identical to the 50 nt upstream of the stop codon (Oligo KIA, green in **Fig. 1**) and 50 nt of sequence 20–30 downstream of the stop codon (Oligo KIB, orange in **Fig. 1**). These two primers are used in a PCR reaction of the following composition: 0.2 µ*M* oligo KIA, 0.2 µ*M* oligo KIB, 25 ng plasmid, 250 µ*M* dNTPs, using a proofreading enzyme and buffer to reduce mutations resulting during PCR amplification *(4)*. Amplification is performed as follows: 25 cycles at 94 °C for 30 s, 55 °C for 3 min, and 72 °C for 4 min. PCR products must be separated from the plasmids, which can autonomously replicate in yeast and are preparatively purified from 1% agarose gels using the Qiagen (or similar) gel purification kit.

Table 1
A Family of Oligonucleotide-Compatible Vectors for Epitope Tagging Genes

Plasmid series	Auxotrophic markers available[a]	Epitope tag	Elution procedure
pPyxxx[b]	H, L, W, K	Polyoma epitope	MEYMPME peptide
pCT-1xxx	H, L, W	CT-1 epitope	GRILTLPRS peptide
pCBPxxx	H, L, W, K	Calmodulin-binding peptide	4 mM EGTA[c]
pHISxxx	H, L, W, K	Polyhistidine	200 m*M* imidazole
pCHPxxx	H, L, W, K	HIS-tag and polyoma epitope	Imidazole + Py peptide
pPrAxxx	H, L, W, K	Protein A (PrA)	None, used for negative selection
pTAPxxx	L, W, U	CBP + PrA; TEV[d] site	TEV cleavage; EGTA

[a]H, HIS3; L, LEU2; W, TRP1; K, KAN (G418[r]).
[b]xxx corresponds to the naming convention of the pRS series of vectors *(6)*.
[c]Ethylene glycol-bis(2-aminoehtylether)-*N,N,N',N'*-tetraacetic acid, pH 8.0.
[d]obacco etch virus protease cleavage site.

3.1.2. Transformation of Yeast With PCR Fragment

High-efficiency yeast transformation is required to obtain sufficient numbers of homologously targeted colonies. Detailed protocols are available elsewhere *(5)*; briefly:

1. The strain of interest is grown in 50–100 mL YPD to an optical density (OD) (A_{600}) of 0.7 to 1.0. Cells must be actively growing and cannot simply be diluted to the desired density. It is preferable to use derivatives of the strain S288C, the genome of which has been sequenced (*see* **Note 2**) *(6)*.
2. Cells are pelleted in 50-mL conical tubes in a clinical tabletop centrifuge, washed once with sterile water, resuspended in 1 mL 100 m*M* LioAc, and transferred to a microcentrifuge tube that has volume markings.

Fig. 1. Design of oligonucleotides and experimental schema for epitope tag insertion into the yeast genome. (**A**) Schematic of the use of gene-specific oligonucleotide primers on a generic epitope tag-containing vector of the series described in the text. (**B**) Reaction scheme for homologous recombination of resulting polymerase chain reaction (PCR) fragment into the genome of yeast. Targeting of Your Favorite Gene (YFG) by design of the PCR primers as described is accomplished by homologous recombination directed by gene-specific sequences preceding the stop codon (green) and downstream of the stop codon (orange). Plasmid-specific sequences common

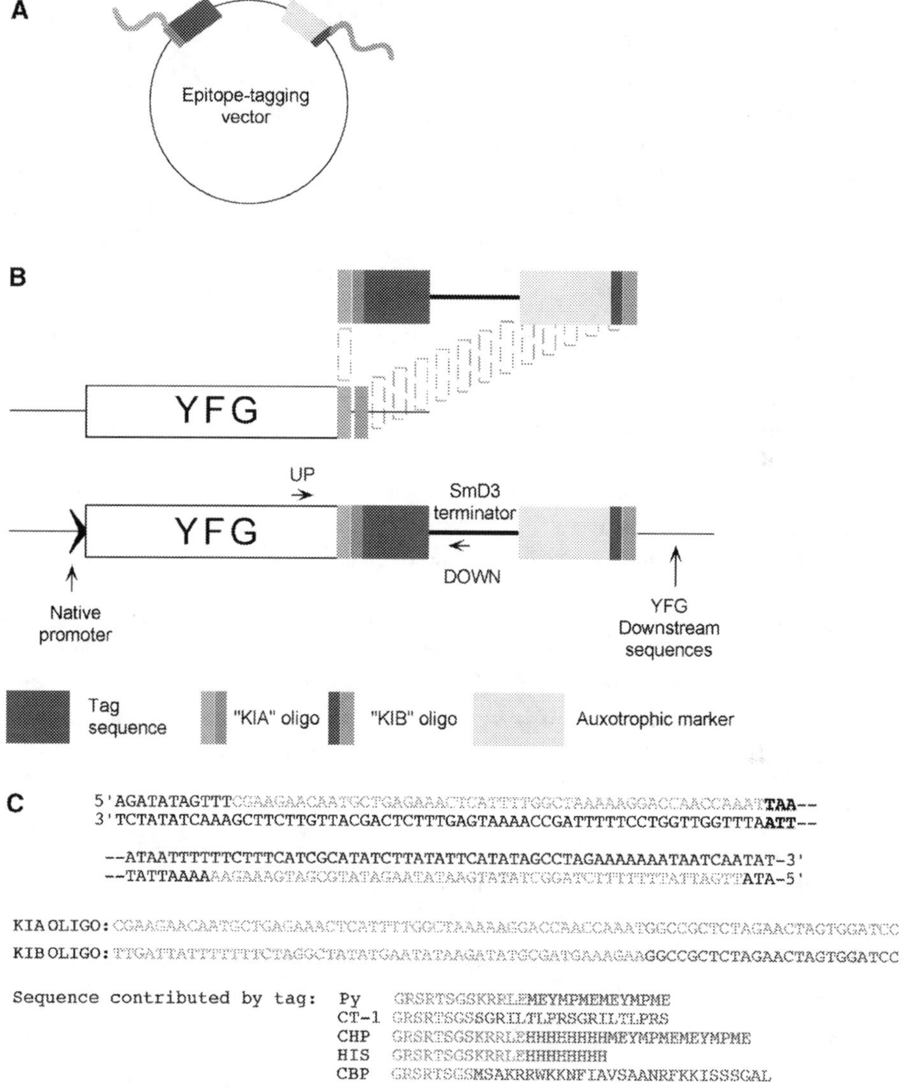

Fig. 1. (continued) to all plasmids are represented in cyan and red. Epitope tag sequences are represented in purple, and the auxotrophic marker sequences are represented in yellow. (**C**) Specifics of the oligonucleotide design for this family of vectors. Top: Example of oligonucleotide design for the *SNU17* gene of yeast. The stop codon is in bold (**TAA**). Colors of the sequences correspond to the descriptions in (**B**). Note the polarity of the sequences, as both strands are shown. Sequence in KIA oligo is in the sense strand; sequence in KIB oligo is in antisense strand. Middle: Examples of sequences in conjunction with the plasmid-specific sequences (cyan and red). Bottom: Expected polypeptide additions to the end of the polypeptide from the targeted gene for comparison to DNA sequencing reactions (*see* text). (*See Color Plate*)

3. Cells are pelleted at 5000 *g* for 1 min in a microcentrifuge and the supernatant removed.
4. Estimate the packed cell volume and add 1.5 volumes 100 m*M* LiOAc.
5. Vortex to resuspend cells, aliquot in 50-μL portions into as many tubes as you require. Two additional tubes should be included, one for a positive control, one for a negative control.
6. Tubes are centrifuged at 5000 *g* for 1 min and the supernatant removed.
7. To each tube, add 240 μL 50% PEG3350, 36 μL 1 *M* LiOAc, 50 μL 2 mg/mL boiled and subsequently chilled salmon sperm DNA.
8. For a positive control, add 34 μL of 1 ng/μL of a vector that will confer growth on the medium dictated by the auxotrophic marker in the plasmid used in **Subheading 3.1.1**.
9. To the negative control, add 34 μL sterile water.
10. To each experimental tube, add 34 μL of the purified PCR fragment from **Subheading 3.1.1**.
11. Vortex all tubes well and incubate at 30 °C for 30 min.
12. Transfer to 42 °C for 45 min.
13. Pellet the cells by microcentrifugation at 5000 *g* for 1 min, remove supernatant, and resuspend cells in 100 μL water.
14. Plate on proper selective SD medium *(7)*. To avoid plating too many colonies on one plate, use two plates per transformation, one with 95 μL of transformation mix and one plate with 95 μL water plus 5 μL transformation mix.
15. Incubate at 30 °C for 3–5 d, when colonies generally appear.

3.1.3. Confirmation of Correct Targeting and Functional Epitope Addition

There are two ways by which one can verify the proper homologously targeted epitope tag. The first is a PCR-based screen, followed by DNA sequencing (useful for all but the protein A and TAP tags). The second is to functionally test for the presence of the epitope tag in a Western blot (useful for all but the HIS and CBP tags).

3.1.3.1. Correct Targeting

To perform a PCR-based screen, one requires oligonucleotide primers designed as shown in **Fig. 1**. One is designed approx 200 nt upstream of the stop codon of the targeted gene (UP oligo in **Fig. 1B**); the other is fixed in the terminator sequence derived from the targeting plasmid (DOWN oligo [5′-CCTACAATGATGAATGCGTTTCTGCCG-3′] in **Fig. 1B**).

1. Colonies (~4–8) resulting from the transformation in **Subheading 3.1.2**. are grown individually in 3 mL YPD overnight at 30 °C with shaking.
2. The next morning, 0.5 mL is aseptically removed from each tube, and cells are pelleted in a microcentrifuge at 14,000 *g* for 1 min.
3. Cells are washed once in 1 mL water and repelleted.
4. Cells are then resusupended in 1 mL of a solution of lyticase solution.

5. Tubes are incubated at 25 °C to 30 °C for 30–60 min on a rotator or a nutator. This treatment will remove the yeast cell wall (*see* **Note 3**).
6. After cells have been spheroplasted, they are gently pelleted by spinning in a microfuge at 5000 *g* for 1 min.
7. After removing the supernatant, they are washed once with 1 mL of spheroplast wash buffer and repelleted.
8. Remove the wash solution and add 100 μL of 1X PCR buffer (*see* **Note 4**) and pipet up and down to mix.
9. Cells are lysed by incubation at 100 °C for 5 min.
10. Cell debris is pelleted in a microcentrifuge for 10 min at 14,000 *g* at 4 °C.
11. Immediately remove the supernatant and transfer into a fresh tube.
12. Use 5 μL of the supernatant in a 100-μL PCR reaction of the following composition: 0.2 μ*M* oligo UP, 0.2 μ*M* oligo DOWN (*see* **Fig. 1B**), 250 μ*M* dNTPs, using a proofreading enzyme and appropriate buffer. Twenty-five cycles of amplification are performed as follows: 94 °C for 30 s, 55 °C for 60 s, and 72 °C for 1.5 min.
13. PCR products are electrophoretically separated from primers and genomic DNA on a 2% agarose gel and purified with the Qiagen gel purification kit. If UP and DOWN oligonucleotides are designed according to the scheme presented in **Fig. 1**, the PCR fragment should be approx 300 bp. To confirm the in-frame incorporation of the epitope tag, compare the theoretical sequence of the gene and the additional sequences contributed from the epitope tag (*see* **Fig. 1C**) against that realized from sequencing the PCR fragment with the UP or DOWN oligonucleotide.

3.1.3.2. Functional Analysis of Epitope Tag Incorporation by Western Blot

An alternative, or perhaps additional, means of demonstrating that the epitope has been properly inserted into the gene of interest is to test by Western blotting. The method of Yaffe and Schatz *(8)* is used to prepare whole-cell protein samples from cells for Western blot analysis, a rapid method amenable to simultaneous processing of several samples.

1. Briefly, 1 mL of cells grown as in **Subheading 3.1.3.1.** is chilled to 4 °C in ice.
2. Add 150 μL ice-cold lysis buffer A solution to each sample.
3. Mix by inverting tube several times.
4. Incubate on ice for 10 min.
5. Add 150 μL ice-cold lysis buffer B; mix by inverting tube several times.
6. Incubate on ice for 10 min.
7. Pellet the precipitated proteins for 2 min at 14,000 *g* in a microcentrifuge.
8. Wash pellet with 1 mL ice-cold acetone.
9. Spin 2 min at 14,000 *g* in a microcentrifuge and aspirate the supernatant.
10. Dry pellet for 10 min uncapped on the benchtop.
11. Resuspend pellet in 100 μL SDS-PAGE sample buffer. The cell material can be very difficult to resuspend. Try adding approx 20 μL and smearing the pellet along the side of the tube with a pipet tip, then add the remaining 80 μL and mix thoroughly. If the color is yellow, quickly add 10 μL 1 *M* Tris, pH 8.0, and vortex to mix.

12. Heat at 95 °C for 5 min.
13. Load 10 μL on an SDS-PAGE gel appropriate for re-solving your proteins of interest.
14. Electroblot the contents of the gel to a nitrocellulose membrane overnight. In our hands, wet transfer is much more efficient than a semidry transfer for this application.
15. The appropriate monoclonal antibody is used at 10–25 ng/mL in a Western blot.
16. Secondary antibodies for the Py and CT-1 epitopes can be HRP- or AP-conjugated antimouse IgG, and detection should be done with an appropriate chemiluminescent detection substrate. The presence of an appropriate size band in experimental strains, but not in a negative control, in addition to a confirmatory PCR reaction should give one confidence to proceed. Anti-HIS antibodies are not sufficiently sensitive enough to detect HIS tags on low- to medium-abundance proteins. TAP-tag and protein A detection is achieved using the rabbit PAP reagent from Sigma (P1291).

3.2. Preparation of Affinity Resin

Proper affinity resin preparation is critical for efficient affinity purification of RNP complexes. If the antibody is not quantitatively covalently linked to the resin, it will slowly leach off the column during purification, reducing your final yield. The polyoma antibody ***(9)*** is available commercially from Covance (GLU-GLU epitope); however, for large-scale applications, it is more economical to grow the hybridoma cells expressing the polyoma antibody. The CT-1 antibody ***(10)*** is not currently commercially available.

3.2.1. Growth of Hybridoma Cells

The hybridoma cell lines described here grow well in protein-free hybridoma media (PFHM II from Invitrogen). It is best to start from a frozen ampule of cells and slowly grow to large volumes. A cryopreserved ampule of the hybridoma cells is thawed and added to 10 mL of PFHM II in a T-25 flask and grown until confluent. Growth is performed in a humidified CO_2 incubator at 37 °C. From there, the culture is expanded into two T-75 flasks. On reaching confluence in those flasks, expand into the necessary number of T-175 flasks and grow until cells *just* begin to die. Another measurement of when the supernatant is ready to be processed is analysis of protein concentration in the cell-free culture supernatant. Since the PFHM II is protein free, when the concentration of antibodies secreted into the medium reaches 50–100 ng/μL, it is ready to harvest.

3.2.2. Purification of Antibodies From Hybridoma Supernatant

1. Centrifuge the cell culture material from **Subheading 3.2.1**. at 10,000 *g* at 4 °C for 20 min to pellet the cell debris.
2. Carefully remove the supernatant and place into a glass beaker slightly more than twice as large as the volume of supernatant at 4 °C.
3. With gentle stirring, *slowly* add an equal volume of cold, saturated ammonium sulfate solution.

4. Cover with plastic wrap and stir overnight at 4 °C.
5. Centrifuge the ammonium sulfate precipitation at top speed in appropriate size centrifuge tubes. With less than 300 mL of ammonium sulfate precipitate, spin in an SS34 (or equivalent) rotor at more than 30,000 *g* for 30 min. If you have more supernatant than an SS34 can accommodate, the precipitate can be centrifuged in a GSA or GS3 rotor (or their equivalents) at the maximum rated speed for the rotor for 2 h at 4 °C.
6. Carefully remove the supernatant and rinse the pellets with 50% ammonium sulfate (*see* **Note 5**).
7. Resuspend the precipitated material in 0.02 volume (original cell culture volume) 25 m*M* sodium phosphate, pH 7.0.
8. Dialyze the antibody three times against more than 500 volumes 25 m*M* sodium phosphate at pH 7.0 for 4 h each for the first two dialyses, with the last dialysis proceeding overnight.
9. Measure the concentration of the antibody with the Bradford assay, using IgG as a standard ***(11)***, and assess the purity of the antibody by SDS-PAGE analysis with and without DTT in the SDS sample buffer (*see* **Note 6**). Antibodies purified by these means are generally more than 95% pure. Freeze at –20 °C to –80 °C in convenient aliquots (e.g., 20 mg/tube).

3.2.3. Coupling of Antibody to Agarose Beads

Orienting the antibody such that the antigen-binding domain is exposed to the solution is preferable to random coupling to the beads via cyanogen-bromide coupling, which will orient the antibodies randomly with respect to the antigen-binding domain. To perform this, use protein G agarose to bind the Fc region of the antibody, thereby exposing the Fab region to the solution.

1. Wash 5 mL protein G slurry twice with 45 mL IPP100 in 50-mL conical tubes.
2. The protein G agarose beads are recovered by centrifugation at approx 1000 *g* for 2 min in a clinical centrifuge at room temperature.
3. Resuspend the washed beads in antibody solution containing 1.5 times the manufacturer's capacity estimation for protein G binding.
4. This slurry is incubated with slow rotation at room temperature for 2 h.
5. After antibody binding, the slurry is centrifuged at about 1000 *g* for 2 min in a clinical centrifuge at room temperature (*see* **Note 7**).
6. Carefully remove the antibody-depleted supernatant, which can be saved and used for Western blotting in the future.
7. Wash the beads three times with 40 mL 0.2 *M* sodium borate, pH 9.0, carefully removing the supernatant after each spin.
8. Resuspend the beads in 40 mL 0.2 *M* sodium borate, pH 9.0.
9. Add 0.233 g DMP solid to this solution to crosslink the antibody to the protein G.
10. Incubate with rotation for 30 min at room temperature.
11. Centrifuge the tube at approx 1000 *g* for 2 min in a clinical centrifuge at room temperature.

12. Carefully remove supernatant and resuspend beads in 40 mL 0.2 *M* ethanolamine, pH 8.0, to quench the crosslinking agent.
13. Incubate with gentle rotation for 2 h at room temperature.
14. Centrifuge the tube at approx 1000 *g* for 2 min in a clinical centrifuge at room temperature.
15. After careful removal of supernatant, wash beads three times with 40 mL 1X PBS.
16. Resuspend in 1 bead volume of 1X PBS plus 0.02% NaN_3 (*see* **Note 8**) and store at 4 °C in a tightly capped container. The efficiency of antibody crosslinking can be monitored by removing samples at each step and at the end by boiling 20 µL of beads in SDS-PAGE sample buffer and looking for the heavy chain (~50 kDa) and light chain (~25 kDa) in an SDS-PAGE gel.

3.3. Preparation of Yeast Whole-Cell Extracts

There are a number of ways to prepare large-scale yeast whole-cell extracts. Among them are (1) blending in liquid nitrogen (***12,13***), (2) continuous-flow high-pressure disruption, and (3) "bead beating" using glass beads in a specialized apparatus (BioSpec). In our experience, the first method is highly variable in breakage efficiency. The second method requires the acquisition of relatively expensive equipment but is a highly efficient means of processing kilogram quantities of yeast (Microfluidics). Method 3 is amenable to quantities from 80 to 500 g of yeast, generally the amount of cells used for preparative RNP purification. In this chapter, experimental details for the bead-beating procedure are provided.

3.3.1. Growth and Harvesting of Cells

If the epitope tag is incorporated in the genome, there is no further need for selection in minimal medium.

1. Grow cells in YPD to mid-log cell concentrations (OD A_{600} ~ 1.5–2.0) in shake flasks at 30 °C or in fermenters to an empirically determined mid-log phase, which can reach optical densities of 10 or greater.
2. Cells are harvested by centrifugation at 8000 *g* for 10 min at 4 °C.
3. After removal of media, cell mass is determined, and cells are resuspended in 1.5 mL ice-cold buffer A per gram of cells. The low-salt conditions allow for optimization of extract procedures (*see* **Subheading 3.3.2.**).
4. Cells are preserved by slow pipetting into a vat of liquid nitrogen, which also aids in the breakage and allows for storage at –80 °C in a convenient form.

3.3.2. Homogenization of Cells

1. The required mass of cryopreserved cells is removed from the freezer and transferred to an appropriate size glass beaker containing a stir bar (*see* **Note 9**).
2. The beaker containing the cells is set in a room temperature water bath and occasionally stirred with a clean spatula to scrape thawing material off the sides.

3. When the material has liquefied to a custard-like consistency, but before it has completely thawed, begin stirring with the stir bar to accelerate the thawing process.
4. Monitor this carefully, and when all of the frozen material has melted, transfer the material to a cold room.
5. With stirring, add PMSF to 0.4 m*M*, leupeptin and pepstatin each to 2 μg/mL, and EDTA to 0.5 m*M*.
6. Add 200 mL of cell suspension (80 g cell mass) to the required numbers of stainless steel bead-beater chambers.
7. Fill to the manufacturer's recommended level with clean, cold, 0.5-mm glass beads.
8. Assemble the chamber into the ice jacket and fill ice jacket with ice water (*see* **Note 10**).
9. Cycle 1 min on and 2 min off for five cycles. A Chrontrol XT timer (or equivalent) is very useful for automating the cycling.
10. After the cycling is complete, the contents of the chamber (including the beads) are transferred to a 1-L centrifuge bottle (or other 1-L capped container).
11. The supernatant is transferred by pipet to a cold glass beaker containing a stir bar.
12. Buffer A10 containing PMSF, leupeptin, pepstatin, and EDTA as described above is added (100 mL/chamber) to rinse the chamber and is added to the beads.
13. The bottle is capped and agitated gently to rinse the beads.
14. After the beads settle, the supernatant is transferred to the glass beaker.
15. Repeat the rinse with an additional 100 mL A10 containing protease inhibitors.

3.3.3. Preparation of Cell Extracts

Depending on the characteristics of the protein/RNP you are attempting to extract, you will need to add extraction agents, such as KCl, NaCl, detergents, or other salts to efficiently extract the complex of interest.

1. Typically, KCl is added to 200 m*M* final concentration, slowly, with stirring from a 2 *M* KCl stock.
2. After stirring for 30 min at 4 °C, cell debris is pelleted in an ultracentrifuge spun at 100,000 *g* for 60 min at 4 °C.
3. Soluble material from the center of the tubes is harvested, avoiding the pellet and the fatty material at the top of the tube (*see* **Note 11**).
4. If the chromatography is going to be performed that day, this material can be filtered through a 0.45-μ*M* filter and applied directly to the column (*see* **Subheading 3.4.1.**). If it is to be preserved or if other adjustments to the solution are warranted, dialysis into an appropriate solution should be performed (*see* **Note 12**).

3.4. Affinity Chromatography

The affinity chromatography protocols will differ slightly depending on the affinity matrix used. Generally, antibody affinity matrices require slow passage of the extracts over the material. Affinity matrices with much greater affinity (such as Ni-NTA) are used with greater flow rates.

3.4.1. Preparation of Affinity Column

The antibody affinity column is assembled using the antibody-agarose conjugate prepared in **Subheading 3.2.3**. An empirically determined amount of antibody slurry is loaded into a small column and allowed to drain (*see* **Note 13**). If the abundance of the material is unknown, or if the material is of low abundance, use 0.5 mL of settled antibody bead volume per 100 g of cells. This provides approx 2–3 mg of antibody on the column under ideal conditions and will capture more than 80% of the desired material with proper running, washing, and elution conditions *(2)*.

3.4.2. Passage of Extracts Over Affinity Matrix

Passage of extracts over the antibody affinity matrix is best done with a slow-pumping peristaltic pump with a flow rate of 5–30 mL /h. The actual flow rate is dictated by a number of variables, such as the volume of extract, a convenient length of time (overnight usually), and the strength of the antibody/antigen interaction. Using the peristaltic pump ensures that the rate remains constant; however, careful calculation must be made to ensure that the column is not pumped dry. When the last of the extract is about to be pumped through the affinity column, disconnect the peristaltic pump and drain the remains (usually ~5 mL) into the column to flow by gravity. Alternatively, the column can be run by gravity by adjusting the height of the extract over the column to properly adjust the flow rate.

3.4.3. Washing Affinity Matrix

After all of the cell extract has passed over the column, rinse the sides of the column with wash buffer and allow to drain by gravity. Wash the column with a buffer suitable for the stability of the complex of interest. Typically, the polyoma and CT-1 antibodies are stable to washing with 100–150 column volumes of wash solution of up to 250 m*M* monovalent salt concentration. If the complexes are sensitive to 250 m*M* salt concentrations, the wash buffer salt concentration can be reduced to as low as 50 m*M*; however, a second affinity step will be required to reduce background binding to the agarose-based affinity matrix.

3.4.4. Affinity Elution of Bound Material

One of the primary advantages of a peptide-elutable affinity matrix is the gentle elution procedure. The free peptide can be easily removed by ultrafiltration, size-exclusion chromatography, or sedimentation through velocity gradients (*see* below). Other affinity matrices sometimes require the use of proteases, which can be difficult to remove after use, or divalent metal ion chelators such as EGTA (ethylene glycol tetraacetic acid), which in some cases can remove structural metals from RNPs, affecting their integrity (unpublished observations). Use 100 ng/mL peptide dissolved in wash buffer (*see* **Note 14**). The competitive elution procedure produces highly concentrated purified material if the flow is interrupted during the elution step. Apply the elution buffer to the top of the column and allow one column volume to drain by gravity. Flow is stopped for 30 min. This is repeated for five cycles to remove as much of the material as possible. The majority of affinity-purified material is generally contained in the second and third fractions. Antibody columns can be regenerated by washing with 20 column volumes of antibody column stripping solution, then by washing with IPP150 containing 0.02% NaN_3. Columns may be stored for over a year at 4 °C.

3.4.5. (Optional) Second Affinity Chromatography Steps

Often, a second affinity step can increase the purity of the complexes of interest (***1,2***). In multicomponent complexes, this can be achieved by incorporation of a second affinity tag from **Table 1** into the same gene as the peptide epitope (e.g., CHP tag, which incorporates a polyhistidine sequence in tandem with a polyoma epitope) or by incorporating a second affinity tag into a separate polypeptide (e.g., polyhistidine into one polypeptide and polyoma into a different polypeptide present in the same complex). Note that Ni-NTA should not be used as a first purification step from yeast extracts. At low-salt concentrations, antibody affinity columns are efficient at capturing the epitope-tagged polypeptide; however, the binding of other, nonspecific material is higher. An example of low-salt-purified SmD3-CHP-associated RNA is shown in **Fig. 2**. Material purified only through polyoma chromatography is presented in **Fig. 2A**, and equal amounts of material further purified by Ni-NTA chromatography is shown in **Fig. 2B**. Ribosomal RNA, which is often the most abundant contaminant in these procedures, is undetectable after Ni-NTA purification when purification is performed at 50 m*M* KCl. Additional tags available are an additional peptide-elutable antibody epitope, polyhistidine, calmodulin-binding peptide, protein A, and TAP ***(1)***. As others have shown, protein A can be used to negatively select against related, but unwanted, complexes that happen to share a polypeptide, an additional utility that has proven very useful ***(14)***. Note that protein A-containing polypeptides will also be captured in the Py and CT-1 columns but not eluted by the peptide. Proper experimental design is required to make use of these tags in conjunction with the peptide-elutable affinity tags.

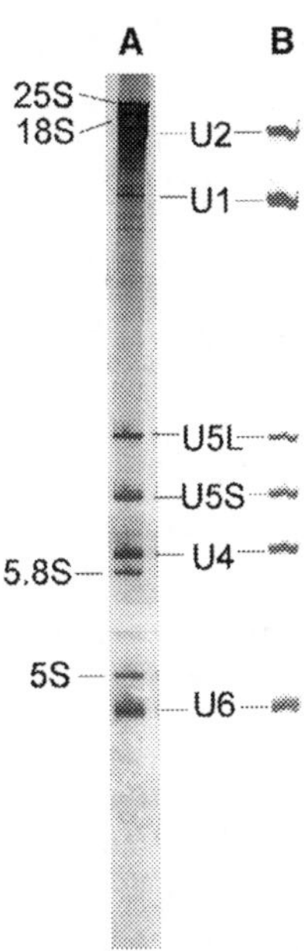

Fig. 2. Resulting RNAs affinity purified from a strain harboring a CHP tag on the SmD3 polypeptide and stained with silver. (**A**) SmD3CHP-associated RNA purified in a single step over an antipolyoma affinity matrix as described in the text. Note the contaminating 25S, 18S, 5.8S, and 5S ribosomal RNAs (rRNAs) in addition to the U1, U2, U4, U5, and U6 small nuclear RNAs. (**B**) Material from the polyoma purification in (**A**) further purified over an Ni-NTA affinity matrix. Note the absence of the rRNAs after the second affinity step.

3.5. Downstream Analysis of Affinity-Purified Material

Even if the material purified by these means were 100% pure, the story is often more complex than is observed by SDS-PAGE analysis of the copurifying polypeptides. This is due to (1) the potential for an extramolar presence of the tagged polypeptide; (2) presence of multiple different, albeit related, complexes in which the tagged polypeptide functions (e.g., the U5 and U4/U6·U5 snRNPs (small nuclear RNPs); ***15–17***); and (3) potentially substrate-engaged RNPs, which will likely be larger and more compositionally complex. To adequately address these issues, physical separation must be achieved, generally by glycerol gradient sedimentation or by size-exclusion chromatography. From there, polypeptides and RNA can be characterized.

3.5.1. Separation of Complexes Based on Size

A tagged polypeptide will often be a component of several different complexes. An example of this is the Sm complex, which binds to each of the spliceosomal U1, U2, U4, and U5 snRNAs. Although functionally related, the U1, U2, and U4/U6·U5 snRNPs are separate entities at high-salt concentrations. To separate complexes based on their molecular weight, one can use size-exclusion chromatography or glycerol gradient sedimentation to separate them. The specifics of the separation techniques are particular to the size of the complex present in the purified material. In our experience, size-exclusion chromatography techniques result in the loss of significant quantities of material, and for precious low-abundance samples may not be the best means of separation. For RNP separation, glycerol gradient sedimentation is preferred because it does not suffer from the losses encountered on gel filtration columns.

In **Table 2**, the conditions for separation of different size complexes are presented. Typically, sedimentation is performed at 4 °C at the indicated speeds and times. All information is based on an 11-mL gradient used in a Beckman SW41 rotor configuration, although smaller tubes and different rotors can be used. Gradients are harvested either from the top with a manual pipetting technique or through the bottom or the top using a gradient fractionator. If there is appreciable precipitated or otherwise pelleted material at the bottom of the tube, harvesting from the bottom leads to contamination of every fraction. Fractions of 420 μL each are harvested from the top of the gradient, which provides good resolution of the resulting material.

RNA and protein from these fractions can be separated by extracting with an equal volume of phenol:$CHCl_3$ (1:1), pH > 6.7. This extraction should be performed by vortexing thoroughly (1–2 min) and centrifugation at 14,000 *g* for 10 min at room temperature (*see* **Note 15**). Pipet the aqueous (top) layer into a separate tube containing 42 μL 3 *M* NaOAc, pH 5.3. Vortex and add 1 mL ice-cold 100% ethanol. Vortex to mix and store at −20 °C overnight. To the organic phase, fill the rest of the tube with acetone (~1.2 mL) that has been prechilled to below −20 °C. Layer the acetone carefully on the organic phase and place the tubes at −20 °C overnight without mixing. Incubation in this manner results in more efficient precipitation than mixing prior to placing at −20 °C.

Table 2
Sedimentation Conditions for Glycerol Velocity Gradients[a]

Size of complex	Glycerol range (%)	Speed (rpm)	Time
<100 kDa	5–20	36,000	24 h
100–1000 kDa	10–20	29,000	18 h
250–3000 kDa	10–30	29,000	18 h
500–6000 kDa	10–40	22,000	16 h
40S[b]–300S	10–50	40,000	100 min

[a]Values given for an 11-ml gradient in SW40 or SW41 rotor or its equivalent at 4 °C.
[b]Svedberg constant.

The next morning, invert all tubes several times and centrifuge both the RNA samples (ethanol precipitations) and the protein samples (acetone precipitations) at 4 °C for 15 min. Wash each pellet carefully with 80% ethanol (–20 °C) and let air dry for 10–20 min. RNA samples are resuspended in RNA sample buffer and are electrophoresed on an appropriately configured urea-PAGE gel (configuration will depend on the size of the RNA to be separated), and protein samples are resuspended in SDS-PAGE sample buffer and electrophoresed on an appropriately configured SDS-PAGE gel. Stain each gel according to the abundance of the RNA and proteins you are expecting, use ethidium bromide staining for abundant RNA ***(18)***, Coomassie blue G250 staining for abundant protein ***(19)***, and silver staining for low-abundance RNA and proteins ***(20)***.

3.5.2. Mass Spectrometric Analysis of Copurifying Polypeptides

Mass spectrometric peptide identification of complex mixtures is a facile and robust means of identifying copurifying polypeptides ***(21–24)***. The services are commercially available, and many institutions have the means to identify proteins in your samples directly from complex mixtures or from isolated polyacrylamide gel bands. Although the techniques for performing these analyses are beyond the scope of this chapter, here are some helpful suggestions for submitting samples that will improve the chances of obtaining good data from the mass spectrometric analysis. First, cleanliness is crucial to avoid contaminating your sample with keratin and other human polypeptides, which can mask the identity of the unknown polypeptides in your samples. Investigators working with RNA typically employ laboratory techniques to avoid RNases, which will also minimize the introduction of human contaminants. The way in which the gels are stained can have an impact on acquisition of interpretable mass spectra. The means and specifics of gel staining can also influence the results of mass spectrometry. Colloidal Coomassie staining of gels ***(19)*** and a particular silver staining technique ***(20)*** will improve the chances of success. Also, preservation of gel slices in a relatively dry state and freezing at –80 °C are recommended.

3.6. Expression of Epitope-Tagged Genes From Plasmids in Escherichia coli or Saccharomyces cerevisiae

Although the expression of genes in yeast from their natural promoter is an optimal and more representative means of studying multisubunit complexes, there may be times when that is not the best way to introduce an epitope-tagged protein into the system under investigation. The vectors described in **Table 1** can be used to drive plasmid-borne epitope-tagged genes in yeast and *E. coli*. For designing yeast constructs, PCR can be performed from yeast genomic DNA, which will include the native promoter, and cloned in-frame with the epitope tags. Complementary DNAs from other organisms can be cloned in-frame to the epitope tags but require the addition of a yeast promoter to drive expression. Choice of plasmid copy number will allow low or high levels of expression depending on the needs of the experiment. Expression in *E. coli* can be achieved by cloning the desired gene in-frame with respect to the epitope tag and including an appropriate promoter and ribosome-binding site in the primer design at the 5' end of the gene. This may be most easily achieved using the T7 RNA polymerase system *(25)*.

3.7. Summary

The use of peptide-elutable epitope tags alone or in combination with other affinity techniques provides a gentle and efficient means of purification of even low-abundance material from yeast or *E. coli* cell extracts. The family of vectors presented here provides great flexibility with respect to auxotrophic markers and tags using a single pair of oligonucleotides per gene.

4. Notes

1. The family of vectors includes both peptide epitopes alone and in conjunction with a polyhistidine epitope tag. Should the genotype of the strain in use allow, multiple tags can be place on multiple genes, or in situations designed to test the higher-ordered organization of a polypeptide (monomer vs dimer, etc.), one can tag each allele in a diploid with a different tag.
2. Using S288C derivatives has several advantages; the sequenced genome is on occasion slightly different from other isolates, such as W303. These small differences can have downstream consequences in strain construction and peptide analysis. Among some of the advantages to using the sequenced strain are (1) increased frequency of homologous recombination in a gene for which there is sequence heterogeneity at the locus of interest, (2) confidence in the copy number of the gene, and (3) comprehensive peptide analysis on mass spectrometric peptide identification.
3. Spheroplasts are rather fragile, and the progress of the treatment should be carefully monitored. After spheroplasting, the cells will clump together, resulting in a stringy appearance. If the reaction is allowed to proceed too far, or if the concentration of lyticase is too high, the yield of spheroplasts will decrease due to cell lysis.
4. The composition of the PCR buffer will be dictated by the enzyme choice.
5. Allow the 50% ammonium sulfate wash solution to perfuse the pellet for approx 10 min before respinning at the same speed for 10 min. If the pellet is "loose" and slides down the side of the tube, it can be transferred to an SS34 tube (should the volume be large enough to have warranted the use of larger buckets for the precipitation) and spun at a higher RCF.
6. In the presence of SDS, DTT, and heat, the antibody will dissociate into the heavy (~50-kDa) and light (~25-kDa) chains. In the absence of DTT, the antibody will migrate in the approx 150-kDa range. The composition of the SDS-PAGE

running and sample buffers depends on the electrophoresis method used in the laboratory.

7. Centrifugation is not the ideal means of harvesting agarose beads as there is a 3–5% loss of beads in every spin. Alternatively, to increase the yield of harvested beads, one may filter at these steps using a chromatography column or a Buchner funnel with fritted disk for the washing and quenching steps.
8. Sodium azide (NaN_3) is highly toxic. Handle with care.
9. If the frozen cells are slowly extruded into the liquid nitrogen, the frozen pellets have approx twice the volume of the thawed cell suspension. This approximation can be used to appropriately apportion the required amount of cells to avoid waste or the need to refreeze cells.
10. Some protocols call for the use of a salt water-ice bath. This often causes freezing of the cell/bead mixture and reduces the efficiency of cell breakage.
11. Typically, one can safely harvest approx 15 mL of extract from a Ti-60 size tube and approx 40 mL from a Ti-45 size tube. If the material is to be used to recover a particular biochemical activity, the extract may need to be harvested from a certain location in the centrifuge tube. Empirical determination of the location of the activity of interest may be required.
12. To cryopreserve the extract, dialysis into an appropriately buffered solution containing at least 20% glycerol is recommended. Generally, dialysis is performed twice for 4 h each into more than 100 extract volumes of buffer at 4 °C with stirring. Equilibration is more rapid and thorough if the dialysis bags are inverted regularly.
13. Bio-Rad Econo-columns 0.5 × 5 cm are generally used.
14. Peptides can be used from the crude synthesis preps (>80% purity) or can be purified through a SEP-PAK C-18 cartridge according to the manufacturer's instructions (to >95% purity). As the peptide is in such high molar abundance under these conditions, as long as the peptide is free from nucleases and proteases and does not affect the pH of the solution, no functional differences are noted using highly purified peptides.
15. Centrifugation at low (4 °C) temperatures results in clouding of the layers in many cases and is not recommended. Also, use of RNA phenol ($pH < 6.5$) is not recommended for glycerol gradient extraction as the density of high concentrations of glycerol does not allow for phase separation at this step.

Acknowledgments

I gratefully acknowledge the support of John Abelson, in whose laboratory these techniques were developed. I thank Christine Guthrie and Amy Kistler for suggesting the polyoma epitope and for the contribution of the hybridoma cells. This work is supported by a grant from the Welch Foundation (F-1564), the National Science Foundation (MCB-0448556), and the American Cancer Society (RSG-05-137-01-MCB).

References

1. Puig, O., Caspary, F., Rigaut, G., et al. (2001) The tandem affinity purification (TAP) method: a general procedure of protein complex purification. *Methods* **24**, 218–229.
2. Stevens, S. W. (2000) Analysis of low-abundance RNPs from yeast by affinity chromatography and mass spectrometry microsequencing. *Methods Enzymol.* **318**, 385–398.
3. Wach, A., Brachat, A., AlbertiSegui, C., Rebischung, C., and Philippsen, P. (1997) Heterologous HIS3 marker and GFP reporter modules for PCR-targeting in *Saccharomyces cerevisiae. Yeast* **13**, 1065–1075.
4. Barnes, W. M. (1994) PCR amplification of up to 35-kb DNA with high fidelity and high yield from lambda bacteriophage templates. *Proc. Natl. Acad. Sci. U. S. A.* **91**, 2216–2220.
5. Gietz, R. D., and Woods, R. A. (2002) Transformation of yeast by lithium acetate/single-stranded carrier DNA/polyethylene glycol method. *Methods Enzymol.* **350**, 87–96.
6. Baker-Brachmann, C., Davies, A., Cost, G. J., et al. (1998) Designer deletion strains derived from *Saccharomyces cerevisiae* S288C: a useful set of strains and plasmids for PCR-mediated gene disruption and other applications. *Yeast* **14**, 115–132.
7. Sherman, F. (1991) Getting started with yeast. *Methods Enzymol.* **194**, 3–21.
8. Yaffe, M. P., and Schatz, G. (1984) Two nuclear mutations that block mitochondrial import in yeast. *Proc. Natl. Acad. Sci. U. S. A.* **81**, 4819–4823.
9. Grussenmeyer, T., Scheidtmann, K. H., Hutchinson, M. A., Eckhart W., and Walter, G. (1985) Complexes of polyoma-virus medium T-antigen and cellular proteins. *Proc. Natl. Acad. Sci. U. S. A.* **82**, 7952–7954.
10. Ganderton, R. H., Stanley, K. K., Field, C. E., et al. (1992) A monoclonal anti-peptide antibody reacting with the insulin-receptor β-subunit. *Biochem. J.* **288**, 195–205.
11. Bradford, M. (1976) A rapid and sensitive method for the quantitation of microgram quantities of protein utilizing the principle of protein dye-binding. *Anal. Biochem.* **72**, 248–254.
12. Ansari, A., and Schwer, B. (1995) SLU7 and a novel activity, SSF1, act during the PRP16-dependent step of yeast pre-mRNA splicing. *EMBO J.* **14**, 4001–4009.
13. Stevens, S. W., and Abelson, J. (2002) Yeast pre-mRNA splicing: Methods, mechanisms, and machinery. *Methods Enzymol.* **351**, 200–220.
14. Bouveret, E., Rigaut, G., Shevchenko, A., Wilm, M., and Séraphin, B. (2000) A Sm-like protein complex that participates in mRNA degradation. *EMBO J.* **19**, 1661–1671.
15. Bach, M., Winkelmann, G., and Lührmann, R. (1989) 20S small nuclear ribonucleoprotein U5 shows a surprisingly complex protein composition. *Proc. Natl. Acad. Sci. U. S. A.* **86**, 6038–6042.

16. Behrens, S. E., and Lührmann, R. (1991) Immunoaffinity purification of a [U4/U6•U5] tri-snRNP from human-cells. *Genes Dev.* **5**, 1439–1452.
17. Stevens, S. W., Barta, I., Ge, H. Y., et al. (2001) Biochemical and genetic analyses of the U5, U6 and U4/U6•U5 small nuclear ribonucleoproteins from *Saccharomyces cerevisiae. RNA* **7**, 1543–1553.
18. Sambrook, J., Fritsch, E. F., and Maniatis, T. (1989) *Molecular Cloning: A Laboratory Manual*, Cold Spring Harbor Laboratory Press, Plainview, NY.
19. Neuhoff, V., Arold, N., Taube, D., and Ehrhardt, W. (1988) Improved staining of proteins in polyacrylamide gels including isoelectric focusing gels with clear background at nanogram sensitivity using Coomassie Brilliant Blue G-250 and R-250. *Electrophoresis* **9**, 255–262.
20. Blum, H., Hildburg, B., and Gross, H. J. (1987) Improved silver staining of plant proteins, RNA and DNA in polyacrylamide gels. *Electrophoresis* **8**, 93–99.
21. Davis, M. T., and Lee, T. D. (1997) Variable flow liquid chromatography tandem mass spectrometry and the comprehensive analysis of complex protein digest mixtures. *J. Am. Soc. Mass Spectrom.* **8**, 1059–1069.
22. Davis, M. T., and Lee, T. D. (1998) Rapid protein identification using a microscale electrospray LC/MS system on an ion trap mass spectrometer. *J. Am. Soc. Mass Spectrom.* **9**, 194–201.
23. Eng, J. K., McCormack, A. L., and Yates, J. R. (1994) An approach to correlate tandem mass-spectral data of peptides with amino-acid sequences in a protein database. *J. Am. Soc. Mass Spectrom.* **5**, 876–989.
24. Link, A. J., Eng, J., Schieltz, D. M., et al. (1999) Direct analysis of protein complexes using mass spectrometry. *Nature Biotechnol.* **17**, 676–682.
25. Studier, F. W., Rosenberg, A. H., Dunn, J. J., and Dubendorff, J. W. (1990) Use of T7 RNA polymerase to direct expression of cloned genes. *Methods Enzymol.* **185**, 60–89.

6

CLIP: Crosslinking and ImmunoPrecipitation of In Vivo RNA Targets of RNA-Binding Proteins

Kirk B. Jensen and Robert B. Darnell

Summary

We present a newly developed method for fixing RNA–protein complexes *in situ* in living cells and the subsequent purification of the RNA targets. Using this approach, complex tissue such as mouse brain can be ultraviolet (UV) irradiated to covalently crosslink RNA–protein complexes. Once covalently bound, RNA–protein complexes can be purified under stringent conditions, allowing a highly specific purification scheme to be employed. After UV irradiation, the tissue is solubilized and the RNA partially digested, allowing a small fragment to remain attached to protein. RNA–protein complexes of interest are partially purified by immunoprecipitation and noncovalently associated RNA removed by sodium dodecyl sulfate polyacrylamide gel electrophoresis (SDS-PAGE). These purified RNA–protein complexes are isolated and treated with proteinase K, which removes protein but leaves intact RNA. This RNA is abundant enough, and competent for, RNA linker ligation, reverse transcriptase polymerase chain reaction (RT-PCR) amplification, and sequencing. Database matching of these short 70- to 100-nt RNA CLIP (crosslinking and immunoprecipitation of RNA–protein complexes) "tags," which mark the native binding sites of RNA binding proteins, potentially allows the entire target repertoire of an RNA binding protein to be determined.

Key Words: CLIP; immunoprecipitation; photocrosslinking; RNA binding protein.

1. Introduction

Those interested in the role RNA binding proteins play in the development and function of organisms have been limited by a paucity of techniques available to study RNA–protein interactions in vivo. Conversely, biochemists have developed potent methods for studying interactions in vitro but have not applied these to living tissues. We describe here a method in which a classical in vitro tool used by RNA biochemists—ultraviolet (UV) crosslinking—is adapted to the study of RNA–protein complexes in living tissues. We have successfully

From: *Methods in Molecular Biology, vol. 488: RNA-Protein Interaction Protocols*
Edited by: Ren-Jang Lin © Humana Press Inc, Totowa, NJ

used this method, termed CLIP (crosslinking and immunoprecipitation of RNA–protein complexes) to identify a number of target RNAs of the Nova family of neuron-specific RNA binding proteins *(1,2)*, and we are piloting its use with several other RNA binding proteins of interest in our laboratories, notably the FMRP and Hu families.

The term CLIP is largely self-explanatory, with the following addendums: We have used a living organ—specifically brain tissue—as the source for in vivo RNA–protein complexes to be CLIPed. This tissue is dissected from the animal (mouse in our experiments), rapidly dissociated on ice by trituration to allow UV light to penetrate the cells, and irradiated immediately thereafter. A series of subsequent biochemical steps is used to partially hydrolyze the RNA, allowing short (~70- to 100-nt) fragments to remain bound to the protein. We have used ribonuclease (RNase) T1, which leaves a 5' –OH and a 2',3'-cyclic phosphate on the RNA chain. The complex is then purified by immunoprecipitation, a step obviously critical to the protocol and dependent on an antibody good for immunoprecipitation. The RNA–protein complex is then treated with T4 polynucleotide kinase (PNK; which both adds a 5'-PO_4^- group and resolves the 2',3'-cyclic phosphate to a 3' –OH), a step that allows the introduction of radioactive tracer and prepares the RNA ends for subsequent directional linker ligation.

The complex is then further purified by boiling in sodium dodecyl sulfate (SDS)-sample buffer and running on denaturing SDS polyacrylamide gel electrophoresis (PAGE) gels. This gel purification step is critical to the success of CLIP because it partitions the RNA–protein covalent complex away from any RNA that has immunoprecipitated with the protein but was not crosslinked to the protein. While this "free" RNA pool may contain *bone fide* RNA targets of the protein, it may also include RNAs that have only bound the protein in vitro during the purification steps of the protocol, and thus it is critical to remove these RNAs *(3)*. A final purification step is achieved by transferring the gel to nitrocellulose; this transfer allows purification of RNA–protein complexes from the membrane (much easier than from the gel matrix itself), and at the same time, any remaining free RNA in the gel passes through the nitrocellulose and toward the positive electrode. Radioactive bands corresponding in size to RNA–protein complexes are cut out from the nitrocellulose, and complexes are eluted and treated with proteinase K to remove the protein component. RNA is isolated, ligated to a 5' linker harboring 5' and 3' –OH groups, and then ligated to a 3' linker harboring a 5' –OH group and a blocked 3' end (we have used puromycin). The fully ligated RNA is then gel purified to obtain only RNAs that contain an insert size greater than 30 nt. This final pool of RNA molecules is used with a linker-specific primer and reverse transcriptase to obtain complementary DNA (cDNA), which is then PCR amplified (again with linker-specific primers), cloned, and individual transformants prepared for sequencing. Finally, the CLIP tags are of sufficient length to easily and unambiguously identify the parental RNAs by database searching.

2. Materials

2.1. Solutions

1. 1X HBSS* buffer: 1X Hank's balanced salt solution, Ca-Mg free (10X from Gibco, cat. no. 14186-012), 10 m*M* HEPES, pH 7.3.
2. 1X PXL buffer: 1X phosphate-buffered saline (PBS; tissue culture grade; no Mg^{2+}, no Ca^{2+}), 0.1% SDS, 0.5% deoxycholate, 0.5% NP-40.
3. 1X PNK^+ buffer: 50 m*M* Tris-Cl, pH 7.4, 10 m*M* $MgCl_2$, 0.5% NP-40.
4. 1X PK buffer: 100 m*M* Tris-Cl, pH 7.5, 50 m*M* NaCl, 10 m*M* ethylenediaminetetraacetic acid (EDTA).
5. 1X PK buffer/7 *M* urea (this buffer must be made fresh): 100 m*M* Tris-Cl, pH 7.5, 50 m*M* NaCl, 10 m*M* EDTA, 7 *M* urea.
6. "RNA phenol": Pure crystalline phenol equilibrated with 0.15 *M* NaOAc, pH 5.2.
7. $CHCl_3$ solution: 49:1 $CHCl_3$:isoamyl alcohol.
8. Nucleic acid elution buffer: 1 *M* NaOAc, pH 5.2, 1 m*M* EDTA.

2.2. Enzymes, Reagents, Kits, Plasticware, Equipment

1. Stericups (Millipore).
2. Stratalinker (Stratagene).
3. RNasin (Promega).
4. RQ1 deoxyribonuclease (DNase) (Promega).
5. Protein A Dynabeads (Dynal).
6. T1 RNase, biochemistry grade, 1 U/μL (Ambion).
7. T4 PNK (Ambion).
8. Novex NuPage Bis-Tris gels (Invitrogen).
9. Novex LDS loading buffer.
10. BA-85 nitrocellulose (S&S).
11. RNA ligase (Fermentas).
12. SpinX colums (Costar).
13. 1-cm filters (Whatman).
14. SuperScript III (Invitrogen).
15. Pfu (Stratagene).
16. Taq (PerkinElmer).
17. TOPO TA cloning kit (Invitrogen).
18. Glycogen (Ambion, 5 mg/mL).

2.3. Linker and Primer Sequences

1. RNA linkers (from Dharmacon): RL5, 5'-OH AGG GAG GAC GAU GCG G 3'-OH; RL3, 5' -P CGA GAU GGC GGC UUC CUG C 3'-puromycin.
2. DNA primers (from Operon): DP5, AGG GAG GAC GAT GCG G; DP3, GCA GGA AGC CGC CAT CTC G.

3. Method

3.1. Ultraviolet Crosslinking of Tissue/Cell Lines

1. For our CLIP experiments with neuronal RNA binding proteins, we harvest brain and spinal cord tissue from postnatal d 8 mice (ICR strain); let the tissue sit in ice-cold HBSS* until the harvest is complete. Other rodent tissues should be harvested the same way. If your tissue source contains a lot of RNase or proteolytic activity, your best bet is to keep the tissue as cold as possible and work quickly.
2. When we use more than 10 brains at a time, we set up a 500-mL Stericup, remove the cellulose filter, and make a conical filter out of a sheet of 200-μm nylon mesh to replace it. The tissue can be disrupted by passing the tissue through the mesh using a flat-tip cell scraper and the remaining tissue washed through the mesh with more HBSS*. For 10 brains, the resulting cell suspension is about 100 mL. If we harvest only a small number of brains, the tissue can be triturated using a 5-mL pipet.
3. Next, transfer the tissue suspension to 50-mL Falcon tubes and spin at 2500 *g* for 5 min at 4 °C. Remove the supernatant and resuspend the tissue in approx 10X the original volume of tissue. Place the suspension in a 150-mm tissue culture dish (around 15–20 mL/plate) and irradiate the suspension for 400 mJ/cm^2 in a Stratagene Stratalinker. Use a dish of ice underneath the tissue suspension as you crosslink to keep the suspension cold.
4. Collect the irradiated suspension in a 50-mL Falcon tube, then wash out each plate with an additional 5 mL of HBSS*, also collecting this wash. Spin down the tissue again at 2500 rpm for 5 min at 4 °C. Resuspend the tissue in 2X volume of the original tissue volume and pipet the solution into Eppendorf tubes. Spin the tubes briefly and take off the supernatant. Freeze at –80 °C until use.
5. For crosslinking adherent cells, grow cells in 150-mm dishes. Wash cells once with 10 mL cold 1X HBSS* and add 10 mL 1X HBSS* to the dish. Irradiate as above and collect the cells by scraping. Wash the dish with an additional 5 mL of 1X HBSS* and spin down the cell suspension as above. Collect and freeze the cells as above.

3.2. Bead Prep

1. We have found that protein A Dynabeads work well for CLIP. The binding capacity of the Dynabeads is not outstanding (25 μg immunoglobulin G [IgG] per 100 μL of bead solution), but this is outweighed by the ease of handling of magnetic beads during the wash steps. For our immunoprecipitations, we generally use about 100 μL of bead solution (about 50–75 μL of beads) per 300 μL of packed, crosslinked tissue. Depending on your particular protein and antibody combination for CLIP as well as the protein's abundance in tissue/cells, these numbers could vary greatly.
2. Pipet 100 μL of protein A Dynabead solution. Place beads in magnetic stand to capture and wash beads with 500 μL of 1X PXL. Repeat the wash two more times. Resuspend beads in 100 μL of 1X PXL and add an appropriate amount of your

anti-RNA binding protein antibody. Rock the tube for 30–45 min at room temperature to bind the antibody to the beads. Wash the beads three times with 1X PXL.

3.3. Crosslinked Lysate Workup

1. Lyse each 100 µL of crosslinked cells using 100 µL of ice-cold 1X PXL. Let the lysates sit on ice for 10 min. Add 10 µL RNasin and 10 µL of RQ1 DNase to each tube; incubate at 37 °C for 15 min. If you have a shaking, heating block, like an Eppendorf Thermomixer, also shake the tubes at 1000 rpm.
2. Add 1 µL of RNase T1 stock (1 U/µL) to the solution; incubate at 37 °C for 10 min, shaking at 1000 rpm. The actual RNase T1 dilution you use will need to be determined empirically (*see* **Note 1**).
3. Next, spin the lysates in a prechilled microultracentrifuge at 90K for 25 min at 4 °C. We use open-topped polycarbonate tubes in a TLA 120.2 rotor (*see* **Note 2**).

3.4. Immunoprecipitation

For immunoprecipitation, carefully remove the supernatant from the pelleted debris and add supernatant to one prepared tube of beads. Again, we use about 100 µL of beads per 300 µL of crosslinked tissue. Rock the beads/lysate for 1 h at 4 °C. Wash three times with 1 mL ice-cold 1X PXL and twice with 1 mL ice-cold 1X PNK$^+$.

3.5. Kinase the Immunoprecipitated RNA

1. Resuspend beads in 80 µL of 1X PNK$^+$ and add 5 µL of P^{32} γ-ATP (γ-adenosine triphosphate) (>3000 Ci/m*M*) and 2 µL of PNK enzyme (*see* **Note 3**). Incubate in a thermomixer at 37 °C and 1000 rpm for 20 min.
2. "Finish" the reaction by adding 5 µL of 1 m*M* ATP. Let the reaction continue another 5 min at 37 °C. Wash the beads four times with 1 mL of ice-cold 1X PNK$^+$.
3. Resuspend the beads in 30 µL of 1X PNK$^+$ and 30 µL of Novex LDS loading buffer (*do not* add any reducing agent). If reducing agent is used, the heavy and light chains of immunoglobulin can interfere with the migration of proteins smaller than about 75 kDa. If you are interested in performing CLIP on a very large protein, you may be best served by using reducing agents to remove intact IgG. Incubate at 70 °C for 10 min. Isolate beads in the magnetic stand and take the supernatant for loading.

3.6. SDS-PAGE Gel and Transfer

1. Load 2 wells/tube of a Novex NuPAGE 10% Bis-Tris gel (*see* **Note 4**). Run the gel at 150 V until the dye front is at the bottom of the gel. After the gel run, transfer gel to a piece of S&S BA-85 nitrocellulose using the Novex wet transfer apparatus (*see* **Notes 5** and **6**).
2. After the transfer, rinse the NC filter in 1X PBS and gently blot on Kimwipes to dry. Wrap the membrane in plastic wrap and expose to film.

3.7. Cut Out Crosslinked RNA–Protein Complex

1. The autoradiography of the immunoprecipitated material usually shows radiolabeled RNA–protein complexes starting at approx the molecular weight of the noncrosslinked protein and extending in a "smear" to higher molecular weights (*see* **Note 7**). In general, the RNA–protein complexes run at approximately the combined molecular weight of the protein and RNA (*see* **Fig. 1**).
2. Since we generally like to work with CLIP tags of about 40–80 nt (or 13,200–26,400 kDa), we try to cut out the crosslinked material that is 10–30 kDa bigger than the migration of the protein itself. If you have titrated your RNase T1 correctly, the majority of radiolabeled signal will fall within this range. Using a scalpel blade, cut out this band and then cut the nitrocellulose into small pieces. Place these pieces into a single, clean Eppendorf tube (*see* **Note 8**).

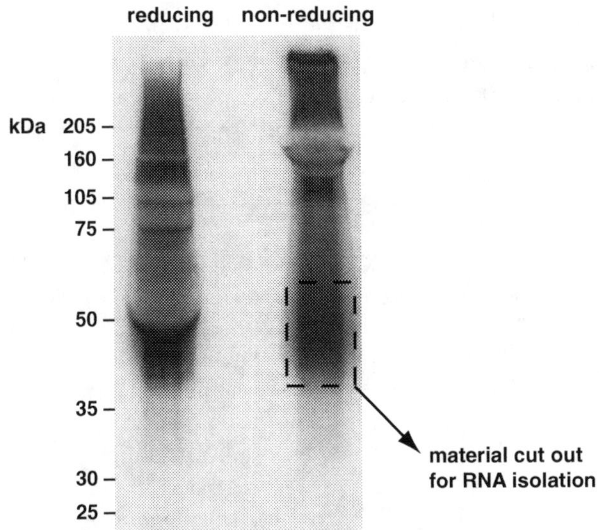

Fig. 1. Autoradiograph of ^{32}P-labeled RNA crosslinked to the neuronal Hu proteins. P4 mouse brain tissue was irradiated with short-wavelength ultraviolet (UV) light. The tissue was collected, and the soluble extracts were treated with ribonuclease (RNase) T1 and immunoprecipitated with antiserum specific for the neuronal Hu proteins (HuB, HuC, and HuD). The purified material was labeled with ^{32}P and run on reducing and nonreducing sodium dodecyl sulfate polyacrylamide gel electrophoresis (SDS-PAGE). The gel contents were transferred to nitrocellulose by wet transfer; the filter was blotted dry and exposed to film. The lane with reducing agent shows the interference caused by the approximately 50-kDa heavy chain "pushing" the gel contents out of its way. In the nonreducing lane, a fraction of the membrane from approximately 40 to 60 kDa was used for purification of the RNA–protein complexes. (The Hu proteins usually run at about 35 kDa.)

3.8. RNA Isolation and Purification

1. Make a 4-mg/mL proteinase K solution in 1X PK buffer; preincubate this stock at 37 °C for 20 min to digest any RNases. Add 200 µL of this proteinase K solution to each tube of isolated NC pieces; incubate 20 min at 37 °C with shaking (1200 rpm on Thermomixer).
2. Add 200 µL PK/7 *M* urea buffer; incubate another 20 min at 37 °C with shaking (1200 rpm on Thermomixer).
3. Add 400 µL RNA phenol and 130 µL of $CHCl_3$ solution. Vortex these tubes and then incubate at 37 °C for 20 min at 1400 rpm with shaking (1200 rpm on Thermomixer). Next, spin tubes at full speed in microcentrifuge at 4 °C or room temperature for 10 min.
4. Put the aqueous phase from each tube in a clean tube and add 50 µL of 3 *M* NaOAc, pH 5.2, and 1 mL of 1:1 EtOH:isopropanol. Do not add any carrier. Precipitate overnight at –20 °C.

3.9. RNA Ligations

1. Spin down RNA at full speed in a cold centrifuge for 30 min. Check with a hand-held Geiger counter to see if you have decent precipitation of counts; if not, you might have to add glycogen (2.5 µg) to the tube and reprecipitate (by mixing and reincubation at –20 °C). Wash the pellet with 100 µL of 75% EtOH and then dry the pellet in a SpeedVac.
2. Count the RNA in a scintillation counter by Cerenkov counts (we just stick in the dry, closed Eppendorf tube.) If you have 100,000 ccpm or more of RNA, you should not have any problem with the ligation step and amplification, and you may want to save about 20% to run as an unligated control for the next gel purification. If you have less, you may want to use all of it for the ligation step. If so, resuspend your RNA in 5.7 µL H_2O.

For the RNA ligation, use the following (10 µL total):

1 µL 10X T4 RNA ligase buffer (Fermentas)
1 µL bovine serum albumin (BSA) (0.2 mg/mL)
1 µL ATP (10 m*M*)
0.3 µL T4 RNA ligase (3 U, Fermentas) (*see* **Note 9**)
1 µL RL5 RNA linker at 20 pmol/µL (*see* **Note 10**)
5.7 µL RNA resuspended in H_2O

Incubate at 16 °C for 1 h, then add the following to the reaction:

1 µL RL3 RNA linker at 40 pmol/µL (*see* **Note 11**)
0.5 µL ATP (10 m*M*)
0.2 µL T4 RNA ligase

Incubate at 16 °C overnight, Then add the following to the reaction (100 µL total):

77 µL H_2O
11 µL 10X DNase I buffer

5 μL RNasin
5 μL RQ1 DNase

Incubate at 37 °C for 20 min, Then add the following to the reaction:

300 μL H_2O
300 μL RNA phenol (*see* **Note 12**)
100 μL $CHCl_3$ solution

Vortex, spin, and take the aqueous layer. Precipitate by adding

50 μL 3 *M* NaOAc, pH 5.2
2 μL glycogen (20 mg/mL stock)
1 mL 1:1 EtOH:isopropanol

Precipitate overnight at −20 °C.

3.10. Size Purification of the Ligated RNA

Spin, wash, and dry RNA pellets as above. Check RNA recovery by Cerenkov counting in a scintillation counter. Pour a 20% denaturing polyacrylamide gel (1:19 bis:acrylamide, 7 *M* urea). Resuspend the entire ligation reaction in 5 μL H_2O plus 5 μL of formamide loading buffer. Heat the resuspended RNA at 95 °C for 5 min and load the entire solution on the gel (along with preligation RNA if you have it); use radiolabeled PhiX markers for sizing the RNA.

After the gel run, place the gel on an old piece of film and wrap with plastic wrap. Place with a film at −80 °C, generally overnight for a good signal. Using the exposed film, cut out the RNA that is greater than about 65 nt (35 nt for the linkers, around 30 nt minimum for the CLIP tag itself). If you have good product past 100 nt, we generally cut out two gel slices: 65–100 nt and 100 nt and above (*see* **Fig. 2**). Place the gel slices in Eppendorf tubes; add 350 μL of nucleic acid elution buffer and crush with a 1-mL syringe plunger; incubate at 37 °C for 30 min at 1200 rpm in a Thermomixer. With a cut off P1000 tip, transfer the gel slurry to a Costar SpinX column to which you have added a 1-cm glass prefilter (Whatman). Spin the columns full speed in the microcentrifuge, take the flowthrough and place it in a clean Eppendorf tube. Add 1 mL 1:1 EtOH:isopropanol and 2 μL of glycogen (20 mg/mL stock) and precipitate overnight at −20 °C. Also, *see* **Notes 13** and **14**.

3.11. cDNA and PCR

Spin, wash, and dry RNA as above; count the RNA again in a scintillation counter to quantitate yield. Resuspend the purified RNA in 9 μL H_2O and add 2 μL of DP3 at 5 pmol/μL. Heat at 65 °C for 5 min; chill and quick spin.

Then, add

2 μL 10 m*M* deoxynucleotide 5'-triphosphates (dNTPs)
2 μL 0.1 *M* dithiothreitol (DTT)

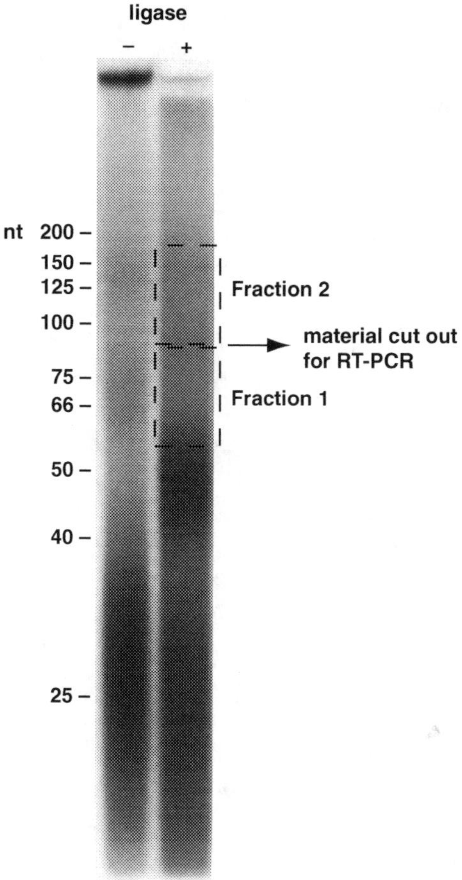

Fig. 2. Gel purification of the CLIP (crosslinking and immunoprecipitation of RNA–protein complexes) RNA after linker ligation. RNA before and after linker ligation was run on a 7 *M* urea/20% polyacrylamide gel. The modal size of this RNA pool before ligation was approximately 30 nt. After ligation, the modal size of the RNA increased to about 50 nt. The postligation pool is probably composed of significant fractions of single-linker ligation (CLIP RNA of 30 nt plus one linker of 16 nt) and smaller amounts of both 5' and 3' linker ligations (CLIP RNA of 30 nt and linkers of 35 nt). Two fractions of RNA were isolated from this gel for reverse transcriptase polymerase chain reaction (RT-PCR). Fraction 1 is from approx 60 to approx 85 nt (with CLIP tag inserts of 25–50 nt), and fraction 2 is from approx 85 to approx 175 nt (with CLIP tag inserts of 50–140 nt).

4 μL 5X SuperScript reverse transcriptase (RT) buffer
0.5 μL RNasin
0.5 μL SuperScript III

Incubate at 55 °C for 30 min, then 90 °C for 5 min. Chill on ice, then use 3 µL of the RT reaction for the PCR step. For the PCR reaction,

4 µL 10X Pfu buffer
4 µL DP5 primer, 5 pmol/µL
4 µL DP3 primer, 5 pmol/µL
4 µL radiolabeled DP5 primer
4 µL 2.5 m*M* dNTPs
1 µL Pfu
16 µL water
3 µL of the RT reaction

Cycle 35 times: 94 °C for 30 s, 58 °C for 30 s, and 72 °C for 30 s.

Pour a 10% denaturing polyacrylamide gel and run about 10 µL of the PCR reaction on the gel; use radiolabeled markers and autoradiography gel (*see* **Fig. 3**). Cut out the major bands of about 60–100 nt (and 100 nt and up if you have it). There will probably be a major contaminant band at approx 40 nt; this is a PCR product from the linker-linker dimer (no insert), and you definitely do not want to clone any of this (*see* **Note 15**). Purify and precipitate the DNA as you did above for the RNA except let the DNA elute from the crushed gel slurry for at least 4 h. After precipitation, resuspend the purified DNA in 5–10 µL of water.

3.12. TOPO Cloning and Sequencing

We used three or four more rounds of PCR (one can use more if your yield is low for the first PCR). Desalt the reaction using a spin column. Then, generate the 3' A end:

3.5 µL desalted PCR reaction
0.5 µL 10X Taq buffer
0.5 µL 10 m*M* deoxyadenosine 5'-triphosphate (dATP)
0.5 µL Taq polymerase (5 U)

Incubate at 72 °C for 20 min; place on ice and use immediately in the TOPO cloning reaction. For the TOPO clone,

2–4 µL of tailed PCR product
H_2O to 4 µL
1 µL salt solution (from TOPO kit)
1 µL pCR4-TOPO vector

Mix gently and incubate 5 min at room temperature (store 3 µL that you do not use in first day cloning at –20 °C for potential subsequent transformation). Transform *Escherichia coli* as suggested by the TOPO cloning kit. Miniprep and sequence individual transformants as you would for your other sequencing reactions (*see* **Note 16**).

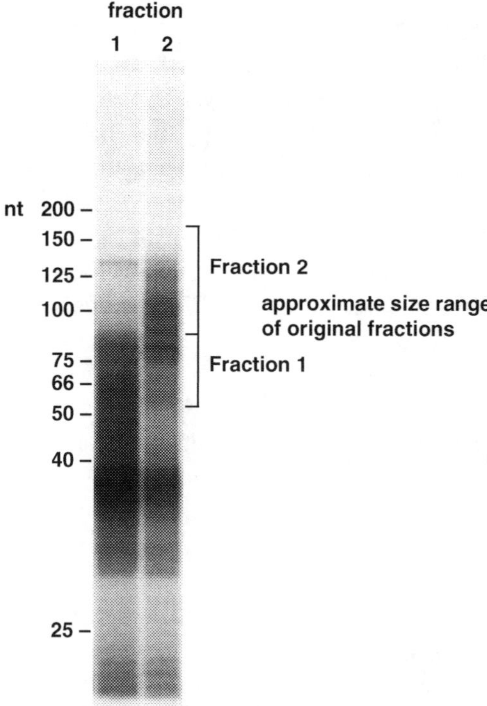

Fig. 3. Reverse transcriptase polymerase chain reaction (RT-PCR) of purified linker-ligated CLIP (crosslinking and immunoprecipitation of RNA–protein complexes) RNA. Fractions 1 and 2 of CLIP RNA gel purified after linker ligation (**Fig. 2**) were used to make complementary DNA for polymerase chain reaction (PCR) amplification. The PCR products obtained from fraction 1 (lane 1) and fraction 2 (lane 2) were run on a 7 *M* urea/20% polyacrylamide gel. Fraction 1 PCR products primarily cover a range from 50 to 80 nt, while those from fraction 2 fall within a range of 70 to 125 nt. In general, these sizes are consistent with the sizes of RNA taken from the gel purification of the linker-ligated CLIP RNAs (both fractions of PCR products are slightly shorter than predicted). To obtain CLIP tag inserts of 30 nt or greater, only PCR products great than 65 nt should be gel purified and used for the final PCR reaction before cloning. Also note the prominent product at about 38 nt. We believe that this is a PCR product resulting from amplification of a linker-linker ligation (a very prominent side reaction of the linker ligation step). While this RNA should be removed during gel purification of the RNA, its abundance makes it very difficult to get rid of it. After gel purification of the correct PCR products, this band will disappear (thus, it is not a PCR-generated "primer-dimer").

3.13. Database Searching

If one cannot identify the parental RNAs from which the CLIP tags are derived, the CLIP experiment is useless. Given the number of finished genomes and the number in progress, it is now unlikely that finding the origin of a CLIP tag poses any difficulty for most. We prefer to use the University of California at Santa Cruz genome entry site (genome.ucsc.edu) and the BLAT search tool for all of our CLIP tag matching. BLAT is much faster than BLAST and the "browser" view of search results is a great way to locate CLIP tags within a gene structure. For those who are doing CLIP with a genome on the University of California at Santa Cruz site, we highly recommend the BLAT search tool. For those doing BLAST searches, the National Center for Biotechnology Information (NCBI) now has a dedicated BLAST search page for those looking for short, nearly exact matches, and it is well suited for CLIP tag identification (*see* **Notes 9** and **17**).

4. Notes

1. The correct dilution of RNase T1 is critical; you will definitely need to do titrations of T1 to figure out which dilution will give you RNA CLIP tags of about 40–60 nt. To determine the correct T1 concentration, we generally perform the steps of the method up until the autoradiography of the SDS-PAGE filter transfer. If you run several reactions with different T1 amounts (and a T1 negative reaction), you should be able to see the effect of the amount of T1 on the migration of the RNA–protein complex on the gel (*see* **Subheading 3.7.** and **Note 7**).
2. This step clears the lysate of all high molecular weight material, like ribosomes, or very large RNPs (ribonucleoproteins).
3. This ensures that all RNA has a 5' phosphate. Also, T4 PNK has a "re-solving" activity that opens up the 2', 3'-cyclic phosphate at the 3' end of the RNA tags to give you some fraction of free 3' –OH *(4)*.
4. The Novex gels are critical. A pour-your-own SDS-PAGE gel (Laemmli) has a pH during the run that can reach approx 9.5 and can lead to alkaline hydrolysis of the RNA. The Novex NuPAGE buffer system is close to pH 7.0.
5. This nitrocellulose is pure NC and is a little fragile, but it works much better for the RNA–protein extraction step.
6. The gel itself should be a lot less hot than the amount of radioactivity that you loaded. Most of the signal should be in the lower buffer. If there is still a lot of signal in the gel, it will likely be below a molecular weight of about 20–15 kDa. You can cut this off the gel before the transfer if your protein is greater than 30 kDa.
7. Use a luminescent sticker (e.g., Glogos from Stratagene) so that you can align the filter back to the autoradiograph. If a strong radioactive band is seen after 1-h exposure, you probably have enough RNA for the CLIP procedure. The protein will migrate higher than normal (and with a greater size distribution) due to the bound RNA tags (which are not of uniform length). A very sharp radioactive band

is a sign of overdigestion, and usually the bound RNAs are too small to be useful for CLIP assay.

8. Use the film to allow you to cut out the corresponding piece of nitrocellulose using a scalpel blade. Cut this piece horizontally into as many smaller pieces as you can manage and place them all into a single Eppendorf tube. If there is signal for a molecular weight range greater than about 20 kDa, we typically divide the material into two or more sections of 20 kDa each; each section of nitrocellulose would then be cut and placed into its own individual Eppendorf tube.
9. Regarding contamination with fungal and bacterial sequences, we believe that some commercial RNA ligases we have tested have nontrivial amounts of nucleic acid in the enzyme preps. If there is insufficient CLIP RNA in the ligation reaction, one can end up ligating linkers to junk RNA. We have overcome this by using the Fermentas ligase (very clean in our hands) and using as much input CLIP RNA as possible. More recent work in the lab is investigating techniques to overcome this issue altogether.
10. It is critical to rigorously gel purify the RNA oligos for CLIP. N-1 or other non-full-length products, most notably the loss of the 3' puromycin or the 5' phosphate from the RL3 RNA linker, will ruin your CLIP experiment. Make sure you have performed this purification step well. We also like to purify the DNA oligos as we find the PCR reactions cleaner than when performed with "desalted"-quality oligos straight from the manufacturer.
11. The ligation reaction is performed sequentially (first 1-h incubation between the CLIP tag RNA and the RL5 linker, followed by the addition of the RL3 linker) to ensure that the CLIP tag RNA has sufficient time to achieve adequate ligation to the 5' RL5 linker before the addition of the 3' RL3 linker. Since the linkers are in vast molar excess to the CLIP RNA, leaving out this sequential step would lead to rapid "dead-end" ligation of the 5' and 3' linkers to each other without any CLIP tag insert. Importantly, neither linker can circularize, and the presence of a 5' –OH on the 5' linker and a 3'-puromycin on the 3' linker prevents higher-order ligation products.
12. *RNA phenol* is pure phenol that has been equilibrated with 0.15 *M* NaOAc, pH 5.2; $CHCl_3$ is chloroform 49:1 with isoamyl alcohol. We find that this phenol gives us more consistent results than a phenol equilibrated with buffers in the pH 4.0 range.
13. As the RL3 linker is blocked at its 3' end with puromycin, it is only competent to ligate at its 5' end. The result of the ligation will be RL5-CLIP tag-RL3 RNA and a prominent side reaction of RL5-RL3 linkers only. These will be removed by gel purification of the ligated products.
14. We do not believe that the presence of a single amino acid (or possibly several) at the site of crosslinking within the RNA generally interferes with reverse transcription. However, the photochemistry of short-wavelength irradiation of nucleic acid and protein is not fully understood; evidence from reverse transcription of RNAs with bromo- and iodosubstituted pyrimidines suggests that crosslinking occurs away from the basepairing faces, which are intact and competent to act as a substrate for RT ***(5,6)***.

15. There should be a sharp band at 36 nt and less sharp from 39 to 48 nt; these are all PCR products of the linker-linker ligation. The good bands should run in the same place as where you cut them out, 70–100 nt and 100–150 nt. It is possible that the larger fraction (the 100- to 150-nt fraction) will show products below 100 nt; this is okay.
16. We generally sequence around 25–50 colonies from the transformations and use this small set to check if the CLIP protocol has worked. In general, we consider a successful experiment one in which there is: (1) very little contaminating sequence in the clones (for example, if the CLIP was from mouse, we do not expect to see sequence matches to fungi, bacteria, or other organisms); (2) a high rate of matches to genomic sequence that is known to be transcribed (works only for sequenced and decently annotated genomes); (3) an absence of any sort of linker/primer contamination (generally happens when there is insufficient RNA in the ligation); and (4) the CLIP tag length is consistent with the gel purification estimates and is long enough to be used for database searching.
17. In our hands, BLAT or BLAST searching of CLIP tags reveals perfect or near-perfect matches to the mouse genome. Greater than 90% of our CLIP tags match the genome exactly or with one mismatch. We have speculated that the mismatches may be due to reverse transcriptase mistakes when the enzyme encounters the crosslinked nucleotide (which is still likely attached to at least a single amino acid), but we have not investigated this further.

References

1. Ule, J., Jensen, K. B., Ruggiu, M., et al. (2003) CLIP identifies Nova-regulated RNA networks in the brain. *Science*, **302**, 1212–1215.
2. Ule, J., Stefani, G., Mele, A., et al. (2006) An RNA map predicting Nova-dependent splicing regulation. *Nature* **444**, 580–586.
3. Mili, S., and Steitz, J. A. (2004) Evidence for reassociation of RNA-binding proteins after cell lysis: implications for the interpretation of immunoprecipitation analyses. *RNA* **10**, 1692–1694.
4. Cameron, V., and Uhlenbeck, O. C. (1977) 3'-Phosphatase activity in T4 polynucleotide kinase. *Biochemistry* **16**, 5120–5126.
5. Jensen, K. B., Atkinson, B. L., Willis, M. C., et al. (1995) Using in vitro selection to direct the covalent attachment of human immunodeficiency virus type 1 Rev protein to high-affinity RNA ligands. *Proc. Natl. Acad. Sci. U. S. A.* **92**, 12220–12224.
6. Meisenheimer, K. M., and Koch, T. H. (1997) Photocross-linking of nucleic acids to associated proteins. *Crit. Rev. Biochem. Mol. Biol.* **32**, 101–140.

7

Quantitative Analysis of Protein–RNA Interactions by Gel Mobility Shift

Sean P. Ryder, Michael I. Recht, and James R. Williamson

Summary

The gel mobility shift assay is routinely used to visualize protein–RNA interactions. Its power resides in the ability to resolve free from bound RNA with high resolution in a gel matrix. We review the quantitative application of this approach to elucidate thermodynamic properties of protein–RNA complexes. Assay designs for titration, competition, and stoichiometry experiments are presented for two unrelated model complexes.

Key Words: Competition binding; equilibrium dissociation constant; free energy; gel mobility shift; RNA binding protein; titration.

1. Introduction

1.1. Gel Mobility Shift Assay

The gel mobility shift assay is a simple yet powerful tool that is commonly used to visualize the interaction between proteins and nucleic acid. This method, also referred to as electrophoretic mobility shift assay or EMSA, relies on the property that nucleic acid will migrate through an agarose or polyacrylamide gel matrix toward an anode on application of an electric field. In a somewhat simplified view, the migration of RNA through a gel is governed by three primary factors: the molecular weight (and hence the charge) of the RNA, its three-dimensional shape, and the physical properties of the gel substrate. Interaction of a protein that modulates the RNA conformation or substantially increases the molecular weight of the ribonucleoprotein particle can lead to differential mobility in the gel. The choice of the gel matrix can amplify or dampen this effect depending on the size and shape of the RNA–protein complex. By this approach, formation of most RNA–protein complexes can be

From: *Methods in Molecular Biology, vol. 488: RNA-Protein Interaction Protocols*
Edited by: Ren-Jang Lin © Humana Press Inc, Totowa, NJ

conveniently monitored by comparing the mobility of the nucleic acid in the presence and absence of the protein.

This method is generalizable and has been used to study a wide variety of protein–RNA complexes. Some of the first examples involved characterization of polyribosomes on bacterial messenger RNA (mRNA) and the interaction of small ribosomal subunit proteins with 16S ribosomal RNA (rRNA) *(1,2)*. Since then, the approach has been used to analyze the interaction of structured and linear RNA with protein and small molecule cofactors in a variety of systems *(3–8)*.

In this overview, we compare and contrast the gel mobility shifts of two systems with distinct binding properties to highlight the useful quantitative analyses that can be performed with this method (**Fig. 1**). In the first example, we review the ability of the translation repressor germline development-1 (GLD-1) from *Caenorhabditis elegans* to shift the mobility of its cognate 28-nt linear RNA target (termed TGE [tra-2 and gli-1 element]) from the 3'-untranslated region of *tra-2* mRNA *(9)*. In the second, we review binding of the small ribosomal subunit protein S15 from the hyperthermophilic bacterium *Aquifex aeolicus* to a fragment of 16S rRNA termed Afr1 (Aquifex 16S rRNA fragment 1) *(10)*. The accompanying protocol is relevant to both systems, even though the mechanism of the shift is different. In chapter 8, the binding properties of two related systems are compared using isothermal titration calorimetry (ITC).

1.2. Determination of the Equilibrium Dissociation Constant

A frequent application of the gel mobility shift assay is to probe the competency of a protein to bind to a specific RNA sequence. In this case, labeled RNA is incubated in the presence or absence of the protein, and the mobility of bound versus free RNA is compared side by side on a native gel. Differential mobility in the presence of protein is indicative of an interaction between the protein and the RNA (*see* **Fig. 1C** for an example of gel retardation and **Fig. 1D** for an example of gel acceleration). The fraction of bound RNA (f) can be determined by measuring the counts present in the bound species and dividing by the total counts present in the lane.

This type of experiment is often performed with a single concentration of protein or cell extract and is sufficient to provide a quick yes or no answer or to corroborate binding observed by a separate method. However, a single concentration is not suitable to make definitive statements concerning the ability of a protein to discriminate between RNA sequences or to compare the binding properties of different RNA binding proteins. This is because the fraction of bound RNA is sensitive to changes in protein concentration for only a narrow range surrounding the equilibrium dissociation constant K_d. For experiments that utilize only one protein concentration, this insensitivity can lead to serious misinterpretations of binding specificity.

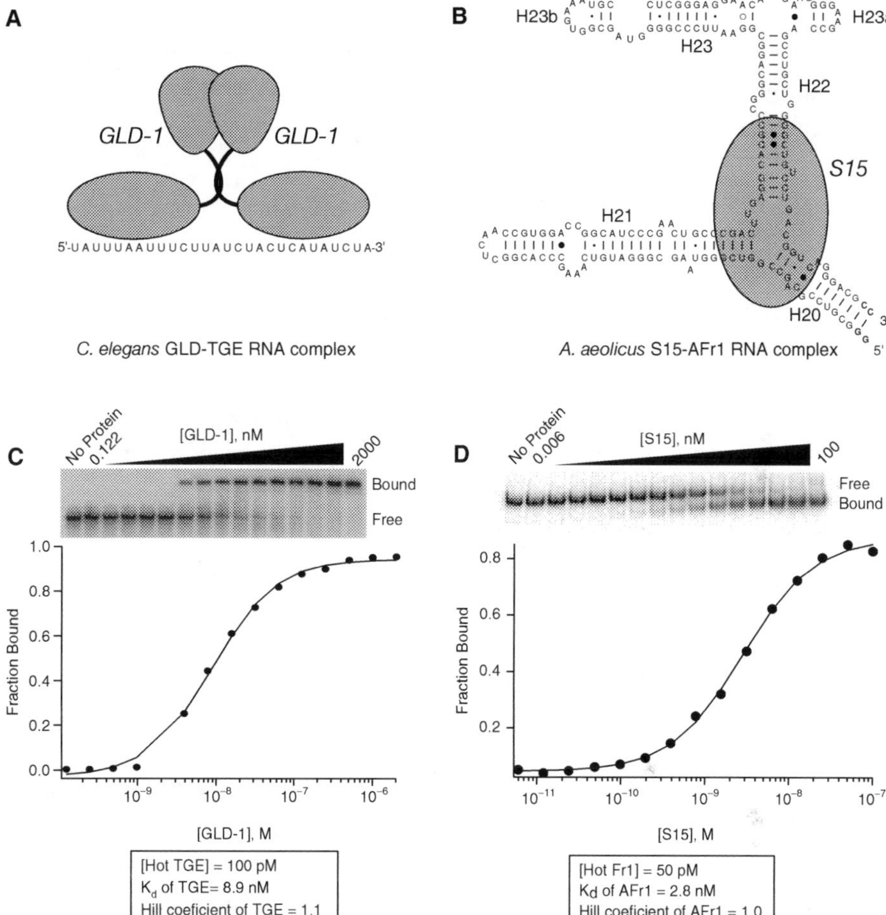

Fig. 1. Measurement of the equilibrium dissociation constant by gel mobility shift assay for two divergent systems. (**A**) Model of the interaction between *Caenorhabditis elegans* GLD-1 and TGE RNA. GLD-1 binds to RNA as a preformed stable homodimer ***(9,20)***. (**B**) Model of the interaction between *Aquifex aeolicus* S15 with Afr1 RNA, derived from the central domain of 16S rRNA. (**C**) The binding of recombinant GLD-1 induces a mobility shift in TGE RNA. GLD-1 RNA-binding domain was expressed and purified as a fusion protein with maltose binding protein (MBP) ***(9)***. In this experiment, the mobility of the TGE RNA (MW = 8.7 kDa) is retarded by the interaction with homodimeric GLD-1 (MW = 125 kDa). Bound and free RNA populations are labeled. The fraction of bound RNA is plotted as a function of the monomeric concentration of GLD-1. A fit to the Hill equation (**Eq. 1**) is shown. The K_d and Hill coefficient for this particular experiment are given. The binding is not cooperative (n = 1), suggesting the GLD-1 dimers bind as a stable, preformed unit. (**D**) Gel mobility shift of Afr1 RNA due to S15 binding. In this case, the RNA species is much larger than the protein. Also, the mobility of the RNA is accelerated, likely indicating that a protein-dependent conformational change in the RNA structure mediates the shift. As in **C**, the plot of the data and a fit are shown.

The equilibrium dissociation constant for the binding reaction in the **Scheme 1** is defined by **Eq. 2**:

$$RP \rightleftharpoons R + P \tag{1}$$

$$K_d = \frac{[R][P]}{[RP]} \tag{2}$$

where $[R]$, $[P]$, and $[RP]$ are the molar concentrations of free RNA, free protein, and bound RNA–protein complex at equilibrium, respectively. Theoretically, if one could monitor the equilibrium concentrations of all three species, then the K_d could be determined from a single equilibration. However, the gel shift experiment monitors the fraction of bound RNA rather than the free protein concentration. The fraction of bound RNA is related to the K_d by **Eq. 3**:

$$f = \frac{[RP]}{[R]+[RP]} = \frac{[P_t]}{[P_t]+K_d} = \frac{1}{1+(K_d/[P_t])} \tag{3}$$

where P_t is the total protein concentration used in the equilibration. This equation assumes that the RNA is in trace, such that $[RP]$ is negligible, and $[P_t]$ approximates the free protein concentration at equilibrium. By determining f for multiple values of P_t, the equilibrium dissociation constant can be derived from a nonlinear least-squares fit. The equilibrium dissociation constant has units of concentration and is a meter of the binding affinity; weak interactions have relatively large values of K_d, while strong interactions have smaller values.

1.3. Relationship Between the Equilibrium Dissociation Constant and Free Energy

The equilibrium dissociation constant is directly related to the Gibbs standard free energy change ($\Delta G°$) for a dissociation reaction via **Eq. 4**:

$$\Delta G^{\circ} = -RT \ln K_d \tag{4}$$

where R is the gas constant, and T is the temperature. Therefore, measurement of the equilibrium dissociation constant defines a thermodynamic parameter necessary to describe the energetics of an interaction. Of particular value, the ability of an RNA binding protein to discriminate between two sequences can be quantified in energetic terms by calculating the change in the standard free energy change ($\Delta\Delta G°$) using **Eq. 5**:

$$\Delta\Delta G^{\circ} = RT \ln \frac{K_{d2}}{K_{d1}} \tag{5}$$

where K_{d1} and K_{d2} represent the equilibrium dissociation constants for each sequence. For example, if a protein binds to sequence 2 fivefold weaker than sequence 1, then the corresponding $\Delta\Delta G°$ is approximately equivalent to 1 kcal/mol at 37 °C, the thermodynamic equivalent of a single hydrogen bond in an RNA–protein complex *(11)*.

Determination of the equilibrium dissociation constant of an RNA–protein interaction eliminates the risk that specificity will be misjudged due to unresponsive assay conditions, defines a useful thermodynamic parameter, and provides a quantitative platform to assess the energetic penalty of defined mutations. We present protocols to derive this parameter from direct titration and competition-binding experiments by gel mobility shift. In addition, we include a stoichiometric-binding protocol to assess the active fraction of protein or the ratio of protein to RNA in the shifted complex.

1.4. Direct Titration Experiments

As mentioned, the radiolabeled RNA concentration used in the equilibration reactions must be trace to satisfy the assumptions in **Eq. 3**. Under ideal conditions, the final molar concentration of RNA in each reaction is 10- to 100-fold less than the K_d. However, the achievable lower limit of the RNA concentration is defined by the specific activity of the radiolabel. In practice, it is unfeasible to utilize less than 10 p*M* RNA in the binding reactions due to the sensitivity of the phosphorimaging screens. Therefore, interactions with K_d values tighter than 0.1 n*M* are difficult to analyze quantitatively by this method.

Conversely, weak interactions are difficult to measure because significant dissociation of the ribonucleoprotein complex can occur during the time that it takes to load and run the native gel. Dissociation can lead to smearing of the bound species and difficulty in quantitation. A good rule of thumb is that equilibrium dissociation constants greater than 1–3 μ*M* typically cannot be accurately determined by standard gel mobility shift. Weaker interactions can be monitored using the competition experiments described below or by using polyacrylamide coelectrophoresis (PACE), a variant of the gel shift in which the ligand is polymerized into the gel matrix. PACE has been reviewed elsewhere and as such is not described further here *(12)*.

For an initial experiment, it is usually safe to design the experiment assuming that the binding constant will fall between 1 and 100 n*M*. However, the assay conditions must eventually be optimized once an initial determination of the K_d has been made. The concentration of protein used in the titration should span at least two orders of magnitude above and below the binding constant. The protein concentration stocks should be prepared by serial dilution to minimize variation and pipet error that can skew the results. The final concentration series used for the measurement should be chosen to maximize the number of data points within the binding transition. Usually, a 1:1 or 2:1 protein-to-buffer dilution series works well. The concentrations used in the protocols here are optimized for the interaction of GLD-1 with TGE RNA. A comparison of the gel shifts observed for GLD-1 binding to TGE and S15 binding to Afr1 can be found in **Fig. 1**.

1.5. Competition-Binding Experiments

Two common goals in the analysis of an RNA binding protein are to find the minimal RNA fragment that binds with high affinity and to define the sequence specificity of that interaction by probing a series of RNA mutants. These goals can be met using direct titration if every RNA fragment or mutation to be tested is individually radiolabeled and the gel shift protocol for each species is optimized. This can be an expensive, arduous, and somewhat time-consuming task.

As an alternative, quantitative competition-binding experiments can be performed. These experiments utilize a well-defined shift as the signal to probe the binding competency of unlabeled RNA variants. One advantage of this approach is that several mutants can be characterized using only one labeled species. A second is that weak interactions can be measured since there is no theoretical limit to the amount of competitor that can be added in the equilibration. Competitor RNA is titrated into the labeled complex, and the ability of the RNA to disrupt the complex is determined by native gel electrophoresis. The efficiency of unlabeled RNA binding is determined by plotting the fraction of bound radiolabeled RNA as a function of competitor concentration.

If the binding conditions are chosen carefully, the equilibrium dissociation constant of the competitor K_c can be determined. The amount of protein included in the equilibration must be below the saturation limit for the K_d of the labeled species. The amount of protein used in the competition experiment should give 60–90% maximal binding of the labeled RNA in the absence of competitor. This is to ensure that competition occurs during the equilibration. Also, calculation of the K_c presumes that the competitor species binds with identical stoichiometry to the same binding site as the labeled species. Before the binding of any mutation or fragment series is measured by competition experiments, self-competition using unlabeled wild-type RNA should be performed to ensure that the K_c compares favorably to the K_d determined by direct titration. The protocol below details experimental conditions for setting up a self-competition-binding experiment using the GLD-1/TGE complex and unlabeled TGE RNA as a competitor. Self-competition plots are compared for GLD-1 and S15 in **Fig. 2**. Detailed fitting procedures are presented in **Subheading 3.5.**

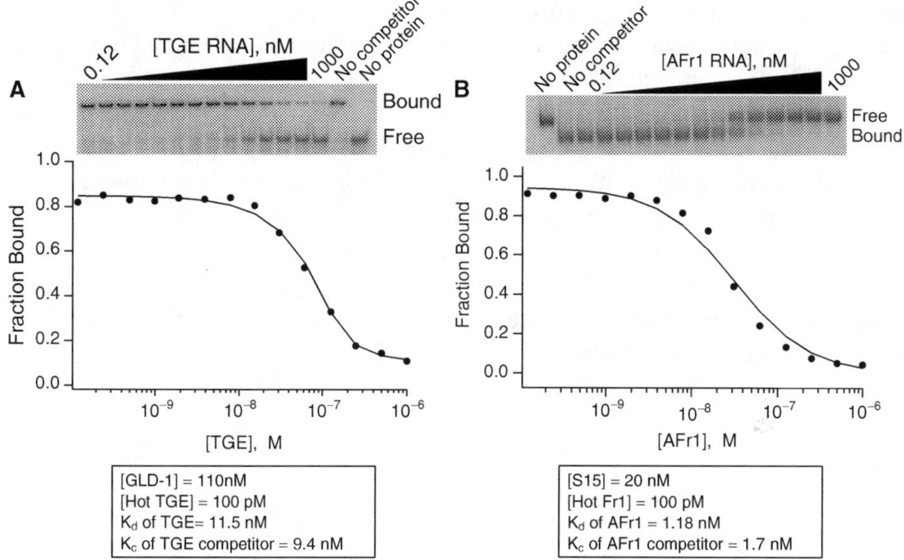

Fig. 2. Competition mobility shift experiments. (**A**) Titration of cold TGE RNA into a GLD-1–TGE complex. A fit of the fraction bound versus competitor RNA concentration is shown. The concentrations of GLD-1, labeled TGE RNA, and the K_d and K_c of GLD-1 for TGE RNA are shown. The K_c was determined of a fit to the data using the Weeks and Crothers solution of the Lin and Riggs equation ***(18,19)***. (**B**) Competition of unlabeled AFr1 RNA into an S15-AFr1 complex. The data are plotted and fit as in **A**. In both cases, the K_c is equivalent within error to the K_d determined by direct titration.

1.6. Stoichiometric-Binding Experiments

If the active concentration of protein is known, then the stoichiometry of an RNA–protein interaction can be inferred from gel mobility shift experiments using conditions in which the RNA concentration is not in trace. In this experiment, the protein is titrated into RNA of known concentration that has been spiked with a trace amount of radiolabeled RNA. The molar equivalent of protein necessary to saturate binding is determined by measuring the fraction of bound radiolabeled RNA and plotting the data as a function of molar equivalents of protein to RNA. A comparison of the results with theoretical saturation curves defines the stoichiometry of binding. For an accurate determination, the concentration of RNA used in the experiment should be 50- to 100-fold greater than the K_d.

If the active protein concentration is not known, then the interpretation of the experiment can be ambiguous. If the stoichiometry of binding is known to be 1:1 by another method, then the same approach gives an approximation of the active protein concentration. For example, stoichiometric binding of GLD-1 to TGE RNA best matches the 2:1 theoretical curve (**Fig. 3**). This can indicate that two molecules of GLD-1 bind to this RNA, or it can mean that the protein preparation is only 50% active. In this case, GLD-1 is known to form stable homodimers, consistent with the 2:1 model. In addition, the stoichiometry was validated with ITC (*see* chapter 8) using this and a shorter RNA construct to distinguish between these two possibilities. In either case, stoichiometric-binding experiments serve as a valuable control in defining a model for any system. A protocol for determining binding stoichiometry is given in **Subheading 3.3**. The concentrations given are relevant to GLD-1 binding to TGE RNA.

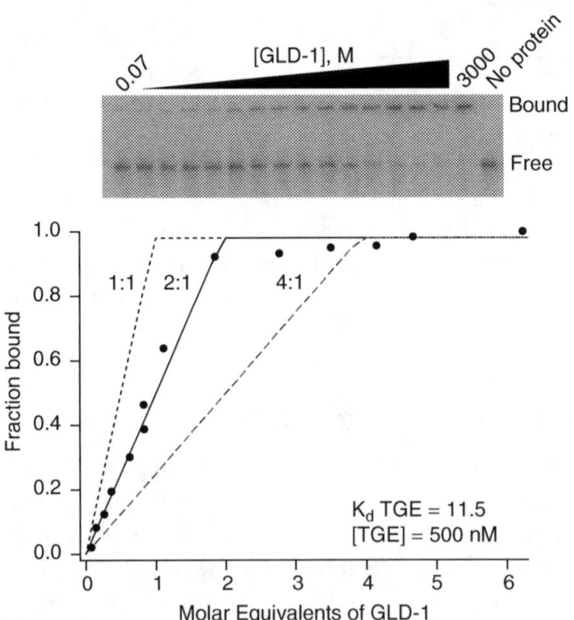

Fig. 3. Stoichiometry shift of GLD-1 binding to TGE RNA. The data are plotted as the fraction of bound labeled RNA versus molar equivalents of GLD-1 to TGE RNA. The total RNA concentration used in the equilibrations and the K_d of GLD-1 for TGE RNA determined by direct titration are given. The data are compared to theoretical saturation curves for 1:1, 2:1, and 4:1 protein-to-RNA stoichiometry. The 2:1 curve most closely approximates the data, indicating that two copies of GLD-1 interact with a single TGE RNA molecule. This is consistent with literature indicating that this protein forms a stable homodimer ***(9,20)***. (Adapted with permission from **ref. *9***.)

2. Materials

1. RNA suitable for quantitative measurements, prepared by chemical synthesis or by in vitro transcription: The molar concentration of the RNA should be measured as accurately as possible with ultraviolet (UV) spectrophotometry using base-hydrolyzed RNA (*see* **Note 1**).
2. ^{32}P-labeled RNA prepared by end or body labeling: The molar concentration of labeled RNA should be estimated from the specific activity of the label using liquid scintillation or Cerenkov counting.
3. RNA binding protein: The protein should be purified as much as possible and concentrated to at least 20 µ*M*. The concentration of protein should be determined by UV spectroscopy. An online calculator to estimate the extinction coefficient of any peptide sequence based on the method of Gill and von Hippel is available at http://us.expasy.org/tools/protparam.html ***(13)***. An estimate of the active protein concentration can be determined by stoichiometric-binding experiments outlined here.
4. Buffers and reagents: All buffers should be prepared with distilled, deionized water (ddH_2O) and filter sterilized prior to use. These buffers may be prepared in advance and stored at –20 °C. To minimize ribonuclease contamination, it is safest to work with gloves and certified ribonuclease (RNase)-free tubes. The contents of the binding buffer, including buffer and pH, ion concentration, detergent, and nonspecific competitors should be optimized for each system (*see* **Note 2**).

   2X GLD-1 binding buffer: 20 m*M* Tris-HCl, pH 8.0, 50 m*M* NaCl, 0.2 m*M* ethylenediaminetetraacetic acid (EDTA).
   2X S15 binding buffer: 40 m*M* K-HEPES, pH 7.6, 660 m*M* KCl, 20 m*M* $MgCl_2$, 0.2 m*M* EDTA.
   1 mg/mL transfer RNA (tRNA), prepared from baker's yeast (Sigma R9001).
   50 µg/mL heparin, sodium salt (USB 16920).
   1% (v/v) IGEPAL CA-630 (Sigma I8896).
   6X loading buffer: 30% (v/v) glycerol, 0.25% (w/v) xylene cyanol, 0.25% (w/v) bromophenol blue.

5. Electrophoresis stock solutions: These solutions should be prepared with distilled, deionized water and stored at room temperature. The gel mix is stable for at least 1 mo if stored in an amber bottle, but we usually prepare it fresh immediately before use.

   10X TBE: 108 g Tris base, 55 g boric acid, and 9.3 g EDTA, tetrasodium salt dissolved into distilled, deionized water to a final volume of 1 L. Dilute a sufficient amount to 0.5X for use as running buffer in a standard vertical electrophoresis apparatus.
   Gel mix (GLD-1): 6% (v/v) 29:1 acrylamide/bis-acrylamide solution (available as a 40% stock from Fisher BP1408-1), 0.5X TBE.
   Gel mix (S15): 15% (v/v) 29:1 acrylamide/bis-acrylamide solution, 0.5X TBE.
   10% (w/v) ammonium persulfate.
   N,N,N',N' tetramethylethylenediamine (TEMED; Fisher BP150–20).

6. Electrophoresis equipment, including a power supply that can be used at 4 °C.
7. Phosphorimager and imaging screens or equivalent.

3. Methods

3.1. Binding Reactions for Direct Titration

1. The final equilibration reactions contain 1X binding buffer, 0.01% IGEPAL CA-630, 0.1 mg/mL of tRNA, 5 μg/mL of heparin, 100 p*M* radiolabeled RNA, and protein concentrations ranging from 2 μ*M* to 120 p*M* (see **Table 1** for an example; *see* **Note 2** for an explanation of additives used in the equilibration reaction). Typical reaction volumes are 20 μL.
2. Prepare a master mix (MM) that contains 1.11X concentration of all of the reagents excluding protein and heparin. Enough mix should be prepared for at least 16 equilibrations. In practice, it is wise to prepare excess MM. The MM should be heated and then allowed to cool to the equilibration temperature to disrupt RNA aggregates and to promote proper folding (*see* **Note 3**).
3. Prepare 30 μL of 20 μ*M* protein in 50 μg/mL heparin to make the master protein (MP) stock. Generate a 1:1 serial dilution series by transferring 15 μL of the MP stock into 15 μL of 50 μg/mL heparin and mixing by gently pipeting up and down. Repeat to generate a series of 10X protein stocks ranging from 20 μ*M* to 1.22 n*M*.

Table 1
Equilibration Reactions for a Direct Titration Experiment

Lane	MM (μL)	Protein stock (μL)	Dilution	Final (protein) (n*M*)	Buffer (μL)	Total volume (μL)
1	18	0	—	—	2	20
2	18	2	1:0	2000	0	20
3	18	2	1:2	1000	0	20
4	18	2	1:4	500	0	20
5	18	2	1:8	250	0	20
6	18	2	1:16	125	0	20
7	18	2	1:32	62.5	0	20
8	18	2	1:64	31.3	0	20
9	18	2	1:128	15.6	0	20
10	18	2	1:256	7.8	0	20
11	18	2	1:512	3.9	0	20
12	18	2	1:1,024	2.0	0	20
13	18	2	1:2,048	1.0	0	20
14	18	2	1:4,096	0.50	0	20
15	18	2	1:8,192	0.25	0	20
16	18	2	1:16,384	0.12	0	20

MM, master mix.

4. Aliquot 18 µL of MM into 16 RNase-free tubes. To 15 of these, add 2 µL of one of the 10X protein stocks and mix by pipeting up and down. To the last tube, add 2 µL of H_2O. This reaction is the no-protein control.
5. Incubate the reactions at a constant temperature until equilibrium is reached. The time necessary to achieve equilibrium depends on the association and dissociation rate constants. Since the apparent association rate depends on the protein concentration, it takes more time for the reactions with low protein concentrations to reach equilibrium.
6. Add 4 µL of 6X loading buffer to each tube. Working as quickly as possible, load 5 µL of each reaction onto a prerun, precleaned native polyacrylamide gel (*see* **Subheading 3.4.**) at 4 °C. Run the gel at 600 V for as long as necessary to achieve separation. The time necessary is highly variable and system dependent. It must be optimized for each RNA or protein construct used. For example, adequate separation of the GLD-1/TGE complex from free TGE RNA is achieved in 30 min on a 6% native gel, while separation of S15-bound Afr1 RNA requires at least 15 h on a 15% gel.
7. Take down the gel, dry it, and expose the gel to a phosphorimager screen. Quantitate and fit the data as detailed in **Subheading 3.5**.

3.2. Competition-Binding Reactions

1. The final equilibration reactions are identical to the direct titration shift, except a constant concentration of protein is included, and variable concentrations of competitor RNA are used (1X binding buffer, 0.01% IGEPAL CA-630, 0.1 mg/mL of tRNA, 5 µg/mL of heparin, 100 p*M* radiolabeled RNA, 100 n*M* GLD-1, and cold RNA concentrations ranging from 1 µ*M* to 120 p*M*). Again, the usual reaction volume is 20 µL.
2. Prepare a MM that contains 1.25X concentration of all of the reagents excluding protein and heparin. Heat and cool the mix to fold the labeled RNA as necessary. Be sure to prepare enough mix for at least 16 reactions.
3. Prepare a MP mix that contains a 10X concentration of the protein dissolved in 50 µg/mL of heparin. Enough MP solution should be prepared for at least 16 equilibrations.
4. Prepare 30 µL of 10 µ*M* competitor RNA to make the master competitor (MC) stock. Heat and cool this mix to fold the competitor RNA as necessary. Prepare a series of 10X stocks by serial dilution of 15 µL of the MC stock into an equal volume of water to generate a range from 10 µ*M* to 1.22 n*M*.
5. Add 0.125 volumes of the MP solution to 1 volume of the MM solution to generate the master mix plus protein (MMP) stock. Be sure to retain at least 16 µL of the MM solution to use in the no-protein control equilibration.
6. Aliquot 18 µL of MMP solution into 15 individual tubes. To a separate tube, add 16 µL of MM solution. Add 2 µL of one of the 10X competitor stocks to 14 of the 15 MMP tubes and mix by pipeting up and down. To the remaining MMP tube, add 2 µL of H_2O and mix. This is the no-competitor control. To the MM tube, add 2 µL of 50 µg/mL heparin stock and 2 µL of H_2O. This is the no-protein control.
7. Follow **steps 5–7** for the direct titration protocol in **Subheading 3.1**.

3.3. Stoichiometric-Binding Reactions

1. Stoichiometric-binding reactions are set up exactly like the direct titration binding reactions except a constant concentration of unlabeled RNA is included in each equilibration. The final concentrations for GLD-1 binding to TGE RNA are 1X binding buffer, 0.01% IGEPAL CA-630, 0.1 mg/mL of tRNA, 5 μg/mL of heparin, 100 p*M* radiolabeled RNA, 500 n*M* cold RNA, and protein concentrations ranging from 4 μ*M* to 0.07 μ*M*. Once again, typical binding reaction volumes are 20 μL.
2. Prepare a MM at 1.33X concentration, including every reagent except for protein and heparin. Fold the RNA by heating and cooling as necessary.
3. Prepare 60 μL of 16 μ*M* (4X) MP stock by diluting the protein into 50 μg/mL heparin. Generate a 3:1 protein-to-buffer serial dilution series by pipeting 45 μL of the MP stock into 15 μL of 50 μg/mL heparin and mixing carefully. This gives a series of 4X stocks that will span the range of 0.14 to 8 molar equivalents of protein to RNA in the final equilibrations
4. Aliquot 15 μL of MM into 16 tubes and add 5 μL of one of the 4X MP stocks to all but 1 of the tubes. To the remaining tube, add 5 μL of 50 μg/mL heparin to serve as the no-protein control.
5. Follow **steps 5–7** from the direct titration protocol in **Subheading 3.1**.

3.4. Native Gel Electrophoresis

Central to the gel mobility shift assay is the native gel. Native gels are differentiated by the lack of denaturant, such as urea, formamide, or sodium dodecyl sulfate (SDS). The primary assumption of the gel mobility shift assay is that the binding reaction is quenched on entrance into the gel. In other words, the concentration of bound and free RNA should not change after loading. The veracity of this assumption depends on the dissociation rate constant of the interaction and the "caging effect," which has been reviewed extensively elsewhere and is not considered further here ***(14)***.

For the purpose of this protocol, we describe the preparation of a 29:1 acrylamide to bis-acrylamide, 0.5X TBE native polyacrylamide gel for use in a vertical gel apparatus. This gel substrate is relevant to many binding interactions, including both GLD-1 and S15, and the separation efficiency can be easily tuned by the modulating the percentage of acrylamide. We typically pour gels between glass plates that are 35 cm wide by 25 cm tall (inner plate; outer plate is 27 cm) using 0.05-cm Teflon spacers and run the gel with an Owl Scientific ADJ3 or equivalent apparatus. The gel can be poured in advance or while the reaction is coming to equilibrium.

1. Clean and assemble the glass plates and the spacers. Use binder clips to hold the plates together.
2. Prepare 50 mL of gel mix using the recipe in the Materials subheading.
3. Add 1/1000 volume of TEMED and 1/100 volume of 10% APS to the gel mix and mix thoroughly.

4. Working quickly, pour the mix into the space between the plates, being careful to avoid air bubbles. If the plates are placed on a flat surface, the gel can be poured without tape by applying the gel mix as a bead across the top of the plate and displacing air from the bottom. Alternatively, the sides and bottom can be taped and the gel poured while at an angle.
5. Prior to polymerization, insert a comb into the top of the gel. We use combs with 36 teeth that are 0.5 cm wide by 1.0 cm tall and are spaced 0.3 cm apart. These combs enable two simultaneous 16-point titrations on a single gel. Wait until polymerization is complete, approximately 15 min.
6. Remove the comb and assemble the gel into the apparatus in a 4 °C room. Using a syringe, clear out bubbles and unpolymerized acrylamide from the wells. Add chilled 0.5X TBE running buffer to the top and bottom buffer chamber and prerun the gel at 600 V until the temperature has equilibrated (at least 1 h).
7. Immediately before loading, turn down the power supply to 100 V and reclean the wells. Be careful to avoid coming into contact with both buffer chambers as current is flowing through the gel. Once all of the lanes have been loaded, turn the power supply back up to 600 V. Run the gel as long as necessary to achieve separation, which unfortunately must be determined empirically.
8. Remove the gel from the apparatus and empty the buffer chambers. Make sure that the lower buffer chamber is not radioactive. If it is, then carefully remove the bottom buffer and dispose of in an approved container. Separate the plates, transfer the gel to Whatman filter paper, cover with plastic wrap, and dry with a gel-drying apparatus.

3.5. Data Analysis

Each of the fitting methods requires determination of the fraction of bound RNA. This is most easily achieved using phosphorimaging plates. Other methods do exist but are less accurate and have been reviewed previously ***(15)***. The dried gel should be exposed to the phosphorimaging plate for at least one overnight period; however, low concentrations of labeled RNA or a low specific activity may necessitate longer exposure times. The fraction of bound RNA is determined by measuring the volume of counts in both the bound and free species using the software associated with the phosphorimager. The fraction bound equals the counts in the bound population divided by the sum of the bound plus free counts.

Direct titration reactions should be plotted as a fraction of bound RNA versus the protein concentration. Displaying the protein concentration axis on the log scale facilitates visual inspection of the data. For bimolecular interactions, the midpoint of the transition is the apparent K_d. The transition should span two orders of magnitude in protein concentration. If this is not the case, then binding may be proceeding by a more complex mechanism or equilibrium may not have been reached during the incubations.

The data can be fit to **Eq. 3** using nonlinear least-squares methods. In practice, it is usually necessary to fit the data to a modified form of the Hill equation (**Eq. 6**) that can compensate for deviations from ideal conditions, including incomplete binding due to an inactive population of RNA, loss of protein sample at low concentrations, cooperative binding, oligomerization, or other more complex mechanisms of binding ***(16)***. The modified Hill equation is as follows:

$$f = b + \left[\frac{m-b}{1+(K_d/[P_t])^n} \right] \tag{6}$$

where m and b are normalization factors that represent the fraction of bound RNA at the upper and lower asymptotes of the titration, respectively, and n is the Hill coefficient. The Hill coefficient is a measure of the cooperativity of binding, and for bimolecular association of protein and RNA, its value is 1. Deviations from unity may indicate cooperative binding of multiple proteins or that the binding reactions have not reached equilibrium. Small deviations from integer values are common and are usually caused by some protein or RNA sticking to the equilibration vessel.

We favor IGOR (Wavemetrics) software, which incorporates the Levenberg-Marquadt algorithm and statistical analysis tools into a convenient graphical rendering program ***(17)***. Using this software, the fit error can be estimated by weighting the data such that the c^2 value is equivalent to $n - 1$ the number of data points used in the fit. The final reported value of the equilibrium dissociation constant should be an average of three independent determinations. If the data are clean, then the standard deviation of three determinations should be greater than the estimated error of each individual fit.

Competition reactions should be plotted as the fraction of bound labeled RNA versus the concentration of unlabeled competitor. The concentration of RNA that disrupts 50% of the bound labeled complex (IC_{50}) can be derived from a fit of the data to a sigmoidal dose-response curve (**Eq. 7**):

$$f = b + \left[\frac{m-b}{1+(IC_{50}/[C])^n} \right] \tag{7}$$

where C is the concentration of unlabeled competitor. If the protein concentration, labeled RNA concentration, and equilibrium dissociation constant of the labeled species are known to a high degree of accuracy, then the IC_{50} can be converted into the equilibrium dissociation constant for the competitor K_c using the Lin and Riggs equation (**Eq. 8**) ***(18)***:

$$K_c = \frac{2K_d IC_{50}}{2P - R - 2K_d} \tag{8}$$

where P is the protein concentration, R is the labeled RNA concentration, and K_d is the equilibrium dissociation constant of the labeled RNA. In practice, it is more efficient to fit the data to a quadratic solution of the Lin and Riggs equation (**Eq. 9**) determined by Weeks and Crothers ***(19)***:

$$f=\frac{m-b}{2R}\left(K_d+(K_d/K_c)C+P+R-\sqrt{[K_d+(K_d/K_c)C+P+R]^2-4RP}\right)+b \qquad (9)$$

where each of the fitted parameters can be held constant, constrained, or allowed to float during the fit as is appropriate. The Lin and Riggs expression assumes that the competitor binds with the same stoichiometry to the same site as the labeled species. If this is not the case, then this correction should not be used.

Stoichiometric-binding data should be plotted as the fraction of bound RNA versus the molar equivalent of protein to RNA in each reaction. On a linear scale, the data should appear as two lines that intersect at a defined point. The value of the x-axis at this point is the apparent stoichiometry of protein to RNA in the complex. For bimolecular association, this value should be 1 molar equivalent. A value of 2 may indicate that two copies of the protein are bound to the RNA in the shifted species, as is the case with homodimeric GLD-1 binding to TGE RNA, or it may indicate that 50% of the protein is not active. If the data appear to form a curved line rather than two distinct lines, then it is likely that insufficient cold RNA was used in the equilibration. The intersection of the lines can be derived from linear fits of each phase of the data or by comparison to theoretical saturation curves as in **Fig. 3**.

4. Notes

1. The concentration of any RNA species can be measured using A_{260} by calculating the molar extinction coefficient based on the content of each of the bases. While this approach is fairly accurate for linear RNA, it is not suitable for RNA that contains significant secondary and tertiary structure. This is due to the hyperchromic effect, where π-orbital stacking of the nucleobases diminishes their capacity to absorb UV light. This limitation can be overcome by hydrolyzing a fraction of the RNA stock by treatment with strong base prior to UV absorption spectroscopy.

   Aliquot 2 μL of the RNA sample into three 0.5-mL microcentrifuge tubes.
   Aliquot 2 μL of ddH_2O each into three more tubes.
   Add 8 μL of 1 *M* NaOH into each tube.
   Incubate the tubes at 37 °C for at least 1 h.
   Add 8 μL of 1 *M* HCl to each tube.
   Add 282 μL of ddH2O.
   Using the water tubes as blanks, take the absorbance of each solution at 260 nm.
   Average the three readings and use that value to determine the concentration.

2. Several additives may be used in equilibration reactions to improve the quality of the shift. It is often useful to include tRNA, heparin, or salmon sperm DNA to inhibit nonspecific binding and serve as a carrier nucleic acid to inhibit sample sticking to tubes, pipet tips, and the wells of the gel. Detergent can help to solubilize the protein at high concentration and may help the ribonucleoprotein particles enter the gel matrix. If the protein has a tendency to stick to surfaces, then bovine serum albumin (BSA) can be included in the titration to minimize this effect. The identity and concentration of the additives used in the equilibration should be optimized for each system. The additives used in this protocol are generally useful and work well for both S15 and GLD-1.
3. Storage of RNA in a –20 °C freezer increases its usable life span. Unfortunately, it also promotes denaturation of RNA structure and aggregation. Therefore, it is usually necessary to refold RNA prior to use in gel mobility shift experiments. There are several protocols for folding RNA, including heating the RNA to 65 °C or warmer followed by slow cooling to assay temperature. Other techniques include heating the RNA and snap cooling on ice and prolonged incubation of RNA at an elevated temperature. The folding of some highly structured RNA species is promoted by monovalent and divalent cations, which may be added to the folding reaction. The state of the folded RNA can be assessed on a native gel. Typically, the unbound RNA should migrate through the gel as a single species. Multiple bands may indicate that the RNA is not properly folded or that it has begun to degrade.

Acknowledgments

We thank Ivan Baxter for the alkaline hydrolysis protocol and Dana Abramovitz for generation of the GLD-1 expression construct. S.P.R. was supported by a Damon Runyon Cancer Research Foundation Fellowship (DRG-1723). M.I.R was supported by a Research Scholar grant from the American Cancer Society (PF-01-087-01-GMC). This research was supported by grants from the National Institutes of Health (NIH GM53320 and GM53757).

References

1. Dahlberg, A. E., Dingman, C. W., and Peacock, A. C. (1969) Electrophoretic characterization of bacterial polyribosomes in agarose-acrylamide composite gels. *J. Mol. Biol.* **41**, 139–147.
2. Schaup, H. W., Green, M., and Kurland, C. G. (1970) Molecular interactions of ribosomal components. I. Identification of RNA binding sites for individual 30S ribosomal proteins. *Mol. Gen. Genet.* **109**, 193–205.
3. Murphy, F. L., Wang, Y. H., Griffith, J. D., and Cech, T. R. (1994) Coaxially stacked RNA helices in the catalytic center of the Tetrahymena ribozyme. *Science* **265**, 1709–1712.
4. Samuels, M. E., Bopp, D., Colvin, R. A., Roscigno, R. F., Garcia-Blanco, M. A., and Schedl, P. (1994) RNA binding by Sxl proteins in vitro and in vivo. *Mol. Cell. Biol.* **14**, 4975–4990.

5. Yakhnin, A. V., Trimble, J. J., Chiaro, C. R., and Babitzke, P. (2000) Effects of mutations in the L-tryptophan binding pocket of the Trp RNA-binding attenuation protein of *Bacillus subtilis*. *J. Biol. Chem.* **275**, 4519–4524.
6. Vargason, J. M., Szittya, G., Burgyan, J., and Tanaka Hall, T. M. (2003) Size selective recognition of siRNA by an RNA silencing suppressor. *Cell* **115**, 799–811.
7. Batey, R. T., and Williamson, J. R. (1996) Interaction of the *Bacillus stearothermophilus* ribosomal protein S15 with 16S rRNA: I. Defining the minimal RNA site. *J. Mol. Biol.* **261**, 536–549.
8. Zamore, P. D., Williamson, J. R., and Lehmann, R. (1997) The Pumilio protein binds RNA through a conserved domain that defines a new class of RNA-binding proteins. *RNA* **3**, 1421–1433.
9. Ryder, S. P., Frater, L., Abramovitz, D. L., Goodwin, E. B., and Williamson, J. R. (2004) RNA target specificity of the STAR/GSG domain post-transcriptional regulatory protein GLD-1. *Nat. Struct. Mol. Biol.* **11**, 20–28.
10. Recht, M. I., and Williamson, J. R. (2001) Central domain assembly: thermodynamics and kinetics of S6 and S18 binding to an S15-RNA complex. *J. Mol. Biol.* **313**, 35–48.
11. delCardayre, S. B., and Raines, R. T. (1995) A residue to residue hydrogen bond mediates the nucleotide specificity of ribonuclease A. *J. Mol. Biol.* **252**, 328–336.
12. Cilley, C. D., and Williamson, J. R. (1999) PACE analysis of RNA-peptide interactions. *Methods Mol. Biol.* **118**, 129–141.
13. Gill, S., and von Hippel, P. (1989) Calculation of protein extinction coefficients from amino acid sequence data. *Anal. Biochem.* **182**, 319–326.
14. Cann, J. R. (1989) Phenomenological theory of gel electrophoresis of protein-nucleic acid complexes. *J. Biol. Chem.* **264**, 17032–17040.
15. Setzer, D. R. (1999) Measuring equilibrium and kinetic constants using gel retardation assays. *Methods Mol. Biol.* **118**, 115–128.
16. Hill, A. V. (1910) The possible effects of the aggregation of the molecules of haemoglobin on its oxygen dissociation curve. *J. Physiol. (London)* **40**, 4–7.
17. Marquardt, D. (1963) An algorithm for least-squares estimation of nonlinear parameters. *J. Soc. Ind. Appl. Math.* **11**, 431–441.
18. Lin, S. Y., and Riggs, A. D. (1972) Lac repressor binding to non-operator DNA: detailed studies and a comparison of equilibrium and rate competition methods. *J. Mol. Biol.* **72**, 671–690.
19. Weeks, K. M., and Crothers, D. M. (1992) RNA binding assays for Tat-derived peptides: implications for specificity. *Biochemistry* **31**, 10281–10287.
20. Chen, T., Damaj, B. B., Herrera, C., Lasko, P., and Richard, S. (1997) Self-association of the single-KH-domain family members Sam68, GRP33, GLD-1, and Qk1: role of the KH domain. *Mol. Cell. Biol.* **17**, 5707–5718.

8

Monitoring Assembly of Ribonucleoprotein Complexes by Isothermal Titration Calorimetry

Michael I. Recht, Sean P. Ryder, and James R. Williamson

Summary

Isothermal titration calorimetry (ITC) is a useful technique to study RNA–protein interactions as it provides the only method by which the thermodynamic parameters of free energy, enthalpy, and entropy can be directly determined. This chapter presents a general procedure for studying RNA–protein interactions using ITC and gives specific examples for monitoring the binding of *Caenorhabditis elegans* GLD-1 STAR domain to TGE RNA and the binding of *Aquifex aeolicus* S6:S18 ribosomal protein heterodimer to an S15–ribosomal RNA complex.

Key Words: Binding; calorimetry; ribonucleoprotein complex; RNA–protein interaction; thermodynamics.

1. Introduction

1.1. *Isothermal Titration Calorimetry*

There are many studies in which isothermal titration calorimetry (ITC) has been used to monitor protein–ligand, protein–protein, and protein–nucleic acid interactions *(1–3)*. ITC measures the heat (q) evolved or consumed as aliquots of one reagent are added to a second reagent in a calorimetric cell. The reaction heat as a function of concentration is analyzed to obtain the complete thermodynamic characterization (ΔH, ΔG, ΔS) of the binding reaction *(4)*. A full titration experiment can usually be completed in 2 h.

As shown in **Fig. 1**, within an adiabatic chamber the calorimeter contains a measurement and a reference cell *(5)*. The measurement cell is filled with a solution of a macromolecule in the buffer of choice. The reference cell is filled with either water or the same buffer in which the macromolecule is dissolved. The tip of a syringe containing the ligand (either a small molecule or another macromolecule) is inserted in the measurement cell. Aliquots of the ligand are

From: *Methods in Molecular Biology, vol. 488: RNA-Protein Interaction Protocols*
Edited by: Ren-Jang Lin © Humana Press Inc, Totowa, NJ

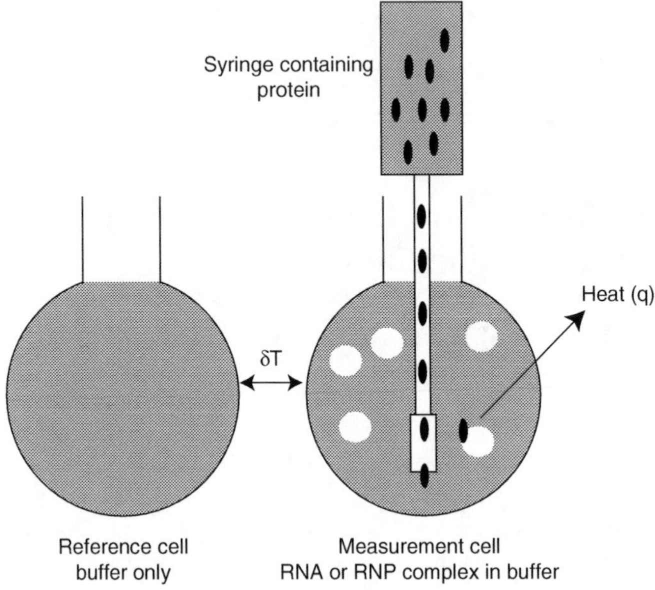

Fig. 1. Schematic of the measurement and reference cells inside the adiabatic housing of the isothermal titration calorimeter. The measurement and reference cells have an active volume of about 1.4 mL. The instrument maintains a constant temperature difference (dT) between the measurement and reference cells. Aliquots of a ligand solution are added to the measurement cell using a syringe controlled by a motorized injection system. Binding of the ligand to the macromolecule releases or consumes heat.

added to the measurement cell. The instrument maintains a small constant temperature difference between the measurement and reference cells. In response to binding of ligand to macromolecule in the measurement cell, which produces or consumes heat, the feedback system compensates to maintain the constant temperature difference between the measurement and sample cells. The signal measured during the titration is the rate of heating as a function of time, with a pulse corresponding to the addition of aliquots of ligand to the measurement cell. Integration of these pulses as a function of time yields a plot of q as a function of injection number or molar ratio of ligand to macromolecule.

1.2. RNA–Protein Interactions

There is a great range in the size and complexity of RNA–protein complexes. These span the simplest case involving a single protein binding to an isolated fragment of RNA to the large ribonucleoprotein (RNP) complexes that involve multiple proteins binding to one or more RNAs.

The great variability observed in RNA–protein interactions makes ITC an ideal choice to study these interactions. In contrast to most other methods of detecting binding of RNA and proteins, there is no inherent size limitation to the macromolecules under investigation. Fluorescence anisotropy ***(6,7)*** requires that one of the two interacting components be much smaller than the other, and surface plasmon resonance ***(8)*** requires that one of the two components be immobilized. Gel mobility shift assays require that the migration of an RNP complex in the gel is sufficiently different from the free RNA, which is less likely when complexes between small proteins and large, highly structured RNAs are studied. Gel shift assays have the requirement that the RNP complex is kinetically stable during electrophoresis, which is often not the case for weak-binding complexes. ITC does not require labeling or immobilization of either macromolecule, is performed under equilibrium conditions, and can be performed under a wide variety of buffer conditions.

1.3. Experimental Design

The amount of material required is an important factor to consider when performing an ITC experiment. Unlike gel mobility shift assays or fluorescence anisotropy measurements, in which the labeled molecule is often present in trace quantities, ITC requires nanomole quantities of each reactant. To obtain a reliable value for the association constant of an interaction, the appropriate concentrations of reactants must be used. As described by Wiseman et al. ***(5)***, the parameter c, which equals the association constant K_a times the total concentration of the reactant in the calorimeter cell M_{tot}, should lie between 1 and 1000. A more reliable estimation of K_a can be obtained if the value of c is kept between 10 and 500. A sufficient concentration of titrant should be used so that its concentration at the completion of the titration will be 1.5–2 × n (the number of binding sites per molecule) times the concentration of reactant in the cell. For a titration in which a total of 250 µL of protein is added to a reaction cell containing approximately 1.5 mL of RNA, the protein should be 10–20n times the concentration of RNA.

1.4. Biological Systems

Two different systems are chosen to demonstrate both a simple protein–RNA interaction and a protein–RNP interaction.

1. Germline development in *Caenorhabditis elegans: C. elegans* GLD-1 (germline development-1) belongs to the highly conserved STAR/GSG (signal transduction and activation of RNA/GLD-1, Sam-68, GRP-33) family of RNA binding proteins, which play a central role in metazoan development. STAR/GSG proteins are composed of a single KH domain flanked by two regions homologous to the murine quaking gene, Qua1 and Qua2. STAR domain proteins form functional homodimers in cells via the Qua1 domain, and the KH and Qua2 regions form an extended RNA interaction surface. GLD-1, a *C. elegans* germline developmental

regulator, binds to a 28-nt sequence (TGE, tra-2 and gli-1 element) in the 3'-UTR (untranslated region) of *tra-2* messenger RNA and recruits a complex that silences translation. The protocol here describes monitoring the binding of the GLD-1 STAR domain homodimer to the 28-nt TGE RNA *(9)*.

2. Assembly of the platform domain of the 30S ribosomal subunit from the hyperthomophilic bacterium *Aquifex aeolicus* ***(10)***: There is ordered binding of five ribosomal proteins (S6, S8, S11, S15, and S18) to the central domain of 16S ribosomal RNA (rRNA) during assembly. S6 and S18 form a heterodimer with a K_d of 8.7 n*M*. The S6:S18 heterodimer binds to an S15–rRNA RNP complex. The protocol here describes monitoring the binding of a preformed S6:S18 heterodimer to an RNP complex containing ribosomal protein S15 bound to a fragment of the central domain of 16S rRNA.

2. Materials

1. 10–20 nmol of purified RNA (2 mL at 5–10 μ*M*) for each titration.
2. 100 nmol of each purified protein (1 mL at 100 μ*M*) (*see* **Note 1**).
3. Dialysis membrane of the appropriate molecular weight cutoff (3,500–30,000).
4. Buffer solutions
   a. For *A. aeolicus* ribosomal proteins and central domain rRNA: 20 m*M* potassium-HEPES, pH 7.6, 330 m*M* potassium chloride, 10 m*M* magnesium chloride.
   b. For *C. elegans* GLD-1 protein and TGE RNA: 20 m*M* Tris-HCl, pH 8.0, 25 m*M* NaCl, 1 m*M* dithiothreitol (DTT).
5. Isothermal titration calorimeter: The protocol was written for the MCS-ITC model calorimeter made by Microcal. (Northampton, MA). This model uses a circulating water bath to allow cooling of the unit below ambient temperature. The newer VP-ITC calorimeter does not require the circulating water bath to provide cooling, and experiments can be performed between 4 °C and 80 °C. These experiments can be performed on the VP-ITC using only slight modifications of the following procedures.
6. 5% solution of Contrad 70 (Decon Labs, Bryn Mawr, PA).

3. Methods

3.1. Sample and Equipment Preparation

Next is a step-by-step protocol for running an ITC experiment to study the binding of a single protein to RNA. To study the binding of a protein to an RNP complex, the only modification required is the formation of the RNP complex prior to degassing the samples (**Subheading 3.1.4.**, following **step 1**).

3.1.1. RNA Synthesis

1. Description of synthesis and purification of the RNA is beyond the scope of this chapter. The RNA can be obtained either by chemical synthesis (for RNAs shorter than 50 nt) or by transcription from a DNA template using T7 RNA polymerase ***(11)*** (for RNAs of any length).
2. For studies of ribosome assembly, the fragments of 16S rRNA were transcribed using T7 RNA polymerase from a plasmid DNA template.

3. TGE RNA for *C. elegans* GLD-1 STAR studies was produced by chemical synthesis (Dharmacon) and deprotected and lyophilized as per the manufacturer's protocol.
4. RNA produced by in vitro transcription methods should be purified on a denaturing polyacrylamide gel. The RNA product is visualized by ultraviolet (UV) shadowing and the band excised and RNA eluted by electroelution (Schleicher and Schuell). Concentrate the RNA by precipitation with ethanol and dissolve in an appropriate volume of dialysis buffer to yield a 20 µ*M* solution.

3.1.2. Protein Purification

1. The protein to be used in the titration should be purified to near homogeneity. At a minimum, it must be free of contaminating proteases and nucleases. Since the value of the fit parameters depends highly on the accurate determination of the active protein concentration, a very pure sample is better for the ITC experiment. GLD-1, S6, S18, and S15 can be purified as previously described ***(9,10)***.
2. Prepare solutions of proteins that are at least 1.5X the final desired concentration to allow for dilution that may occur during dialysis.

3.1.3. Dialysis of Proteins and RNA

1. Prepare 4 L of the appropriate buffer (A or B) and chill to 4 °C. Dialyze RNA and protein against this buffer overnight at 4 °C (*see* **Note 2**).
2. Equilibrate the calorimeter at the experimental temperature by first setting the circulating water bath 5 °C to 10 °C below the desired temperature for the experiment. Set the thermostat to the desired temperature (30 °C or 40 °C) and allow to equilibrate at least 12 h.
3. Following dialysis of the proteins and RNA, retain the buffer for washing of the injection syringe, reference, and measurement cells of the ITC and for control titrations and dilution of RNA or proteins (as necessary).
4. Determine the concentration of RNA and proteins and dilute as necessary using dialysis buffer (*see* **Note 3**).

3.1.4. Loading Samples in Calorimeter

1. Anneal the RNA by heating to 95 °C for 2 min and place on ice for 5 min (*see* **Note 4**).
2. Degas 10 mL of dialysis buffer and the RNA sample for 7–10 min with a vacuum pump.
3. Wash the reaction and reference cells with a 5% Contrad 70 solution and rinse thoroughly with water. Fill the reference cell with buffer or water.
4. Rinse the measurement cell with buffer before adding the RNA sample slowly to avoid introducing any air bubbles to the sample.
5. Fill the 250-µL injection syringe with either GLD-1 or S6:S18 heterodimer solution (*see* **Note 5**).
6. Insert the injection syringe into the reaction cell, taking care to avoid bending the needle.

3.2. Titration Experiment

1. Set the run parameters for the experiment: temperature (30 °C for GLD-1 STAR/TGE RNA or 40 °C for S6:S18/S15–RNA complex), number of injections (26), reference power (10 µcal/s), initial delay (60 s), time between injections (240 s), stirring speed (400 rpm), concentration in the cell (5 µ*M* for TGE RNA, 10 µ*M* for S15–RNA complex) and the syringe (110 µ*M* for GLD-1 STAR, 100 µ*M* for S6:S18), and the injection volume (first injection 2 µL, next 25 injections 10 µL) (*see* **Note 6**) (**Fig. 2**).
2. Start the experiment. The MCS-ITC can be set up to automatically proceed through all steps of the experiment. The instrument will proceed through a thermal equilibration step that, depending on how closely the temperature of the sample placed in the reaction cell matches the experimental temperature, generally takes between 10 and 30 min. Following thermal equilibration, the injection mechanism engages the syringe, and stirring commences. This is followed by a mechanical equilibration until a stable baseline is achieved, which generally takes an additional 10 to 15 min.

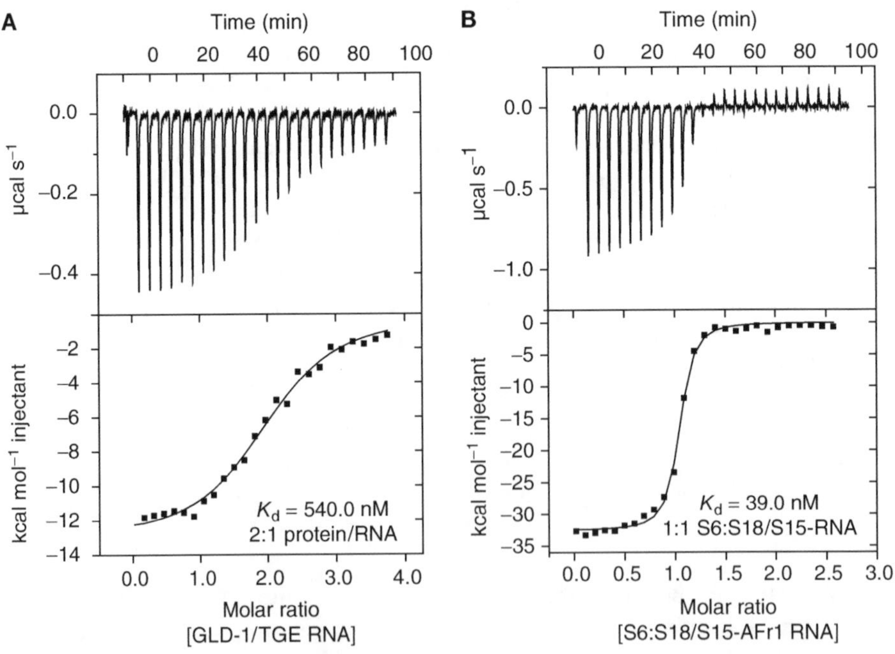

Fig. 2. (**A**) Titration of TGE RNA (4.9 µ*M*) with GLD-1 STAR (95 µ*M*) at 30 °C. The data were fit to a single binding site model, yielding a ΔH_{obs} of −11.7 kcal mol^{-1}, K of 1.81 × 10^{6}, and n of 2.3. (**B**) Titration of S15–ribosomal RNA (10 µ*M*) complex with S6:S18 heterodimer (100 µ*M*) at 40 °C. The data were fit to a single binding site model, yielding a ΔH_{obs} of −32.6 kcal mol^{-1}, K of 2.56 × 10^{7}, and n of 1.0.

3. On completion of the experiment, the titrated RNA–protein complex should be removed from the reaction cell, and the cell should be cleaned thoroughly.
4. Perform an additional experiment in which protein at the same concentration used in **step 1** is titrated into a measurement cell containing buffer. The heat evolved by dilution of the concentrated protein solution will be subtracted from the data for titration of protein into RNA (*see* **Note 7**).

3.3. Data Analysis

1. The Origin software provided by Microcal (Northampton, MA) can be used for data analysis. For both examples presented here, a model describing a single set of identical sites is adequate to describe the data ***(5,12)***. Assuming a 1:1 interaction in which a macromolecule M binds to a ligand X, the following binding equilibrium exists:

$$M+X \leftrightarrow MX \tag{1}$$

$$K=\frac{[MX]}{[M][X]} \tag{2}$$

It follows that

$$X_t=[X]+[MX], \tag{3}$$

and

$$M_t=[MX]+[M]=[MX]+\frac{[MX]}{K[X]} \tag{4}$$

At the start of the titration, the macromolecule concentration is M_t in a volume V_0. As aliquots of ligand are added during the titration, the total volume will be $V_0 + \Delta V$. The concentration of macromolecule is reduced during the titration, so that

$$M_t=M_t^o\left(\frac{1-\frac{V}{2V_0}}{1+\frac{V}{2V_0}}\right) \tag{5}$$

and similarly,

$$X_t = X_t^o \left(1 - \frac{V}{2V_0}\right) \quad (6)$$

The total heat content Q of the solution contained in V_0 is

$$Q = n\Theta M_t \Delta H V_0 \quad (7)$$

in which Θ is the fraction of sites on M occupied by X, n is the number of sites, and V_0 is the active cell volume.

$$Q = \frac{nM_t \Delta H V_0}{2}\left[1 + \frac{X_t}{nM_t} + \frac{1}{nK_a M_t} - \sqrt{\left(1 + \frac{X_t}{nM_t} + \frac{1}{nK_a M_t}\right)^2 - \frac{4X_t}{nM_t}}\right] \quad (8)$$

in which K_a is the association constant, and Mt and Xt are the bulk concentration of macromolecule (in V_0) and ligand, respectively.

The data are plotted as the differential heat evolved for each injection of an aliquot of ligand X to the sample in the measurement cell versus the molar ratio of X_t/M_t. The differential heat is described by

$$\frac{dQ}{dX_t} = \Delta H^0 V_0 \left(\frac{1}{2} + \frac{1 - \left(\frac{X_t}{nM_t}\right) - \left(\frac{1}{nK_a M_t}\right)}{2\sqrt{\left(\frac{X_t}{nM_t} + \left(\frac{1}{nK_a M_t}\right) + 1\right)^2 - 4\left(\frac{X_t}{nM_t}\right)}}\right) \quad (9)$$

The heat released from the *i*th injection is

$$\Delta Q(i) = Q(i) + \frac{dV_i}{V^0}\left[\frac{Q(i) + Q(i-1)}{2}\right] - Q(i-1) \quad (10)$$

In fitting data with the single set of identical sites model, initial guesses are made for the values of K_a, ΔH and n. ΔQ(i) is calculated for each injection and compared to the measured values. The values of K_a, ΔH, and n are varied until the best fit of the data is obtained using standard Marquardt methods ***(13)*** (*see* **Note 8**). The free-energy change for the interaction is calculated using the relationship

$$\Delta G = -RT \ln K_a \quad (11)$$

The entropy change (ΔS) is obtained using the standard thermodynamic expression

$$\Delta G = \Delta H - T\Delta S \quad (12)$$

2. The value of n obtained using this model should be close to the stiochiometry of the interactions. For interactions with a 1:1 stiochiometry, n should typically lie between 0.9 and 1.1 (or 1.8–2.2 for 2:1 stiochiometry). Large deviations from these values indicate one of two possibilities:
   a. For $n < 1$, the concentration of RNA in the cell has been overestimated, the concentration of protein has been underestimated, or not all of the RNA is in an active state.

For $n > 1$, the concentration of protein in the syringe has been underestimated, the concentration of RNA has been underestimated, or not all the protein is active.

4. Notes

1. Only 500 μL of this sample will be used for the experiment and control titrations, but additional protein is needed so that the injection syringe can be filled without introducing bubbles. The needle contains two holes, one at the tip and another on the side. The hole on the side must be covered by protein sample during filling, necessitating the additional volume.
2. Depending on the buffer in which the proteins are stored, it may be necessary to prepare an additional 4 L of dialysis buffer and change the buffer approximately 8 h into the dialysis. This is particularly important if the protein is stored in a salt concentration that is much higher than in the measurement buffer or if a large volume of protein (>10 mL) is being dialyzed.
3. The concentration of the proteins used in this example was determined using the calculated extinction coefficient at 280 nm ***(14)***. Other methods may be used to determine concentrations of proteins, but inaccurate concentration determination will affect all parameters in the subsequent determination of ΔH, ΔG, and n.
4. The RNA samples should be annealed prior to degassing as the heating process will liberate more dissolved air. If an *Aquifex aeolicus* S15–RNA complex is to be titrated, form the complex before degassing by incubating the annealed RNA with an equimolar amount of protein for 5 min at 65 °C. The *A. aeolicus* protein is stable under these conditions.
5. It is possible, although not recommended, to perform the titration with protein in the measurement cell and RNA in the syringe. The initial injections will create a situation in which protein is in great excess over RNA, which can produce complexes that are prone to aggregation/precipitation.
6. The injection parameters will vary based on the specific interaction measured. The values given have yielded good results with both systems described here. The first injection is kept small because there is the possibility of some mixing of protein solution on the outside or in the tip of the needle during the equilibration process, resulting in an erroneous data point. The heat evolved during the first injection will not be included in the data fitting, but the amount of protein in this injection will be used to calculate the total ligand concentration in the reaction cell.
7. A control titration such as that described is not always necessary. If there is only a single binding site for the protein on the RNA, and it is fully saturated by the end of the titration, the heat observed following saturation can be used as the dilution reference.

8. For particularly high-affinity interactions ($K_a \geq 1 \times 10^9$), the value of K_a may need to be held constant as it will not be well defined unless the value of c can be kept between 10 and 500. If a competitive inhibitor binds to the same site as a high-affinity binding ligand, one can determine the K_a of the high-affinity ligand using displacement isothermal titration calorimetry ***(15)***.

Acknowledgments

M.I.R. was supported by Research Scholar Grant PF-01-087-01-GMC from the American Cancer Society. S.P.R. was supported by the Damon Runyon Cancer Research Foundation Fellowship (DRG-1723). This research was supported by grants from the Skaggs Institute for Chemical Biology and the National Institutes of Health (NIH, GM53320 and GM53757).

References

1. Ladbury, J. E., and Chowdhry, B. Z. (1996) Sensing the heat: the application of isothermal titration calorimetry to thermodynamic studies of biomolecular interactions. *Chem. Biol.* **3**, 791–801.
2. Leavitt, S., and Freire, E. (2001) Direct measurement of protein binding energetics by isothermal titration calorimetry. *Curr. Opin. Struct. Biol.* **11**, 560–566.
3. Weber, P. C., and Salemme, F. R. (2003) Applications of calorimetric methods to drug discovery and the study of protein interactions. *Curr. Opin. Struct. Biol.* **13**, 115–121.
4. Tame, J. R. H., O'Brien, R., and Ladbury, J. E. (1998) Isothermal titration calorimetry of biomolecules. In *Biocalorimetry: Applications of Calorimetry in the Biological Sciences* (Ladbury, J. E., and Chowdhry, B. Z., eds.), Wiley, Chichester, U.K., pp. 27–38.
5. Wiseman, T., Williston, S., Brandts, J. F., and Lin, L. N. (1989) Rapid measurement of binding constants and heats of binding using a new titration calorimeter. *Anal. Biochem.* **179**, 131–137.
6. LeTilly, V., and Royer, C. A. (1993) Fluorescence anisotropy assays implicate protein–protein interactions in regulating trp repressor DNA binding. *Biochemistry* **32**, 7753–7758.
7. Wilson, G. M., Sutphen, K., Chuang, K., and Brewer, G. (2001) Folding of A+U-rich RNA elements modulates AUF1 binding. Potential roles in regulation of mRNA turnover. *J. Biol. Chem.* **276**, 8695–8704.
8. Katsamba, P. S., Park, S., and Laird-Offringa, I. A. (2002) Kinetic studies of RNA–protein interactions using surface plasmon resonance. *Methods* **26**, 95–104.
9. Ryder, S. P., Frater, L. A., Abramovitz, D. L., Goodwin, E. B., and Williamson, J. R. (2004) RNA target specificity of the STAR/GSG domain post-transcriptional regulatory protein GLD-1. *Nat. Struct. Mol. Biol.* **11**, 20–28.
10. Recht, M. I., and Williamson, J. R. (2001) Central domain assembly: thermodynamics and kinetics of S6 and S18 binding to an S15–RNA complex. *J. Mol. Biol.* **313**, 35–48.

11. Milligan, J. F., and Uhlenbeck, O. C. (1989) Synthesis of small RNAs using T7 RNA polymerase. *Methods Enzymol.* **180**, 51–62.
12. Indyk, L., and Fisher, H. F. (1998) Theoretical aspects of isothermal titration calorimetry. *Methods Enzymol.* **295**, 350–364.
13. Marquardt, D. (1963) An algorithm for least-squares estimation of nonlinear parameters. *J. Soc. Indust. Appl. Math.* **11**, 431–441.
14. Gill, S. C., and von Hippel, P. H. (1989) Calculation of protein extinction coefficients from amino acid sequence data. *Anal. Biochem.* **182**, 319–326.
15. Sigurskjold, B. W. (2000) Exact analysis of competition ligand binding by displacement isothermal titration calorimetry. *Anal. Biochem.* **277**, 260–266.

9

Characterization of RNA–Protein Interactions by Phosphorothioate Footprinting and Its Applications to the Ribosome

A. Özlem Tastan Bishop, Ulrich Stelzl, Markus Pech, and Knud H. Nierhaus

Summary

Analogs of naturally occurring substances obtained by chemical modifications are powerful tools to study intra- and intermolecular interactions. We have used the phosphorothioate technique to analyze RNA–protein interactions, here the interactions of transfer RNAs (tRNAs) with the three ribosomal binding sites. We describe preparation and purification of thioated tRNAs, formation of functional complexes of programmed ribosomes with tRNAs, and the evaluation of the observed phosphorothioate footprints on the tRNAs.

Key Words: Phosphorothioate technique; ribosomal tRNA contact sites; RNA–protein interactions.

1. Introduction

Chemical modifications of biological molecules at specific functional groups play an important role to test and establish structure–function relationships. For example, chemical modifications in the phosphate group of the ribonucleoside phosphates can be applied to obtain information on questions concerning nucleic acid structure and function. Phosphates in the nucleic acid backbone often have an essential role in nucleic acid–protein interactions ***(1–3)*** since phosphate oxygens are hydrogen bond acceptors and also participate in electrostatic interactions with positively charged amino acid side chains ***(4)***. Eckstein's group introduced the phosphorothioate backbone modification of nucleic acids and demonstrated its usefulness for RNA sequencing ***(5)*** and for analysis of RNA–protein interactions ***(6)***. The phosphorothioate technique has advanced and developed into an important tool for analyzing RNA structure

From: *Methods in Molecular Biology, vol. 488: RNA-Protein Interaction Protocols*
Edited by: Ren-Jang Lin © Humana Press Inc, Totowa, NJ

and its interaction with ligands *(7–9)*. It was successfully used for analysis of transfer RNAs (tRNAs) complexed with their respective synthetases *(6,7)*. Thioate contact patterns of tRNAs on the ribosome provided insight to ribosomal function and led to the development of the alpha-epsilon model for the ribosomal elongation cycle *(10–12)*. Phosphorothioate footprinting of functional ribosomal complexes showed that the messenger RNA (mRNA) is fixed on the ribosome only in the region of codon–anticodon interaction *(13)*. The phosphorothioate technique was also applied for the characterization of binding sites for ribosomal proteins on the large ribosomal RNAs, providing clues to the assembly and structure of the ribosome *(14–19)*. Although the technique is equally applicable to both DNA and RNA, in this chapter we mainly talk about RNA and RNA-based applications.

1.1. What Is a Phosphorothioated RNA?

The exchange of one of the two nonbridging phosphate oxygens with a sulfur atom in the ribonucleoside phosphates changes them into ribonucleoside phosphorothioates. Phosphorothioated RNA can be synthesized either by chemical synthesis, with the use of a sulfurizing reagent in the oxidation step, or by enzymatic incorporation of modified nucleotide triphosphates (NTPs) *(7)* during in vitro RNA transcription. In this chapter, we describe in vitro transcription for enzymatic incorporation of the phosphorothioated ribonucleotides. Briefly, for the in vitro synthesis of the thioated RNA, one of the four ribonucleoside triphosphates is added as a mixture of this triphosphate and the corresponding ribonucleoside 5′-*O*-(1-thiotriphosphate) [NTP(αS)] to the transcription reaction. As a result, the phosphorothioate group is randomly incorporated by the polymerase at the positions of the corresponding ribonucleoside triphosphates. The most commonly used enzyme for this purpose is the T7 RNA polymerase, incorporating NTP(αS) with an efficiency of about 25% of that of nonthioated NTPs.

1.2. Features of the Thioated RNA

Substitution of nonbridging oxygen by sulfur is conservative since the size of the sulfur atom is only slightly larger than the oxygen atom, and the negative charge of the phosphate group is retained.

However, the exchange of nonbridging oxygen in the phosphorothioate modification changes the metal ion coordination. Sulfur has affinity for soft ions such as Mn^{2+}, Co^{2+}, Zn^{2+}, or Cd^{2+}, whereas oxygen prefers hard ones, such as Mg^{2+} *(7,8)*. Assume an oxygen in a functionally important position concerning Mg^{2+} coordination. If such an oxygen is replaced by a sulfur atom in the phosphorothioated RNA, the replacement will weaken or block the respective function (*see* **Subheadings 1.4.** and **1.5.**, interference concept). Sometimes, the activity can be restored using soft ions.

In addition, the oxygen-to-sulfur exchange makes the phosphorous a chiral center. There are two diastereomers of the phosphorothioate internucleotide linkage, S_p and R_p configurations (**Fig. 1** shows as an example the diastereomers of adenosine α-thiotriphosphates). During the enzymatic incorporation of the NTP(αS), T7 RNA polymerase can only accept S_pNTP(αS) diastereomers and inverses them during incorporation to the R_p isomers, with the result that the transcribed thioated ribonucleotides only contain R_p isomers. S_p-phosphorothioated RNAs are usual obtained by chemical synthesis.

1.3. Mechanism of the Cleavage

Phosphorothioate incorporation was initially used to characterize the stereochemistry of nucleic acid phosphoryl transfer reactions. Later, it was combined with the cleavage of the phosphorothioate linkage by molecular iodine for RNA and DNA sequencing *(5)*. Small and chemically inert iodine (I_2) can trigger a cleavage of the phosphate-sugar backbone at the modified phosphate residue. The reaction starts with the nucleophilic attack of the sulfur atom toward iodine or its derivatives (e.g., 2-iodoethanol; **ref.** ***20***; **Fig. 2**). The nucleophilic interaction of the 2′-hydroxyl (2′-OH) group might lead to a direct cleavage (pathway a). The other possibility is the formation of a phosphor-triester intermediate (pathway b). This intermediate has two different routes to follow. It is hydrolyzed either to induce cleavage (route 1) or to restore the phosphordiester bond (route 2 or 3). The cleavage reaction is highly specific, albeit

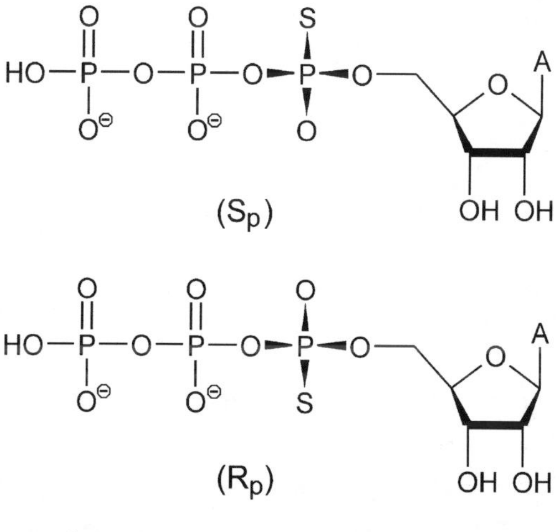

HO—P(=O)(O⁻)—O—P(=O)(O⁻)—O—P(S)(O)—O—[ribose, A; OH OH]

(S_p)

HO—P(=O)(O⁻)—O—P(=O)(O⁻)—O—P(O)(S)—O—[ribose, A; OH OH]

(R_p)

Fig. 1. Absolute configurations of the diastereomers of adenosine α-thiotriphosphates. S_p isomer: adenosine 5′-*O*-(1-thiotriphosphate); R_p isomer: adenosine 5′-*O*-(2-thiotriphosphate).

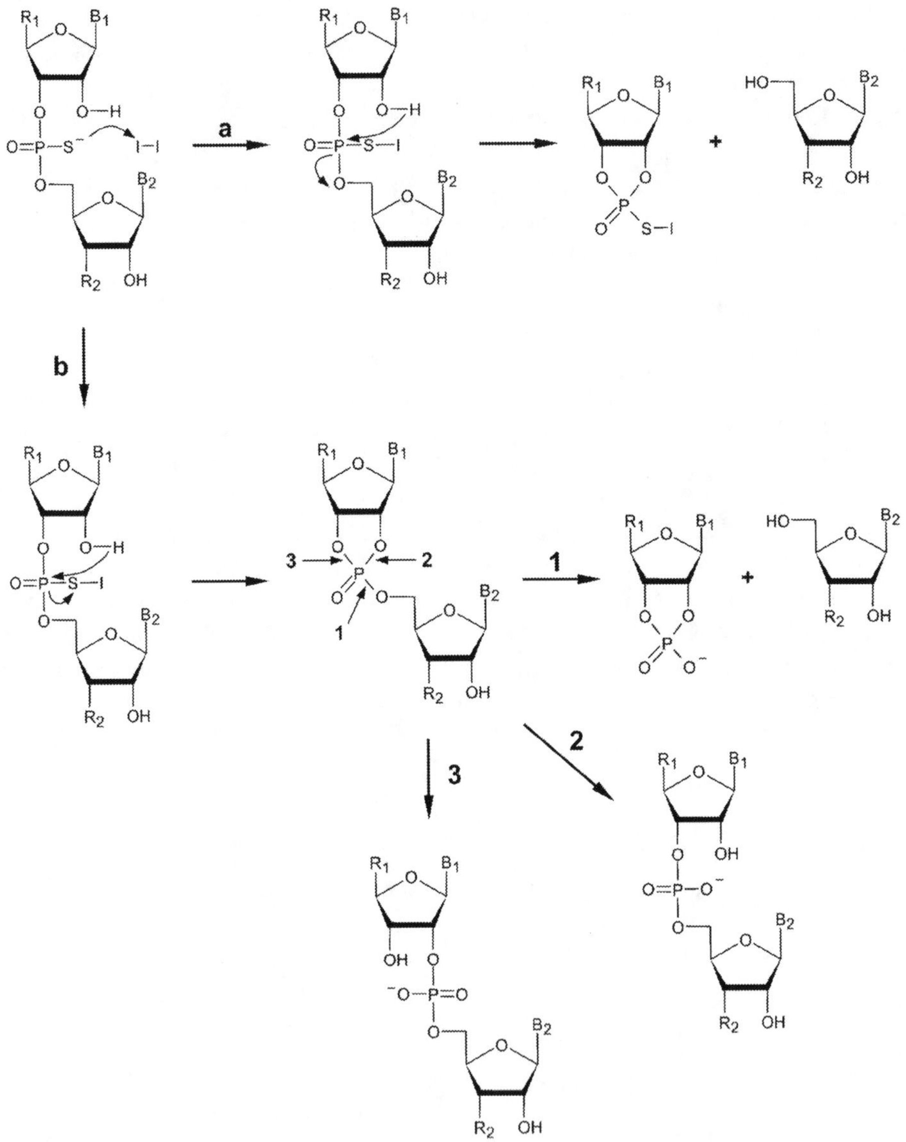

Fig. 2. Reaction scheme of the iodine cleavage. (Modified after **ref. 5**) For details, see **Subheading 1.3**.

with a low efficiency (approximately 5%) ***(21)***. It is largely independent of the secondary structure of the RNA; that is, thiophosphorothioates are attacked in both single- and double-stranded RNA but are sensitive to tertiary structure.

1.4. Experimental Outline

As mentioned in the introduction, our laboratory has used the phosphorothioate technique for the analysis of the interactions of tRNAs with the three ribosomal-binding sites ***(10–13)*** and ribosomal RNA (rRNA) interactions with ribosomal proteins ***(15–18)***. It was shown that in vitro-transcribed phosphorothioated tRNAs containing up to one phosphorothioate per molecule exhibit an activity of at least 80% of that of nonthioated tRNAs ***(10)***. Phosphorothioated tRNAs bound by the ribosome produce specific cleavage patterns that are characteristic for the particular ribosomal binding site as well as dependent on the functional state of the ribosome ***(10–12)***. The rationale behind the analyses of RNA–protein or RNA–ribosome interactions is the following (summarized in **Fig. 3**):

After transcribing and purifying the thioated RNAs (tRNA or a fragment of rRNA), the RNAs are 5′ labeled with ^{32}P, and the complex (ribosome–tRNA or ribosomal protein–rRNA) is formed. Complexes are first cleaned by removal of unbound RNAs from the reaction mixture. A variety of techniques such as gel filtration (e.g., in case of a ribosome–tRNA complex) ***(10,12)***, sucrose gradient centrifugation ***(14,19)***, nitrocellulose filtration ***(16,18)***, or native gel electrophoresis (e.g., in case of ribosomal protein–rRNA) ***(17)*** can be applied.

Footprinting experiments are done in two different ways. In the course of an *interference experiment*, the RNA is extracted from the complex and subsequently cleaved with iodine. A band that is weakened (or even sometimes missing) indicates that a thio modification at the corresponding phosphate of the RNA impairs (prevents) formation of the RNA–protein complex. Sometimes, a distinct band is even stronger, indicating that the modification at the corresponding position facilitates complex formation (*see also* **Subheading 1.5.**). In a *protection experiment* the RNA is cleaved in the complex under native conditions. The RNA cleavage products are extracted from the complex after the cleavage reaction. A *weakened or absent* band in this experiment indicates that iodine does not have free access to the corresponding phosphate of the RNA; that is, the position is protected from iodine cleavage. An example for protection experiments, and evaluation of the protection pattern from the gel, are given in **Fig. 4**, and **Table 1**. For color codes, see **Fig. 5**.

In both types of experiments, RNA cleavage products are separated on a sequencing gel and compared to the cleavage efficiency of the free RNA in the absence of the binding substrate. In a control experiment, the iodine, which triggers cleavage, is omitted to control for cleavage specificity. Provided that the ribosome or protein preparation is essentially ribonuclease (RNase) free, this experiment does not produce cleavage bands.

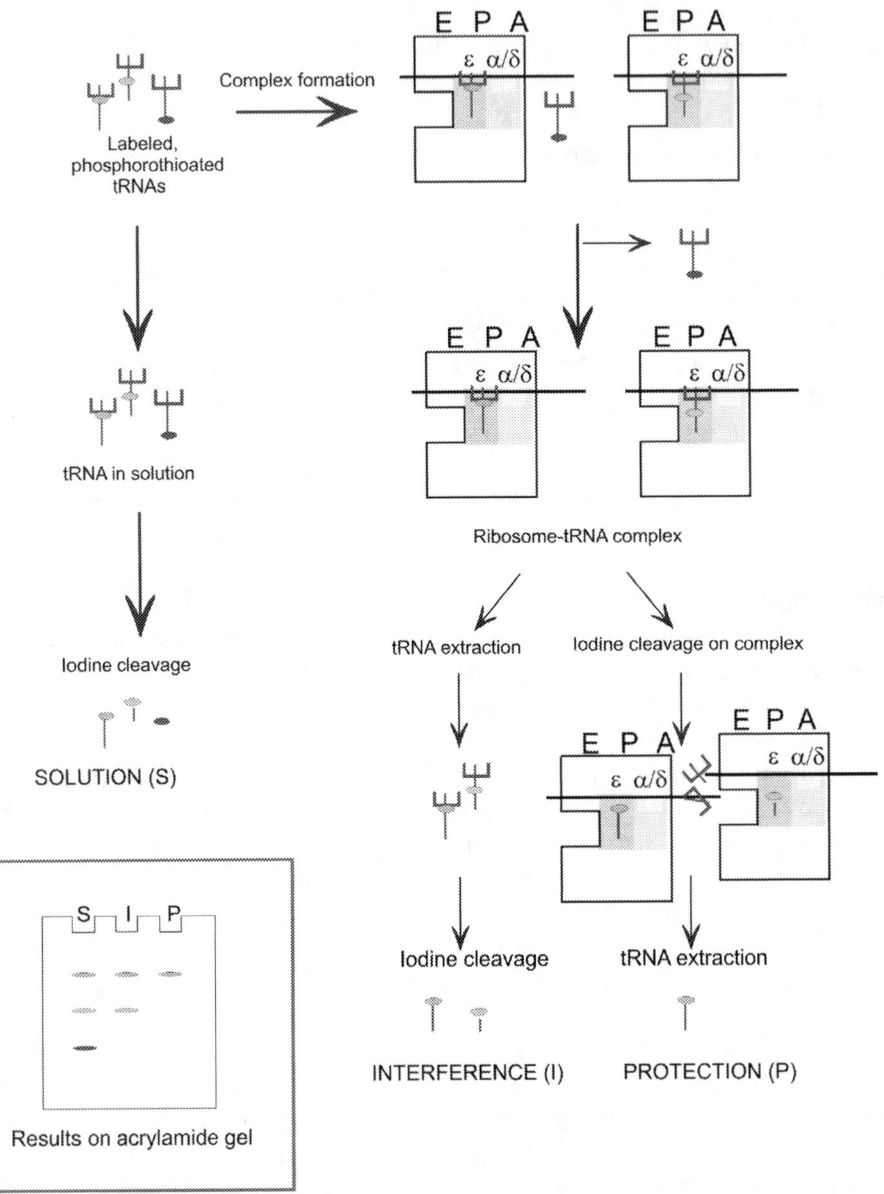

Fig. 3. Outline of a phosphorothioate cleavage experiment on ribosomes. The cleavage products are analyzed on a sequencing gel. S, cleavage products of thioated tRNAs in solution; three bands are shown (orange, blue, and violet). I, cleavage products from an interference experiment; the violet band is missing since the corresponding phosphorothioate prevented complex formation. Even if this band is weakened, it should not be considered in the protections experiment. P, cleavage products from a protection experiment. The blue band is missing, indicating that the access of iodine to the corresponding phosphorothioate is sterically hindered. For data quantification, intensities of the corresponding bands are compared: Interference = I/S; Protection = P/I. (*See Color Plate*)

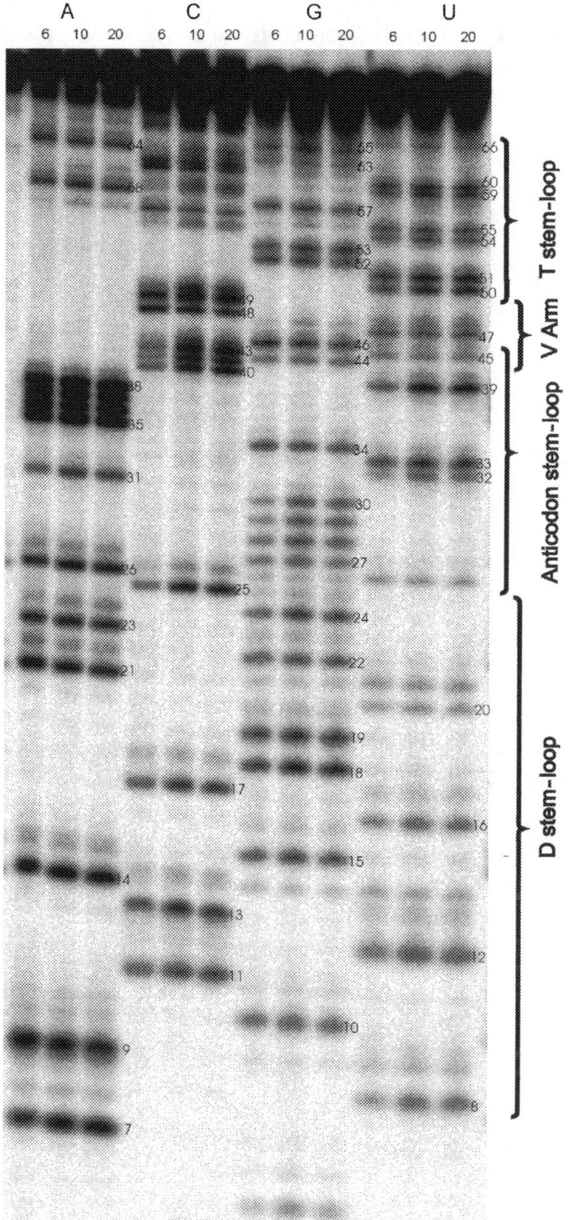

Fig. 4. Sequencing gel of protection experiments with thioated tRNAPhe in solution. tRNAPhe is either in a buffer containing 6 m*M* Mg^{2+} and polyamines (spermine and spermidine) or in conventional buffers with 10 and 20 m*M* Mg^{2+} without polyamines (lanes marked 6, 10, and 20, respectively). The number next to the bands indicates the nucleotide position; structural elements of the tRNA are indicated on the right. For example, tRNAs in the 6 m*M* Mg^{2+}/polyamine buffer are protected at positions 40 to 43.

Table 1
Evaluation of the Protection Patterns of Thioated tRNAPhe Under Various Buffer Conditions

	A6/A10	A6/A20	A10/A20
1G	n.d.	n.d.	n.d.
2C	n.d.	n.d.	n.d.
3C	n.d.	n.d.	n.d.
4C	n.d.	n.d.	n.d.
5G	n.d.	n.d.	n.d.
6G	n.d.	n.d.	n.d.
7A	**0.43**	**0.39**	0.90
8U	**0.31**	**0.34**	1.10
9A	*0.76*	*0.75*	0.98
10G	*0.70*	0.82	1.17
11C	**0.44**	0.63	1.45
12U	**0.41**	**0.37**	0.92
13C	**0.45**	*0.54*	1.22
14A	*0.54*	*0.54*	1.00
15G	*0.61*	*0.72*	1.17
16U	**0.38**	**0.36**	0.94
17C	*0.58*	*0.63*	1.09
18G	*0.63*	*0.70*	1.12
19G	*0.65*	*0.75*	1.14
20U	*0.61*	*0.57*	0.94
21A	0.87	0.91	1.04
22G	*0.64*	*0.72*	1.13
23A	*0.63*	*0.57*	0.91
24G	**0.36**	**0.49**	1.35
25C	**0.29**	**0.35**	1.22
26A	**0.46**	**0.43**	0.94
27G	*0.55*	*0.67*	1.23
28G	**0.36**	**0.42**	1.16
29G	**0.33**	**0.35**	1.06
30G	**0.27**	**0.27**	1.02
31A	**0.31**	**0.34**	1.08
32U	**0.34**	**0.39**	1.14
33U	**0.45**	*0.50*	1.12
34G	*0.70*	0.94	1.34
35A	*0.72*	0.82	1.14
36A	0.85	0.85	1.00
37A	*0.76*	*0.79*	1.03
38A	*0.67*	*0.68*	1.02
39U	**0.29**	**0.31**	1.05
40C	**0.46**	*0.59*	1.27
41C	**0.38**	*0.55*	1.43

Table 1
Continued

	A6/A10	A6/A20	A10/A20
42-43C	**0.39**	*0.57*	1.49
44G	*0.78*	0.92	1.18
45U	1.07	0.96	0.90
46G	1.01	1.18	1.17
47U	0.88	0.80	0.90
48C	1.16	1.28	1.10
49C	*0.70*	0.91	1.30
50U	**0.47**	**0.45**	0.96
51U	*0.51*	*0.51*	1.00
52G	*0.77*	1.13	1.47
53G	*0.60*	*0.78*	1.30
54U	*0.63*	*0.71*	1.13
55U	0.85	0.84	0.99
56C	1.12	1.31	1.17
57G	*0.51*	*0.67*	1.31
58A	*0.64*	*0.66*	1.04
59U	*0.58*	*0.56*	0.96
60U	*0.60*	*0.60*	1.02
61C	1.04	1.13	1.09
62C	0.88	0.95	1.09
63G	*0.61*	*0.69*	1.13
64A	*0.77*	*0.71*	0.93
65G	**0.46**	*0.58*	1.25
66U	**0.45**	**0.48**	1.06
67C	n.d.	n.d.	n.d.
68C	n.d.	n.d.	n.d.
69G	n.d.	n.d.	n.d.
70G	n.d.	n.d.	n.d.
71G	n.d.	n.d.	n.d.
72C	n.d.	n.d.	n.d.
73A	n.d.	n.d.	n.d.
74C	n.d.	n.d.	n.d.
75C	n.d.	n.d.	n.d.
76A	n.d.	n.d.	n.d.

A6, accessibility under polyamine conditions (6 m*M* Mg^{2+}); A10 and A20, accessibility under conventional buffer conditions without polyamines at 10 and 20 m*M* Mg^{2+}, respectively; n.d., not determined.

Bold numbers: The intensity ratio of the corresponding bands is below 0.49 (e.g., A7, 0.43 for A6/A10 means that the accessibility of phosphate at position A7 is protected more under polyamine buffer conditions than under conventional buffer conditions). *Italicized* numbers: The ratio is between 0.5 and 0.79. Underlined numbers: The ratio is above 1.15. Ratios between 0.8 and 1.14 are not considered to differ significantly. The numbers are averages of up to four experiments; the standard deviation was below 10%.

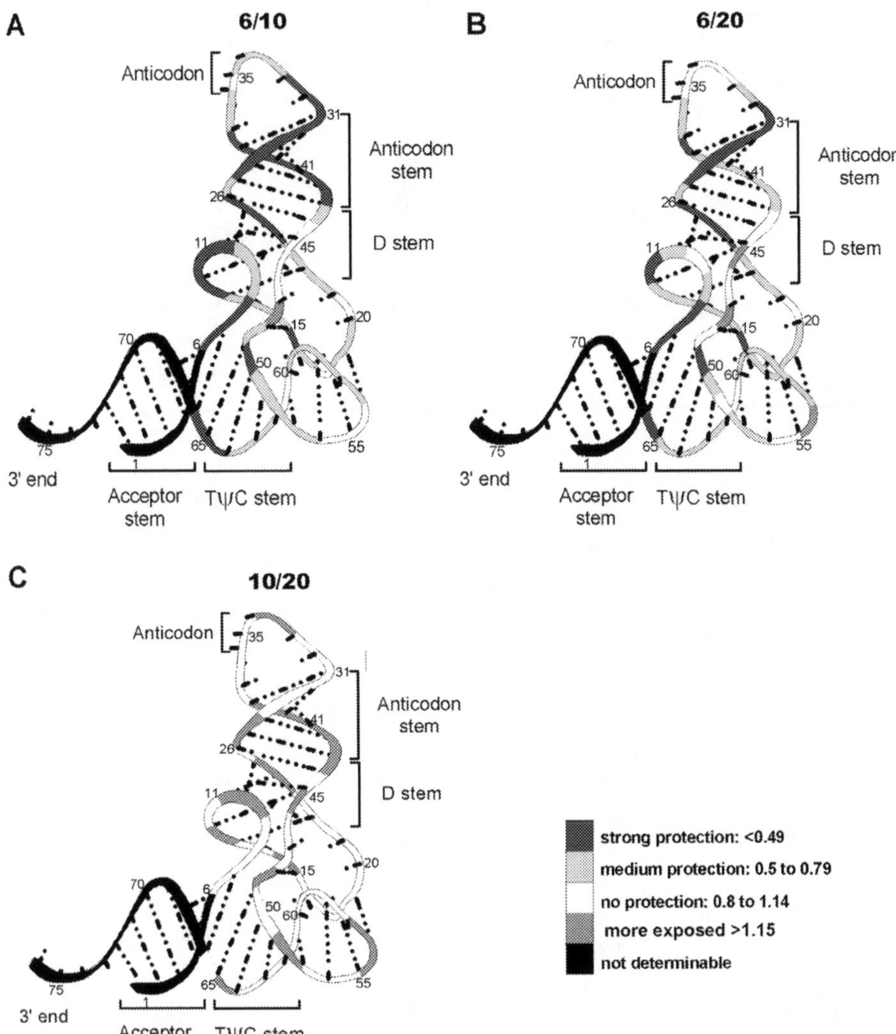

Fig. 5. Color code representation of the protection patterns shown in **Table 1. 6**, accessibility under polyamine conditions (6 m*M* Mg^{2+}); 10 and 20, accessibility under conventional buffer conditions without polyamines at 10 and 20 m*M* Mg^{2+}, respectively. (*See Color Plate*)

1.5. Interpretation of the Results

The phosphorothioate method provides a tool for the analysis of the RNA structure with atomic resolution since every single substitution can be mapped by separation of the iodine cleavage products on a sequencing gel. Interpretation of the results, however, is not trivial. Each band in the sequencing gel is compared with the corresponding band of iodine-treated RNA in the absence of ligands. A band in both the interference and the protection experiment can either be weakened or missing or can be stronger than in the control experiment.

In the interference experiments, a weakened or missing band is an indicator for the absence of a certain modification in the RNA–protein complex. In other words, the presence of a thioate at a particular position interferes with binding. In the protection experiments, a weakened or missing band means that iodine does not have free access to the backbone of the RNA in that region, so the position is protected from iodine cleavage by the ligand or the RNA itself, which has changed its conformation on ligand binding.

A band that appears stronger in the interference experiments indicates that a phosphorothioate at this position is present in a larger fraction within the complex than in the initial αS-RNA pool; that is, this phosphorothioate increases the affinity between the rRNA fragment and the protein, thus leading to an enrichment of this fragment within the complex. A *stronger* band in the protection experiment also occurs, meaning that the phosphorothioate at this position is more exposed (more reactive) to the iodine within the complex than in the naked fragment.

Strictly speaking, it cannot be differentiated whether signals are due to direct contacts of the ligand and the RNA or whether signals reflect changes in the RNA structure on ligand binding (*see*, e.g., **ref.** ***18***).

2. Materials

While working with RNA, caution should be applied to prevent RNase contamination. All glassware should be incubated at 180 °C for 3 h. Buffer aliquots should be kept at −20 °C and used only a limited number of times.

2.1. In Vitro Transcription of Thioated and Normal tRNAs

The reaction volume per transcription reaction is not larger than 400 µL in double-distilled water (ddH$_2$O). Standard compositions for in vitro transcription assay and the order of the component additions are as follows:

1. 10X transcription buffer: 400 m*M* Tris-HCl (pH 8.0 at 25 °C), 220 m*M* $MgCl_2$, 50 m*M* 1–4-dithioerythritol (DTE), 10 m*M* spermidine (*see* **Note 1**).
2. NTP mix (adenosine triphosphate [ATP], cytosine triphosphate [CTP], guanosine triphosphate [GTP], uridine triphosphate [UTP]): 3.75 m*M* each (*see* **Note 2**).
3. [αS]-NTP (add if you are transcribing thioated RNA) (*see* **Note 3** and **Subheading 3.1.**)
4. 100 µg/mL bovine serum albumin (BSA) (RNase and DNase [deoxyribonuclease] free).
5. 1 U/µL RNasin.
6. 5 U/mL inorganic pyrophosphotase.

7. 20 pmol/mL linearized plasmid DNA or DNA oligonucleotides (*see* **Note 4**).
8. 40 µg/mL T7 RNA polymerase (*see* **Note 5**).

2.2. Purification of Transcripts

2.2.1. Purification via Polyacrylamide Gel Electrophoresis

1. 13% polyacrylamide sequencing gel (e.g., from 40% stock: 19:1 acrylamide/bis-acrylamide).
2. RNA gel loading buffer: 10 m*M* Tris-HCl, pH 7.5, 1 m*M* ethylenediaminetetraacetic acid (EDTA), 7 *M* urea, 0.05% (w/v) xylenecyanol, 0.05% (w/v) bromophenol blue.
3. RNA extraction buffer (from acrylamide gel): 10 m*M* Tris-HCl, pH 7.8, 1 m*M* EDTA, 1% sodium dodecyl sulfate (SDS), 100 m*M* NaCl, 1 m*M* DTE.
4. 70% phenol (diluted with double-distilled water).

2.2.2. Purification via Anion Exchange

For purification via anion exchange, use a Qiagen-tip-100 and the manufacturer's brochure.

2.3. 5′-Labeling and Purification of tRNAs

2.3.1. Dephosphorylation Reaction Mix

The dephosphorylation reaction mix (total volume 50 µL) is as follows:

1. 50 m*M* Tris-HCl, pH 8.3 (0 °C).
2. 1 m*M* $MgCl_2$.
3. 1 m*M* $ZnCl_2$.
4. 0.5 m*M* EDTA.
5. 500–1500 pmol tRNA.
6. 2.5–5 U alkaline phosphatase (from calf intestine).

2.3.2. Phosphorylation Reaction Mix

The phosphorylation reaction mix (total volume 40 µL) is as follows:

1. 50 m*M* Tris-HCl, pH 7.5 (0 °C).
2. 1 m*M* EDTA.
3. 10 m*M* $MgCl_2$.
4. 6 m*M* β-mercaptoethanol.
5. 2.5 pmol/µL dephosphorylated tRNA.
6. 0.4 pmol/µL [γ-^{32}P]ATP (~3.5 µCi/pmol).
7. 0.5 U/µL T4 polynucleotide kinase.

2.4. Aminoacylation of the 5′-[^{32}P]-Labeled tRNA

For the aminoacylation reaction mixture (total volume 2 mL), use the following:

1. 50 m*M* HEPES KOH, pH 7.5 (0 °C).
2. 100 m*M* $MgAc_2$.

3. 10 m*M* KCl.
4. 3 m*M* ATP.
5. 5 m*M* β-mercaptoethanol.
6. 10–50 A_{260} units of tRNA (1 A_{260} unit ≈ 1500 pmol).
7. [^{14}C] or [^{3}H] amino acid ≥ 1000 dpm/pmol, concentration should be 3 *M* to 5 *M* more than that of tRNA.
8. Optimized amount of S-100 supernatant (containing aminoacyl synthetases), which is usually about 40 μg of it per 10 A_{260} units of tRNA.

2.5. Acetylation of tRNAs

1. 5 nmol aminoacyl-tRNA dissolved in 500 μL 0.3 *M* NaAc, pH 5.5.
2. Acetic acid anhydride.

2.6. Separation of Charged and Noncharged tRNA Species via Reversed-Phase High-Performance Liquid Chromatography

1. Reversed-phase high-performance liquid chromatographic (HPLC) C4 column (Nucleosil 300-5).
2. Buffer A: 400 m*M* NaCl, 10 m*M* $MgAc_2$, 20 m*M* NH_4Ac, pH 5.0.
3. Buffer B: 60% uvasol-methanol in buffer A.

2.7. Binding of tRNAs to the Different Ribosomal Binding Sites

1. Ribosomal binding buffer: 20 m*M* HEPES KOH, pH 7.6 (0 °C), 6 m*M* $MgAc_2$, 150 m*M* NH_4Ac, 2 m*M* spermidine, 0.05 m*M* spermine (*see* **Note 6**).
2. Concentration of the components (e.g., 70 S ribosome, mRNA, tRNAs, etc.) is given in **Subheading 3.7.**
3. EF-G in 20 m*M* HEPES KOH, pH 7.5, 10 m*M* $MgCl_2$, 100 m*M* KCl, 20% glycerol.
4. Puromycin solution (6 mg/mL) in binding buffer, prepared freshly and adjusted to pH 7.5 by adding approximately 1/100 volume 1 *M* KOH.
5. 65 μL 0.3 *M* NaAc (pH 5.5) saturated with $MgSO_4$.
6. Nitrocellulose filters.

2.8. Footprinting Experiments

1. Iodine: 50 m*M* solution in ice-cold EtOH.
2. Sephacryl S-300 spun column.
3. Polyacrylamide sequencing gel (from 40% stock: 19:1 acrylamide/bis-acrylamide).
4. Native gel (6% polyacrylamide gel, 90 × 60 × 0.7 mm, buffered with the binding buffer).

2.9. Image Quant Analysis of Gels

1. PhosphorImager™ (Molecular Dynamics™).
2. Package program: Image Quant version 3.3™ (Molecular Dynamics).

3. Methods

Although we explain the methodology mainly with tRNA-based experiments, it is equally applicable to any RNA molecule.

3.1. Preparative In Vitro Transcription of Thioated and Normal tRNAs

The in vitro transcription system with T7 RNA polymerase follows **ref.** *22*. The main steps in transcribing the phosphorothioated and nonmodified tRNAs are the same. In **Subheading 2.1.**, the components and the order of component additions are given. In the preparative assay, the maximal reaction volume per 1.5-mL Eppendorf tube is 400 μL. Larger volumes decrease the efficiency of the reaction, especially when thioated RNA is transcribed. Pipeting order is very important since the enzyme and the RNase inhibitor can be inactivated if the ionic environment changes drastically. After the addition of T7 RNA polymerase, the reaction is incubated at 37 °C for 8–10 h and stopped by adding EDTA to a final concentration of 25 m*M*. After stopping the reaction, products can be analyzed in an acrylamide gel in case of optimization (analytic transcription) experiments or directly purified by one of the suitable methods explained in **Subheading 3.2.** in the case of a preparative assay.

The thioation of the tRNA at A, C, G, or U positions, respectively, is obtained by replacing 20% of the nucleotide of interest in the reaction with the corresponding [αS]-NTP of the S_p isomers. Transcription resulted in 1.1–1.3 thioated nucleotide per RNA ***(10,12,23)***.

3.2. Purification of Transcripts

Purification of the transcribed RNAs can be done either via denaturing acrylamide gel electrophoresis or by anion exchange via spun columns, depending on the product. However, the best purification results from gel purification as transcription side products are removed efficiently.

3.2.1. Purification via Polyacrylamide Gel Electrophoresis

1. Purify the transcripts via phenol-chloroform extraction of the samples.
2. Precipitate the transcripts from the upper phase with EtOH by adding 1/10 volume 3 *M* NaAc pH 5.0 and 2.5 volumes EtOH. After washing the pellet with 70% EtOH, briefly dry and resuspend in double-distilled water (e.g., in 150 μL of H_2O) and add an equal volume of RNA gel loading buffer.
3. Sterilize the glass plates (14 × 16 cm) for RNase inhibition in 180 °C for 3 h before pouring the gel. Incubate the samples 2 min at 80 °C just before loading to the 13% acrylamide-urea gel (*see* **Notes 7** and **8**).
4. Prerun the gel at 8 W for 30 min and rinse the wells with electrophoresis buffer (1X TBE) to eliminate the excess of urea diffusing from the gel prior to the application of the sample.
5. After loading the sample, run the gel at 8 W until the xylene-cyanol marker migrates to 8–9 cm from the bottom of the well (*see* **Note 8**).
6. Use ultraviolet (UV) shadowing (240–280 nm) to localize the RNA band. Cut the portion of gel containing the RNA of the expected length with a sterile blade and crush into small pieces inside a sterile tube.

7. Cover the disrupted gel pieces with 3 mL of RNA extraction buffer, 3 mL of 70% phenol, and vortex 12 h vigorously.
8. Separate the phases by 45-min centrifugation at 6000 *g* (e.g., 6000 rpm in a Sorvall HB-4 rotor).
9. Recover the aqueous phase and reextract the phenol phase one more time with 2 mL of RNA extraction buffer by vortexing another 2 h.
10. Combine the aqueous phases and extract once with chloroform/isoamylalcohol (24:1).
11. Precipitate the RNA overnight by adding 2.5 volumes of ice-cold EtOH. There is no need to add salt since extraction buffer contains 200 m*M* NaCl.
12. After recovery of the precipitate by centrifugation for 1 h at 6000 *g* (6000 rpm in a Sorvall HB-4 rotor), wash with 70% ethanol, dry at room temperature for 10 min, and dissolve the RNA in double-distilled water. Determine the concentration by measuring the optical absorbance (A_{260}), make small aliquots, freeze in liquid nitrogen, and store at –80 °C.

3.2.2. Purification via Anion Exchange

If the transcription reaction yielded a single band on an analytical gel, the RNA can be purified over a Qiagen ion exchange column (Qiagen-tip-100 for the tRNA purification) directly. Please follow instructions in the manufacturer's brochure.

3.3. 5′ Labeling and Purification of 5′-Labeled tRNAs

The introduction of a radioactive phosphate at the 5′ end is performed using the method described by Chaconas and van de Sande ***(24)*** with slight modifications. The 5′ labeling is a two-step procedure. The first step is the dephosphorylation (hydrolysis of the 5′ phosphate group) with alkaline phosphatase. The second step is the phosphorylation with [γ-^{32}P]ATP in the presence of T4 polynucleotide kinase.

3.3.1. Dephosphorylation of RNA With Alkaline Phosphatase

Generally, the removal of 5′ phosphates from nucleic acids is used to enhance subsequent labeling with [γ-^{32}P]ATP.

1. Carry out the dephosphorylation reaction by incubating the reaction mix given in **Subheading 2.3.1.** at 50 °C for 45 min. This temperature is important for partial denaturation of the tRNA acceptor stem, thus increasing the dephosphorylation reaction. Pipeting order should be as in **Subheading 2.3.1**.
2. Stop the reaction by addition of 1/10 volume of 3 *M* NaAc, pH 5.0.
3. Perform a phenol chloroform extraction. This step is extremely important to remove the enzyme for an efficient phosphorylation.
4. Precipitate by addition of 2.5 volumes of EtOH, wash with 70% EtOH, and resuspend in a suitable volume of double-distilled water (e.g., 10 µL ddH$_2$O for 500 pmol initial tRNA).

3.3.2. Phosphorylation of RNA

The removal of 5′ phosphates from nucleic acids with phosphatases and their readdition in radiolabeled form by bacteriophage T4 polynucleotide kinase is a widely used technique for generating 5′ ^{32}P-labeled probes.

In **Subheading 2.3.2.**, the components were given. The end volume of the reaction is 40 µL. Remember, pipeting order is very important since the enzyme can be inactivated if the ionic environment changes drastically.

1. Carry out the phosphorylation reaction by incubating the reaction mix either at 4 °C for 15 h or at 37 °C for 1 h.
2. Lyophilize the reaction mixture to decrease the volume to roughly 5 µL to be able to load to one slot of gel.
3. Add 5 µL of RNA gel loading buffer and denature it at 80 °C for 1 min before applying to a 13% polyacrylamide-urea sequencing gel.
4. Run the gel at 50 W until xylene-cyanol dye migrates to 8–10 cm from the top (one nucleotide resolution). The important feature of this gel is that it contains 30% acrylamide on the bottom part (~10 cm height) to collect the nonincorporated radioactive nucleotides and decrease the possibility of radioactive contamination while working. This part should be cut and discarded right after stopping the gel.
5. Transfer the remaining part of the gel, which contains the labeled RNAs, to a used film, label the corners of the gel with radioactive ink and cover with transparent plastic wrap.
6. First, a short (1-min) autoradiography is performed to localize the labeled product. Cut the radioactive tRNA band and extract the tRNA as described in **Subheading 3.2.** (purification via polyacrylamide gel electrophoresis).
7. Right after the extraction of labeled RNA, take two 1-µL samples for radioactivity determination and immediately dilute the labeled RNA with the corresponding nonlabeled material to get specific activities ranging between 5,000 and 20,000 dpm/pmol depending on the experimental purpose (*see* **Note 9**).

3.4. Aminoacylation of the 5′-[^{32}P]-Labeled tRNA

In **Subheading 2.4.**, the components and the order of component additions are given.

1. Adjust the pH of the reaction mix to 7.5 with 1 *N* KOH before addition of the S-100 enzymes.
2. Carry out the reaction by incubating the reaction mix at 37 °C for 15 min.
3. Extract the reaction mixture with 1 volume of 70% phenol and recover the radiolabeled aminoacyl-tRNA by EtOH precipitation before the final purification step, which is reversed-phase HPLC.

3.5. Acetylation of tRNAs

With the addition of acetic anhydride, a nonpurified aminoacyl-tRNA is converted into its *N*-acetyl-aminoacyl-tRNA derivative with a yield of above 90%. A modified method of Haenni and Chapeville is used for this purpose *(25)*.

1. Add acetic anhydride (1/30 of the sample volume) and incubate the mixture at 0 °C for 15 min. Repeat the procedure three times.
2. Precipitate the *N*-acetyl-aminoacyl-tRNA with EtOH and resuspend in double-distilled water.

3.6. Separation of Charged and Noncharged tRNA via Reversed-Phase HPLC

The *N*-acetyl-aminoacyl-tRNA and aminoacyl-tRNA can be separated from deacyl-tRNA using reversed-phase HPLC (*see* **Note 10**). The order of elution is first deacyl-tRNA, then aminoacyl-tRNA, and finally *N*-acetyl-aminoacyl-tRNA.

1. Equilibrate the column in buffer A.
2. Apply the tRNA mixture to the column (up to 50 nmol) and elute by a gradient composed of buffer A and an increasing concentration of buffer B starting with 0% buffer B and ending with 40% buffer B.
3. Collect the elute in 1000-μL fractions.
4. Pool the fractions containing *N*-acetyl-aminoacyl-tRNA and aminoacyl-tRNA and precipitate the tRNAs.

3.7. Footprinting Experiments With Thioate tRNAs Bound to the Ribosome

As explained in **Subheadings 1.4.** and **1.5.**, footprinting experiments should be done in two different ways to analyze interference and protection effects. Here, we mainly talk about protection experiments. However, interference experiments can be done in a similar way except with a slight experimental ordering difference (*see* **Fig. 3**).

3.7.1. Functional Complex Formation

To carry out footprinting of the ribosome on thioated tRNA species occupying different ribosomal sites, the following complexes can be defined as representing the main states of the elongating ribosome in vitro ***(26)***:

1. Deacyl-tRNA in the P site (a state similar to a 70S initiation complex that also carries only one tRNA—the Pi state).
2. Ac-aminoacyl-tRNA in the P site (Pi state).
3. Deacyl-tRNA in the P site and Ac-aminoacyl-tRNA in the A site (pretranslocational state—PRE state).
4. Deacyl-tRNA in the E site and Ac-aminoacyl-tRNA in the P site (posttranslocational state—POST state).
5. Ac-aminoacyl-tRNA or deacyl-tRNA in the P site and aminoacyl-tRNA in the A site (PRE state).
6. The investigation of the protection patterns of tRNAs bound to the ribosome can be performed for all ribosomal-binding sites with these defined states. The 5′-[^{32}P]-thioated tRNA is bound to a distinct site chosen for investigation. It is very

important to include control reactions, in which the corresponding thioated tRNA is free in solution.

7. In the following subheadings, a single aliquot for functional experiments containing initially 12.5µL is described. However, for a typical thioate experiment the volume should be increased. For example, for the spun column the total volume should be between 100 and 125µL, for the sucrose cushion between 100 and 200µL. Increasing the volume requires increasing the amount of ribosome (e.g., 120pmol [5 A_{260} units] instead of 5–10pmol) as well as the other components accordingly.

3.7.1.1. P Site Binding: Construction of the Pi Complex

Incubate 5–10pmol of 70S ribosomes in a volume of 12.5µL with 6–10 molar excess of either poly(U) or heteropolymeric mRNA over ribosomes and deacyl tRNA in a 1.5- to 2-fold molar ratio to ribosomes for 15min at 37°C (*see* **Note 11**).

3.7.1.2. A Site Binding: Construction of PRE Complex

A site binding is divided into two types: enzymatic and nonenzymatic. While constructing a PRE complex, the first site to be occupied with a tRNA is the P site.

In case of nonenzymatic A-site binding, add an aminoacyl-tRNA in a 0.8- to 2-fold molar ratio to ribosomes.

For enzymatic A-site binding, form a ternary complex (aminoacyl-tRNA·EF-Tu·GTP) immediately before its addition to the binding assay: Preincubate aminoacyl-tRNA (1–2pmol per pmol of 70S ribosomes), 0.5m*M* GTP, EF-Tu (1.2pmol per pmol of aminoacyl-tRNA), and EF-Ts (up to 2pmol per pmol EF-Tu) for 2min at 37°C under the ionic conditions of the binding buffer

Add the ternary complex to the reaction mixture and then incubate for 30min at 37°C for binding. Final volume is 25µL.

3.7.1.3. Translocation: Construction of POST Complex

POST complexes are constructed via an EF-G- (0.3pmol/pmol 70S) dependent translocation of the PRE complex.

Add 2.5µL of HMK buffer (20m*M* HEPES KOH, pH 7.5, 10m*M* $MgCl_2$, 100m*M* KCl) containing EF-G (*see* **Subheading 2.7.**) and incubate for 10min at 37°C.

3.7.1.4. Puromycin Reaction

Generally, for one puromycin reaction six samples are used:

1. Add 2.5µL of binding buffer to two control samples (without puromycin and either plus or minus EF-G) as background.
2. Next, add 2.5µL of puromycin stock solution (10m*M* in binding buffer, final concentration of puromycin should be 0.7m*M*) to four remaining samples (two

with and two without EF-G) to determine the amount of A-site occupation. Note: Store the puromycin stock solution in small aliquots at −80 °C.
3. Incubate the reaction mixture either at 37 °C for 10–15 min or at 0 °C for about 12 h and stop the reaction by adding 32.5 μL of 0.3 *M* sodium acetate, pH 5.5, saturated with $MgSO_4$.
4. Determine the amount of aminoacyl-puromycin formed by extraction with 1 mL of cold ethylacetate by vortexing the samples for 1 min.
5. Carry out a low-speed centrifugation to get phase separation.
6. Withdraw 800 μL of the ethylacetate for scintillation counting (*see* **Note 12**).

3.7.1.5. Filter-Binding Assay

The nitrocellulose filter assay is a method to measure the tRNAs in the complex.

1. Mix the complexes with 2 mL ice-cold ribosomal binding buffer and immediately filtrate through nitrocellulose filters.
2. Wash the filters twice with ice-cold binding buffer and determine the radioactivity retained on the filter by liquid scintillation counting. Binding assays should include reactions without ribosomes as controls to determine the filter background.

3.8. Iodine Cleavage

As explained in **Subheading 3.7.1.**, the total volume of the complex reaction should be increased at least to 100 μL for the thioate experiments.

1. Take a 5-μL aliquot from the complex reaction for filter-binding assay (see **Subheading 3.7.1.5.**). Keep the rest on ice for iodine cleavage.
2. Add 1/50 volume freshly prepared iodine solution (50 m*M* in EtOH) to the ribosome-binding reaction aliquot saved from **step 1**.
3. Incubate the reaction 1 min on ice and stop the reaction with 1,4-dithiothreitol (DTT) (final concentration 5 m*M*) when orange color of iodine disappears.
4. Purify the complexes immediately with one of the methods discussed next.

3.9. Purification of the Complexes

Ribosomal complexes can be separated from free ligands by several methods, such as spun columns, centrifugation of the complexes through a sucrose cushion, or gel filtration by gravity flow. The best method for the protein–RNA complexes is purification via native gel.

3.9.1. Isolation of Ribosome Complexes via Spun Column

1. Load 100–125 μL sample containing 0.5–5 A_{260} of ribosomal complex immediately after iodine cleavage onto a complementary DNA spun column S300 that was pre-equilibrated by washing three times with 2 mL binding buffer using gravity flow.
2. After loading, immediately centrifuge the column for 1 min at 400 *g* (e.g., 1500 rpm in a Sorvall HB4 rotor) at 4 °C.

3. Collect the flowthrough (fraction 1) and repeat the centrifugation by loading 100–125 µL of binding buffer one more time. Combine both fractions. Recovery is about 60%.

3.9.2. Sucrose Cushion Centrifugation

1. Load 100–200 µL sample containing 1–5 A_{260} of ribosome in binding buffer on a 1–2 mL 10% sucrose cushion in binding buffer. A polycarbonate centrifuge tube (Beckman) can be used.
2. After centrifugation for 18 h at 4 °C and 90,000 *g* (e.g., 40,000 rpm in a Beckman TLA 100.3 rotor), resuspend the pellet in 100 µL of binding buffer. The yield is about 50–60% of input.

3.9.3. Native Gel

1. Prepare a 15-µL sample of an RNA–protein complex containing 75 pmol of [^{32}P]RNA and the amount of protein to achieve approx 75% binding by incubating for 15 min at 37 °C (binding buffer, e.g., 20 m*M* HEPES KOH (pH 7.4), 100 m*M* NH_4Cl, 4 m*M* $MgCl_2$) and then chill on ice.
2. Add 78% glycerol to a final concentration of 10% to the preformed complex, load on a 6% PAA gel (90 × 60 × 0.7 mm). The gel is also buffered with the binding buffer.
3. Run for 2 h at 80 V. Use a pump to keep the ionic conditions of the running buffer constant in case high-salt buffers are used.
4. Localize bands by exposing on an X-ray film (Fuji) for 25–35 min and cut out the bands accordingly (*see* **Note 13**).

3.10. Phenol-Chloroform Extraction

1. Do overnight phenol extraction in the cold room with strong vortex. It is very important to remove all ribosomal proteins from the sample to get clearer and sharp bands.
2. Repeat the phenol extraction one more time (30 min). After a subsequent chloroform extraction, precipitate the samples with EtOH and wash with 70% EtOH (*see* **Note 14**).
3. Dissolve the pellet in 6 µL H_2O and 6 µL loading buffer.
4. Take two 1-µL samples to measure in the scintillation counter to adjust the amount to load to the gel.

3.11. Gel Electrophoresis

Run a 13% denaturing polyacrylamide sequencing gel (acrylamide/bis-acrylamide 19:1, 7 *M* urea; 3,000–10,000 dpm/lane).

3.12. Image Quant Analysis of Gels

Expose the 13% denaturing polyacrylamide sequencing gels (*see* **Subheading 3.11.**) for 12–16 h on a PhosphorImager. Evaluate the scanned gel with the program ImageQuant.

Repeat the experiments up to four times and normalize the data as follows:

Assuming that the results of two experiments performed under identical conditions are considered, the first normalization concerned variation in loading between respective thio-A-lanes (input normalization that yields a normalization factor with which the individual bands of one of the two lanes have to be multiplied). Here, the total counts of the two lanes are normalized, and the intensity of each band of one of the lanes is multiplied by the normalization factor. Input normalization is performed for both complex and solution cleavage experiments individually. The second normalization is between the two sets of data (bound tRNA vs. tRNA in solution). The protection value regarding the amount of $tRNA_{bound}/tRNA_{solution}$ for a distinct band is calculated using the normalized intensities of corresponding bands derived from a tRNA in a complex and a tRNA in solution, respectively.

3.13. An Example of Footprinting Experiments

Figure 4 is an example to iodine cleavage of the tRNAPhe on different buffer systems. **Table 1** shows the calculation of the gels of this experiment, and **Figure 5** represents the color code of the calculations.

4. Notes

1. Final concentration in the transcription reaction mixture should be 1X.
2. The components of the NTP mix were prepared as 100 m*M* stock solutions adjusted to pH 5.5–6.0 with 1 *M* KOH (to minimize spontaneous hydrolysis) and stored at −80 °C in very small aliquots. NTPs are obtained as sodium or lithium salts with an undefined stoichiometry. Dissolve the NTPs at higher concentration first and determine the real concentration by UV photometric measurements before diluting to 100 m*M*. Concentration of the nucleotides changes on addition of [αS]-NTPs in the thioated RNA transcription.
3. [αS]-NTPs were obtained from NEN or Glen Research with 20 m*M* stock concentration.
4. If synthetic oligonucleotides are employed, use 500 pmol/mL of the tRNA template and 1000 pmol/mL T7 primer oligonucleotide.
5. T7 RNA polymerase was overproduced in *Escherichia coli* and purified via affinity chromatography *(27)*. T7 RNA polymerase is also available from various suppliers.
6. Since reducing agents such as β-mercaptoethanol or DTT are interfering with the iodine cleavage, these reagents have to be omitted from the original binding buffer, which contains 4 m*M* β-mercaptoethanol.
7. In our laboratory, a special sample well 9 × 1 cm was made with an additional 2-mm thick spacer prepared for this purpose.
8. A 13% acrylamide gel is used for the tRNA purification. The percentage of acrylamide used in the gel should be chosen to ensure that the RNA with the expected length migrated above this marker.

9. Immediate dilution and aliquoting into small fractions are important to reduce radiolysis of the RNA molecules.
10. Due to its hydrophobic properties, this column is especially suitable for separation of tRNAs carrying hydrophobic amino acids like phenylalanine.
11. Reassociated 70 ribosomes should be used. Preparation is as described in **ref. 26**.
12. The samples without both EF-G and puromycin allow the determination of the background activity. The samples without EF-G but with puromycin determine the direct P-site binding of Ac-aminoacyl-tRNA or its spontaneous translocation.
13. For the analysis of rRNA–protein complexes, the iodine cleavage can occur before purification of the complexes. The complexes remain in their native state during native gel electrophoreses. Thus, in protection experiments the complexes can be subjected to iodine treatment after the corresponding bands have been cut out by soaking the gel piece with iodine for 5 min.
14. Add RNase-free glycogen to assist precipitation of rRNA fragments from rRNA–protein complexes. For tRNA–ribosome complexes, rRNA serves as precipitation support.

References

1. Jen-Jacobson, L. (1997) Protein–DNA recognition complexes: conservation of structure and binding energy in the transition state. *Biopolymers* **44**, 153–180.
2. Mattaj, I. W. (1993) RNA recognition: a family matter? *Cell* **73**, 837–840.
3. Kenan, D. J., Query, C. C., and Keene, J. D. (1991) RNA recognition: towards identifying determinants of specificity. *Trends Biochem. Sci.* **16**, 214–220.
4. Dertinger, D., Behlen, L. S., and Uhlenbeck O. C. (2000) Using phosphorothioate-substituted RNA to investigate the thermodynamic role of phosphates in a sequence specific RNA–protein complex. *Biochemistry* **39**, 55–63.
5. Gish, G., and Eckstein, F. (1988) DNA and RNA sequence determination based on phosphorthioate chemistry. *Science* **240**, 1520–1522.
6. Schatz, S., Leberman, R., and Eckstein, F. (1991) Interaction of *Escherichia coli* $tRNA^{Ser}$ with its cognate aminoacyl-tRNA synthetase as determined by footprinting with phosphorothioate-containing tRNA transcripts. *Proc. Natl. Acad. Sci. U. S. A.* **88**, 6132–6136.
7. Verma, S., and Eckstein, F. (1998) Modified nucleotides: synthesis and strategy for users. *Annu. Rev. Biochem.* **67**, 99–134.
8. Claus, L., Vörtler, S., and Eckstein, F. (2000) Phosphorothioate modification of RNA for stereochemical and interference analyses. *Methods Enzymol.* **317**, 74–91.
9. Strobel, S. A. (1999) A chemogenetic approach to RNA function/structure analysis. *Curr. Opin. Struct. Biol.* **9**, 346–352.
10. Dabrowski, M., Spahn, C. M. T., and Nierhaus, K. H. (1995) Interaction of tRNAs with the ribosome at the A and P sites. *EMBO J.* **14**, 4872–4882.
11. Dabrowski, M., Spahn, C. M. T., Schäfer, M. A., Patzke, S., and Nierhaus, K. H. (1998) Contact patterns of tRNAs do not change during ribosomal translocation. *J. Biol. Chem.* **273**, 32793–32800.
12. Schäfer, M. A., Tastan, A. O., Patzke, S., et al. (2002) Codon–anticodon interaction at the P site is a prerequisite for tRNA interaction with the small ribosomal subunit. *J. Biol. Chem.* **277**, 19095–19105.

13. Alexeeva, E. V., Shpanchenko, O. V., Dontsova, O. A., Bogdanov, A. A., and Nierhaus, K. H. (1996) Interaction of mRNA with the *Escherichia coli* ribosome: accessibility of phosphorothioate-containing mRNA bound to ribosomes for iodine cleavage. *Nucleic Acids Res.* **24**, 2228–2235.
14. Shpanchenko, O. V., Dontsova, O. A., Bogdanov, A. A., and Nierhaus, K. H. (1998) Structure of 5S rRNA within the *Escherichia coli* ribosome: application of iodine cleavage of phosphorothioate derivatives. *RNA* **4**, 1154–1164.
15. Shpanchenko, O. V., Zvereva, M. I., Dontsova, O. A., Nierhaus, K. H., and Bogdanov, A. A. (1996) 5S rRNA sugar-phosphate backbone protection in complexes with specific ribosomal proteins. *FEBS Lett.* **394**, 71–75.
16. Stelzl, U., Spahn, C. M. T., and Nierhaus, K. H. (2000) Selecting rRNA binding sites for the ribosomal proteins L4 and L6 from randomly fragmented rRNA: application of a method called SERF. *Proc. Natl. Acad. Sci. U. S. A.* **97**, 4597–4602.
17. Stelzl, U., and Nierhaus, K. H. (2001) A short fragment of 23S rRNA containing the binding sites for two ribosomal proteins, L24 and L4, is a key element for rRNA folding during early assembly. *RNA* **7**, 598–609.
18. Stelzl, U., Zengel, J. M., Tovbina, M., et al. (2003) RNA-structural mimicry in *Escherichia coli* ribosomal protein L4-dependent regulation of the S10 operon. *J. Biol. Chem.* **278**, 28237–28245.
19. Maiväli, Ü., Pulk, A., Loogväli, E.-L., and Remme, J. (2002) Accessibility of phosphates in domain I of 23S rRNA in the ribosomal subunit as detected by Rp phosporothioates. *Biochim. Biophys. Acta* **1579**, 1–7.
20. Dabrowski, M., Junemann, R., Schafer, M. A., et al. (1996) Contact patterns of RNA ligands with the ribosome in defined functional states as determined by protection against cleavage at phosphorothioated residues: a review. *Biochem. (Moscow)* **61**, 1402–1412.
21. Dabrowski, M., and Nierhaus, K.H. (1998) Synthesis and site-specific binding of thioated tRNAs to probe ribosome–tRNA interactions. *Methods Mol. Biol.* **77**, 413–426.
22. Milligan, F., and Uhlenbeck, O. C. (1989) Synthesis of small RNAs using T7 RNA polymerase. *Methods Enzymol.* **180**, 51–62.
23. Tastan, O. A. (2003) How lazy are the tRNAs on the ribosome? New insights for the alpha-epsilon model. PhD thesis, Free University, Berlin, Germany.
24. Chaconas, G., and van de Sande, J. H. (1980) 5′ 32P labeling of RNA and DNA restriction fragments. *Methods Enzymol.* **65**, 75–88.
25. Haenni, A.-L., and Chapeville, F. (1966) The behavior of acetylphenylalanyl soluble ribonucleic acid in polyphenylalanine synthesis. *Biochim. Biophys. Acta* **114**, 135–148.
26. Blaha, G., Stelzl, U., Spahn, C. M. T., Agrawal, R. K., Frank, J., and Nierhaus, K. H. (2000) Preparation of functional ribosomal complexes and the effect of buffer conditions on tRNA positions observed by cryoelectron microscopy. *Methods Enzymol.* **317**, 292–309.
27. Davanloo, P., Rosenberg, A. H., Dunn, J. J., and Studier, F. W. (1984) Cloning and expression of the gene for bacteriophage T7 RNA polymerase. *Proc. Natl. Acad. Sci. U. S. A.* **81**, 2035–2039.

10

In Vivo Analysis of Ribonucleoprotein Complexes Using Nucleotide Analog Interference Mapping

Lara B. Weinstein Szewczak

Summary

Multicomponent RNA–protein complexes are essential for eukaryotic gene expression. Some, like the spliceosome, have been studied successfully in vitro using biochemical and structural approaches, but many have not been reconstituted in cell-free systems. Nucleotide analog interference mapping (NAIM) can report detailed atomic information about requirements for ribonucleoprotein particle assembly and function in living cells, providing a method to study complexes in a cellular context at a level of detail comparable to many biochemical assays. The method relies on incorporation of phosphorothioate-tagged nucleotide analogs during in vitro transcription, followed by a selection for the active population of molecules and analysis of the selected RNA sequence composition. *Xenopus* oocytes provide a cellular environment for selecting active molecules based on particle assembly or function. Functional group analysis of complexes assembled in vivo provides predictive models for further investigation either in vivo or in vitro as well as benchmarks for evaluating and refining biochemical and structural models.

Key Words: In vivo selection; NAIM; nucleotide analog; phosphorothioate; RNP.

1. Introduction

Many of a cell's most abundant and essential machines are ribonucleoprotein (RNP) complexes. Some RNPs, like the 50 S ribosomal subunit *(1)*, assemble to form stable functional structures, while others, like the eukaryotic spliceosome *(2)*, are spectacularly dynamic, assembling and disassembling as a part of their functional cycle. Still others, small nucleolar RNPs (snoRNPs), individual small nuclear RNPs, and miRNPs may lie somewhere in the mechanistic middle. To understand the function of these particles, it is essential to examine the RNA–protein interactions that allow them to assemble and work.

One of the most successful biochemical approaches to examining RNA–protein interactions is modification interference. In such experiments, the RNA is

From: *Methods in Molecular Biology, vol. 488: RNA-Protein Interaction Protocols*
Edited by: Ren-Jang Lin © Humana Press Inc, Totowa, NJ

modified with a chemical reagent and then used in protein binding or complex formation. Where sites of modification inhibit binding or assembly, the RNA is excluded from the purified "active" population. RNA sequence analysis by direct or indirect methods allows the identification of important sites. In the late 1980s, Eckstein and coworkers described a method for RNA sequence analysis using phosphorothioate substitutions in an otherwise phosphodiester backbone *(3)*. Their observation that the phosphorothioate linkage could be severed selectively by treatment of the RNA with iodine (I_2) in ethanol paved the way for use of this technique in modification interference studies, examining positions where replacement of the pro-R_P oxygen with sulfur impaired binding or activity *(4)*.

Strobel and coworkers expanded the phosphorothioate interference technique to enable analysis of atomic mutations to ribose sugar and heterocyclic base functional groups *(5)*. This approach, termed nucleotide analog interference mapping (NAIM), has been applied with great success to mechanistic studies of RNA catalysis, RNA folding, and RNA–protein interactions (see **refs.** ***6*** and ***7*** for reviews and basic in vitro protocols). NAIM is a quantitative method that enables rigorous analysis of individual functional group contributions to the interaction or process of interest.

NAIM relies on four key steps: (1) incorporation of phosphorothioate-containing nucleoside analog triphosphates into RNA molecules by in vitro transcription; (2) a selection for functional molecules; (3) physical separation of the active subpopulation from the inactive one; and (4) direct comparison of the sequence composition of the active and unselected RNA pools.

1. All of the analogs used in a NAIM experiment are derivatives of 5′-*O*-(1-thio)-nucleoside triphosphates. Substitution of sulfur for oxygen within the RNA backbone not only tests the functional contributions of that atom, but also provides a tag for all subsequent functional group alterations. The phosphorothioate linkage is susceptible to cleavage in the presence of iodine, while phosphodiester bonds are unaffected. Parental analogs (NαTPs) are the α-thiophosphate derivatives of ATP (adenosine triphosphate), GTP (guanosine triphosphate), CTP (cytosine triphosphate), and UTP (uridine triphosphate). Nucleotide analogs (δαTPs) contain the α-phosphorothioate and an additional functional group modification (δ) to the heterocyclic base or to the sugar (**Fig. 1**). The parent or nucleotide analog is added to a transcription reaction containing unmodified nucleoside triphosphates (NTPs) under conditions designed to achieve 5% analog incorporation. This doping procedure produces a pool in which each position of a canonical nucleotide (e.g., A) is substituted with an analog (e.g., AαS or PurαS, purine removes the exocyclic amine of A; *see* **Fig. 1**) such that each RNA molecule within the pool contains only one substitution (NαS or δαS) on average. Activity-based selection of this pool thus allows screening for the effects of substitutions at each position simultaneously and independently.

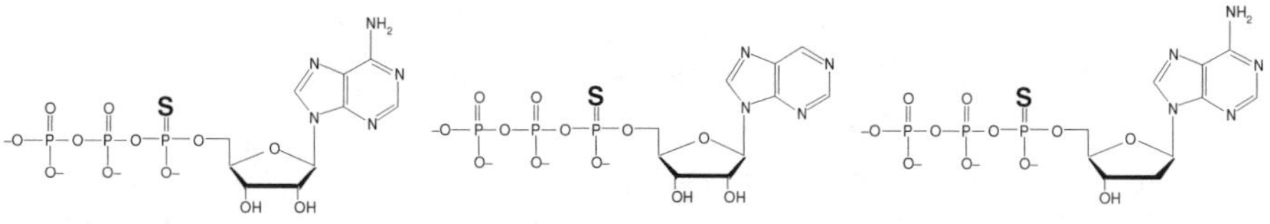

ATPαS PurTPαS dATPαS

Fig. 1. Structures of three adenosine triphosphate (ATP) analogs. Parental analog adenosine α-thiotriphosphate (ATPαS), base analog purine α-thiotriphosphate (PurTPαS), and sugar analog 2′-deoxyadenosine α-thiotriphosphate (dATPαS) are shown. The nonbridging sulfur substitution is enlarged in bold type. The base and sugar functional group alterations are boxed.

2. Many different assays have been employed as selection steps in NAIM experiments. These include RNA catalysis, RNA folding, and RNA–protein interaction both in vitro and in vivo ***(6)***. In each assay, the RNAs are radiolabeled at a single position, enabling direct sequence analysis following I_2 treatment. The label may be introduced before, during, or after selection.
3. Isolation of the active subpopulation of RNAs (the selected pool) from the inactive subpopulation is key for the success of the NAIM technique. Approaches employed for this separation are wide ranging and have included selective radiolabeling, excision of a particular species from polyacrylamide gels, retention on nitrocellulose filters, and accurate processing by cellular pathways ***(6)***.
4. Ultimately, the selected RNAs are treated with iodine, and the resulting fragments are re-solved by denaturing polyacrylamide gel electrophoresis (PAGE). Positions where a functional group change interfered with the selection leave gaps in a sequencing ladder. It is essential to compare this experimental ladder to one generated from unselected RNAs. Not every NαS or δαS incorporates equally well at all positions during in vitro transcription. Consequently, gaps in the selected ladder may arise from gaps present in the initial transcribed pool. The unselected control ladder identifies such sites.

In contrast to conventional mutagenesis, NAIM allows the investigation of specific base alterations as well as providing information about the ribose-phosphate backbone. The conclusions derived from a NAIM experiment thus provide a basis for developing a truly atomic-level model for RNP assembly or function. While most NAIM experiments have been performed on in vitro systems, the method has been extended to allow examination of RNPs in living cells (***8,9***).

Xenopus laevis oocytes provide a cellular system for examining exogenous molecules in the context of endogenous pathways. These large cells survive for 2–3 d once separated from ovary tissue, are amenable to microinjection of macromolecules, and can be fractionated into nuclear and cytoplasmic compartments with straightforward manipulation. Microinjection of *Xenopus* oocytes facilitates investigation of the assembly of a composite box C/D small nucleolar RNA (snoRNA) into a functional RNP by NAIM *(8)*.

The vast majority of box C/D snoRNAs function in the methylation of specific ribose 2′-hydroxyls in ribosomal and small nuclear RNAs. Most vertebrate box C/D snoRNAs are contained within introns of protein-coding genes. The biogenesis pathway for these RNAs involves splicing of the pre-mRNA (messenger RNA), debranching of the intron-lariat, and exonucleolytic trimming to produce the mature snoRNA termini (reviewed in **refs.** ***10–14***). Assembly with snoRNP-specific proteins occurs during splicing and protects the snoRNA from complete degradation (***10,11,15***). Based on this pathway, the selection in this experiment was two-fold. First, the RNAs had to be stable in the cells, indicative of their association with snoRNP proteins. Second, the RNAs were 3′ end labeled using a DNA splint such that only accurately processed molecules were visualized. This approach identified specific atoms and functional groups comprising a structured protein-binding site. Generalized methods for conducting NAIM analysis of in vivo RNP assembly or function derived from that study follow.

2. Materials

2.1. In Vitro Transcription

1. Distilled, deionized water; henceforth referred to as water (H_2O).
2. Plasmid- or oligonucleotide-based linear DNA transcription template.
3. NTP 10X stocks: 10 m*M* ATP, CTP, and UTP; 1 m*M* GTP.
4. 5′-*O*-(1-Thio)-nucleoside triphosphates (nucleotide analogs; *see* **Note 1**).
5. T7 RNA polymerase and T7 Y639F RNA polymerase.
6. (5′)ApppG(5′) (10 m*M*; New England Biolabs).
7. 10X transcription buffer: 400 m*M* Tris-HCl, pH 7.5, 100 m*M* dithiothreitol (DTT), 40 m*M* spermidine, 150 m*M* $MgCl_2$, 0.5% Triton X-100.
8. PAGE equipment.
9. Elution buffer: 10 m*M* Tris-HCl, pH 7.5, 1 m*M* ethylenediaminetetraacetic acid (EDTA), 0.28 *M* NaCl.
10. Ethanol (absolute; stored at −20 °C).

2.2. Oocyte Manipulation

1. Dissecting microscope.
2. Injection apparatus (e.g., Drummond Nanoject).
3. Glass capillary needles.
4. Forceps (#4 or #5).
5. OR3 media: 50% Leibovitz L-15 media (GibcoBRL), 5% penicillin/streptomycin in 0.9% NaCl (Sigma), 1% L-glutamine, 5 m*M* HEPES OH, pH 7.5.
6. Transfer pipets (Falcon, Becton Dickinson).
7. Mineral oil.

8. Germinal vesicle (GV) isolation buffer at 4 °C: 6.5 m*M* Na_2HPO_4, 3.5 m*M* KH_2PO_4, 83 m*M* KCl, 17 m*M* NaCl, 10 m*M* $MgCl_2$, 1 m*M* DTT.

2.3. Nuclear RNA Preparation

1. TE: 10 m*M* Tris-HCl, pH 7.5, and 1 m*M* EDTA
2. Proteinase K mix: 50 m*M* Tris-HCl, pH 7.5, 5 m*M* EDTA, 1.5 % sodium dodecyl sulfate (SDS), 300 m*M* NaCl, 1.5 mg/mL proteinase K (Amresco).
3. PCA: Phenol/chloroform/isoamyl alcohol (50:49:1).

2.4. Splint Labeling

1. [^{32}P]-dNTP (deoxynucleotide 5′-triphosphate; 3.3 μ*M*, 3000 Ci/mmol).
2. dNTPs (if needed, 5 μ*M*).
3. 10X splint-labeling buffer: 500 m*M* Tris-HCl, pH 7.5, 100 m*M* $MgCl_2$, 10 m*M* DTT.
4. DNA oligonucleotide splint.
5. T7 Sequenase v. 2.0 (U.S. Biochemicals).
6. 10X TBE: 890 m*M* Tris base, 890 m*M* boric acid, 20 m*M* EDTA.
7. Formamide dyes: 85% deionized ultrapure formamide, 1X TBE, 10 m*M* EDTA, 0.05% bromophenol blue, 0.05% xylene cyanole.

2.5. Nucleotide Analog Interference Mapping

1. 25 m*M* I_2 in ethanol; store at 4 °C (stable for 1–3 mo).
2. Sequencing gel apparatus.
3. Whatman 3MM paper.
4. Imaging system (e.g., PhosphorImager, Molecular Dynamics).

3. Methods

This section describes methods for (1) transcription of RNAs containing nucleotide analogs, (2) injection of *Xenopus* nuclei, (3) isolation of nuclear RNA, (4) 3′ end labeling, and (5) NAIM analysis of the isolated (selected) RNAs.

3.1. Analog Incorporation by In Vitro RNA Transcription

NAIM relies on the analysis of RNA transcripts containing randomly incorporated nucleotide analogs. Comprehensive lists of analogs used in in vitro NAIM experiments have been published ***(6,7)***. In separate transcription reactions, parental nucleoside α-thiotriphosphates (e.g., ATPαS,) and analog α-thiotriphosphates (e.g., PurTPαS) are added at concentrations to achieve one substitution per RNA molecule. Both linearized plasmid ***(5)*** and oligonucleotide templates ***(16)*** containing T7 RNA polymerase promoters have been used to synthesize analog-containing RNAs. For in vivo assays, the 5′ end of most RNAs must be blocked to prevent degradation by 5′ → 3′ exonucleases (*see* **Note 2**).

1. Assemble transcription reactions (*see* **Note 3**) containing 1X transcription buffer, 0.05 µg/µL linear transcription template, 1X NTPs, 1 m*M* (5′)ApppG(5′), the appropriate concentration of nucleotide analog (reduce concentrations of G analogs 10-fold), and 0.05 µL/mL T7 RNA polymerase.
2. Incubate the reactions at 37 °C for 2 h.
3. Add 50 µL of formamide dyes and purify by denaturing PAGE.
4. Visualize the RNA transcripts by ultraviolet (UV) shadowing, excise the full-length transcripts from the gel, and elute the RNAs from the gel slices into 400 µL of elution buffer by rotating the tubes at 4 °C overnight.
5. Recover the RNAs by transferring the elution buffer into clean 1.6-mL Eppendorf tubes.
6. Ethanol precipitate the RNAs by adding 1 mL of absolute ethanol, mixing well, and freezing the tubes on dry ice for 15–30 min or at –20 °C overnight. Without thawing, spin the tubes at 16,000 *g* for 15 min and decant the EtOH. To remove the last drops of EtOH, spin the tubes at maximum speed for a few seconds and draw off the remaining EtOH without disturbing the RNA pellets. Allow the pellets to air dry for a minimum of 10 min.
7. Dissolve each RNA pellet in 10 µL H_2O and use 1 µL to determine the RNA concentration spectrophotometrically. RNA transcripts can be stored at –80 °C for months.

3.2. Oocyte Manipulation

3.2.1. Oocyte Isolation

1. Harvest ovary tissue from a *Xenopus laevis* using techniques in compliance with individual institutional policies. Store tissue in OR3 medium ***(17)*** at 18 °C for up to 48 h.
2. Working under a dissecting microscope, isolate individual stage 5 and 6 oocytes from the tissue using forceps. Transfer isolated oocytes to fresh OR3 using a transfer pipet. Store at 18 °C or inject immediately (*see* **Note 4**).

3.2.2. Injection

Parental and related nucleotide analog containing RNAs (e.g., AαS, PurαS, and dAαS) must be compared within a single experiment.

1. Dilute an aliquot of each RNA transcript in water to give a final concentration between 0.5 and 5 µ*M* (*see* **Note 5**).
2. Fit a mineral oil-filled glass capillary needle onto the injector. Load one diluted RNA transcript into the needle and inject 9–14 nL into the animal pole of each individual oocyte. Inject 30–40 oocytes per RNA transcript.
3. Incubate the injected oocytes at 18 °C in OR3 for 4–24 h.

3.2.3. Germinal Vesicle Isolation

1. Transfer 5–10 oocytes from OR3 into cold GV isolation buffer. Use a sharp forceps (#5) or needle to puncture the oocyte and gently squeeze the GV into the buffer solution using two forceps (#4).

2. Remove the GVs using a P-20 pipet to a precooled Eppendorf tube on dry ice. Keep the amount of buffer transferred to a minimum. Continue to isolate GVs, adding to the tube and keeping track of the number isolated. Store isolated GVs at −80 °C.
3. Repeat for each set of injected oocytes.

3.3. Nuclear RNA Preparation

Total nuclear RNA is prepared from isolated GVs by removing nuclear proteins using proteinase K digestion ***(18)***.

1. Add 30 µL/GV of proteinase K mix to each tube containing frozen GVs and vortex at maximum speed for 30 s (*see* **Note 6**).
2. Add TE, if necessary, to bring the volume to a convenient amount for ethanol precipitation. Split into multiple tubes if required.
3. Extract each tube with 1 volume of PCA followed by extraction with 1 volume of chloroform.
4. Precipitate each aqueous layer by addition of 2.5–3 volumes of EtOH as described in **Subheading 3.1., step 6**.
5. Dissolve the nuclear RNAs in a small volume (5–10 µL) of water or, if necessary, combine the RNA samples in a larger volume (200–400 µL) and reprecipitate by addition of NaCl to 0.25–0.3 *M* and 2.5–3 volumes of EtOH.

3.4. 3′ End Labeling

The 3′ end labeling using a DNA splint ***(19)*** provides a convenient way to selectively radiolabel injected RNAs within the pool of total nuclear RNA (**Fig. 2A**). If the 3′ end of the RNA is processed in vivo as a consequence of RNA maturation, it also provides a selection for those RNAs that follow the endogenous biogenesis pathway. The splint is complementary to 10–15 nucleotides of the 3′ end sequence of the RNA and carries a short 5′ overhang to direct the incorporation of radiolabel(s). Splint labeling creates a homogeneous 3′ end that is necessary for sequence analysis following I_2 treatment. A splint specific for the in vitro transcript is used to label the unselected RNA.

3.4.1. Annealing

The nuclear RNA pellet (or the amount of unselected transcript comparable to that injected; e.g., 10 GVs injected with 9 nL of 2 µ*M* RNA would maximally yield 0.18 pmol) is annealed with an excess of the DNA splint (20 pmol).

1. Dissolve the nuclear pellet in 5 µL containing the DNA splint and 2X splint-labeling buffer.
2. Heat to above 90 °C for 1–2 min and then slowly cool in a rack at room temperature for 25–30 min.

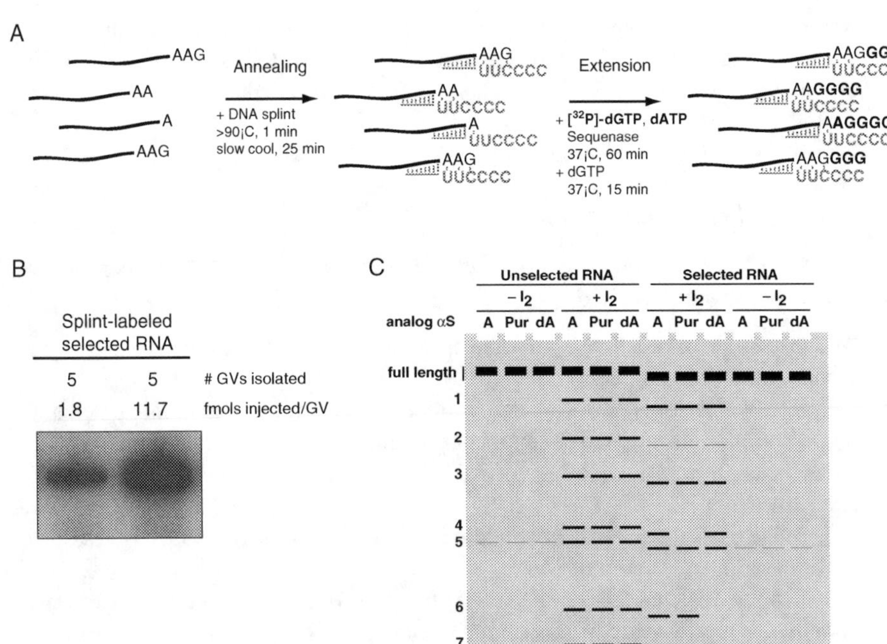

Fig. 2. 3′ end labeling and **nucleotide analog interference mapping** (NAIM) schematic. (**A**) Following incubation in oocytes, the selected RNA pool is 3′ end labeled by annealing a sequence-specific DNA oligonucleotide splint (gray) complementary to the 3′ end of the RNA, leaving a 5′ overhang. The overhang is filled in by Sequenase using a radiolabeled deoxynucleoside triphosphate (e.g., [^{32}P]-dGTP, bold) and any other dNTPs required to produce a homogeneous 3′ end (e.g., dATP, bold). (**B**) A composite **small nucleolar RNA** (snoRNA) pool was injected into *Xenopus* oocytes and subjected to splint labeling after overnight incubation at 18 °C. The sequences at the 3′ end of this RNA are shown in panel **A**. (**C**) Schematic of a sequencing gel re-solving the I_2-cleaved fragments. The 3′-end-labeled RNAs from unselected and selected pools are treated with iodine, and the resulting fragments are separated on a polyacrylamide gel to produce a sequence ladder. If the RNA is trimmed in vivo, the length of the selected population may differ from the unselected, as shown. Untreated samples are run on the same gel to identify positions of I_2-independent hydrolysis (band 5). Comparison of the selected and unselected lanes for a particular analog reveals positions of phosphorothioate interference (band 2) as well as base- (band 4) and sugar-specific (band 6) effects.

3.4.2. Extension

1. Add 0.5–1 μ*M* of the radiolabeled nucleotide and 19.5 U of Sequenase in a volume of 5 μL to the annealed RNA–DNA hybrid. Include other nucleotides if necessary to fill in the overhanging sequence on the DNA splint (**Fig. 2A**).
2. Incubate the extension reaction at 37 °C for 60–90 min.

3. Add an excess (200 pmol) of cold labeling nucleotide and incubate at 37 °C for 15 min to ensure homogeneous 3′ ends.
4. Add 10 μL formamide dyes to quench the reaction.

3.4.3. Purification and Recovery of Labeled RNAs

1. Re-solve the labeled RNAs by denaturing PAGE.
2. Visualize the labeled RNAs by autoradiography (**Fig. 2B**; exposure time may vary from 30 min to 2 h).
3. Excise, elute, and precipitate the RNAs as described above.
4. Determine the amount of RNA in each pellet by Cerenkov scintillation counting.
5. Store the pellets at –80 °C.

3.5. Interference Mapping Analysis

Analysis by NAIM requires that both the selected and unselected end-labeled RNAs be fragmented by the addition of I_2 in EtOH and separated to single-nucleotide resolution on a sequencing gel (**Fig. 2C**). While the selected and unselected RNAs do not necessarily need to be run on the same gel, it is imperative that each parent phosphorothioate-containing RNA (e.g., AαS) is run alongside the corresponding analog-containing RNAs (e.g., A, Pur, and dA; schematized in **Fig. 2C**).

3.5.1. Sample Preparation

1. Dissolve the selected RNAs (parent and analog containing) in formamide dyes to give a minimum of 10,000 cpm/μL.
2. Remove 5- and 2-μL aliquots of each RNA to separate Eppendorf tubes.
3. Combine 2 μL of 25 m*M* I_2/EtOH and 25 μL of formamide dyes (I_2/dyes).
4. Add 1 μL of I_2/dyes to the 5-μL sample (*see* **Note 7**).
5. Add 4 μL of formamide dyes to the 2-μL sample (*see* **Note 8**).

3.5.2. Electrophoresis (see **Note 9**)

1. Prerun a denaturing polyacrylamide gel (4–20% depending on the size of the RNA and the region of interest) for at least 30 min.
2. Heat all of the RNA samples to above 90 °C for 1 min.
3. While the samples are heating, wash all urea and gel debris from the gel wells.
4. Load 4–5 μL of each sample (more if well size permits).
5. Electrophorese the samples to re-solve the region of interest.
6. Transfer the gel to Whatman 3MM paper, cover with plastic wrap, and dry under vacuum (*see* **Note 10**).
7. Expose the dried gel to an imaging plate (e.g., PhosphorImager, Molecular Dynamics).

3.5.3. Analysis

1. Using the imager software (e.g., ImageQuant, Molecular Dynamics), quantify the intensities of each band for the parental- and analog-containing RNAs using area

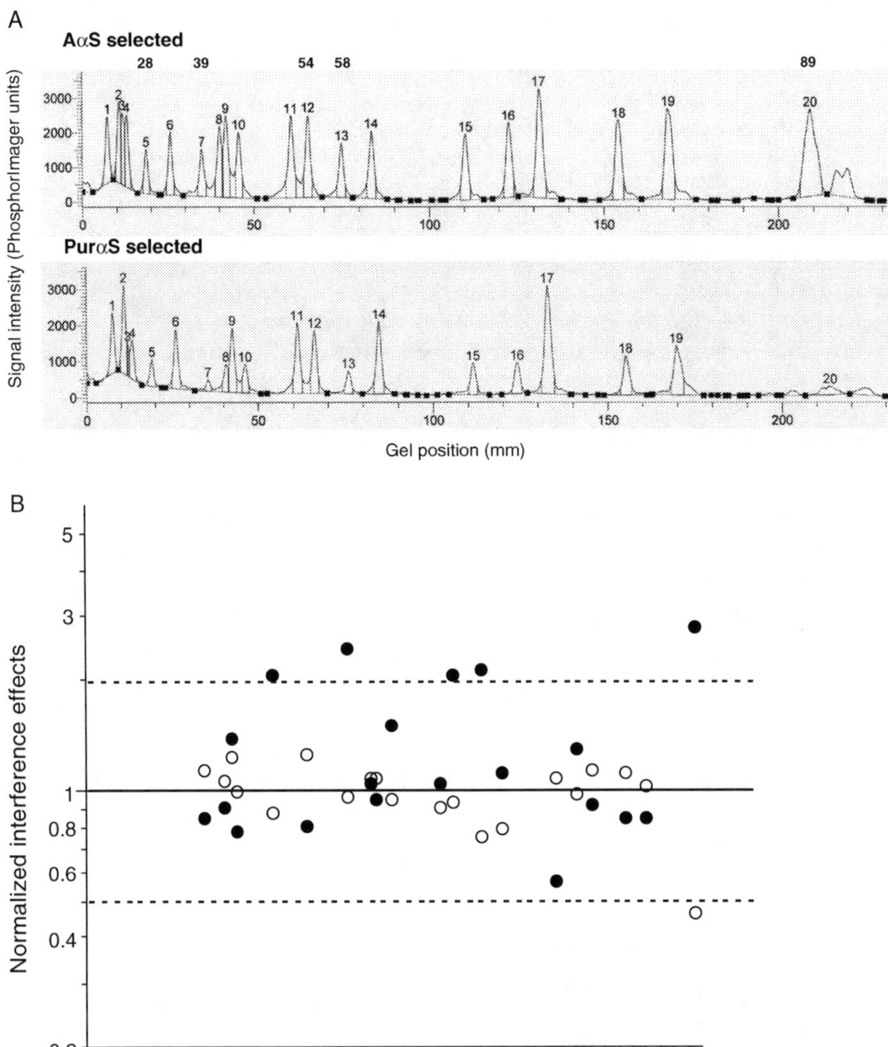

Fig. 3. Analysis of nucleotide analog interference mapping (NAIM) results. (**A**) Area traces for selected composite small nucleolar RNAs (snoRNAs) containing AαS or PurαS. Baseline points (■) were adjusted manually, as were individual peak definitions (vertical lines). The peaks corresponding to A nucleotides are numbered sequentially with the corresponding sequence positions shown in bold (*see* **Note 13**). (**B**) Normalized phosphorothioate and analog interference effects for AαS (○) and PurαS (●) are shown for a composite box C/D snoRNA *(8)*. Values greater than 2 are sites of interference and less than 0.5 are enhancements (dashed lines). These data represent values from a single experiment. A minimum of three independent measurements was averaged to produce final PS and κ values (*see* **Note 14**).

integration (**Fig. 3A**). First, define the baseline so that it is consistent between every lane on the gel. Second, manually adjust the peak definitions to exclude background bands and to delineate peaks that may overlap (*see* **Note 11**).

2. Calculate the phosphorothioate effects (PS) at each position using the equation

$$\text{PS effect} = \frac{\text{N}\alpha\text{S unselected}}{\text{N}\alpha\text{S selected}} \times \text{NF}$$

where NαS is the peak intensity for each position in the parental nucleotide-containing RNA in the unselected and selected lanes, respectively, and NF is a normalization factor adjusting for lane-to-lane differences in loading and the extent of reaction. NF = 1/(Average of all peaks with intensities within 3 standard deviations of the mean).

3. Calculate the interference effect κ for each nucleotide position using the equation

$$K = \frac{(\text{N}\alpha\text{S selected} / \delta\alpha\text{S selected})}{(\text{N}\alpha\text{S unselected} / \delta\alpha\text{S unselected})} \times \text{NF}$$

where selected δαS is the peak intensity for each position in the nucleotide analog-containing RNA in the unselected and selected lanes, respectively.

4. Plot PS and κ values as a function of nucleotide position (**Fig. 3C**; *see* **Note 12**).

4. Notes

1. A number of nucleotide analogs are available commercially (see **ref.** ***6*** for a comprehensive list and for transcription conditions). Incorporation of some analogs, particularly those with modified sugar moieties, requires the use of a mutant polymerase, T7 Y639F *(20)*.
2. The most straightforward blocking agent is a cap structure related to those installed on RNA polymerase II transcripts, (5′)GpppG(5′). However, the biogenesis pathways of many RNAs (e.g., U small nuclear RNAs) involves recognition of the cap structure. If cap recognition and subsequent localization events are not part of the selection, the cap analog (5′)ApppG(5′) can be used. Reducing the final concentration of GTP in the transcription reaction by 10-fold facilitates high levels of cap analog incorporation.
3. Transcription volumes of 50–100 µL produce enough RNA to visualize by UV shadowing in polyacrylamide gels. The amount of RNA recovered after elution and precipitation (5–100 pmol) is sufficient for numerous oocyte injections.
4. A greater number of oocytes survive the injection process when they are injected within a few hours of isolation.
5. Many cellular processes (e.g., snoRNA processing; ***21***) are saturable. The amount of RNA injected should be below the saturation threshold and can be determined empirically by injecting several concentrations of RNA in a pilot experiment.
6. There is no need to incubate the proteinase K reaction. Elevated temperatures and extended times typically result in measurable RNA degradation and thus loss of processed transcript recovery.

7. Many applications of the NAIM method utilize direct addition of I_2/EtOH to the samples. I_2/EtOH is effective over a wide concentration range, but excess EtOH can cause smearing of the radioactive bands, impeding quantitative analysis. Due to the minimal volumes employed with the in vivo samples, the I_2/EtOH solution was further diluted into formamide dyes to reduce the amount of EtOH loaded onto the gel.
8. It is important to run untreated samples of the selected and unselected RNAs on the same gel as the I_2-treated samples. The untreated samples control for I_2-independent degradation. Intense background bands obscure the NAIM signal and render that position uninformative.
9. To achieve single-nucleotide resolution, it is preferable to resolve the RNAs on DNA sequencing gels (0.4-mm thick). However, depending on the size and sequence of the RNA, standard PAGE may suffice.
10. Gels containing more than 8% polyacrylamide will not adhere to the Whatman 3MM paper and so must first be transferred onto plastic wrap and then covered with the paper for drying.
11. Very intense, I_2-independent background bands may prevent analysis of interference values at particular positions. If the intensity of the background band approaches that of the sequence-specific bands, that position should be excluded from all data manipulation. The background band drawn in **Fig. 2C** (band 5) is minor with respect to the intensity of the sequence-specific bands; consequently, the interference value calculated with those peak intensities would still be meaningful. Adenosines in RNA sequences seem to produce a greater number of and more intense background bands than do analogs of the other nucleotides.
12. Typically, κ values greater than 2 are reported as interference effects. κ values less than 0.5 represent enhancements, while $\kappa = 1$ indicates positions unaffected by substitution. The choice of κ greater than 2 identifying sites of interference was made arbitrarily for in vitro studies of the group I ribozyme, which showed abundant sites of analog interference *(5)*. With several repetitions, values as low as 1.5 can be reported confidently. The largest effect observed for in vivo RNP assembly was $3 < \kappa < 4$ *(8)*. In contrast, effects measured in in vitro assays can be much larger ($\kappa > 10$). The differences may arise from several factors, including the ability to manipulate selection conditions in vitro, which is not typically feasible with an in vivo selection, and greater signal-to-noise ratios achieved with in vitro selections. The magnitude of an individual interference value does not directly correlate with the energetic contribution made by a particular functional group in most selection experiments. An advance in the NAIM technique termed quantitative NAIM (QNAIM) was developed to facilitate that comparison *(22)*.
13. The traces shown comprise only half of those needed to calculate interference values. The selected peak areas (band intensities) must be scaled by those for unselected RNAs to eliminate any effects due to variable analog incorporation.
14. The magnitudes of PS and κ values varied more substantially in the in vivo analyses than for an in vitro study using the same RNA *(8)*. Although five sites appeared

to show interference with the purine analog in the data set shown in **Fig. 3**, only three of them (positions 39, 58, and 89) were reproducibly greater than 2.

References

1. Ban, N., Nissen, P., Hansen, J., Moore, P. B., and Steitz, T. A. (2000) The complete atomic structure of the large ribosomal subunit at 2.4Å resolution. *Science* **289**, 905–919.
2. Staley, J. P., and Guthrie, C. (1998) Mechanical devices of the spliceosome: motors, clocks, springs and things. *Cell* **92**, 315–326.
3. Gish, G., and Eckstein, F. (1988) DNA and RNA sequence determination based on phosphorothioate chemistry. *Science* **240**, 1520–1522.
4. Verma, S., and Eckstein, F. (1998) Modified oligonucleotides: synthesis and strategy for users. *Annu. Rev. Biochem.* **67**, 99–134.
5. Strobel, S. A., and Shetty, K. (1997) Defining the chemical groups essential for *Tetrahymena* group I intron function by nucleotide analog interference mapping. *Proc. Natl. Acad. Sci. U. S. A.* **94**, 2903–2908.
6. Cochrane, J. C., and Strobel, S. A. (2004) Probing RNA structure and function by nucleotide analog interference mapping. In *Current Protocols Nucleic Acid Chemistry* (Beaucage, S. L., Bergstrom, D. E., Glick, G. D., and Jones, R. A., eds.), Wiley, Hoboken, NJ, pp. 691–6921.
7. Basu, S., Pazsint, C., and Chowdhury, G. (2004) Analysis of ribozyme structure and function by nucleotide analog interference mapping. In *Ribozymes and siRNA Protocols*, 2nd ed. (Sioud, M., ed.), Humana Press, Totowa, NJ, pp. 57–75.
8. Szewczak, L. B. W., DeGregorio, S. J., Strobel, S. A., and Steitz, J. A. (2002) Exclusive interaction of the 15.5 kD protein with the terminal box C/D motif of a methylation guide snoRNP. *Chem. Biol.* **9**, 1095–1107.
9. Kolev, N., and Steitz, J. (2006) In vivo assembly of functional U7 snRNP requires RNA backbone flexibility within the Sm binding site. *Nat. Struct. Mol. Biol.* **13**, 347–353.
10. Matera, A. G., Terns, R. M., and Terns, M. P. (2007) Non-coding RNAs: lessons from the small nuclear and small nucleolar RNAs. *Nat. Rev. Mol. Cell Biol.* **8**, 209–220.
11. Terns, M. P., and Terns, R. M. (2002) Small nucleolar RNAs: versatile *trans*-acting molecules of ancient evolutionary origin. *Gene Expr.* **10**, 17–39.
12. Weinstein, L. B., and Steitz, J. A. (1999) Guided tours: from precursor snoRNA to functional snoRNP. *Curr. Opin. Cell Biol.* **11**, 378–384.
13. Filipowicz, W., and Pogacic, V. (2002) Biogenesis of small nucleolar ribonucleoproteins. *Curr. Opin. Cell Biol.* **14**, 319–327.
14. Kiss, T. (2001) Small nucleolar RNA-guided post-transcriptional modification of cellular RNAs. *EMBO J.* **20**, 3617–3622.
15. Hirose, T., Shu, M. D., and Steitz, J. A. (2003) Splicing-dependent and -independent modes of assembly for intron-encoded box C/D snoRNPs in mammalian cells. *Mol. Cell* **12**, 113–123.

16. Ryder, S. P., and Strobel, S. A. (1999) Nucleotide analog interference mapping of the hairpin ribozyme: implications for secondary and tertiary structure formation. *J. Mol. Biol.* **291**, 295–311.
17. Romero, M. F., Kanai, Y., Gunshin, H., and Hediger, M. A. (1998) Expression cloning using *Xenopus laevis* oocytes. *Meth. Enzymol.* **296**, 17–52.
18. Hamm, J., Dathan, N. A., and Mattaj, I. W. (1989) Functional analysis of mutant *Xenopus* U2 snRNAs. *EMBO J.* **59**, 159–169.
19. Hausner, T. P., Giglio, L. M., and Weiner, A. M. (1990) Evidence for base-pairing between mammalian U2 and U6 small nuclear ribonucleoprotein particles. *Genes Dev.* **4**, 2146–2156.
20. Sousa, R., and Padilla, R. (1995) A mutant T7 RNA polymerase as a DNA polymerase. *EMBO J.* **14**, 4609–4621.
21. Terns, M. P., Grimm, C., Lund, E., and Dahlberg, J. E. (1995) A common maturation pathway for small nucleolar RNAs. *EMBO J.* **14**, 4860–4871.
22. Cochrane, J. C., Batey, R. T., and Strobel, S. A. (2003) Quantitation of free energy profiles in RNA-ligand interactions by nucleotide analog interference mapping. *RNA* **9**, 1282–1289.

11

T7 RNA Polymerase-Mediated Incorporation of 8-N_3AMP Into RNA for Studying Protein–RNA Interactions

Rajesh K. Gaur

Summary

Ultraviolet (UV)-dependent photochemical crosslinking is a powerful approach that can be used for the identification of RNA–protein interactions. Although 8-azidoATP (8-N_3ATP) has been widely used to elucidate the ATP binding site of a variety of proteins, its inability to serve as an efficient substrate for bacteriophage RNA polymerases apparently restricted its actual potential as a photocrosslinking agent. In this chapter, in vitro transcription conditions that allow for template-dependent incorporation of 8-N_3AMP into RNA are described. In addition, it is shown that a high-affinity MS2 coat protein binding sequence, in which adenosine residues were replaced by 8-azidoadenosine, crosslinks to the coat protein of the *Escherichia coli* phage MS2. This approach can be extended to identify almost any RNA binding protein.

Key Words: In vitro transcription; modified transcript; 8-N_3ATP; RNA–protein interaction; UV crosslinking.

1. Introduction

RNA–protein interactions play a key role in a wide variety of cellular processes, including messenger RNA (mRNA) biogenesis. It is well established that these interactions influence various aspects of cellular and viral gene expression ranging from RNA splicing to mRNA translation. Identification and characterization of RNA–protein interactions is therefore central to understanding the molecular basis of gene expression.

Ultraviolet (UV)-induced crosslinking has proved to be a powerful approach for understanding the structural topography of RNA–RNA and RNA–protein assemblies ***(1,2)***. However, irradiation of biological samples with far-UV light (254 nm) can cause damage to both protein and RNA. Thus, photochemical

From: *Methods in Molecular Biology, vol. 488: RNA-Protein Interaction Protocols*
Edited by: Ren-Jang Lin © Humana Press Inc, Totowa, NJ

agents that are activated with near-UV light (300–360 nm) and can crosslink with efficiencies greater than standard UV-induced crosslinking have drawn considerable attention *(3–7)*. When inserted into RNA and irradiated, such agents not only identify interacting partners by the presence of site-specific crosslinking, but also provide detailed information about the molecular environment of ribonucleoprotein assemblies.

The last two decades have witnessed the development of new photocrosslinking agents as well as the methods for the incorporation of such agents into RNA *(8–12)*. However, DNA template-dependent in vitro transcription in which one of the wild-type nucleoside triphosphates is replaced by a photoactivatable nucleotide analog remains the most popular method for synthesizing RNAs for protein–RNA crosslinking *(13,14)*. Unfortunately, there are very few photoactivatable nucleotide analogs that function as substrates for bacteriophage RNA polymerases, and most of them are pyrimidine based. A pyrimidine-based photocrosslinking nucleotide analog may therefore be of limited use if an adenine base or the purine-rich region of the RNA mediates protein–RNA interaction or if the replacement of a purine base by a pyrimidine-based photocrosslinking nucleotide abolishes the protein–RNA interaction.

A number of adenine-based photoactivatable nucleotides have been described in the literature *(15,16)*. Among them, 8-azidoATP (8-N_3ATP) has been widely used to map the active site of a variety of enzymes and the nucleotide-binding domains of several proteins (*17* and references therein) (**Fig. 1**). Due to the direct attachment of an azide group on the base (at position 8), this nucleotide functions essentially as a zero-length photoaffinity-labeling agent (**Fig. 1**). Moreover, it can be activated with far-UV light (300–360 nm) that is less damaging to proteins and nucleic acids *(8,17)*. Although 8-N_3ATP functions as an elongation substrate for *Escherichia coli* RNA polymerase *(18)*, it appears to act as an inefficient substrate for template-dependent in vitro transcription with T7 RNA polymerase *(19)*. Here, I provide in vitro transcription conditions that allow template-dependent insertion of 8-N_3AMP into RNA. In addition,

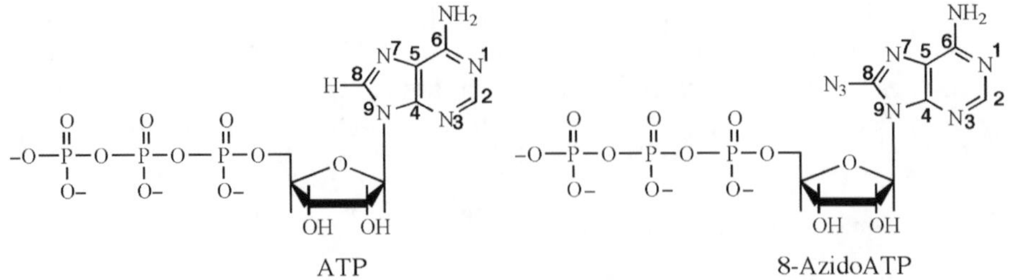

Fig. 1. Structures of adenosine triphosphate (ATP) and 8-N_3ATP (8-azidoATP).

the utility of 8-N_3ATP as a photocrosslinking agent for the identification of RNA–protein interaction is provided.

2. Materials

1. Plasmids containing bacteriophage RNA polymerase promoters are commercially available (e.g., pGEM and pSP series from Promega, Madison, WI).
2. Restriction enzymes can be purchased from New England Biolabs (NEB, Beverly, MA) and should be stored at −20 °C. 10X restriction enzyme buffers are supplied by the manufacturer with the enzyme and should also be stored at −20 °C.
3. T3, T7, and SP6 RNA polymerases (NEB).
4. T7 R&DNA polymerase (Epicenter, Madison, WI).
5. RNasin (Promega, Madison, WI).
6. MMLV reverse transcriptase (Stratagene, La Jolla, CA).
7. Taq polymerase (Roche, Indianapolis, IN).
8. 8-N_3ATP (ALT, Lexington, KY).
9. RNase*Zap* (Ambion, Austin, TX).
10. 10X transcription buffer: 400 m*M* Tris-HCl, pH 8.0, 20 m*M* spermidine.
11. 10X reverse transcriptase (RT) buffer: 500 m*M* Tris-HCl, pH 8.3, 500 m*M* KCl, 60 m*M* $MgCl_2$, 50 m*M* β-mercaptoethanol.
12. 10X polymerase chain reaction (PCR) buffer: 500 m*M* KCl, 100 m*M* Tris-HCl, pH 8.3, 15 m*M* $MgCl_2$.
13. GS buffer: 10 m*M* Tris-HCl, pH 7.5, 50 m*M* KCl, 1 m*M* ethylenediaminetetraacetic acid (EDTA), 3 μg transfer RNA (tRNA).
14. 2X sodium dodecyl sulfate polyacrylamide gel electrophoresis (SDS-PAGE) loading buffer: 100 m*M* Tris-HCl, pH 6.8, 200 m*M* dithiothreitol (DTT), 6% SDS, 0.002% bromophenol blue, 20% glycerol.
15. RNA elution buffer: 0.5 *M* Tris-HCl, pH 7.5, 0.1 m*M* EDTA, 0.1% SDS.
16. TE buffer: 10 m*M* Tris-HCl, pH 8.0, 1 m*M* EDTA.
17. 0.5 *M* Na_2HPO_4 buffer.
18. Sterile water: Diethylpyrocarbonate (DEPC)-treated nuclease-free distilled water (*see* **Note 1**).
19. 100 m*M* solutions of nucleotide triphosphate (NTP) and deoxynucleotide 5′-triphosphate (dNTP; GE Healthcare, Piscataway, NJ).
20. [∝-^{32}P]-UTP (uridine triphosphate) 3000 Ci/mmol (MP Biochemicals, Irvine, CA).
21. Prepare a solution of 0.1% Triton X-100 in sterile water and store at 4 °C.
22. Proteinase K (Sigma-Aldrich, St. Louis, MO): 10 mg/mL; store at −20 °C.
23. Phenol/chloroform/isoamyl alcohol (25:24:1), pH 7.9 (Ambion); store at 4 °C.
24. Chloroform/isoamyl alcohol (24:1).
25. 3 *M* sodium acetate, pH 5.2.
26. 70% ethanol; 100% ethanol.
27. Denaturing polyacrylamide gels (ready-to-use solutions available from National Diagnostic, Atlanta, GA).
28. Formamide loading dye: 80% formamide (v/v), 10 m*M* EDTA, pH 8.0, 1 mg/mL xylene cyanol FF, 1 mg/mL bromophenol blue.
29. Amber microfuge tubes (Sigma-Aldrich).

30. Whatman DE-81 filter disks.
31. UV lamp: Spectroline; model ENF-240C.
32. Phosphorescent markers (RPI, Mount Prospect, IL).
33. Kodak X-OMAT (Rochester, NY).

3. Methods

Before working with RNA, clean the laboratory bench and pipets with a ribonuclease (RNase) decontamination solution (e.g., Ambion's RNase*Zap*). To prevent RNA degradation, all glassware, microfuge tubes, pipet tips, water, and buffers should be autoclaved. In addition, wear laboratory gloves at all times and change them frequently. Buffers and solutions containing nucleoside triphosphates, DTT, spermidine, and phenylmethylsulfonyl fluoride (PMSF) should not be autoclaved. Sterile weighing boats should be used for weighing solid chemicals and reagents. Teflon-coated magnetic stir bars should be soaked overnight in 50% aqueous ethanol containing 20% (w/v) KOH and rinsed thoroughly with sterile water. Glass plates for PAGE should be cleaned with ethanol followed by rinsing with sterile water. Finally, all experiments involving 8-N_3ATP should be performed in dim light using amber microfuge tubes.

3.1. Synthesis, Purification, and Isolation of RNA

3.1.1. Preparation of Template for In Vitro Transcription

1. To generate DNA template for runoff transcription using T7 RNA polymerase, clone the sequence of interest downstream of a T7 promoter in a suitable plasmid.
2. Confirm the authenticity of the insert by sequencing.
3. In a sterile microfuge tube, add 10X restriction enzyme buffer (5 μL), plasmid DNA (5 μg), restriction endonuclease (20–40 U), and sterile water to a volume of 50 μL. Incubate at the recommended temperature for 2–3 h. To confirm the completion of the digestion, load an aliquot of the reaction mixture on a 0.75% agarose gel. Since transcription from the plasmids containing 3′ overhanging ends has been shown to generate poor yield, enzymes that cleave DNA leaving a 3′overhang (for example, Kpn I or Pst I) should be avoided (*see* **Note 2**).
4. To the above reaction mixture, add proteinase K (2–4 μL, 10 mg/mL) and incubate at 37 °C for 1 h.
5. Add 1 volume of phenol/chloroform/isoamyl alcohol, vortex for 15–20 s, and centrifuge for 5 min at 13,000 *g* in a microfuge.
6. Transfer the aqueous phase (top layer) to a new tube and repeat the extraction as in **step 5**.
7. Extract the aqueous phase with chloroform/isoamyl alcohol and add 0.1 volume of 3 *M* sodium acetate, pH 5.2, 2.5 volumes of cold 100% ethanol, followed by incubation on dry ice for 10 min (or 1 h at –20 °C). Collect the DNA by centrifugation at 13,000 *g*; wash the pellet with 70% cold ethanol. Dissolve the dry pellet in TE to a final concentration of 1 μg/μL.

3.1.2. Protocol for Standard Transcription Reaction

The transcription conditions provided below can be used for the synthesis of RNA from a variety of templates, such as linearized plasmids, double-stranded PCR products, or synthetic oligonucleotides in which only the promoter region is double stranded. In addition, the same conditions can be used for the incorporation of a variety of photoactivatable nucleotide analogs such as 4-thioUTP, 5-bromo-UTP, and 5-iodo-CTP.

1. For a 50-μL transcription reaction, mix the following components in a sterile microfuge tube (*see* **Notes 3–7**): 10X transcription buffer (5 μL), 0.1 *M* $MgCl_2$ (10 μL), linearized plasmid (2 μL; 0.5 μg/μL), 100 m*M* DTT (2.5 μL), 4 m*M* NTPs (mixture of ATP, CTP, GTP, and UTP) (5 μL), 40 U/μL RNasin (1 μL), 0.1% Triton X-100 (5 μL), 10 μCi/μL [α-^{32}P]-UTP (0.5 μL), 20 U/μL T7 RNA polymerase (2 μL), and sterile water (17 μL).
2. Mix well by pipeting the reaction mixture up and down several times.
3. Incubate for 2 h at 37 °C.
4. Add RNase-free deoxyribonuclease (DNase) I (1 μL) and incubate for 15 min at 37 °C.
5. Extract with 1 volume of phenol/chloroform/isoamyl alcohol followed by extraction with chloroform/isoamyl alcohol.
6. To the aqueous phase, add 0.1 volume of 3 *M* sodium acetate, pH 5.2.
7. Add 2.5 volumes of 100% cold ethanol, vortex briefly, and incubate for 10 min on dry ice or in a –20 °C freezer (1 h) (*see* **Note 8**).
8. Centrifuge for 10–15 min at 13,000 *g*. Remove supernatant.
9. Wash the pellet with ice-cold 100–200 μL 70% ethanol and centrifuge for 10 min, remove the supernatant carefully, and dry the pellet under vacuum.
10. Dissolve the pellet in 5–10 μL of formamide loading dye, heat at 65 °C for 5 min, and immediately place in the ice. Spin down briefly.
11. Run the RNA sample on a denaturing polyacrylamide gel (0.5-m*M* thickness). Use 20% gels for transcripts up to 50 nt, 8% for up 200 nt, and 4% for longer transcripts (*see* **Note 9**).
12. When done, remove the top plate and spacer, cover the gel with plastic wrap, apply radioactive or phosphorescent markers, and expose the gel in dark to Kodak X-OMAT or similar film for 5–30 min as needed to visualize the bands.
13. Using markers as guide, align film over the gel and identify the desired band. Excise the gel slice with a sterile scalpel and transfer the gel slice to a sterile microfuge tube. Add 200 μL RNA elution buffer and leave 4–5 h or overnight for elution at room temperature.
14. After brief centrifugation, transfer the eluted RNA into a fresh microfuge tube, extract the gel slice with 100 μL of fresh elution buffer and combine.
15. Follow **steps 6–9** for the isolation of RNA.
16. RNA can be dissolved in sterile water or TE and stored at –20 °C.

3.1.3. Quantification of RNA

Assuming equal representation of all four nucleotides, the following protocol can be used to estimate the yield of the transcript *(20)*.

1. Add 2 μL of transcription mixture (after **step 3** at the end of transcription in **Subheading 3.1.2.**) to a fresh microfuge tube containing water (18 μL).
2. Withdraw 8-μL aliquots and spot onto two Whatman DE-81 filter disks.
3. After brief air drying, place one filter in a scintillation vial and count (count A).
4. Incubate the second filter for 3–5 min with 25–30 mL of 0.5 *M* Na_2HPO_4 buffer (nontitrated) with gentle swirling. Remove the buffer.
5. Repeat wash two to three times followed by two brief washings with water.
6. Rinse the filter with ethanol, air dry, and count (count B). Yield (μg) = 1.24 × (Count B/Count A) × (Amount of cold UTP [nmol] in the reaction)

3.1.4. Protocol for the Synthesis of RNA Containing 8-N_3AMP

Under standard reaction conditions in the presence of Mg^{2+}, 8-N_3ATP does not serve as an efficient substrate for T7 RNA polymerase (**Fig. 2**). In addition, replacement of T7 by T3, SP6, or a mutant T7 polymerase (T7 R&DNA polymerase; *see* **Note 10**) failed to improve the substrate properties of 8-N_3ATP (**Fig. 2**). However, replacement of Mg^{2+} by 2.5 m*M* Mn^{2+} enables T7 RNA polymerase to catalyze the polymerization of 8-N_3ATP into RNA (**Fig. 3**). Significantly, transcription reactions containing a mixture of 2.5 m*M* Mn^{2+}/2.5 m*M* Mg^{2+} (*see* **Note 11**) further improve the yield of 8-N_3AMP-containing RNA (**Fig. 3**). To synthesize RNAs containing 8-N_3AMP, following modifications should be made in the protocol in **Subheading 3.1.2.**:

1. Replace the standard NTP mixture by 4 m*M* NTPs-A (4 m*M* of CTP, GTP, and UTP) (5 μL).
2. Substitute 4 m*M* 8-N_3ATP (5 μL) for ATP.
3. Replace DTT with 50 m*M* β-mercaptoethanol (5 μL).
4. Replace 0.1 *M* $MgCl_2$ (10 μL) by 0.1 *M* $MnCl_2$ (1.25 μL) and add water to bring the final volume to 50 μL.
5. Follow **steps 2–16** from **Subheading 3.1.2.**

3.1.5. Synthesis of High-Specific-Activity RNA

For the identification of RNA–protein crosslinking, it is desirable to synthesize RNAs with high specific activity. To achieve this objective, the following changes should be made to the transcription reaction. These changes can be applied for the synthesis of normal RNAs as well as for the RNAs transcribed in the presence of 8-N_3ATP.

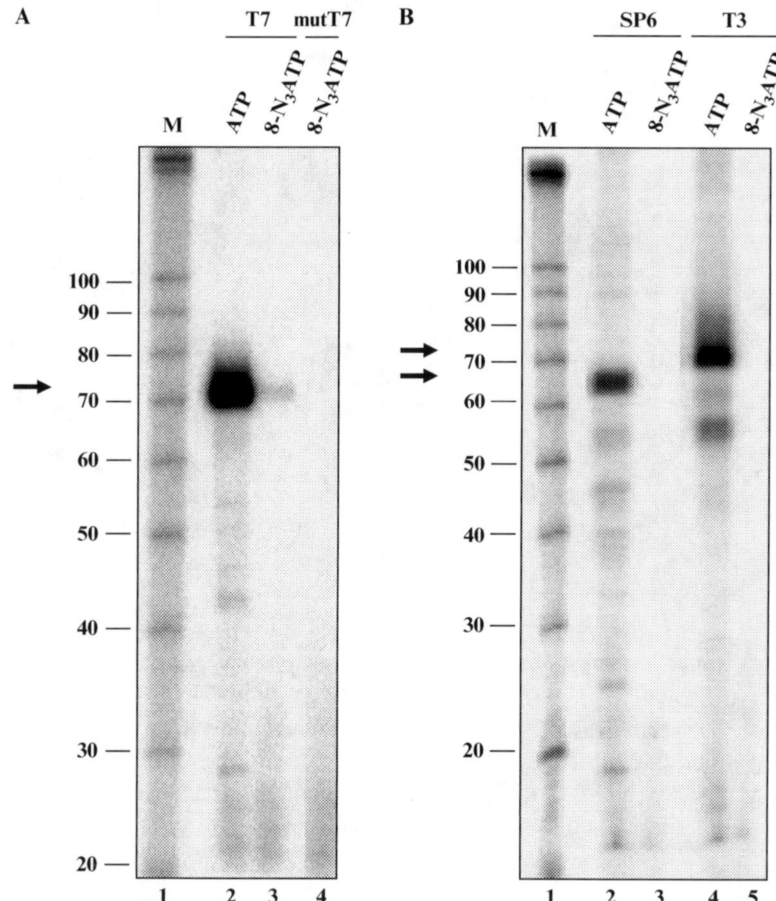

Fig. 2. 8-AzidoATP (8-N_3ATP) does not function as a substrate for SP6, T3, T7, or mutant T7 (T7 R&DNA) polymerase under standard in vitro transcription conditions. (**A**) RNAs synthesized from a *BamH I*-digested plasmid pPIP85.B *(27)*, which directs the synthesis of 68-mer RNA, with indicated polymerase were resolved on a 10% polyacrylamide denaturing gel. M, RNA size marker (Decade Marker™, Ambion). (**B**) As in panel **A**. The *Sal I*-linearized plasmid pAdML *(28)* was used as template for SP6 and T3 transcriptions. The expected products of SP6 (62-mer) and T3 (72-mer) catalyzed transcription reactions are indicated.

1. Replace the standard NTP mixture by NTPs-U (4 m*M* CTP, GTP, ATP, or 8-N_3ATP) (5 µL) and 1 m*M* UTP (5 µL).
2. Use approx 50 µCi of 3000 mCi/mmol [∝-^{32}P]-UTP.
3. Add water to bring the final volume to 50 µL.
4. Follow **steps 2–16** from **Subheading 3.1.2**.

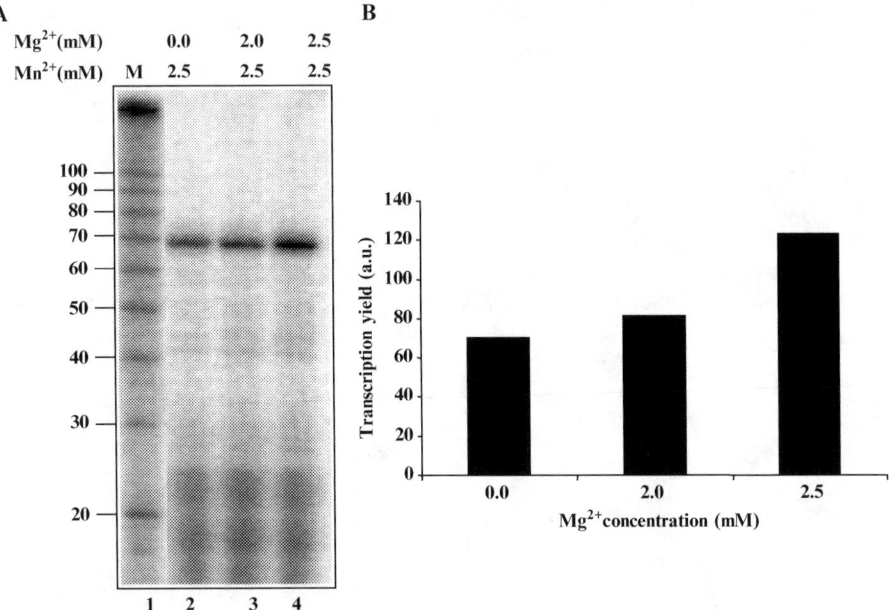

Fig. 3. Synthesis of 8-azidoATP (8-N_3AMP) containing RNA. (**A**) In the presence of 2.5 m*M* Mn^{2+}, 8-N_3ATP serves as a substrate for T7 RNA polymerase (lane 2), and addition of Mg^{2+} improves the yield of 8-N_3AMP-containing RNA (lanes 3–4). The *BamH I*-digested plasmid (pPIP85.B) was used as a template. M, RNA size marker (Decade Marker, Ambion). (**B**) Histogram representing the yield of RNA (from panel **A**) as a function of Mg^{2+} concentration. a.u., arbitrary unit.

3.2. Confirming the Authenticity of 8-N_3AMP-Containing RNA

Transcripts produced by T7 RNA polymerase often contain both 5′ and 3′ heterogeneity *(13,21,22)*. In addition, substitution of Mn^{2+} for Mg^{2+} has been reported to compromise the fidelity of DNA as well as RNA polymerases *(23–25)*. Thus, it is important that RNAs transcribed in the presence of Mn^{2+} be analyzed for their authenticity. Reverse transcription followed by PCR can be used to determine whether T7 RNA polymerase-mediated incorporation of 8-N_3AMP did occur in a template-dependent manner. The results shown in **Fig. 4** demonstrate that under modified transcription conditions (in the presence of 2.5 m*M* Mn^{2+}/2.5 m*M* Mg^{2+}) T7 RNA polymerase can catalyze the incorporation of 8-N_3AMP in a template-dependent manner. Moreover, RNAs containing 8-N_3AMP can be faithfully reverse transcribed.

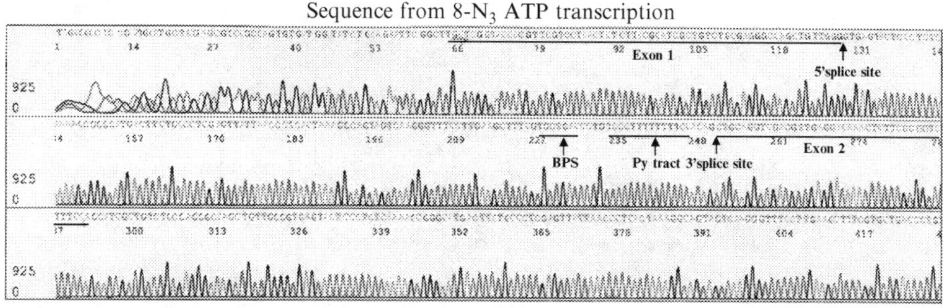

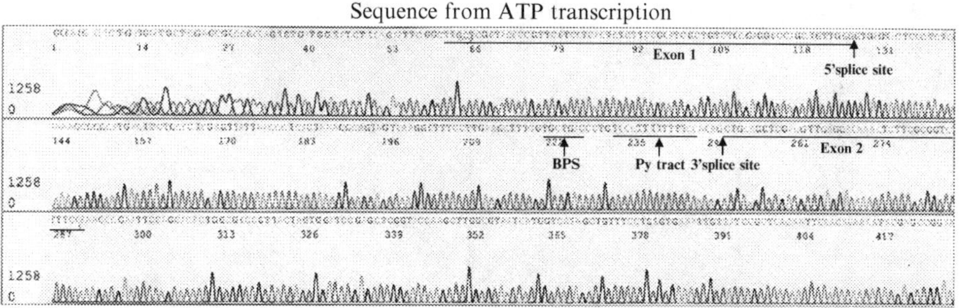

Fig. 4. Sequencing results from adenosine triphosphate (ATP) and 8-azidoATP (8-N$_3$ATP) transcription reactions. RNAs were transcribed in the presence of ATP or 8-N$_3$ATP using *Bam HI*-digested plasmid pRG1 ***(29)*** as the template. Gel-purified RNA samples containing adenosine monophosphate (AMP) or 8-azidoAMP (8-N$_3$AMP) were reverse transcribed and polymerase chain reaction (PCR) amplified. The PCR product was subjected to DNA sequencing. Underlining highlights functionally important regions of pre-mRNA.

3.2.1. RT-PCR and Sequencing

1. In a 10-μL reaction, mix approx 5 pmol of RNA and 50 pmol reverse primer.
2. Incubate for 5 min at 65 °C and chill on ice.
3. Add 10 μL of 5X RT buffer, 1 μL of RNasin, 2.5 μL of 10 m*M* dNTPs, 2.5 μL of MMLV reverse transcriptase, and water to 50 μL (*see* **Note 12**).
4. Incubate at 42 °C for 1 h.
5. For PCR amplification, add 20 μL of the complementary DNA product (from **step 4**) to 80 μL of PCR mixture containing 50 pmol forward and reverse primers, 2.5 μL of 10 m*M* dNTPs, 10 μL of 10X PCR buffer, and 0.5 U Taq polymerase
6. Set the desired cycle conditions of denaturing, annealing, and extension, taking into account the theoretical melting temperature of the primers. In general, a total of 25 cycles is sufficient, but it may be desirable to optimize the number of cycles

A

MS2 protein binding RNA

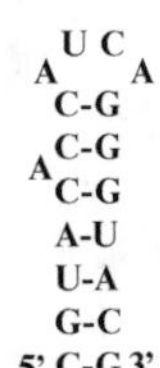

B

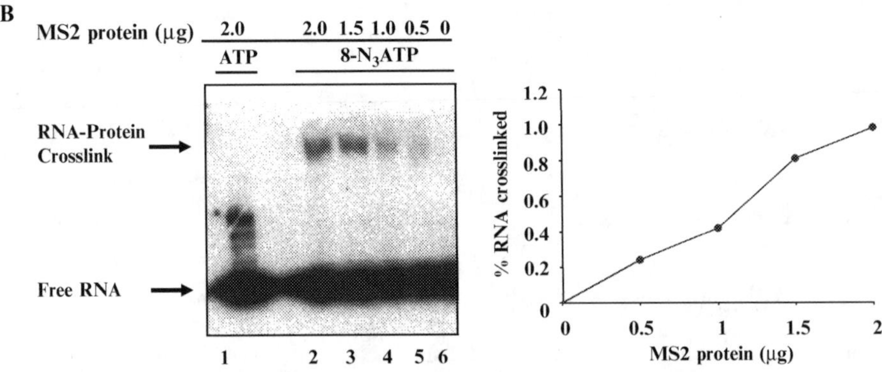

Fig. 5. MS2 protein–RNA crosslinking. **(A)** Sequence and proposed secondary structure of hairpin RNA known to bind MS2 protein. **(B)** Analysis of MS2 protein–RNA crosslinking. [α-^{32}P]UTP-labeled RNA containing adenosine monophosphate (AMP) (lane 1) or 8-azidoAMP (8-N_3AMP) (lanes 2–6) was incubated with (lane 1, 2–5) or without MS2 protein (lane 6) and the mixture was irradiated for 15 min with long-wavelength ultraviolet (UV) light. After ribonuclease (RNase) A digestion, the crosslinked RNA was separated from free probe on a 12.5% sodium dodecyl sulfate polyacrylamide gel.

necessary by analyzing the relative intensities of ethidium bromide staining for each amplification product. The cycle number that produces product within the exponential range should be selected.

7. To the PCR mix, add 1 volume of phenol/chloroform/isoamyl alcohol, vortex for 20 s, and centrifuge for 1 min to separate the phases.
8. Transfer the top layer to a clean tube and repeat the extraction with 1 volume of chloroform/isoamyl alcohol.
9. To the aqueous phase containing the DNA, add 0.1 volume of 3 *M* sodium acetate, pH 5.2, and vortex briefly.
10. Add 2.5 volume of 100% ethanol, vortex, and incubate on dry ice for 10 min.
11. Centrifuge at 13,000 *g* in a microfuge for 10–15 min. Discard the supernatant and wash the pellet with 70% ethanol (500 µL).

12. Centrifuge for 10 min, remove the supernatant, and dry the pellet under vacuum.
13. Dissolve the pellet in nuclease-free water or TE.
14. Use appropriate primer to determine the sequence of the amplified PCR product by using an ABI DNA sequencer.

3.3. Crosslinking of 8-N_3AMP-Containing RNA to MS2 Protein

While performing a UV crosslinking assay, be sure to wear UV safety goggles. A typical protein–RNA crosslinking assay is performed in a volume of 20 µL (*see* **Note 13**). An example of the use of 8-N_3AMP-containing RNA for photocrosslinking is shown in **Fig. 5**.

1. In a 20 µL reaction, incubate approx 10^5 to 10^6 cpm of ^{32}P-labeled RNA (10–20 fmol) and different amounts of MS2 protein (*see* **Note 14**) in GS buffer on ice for 30 min (*see* **Note 15**).
2. Transfer the reaction mixture to individual wells of a microtiter plate on ice. Alternatively, the reaction mixture can be placed as a drop on a Parafilm-covered metal plate on ice.
3. Place the handheld UV lamp directly on top of the microtiter plate and irradiate for 15–20 min with long-wavelength UV light (Spectroline model ENF-240C).
4. Transfer the reaction mixture to the original microfuge tubes.
5. Add RNases A (6 µg) and continue the incubation at 37 °C for 20 min.
6. Add SDS loading buffer, boil 3–5 min, and load samples directly onto an SDS polyacrylamide gel.
7. Dry the gel and analyze the efficiency of crosslinking by PhosphorImager scanning.

4. Notes

1. Treat double-distilled water with DEPC (1 mL/L water) overnight at 37 °C with stirring, followed by autoclaving. Since DEPC is a suspected carcinogen, wear gloves at all times while working with this chemical.
2. If a suitable restriction site is not available, the 3′ overhang can be made blunt by treatment with Klenow fragment (5–10 U/µg DNA) in the presence of 0.25 m*M* dNTPs and incubating for 20 min at 37 °C. After Klenow treatment, isolate DNA by following **Subheading 3.1.1., steps 5–7**. Alternatively, a synthetic oligonucleotide containing the T7 promoter can be used *(13)*.
3. Although the same protocol can be used for three different types of templates, due to the size difference between the plasmid and an oligonucleotide-based template the amount of the template must be adjusted. For example, 1 pmol of a 3-kb plasmid DNA corresponds to 2 µg, whereas the same molar amount of 100-bp template will be equal to 66.66 ng.
4. The yield of the transcription reaction from a plasmid template is normally higher than a synthetic template. However, this problem can be overcome by increasing the amount of the synthetic template. It is recommended to determine the optimal concentration of the template.

5. Standard transcription reactions can be performed with low specific activity [∝-^{32}P]-UTP (800 Ci/mmol). For the synthesis of high-specific-activity RNAs, [∝-^{32}P]-UTP (3000 Ci/mmol) can be used.
6. The presence of spermidine in the transcription buffer may precipitate DNA; thus, transcription reactions should be assembled at room temperature.
7. It is recommended that the Mg^{2+} concentration should exceed the total nucleotide concentration by approximately 4 m*M*. Since the concentration of Mg^{2+} in this protocol is already 20 m*M*, it is not necessary to add additional Mg^{2+} even if a large-scale transcription reaction is assembled (e.g., each NTP 4 m*M*).
8. To minimize the loss of RNA, 5–10 μg glycogen can be added as a carrier.
9. The presence of tracking dyes in the loading buffer also serves as a reference for RNA mobility. For example, on 20%, 8%, and 4% denaturing polyacrylamide gels, RNAs 7, 20, and 75 nt long, respectively, comigrate with bromophenol blue. Similarly, xylene cyanol comigrates with 27, 70, and 200 nt long RNAs.
10. T7 R&DNA polymerase is a recombinant enzyme that is encoded by a DNA sequence having a single-base active-site mutation in the T7 polymerase gene. This allows the enzyme to incorporate 2′-deoxyribonucleoside triphosphates (dNTPs) into full-length transcripts without compromising the catalytic activity for incorporation of canonical NTPs.
11 A mixture of 2.5 m*M* Mn^{2+} and 2.5 m*M* Mg^{2+} has been found to be optimal. However, results may vary depending on the template. Therefore, it is recommended to optimize the concentration of metal ions and template.
12. To monitor the efficiency of the RT reaction, a small amount of [∝-^{32}P]-dNTP can be added to the reaction. The incorporation of radioactive nucleotide can be analyzed by gel electrophoresis and subsequent exposure to X-ray film.
13. Because transcription randomly incorporates 8-N_3AMP, precise identification of the site of protein–RNA contact may not be possible. However, T4 DNA ligase-based RNA ligation can be used to assemble an RNA containing 8-N_3AMP at a specific site ***(26)***.
14. The expression and isolation of MS2 protein was performed according to the published protocol ***(19)***.
15. DTT can reduce the azide group; thus, 8-N_3ATP or RNA containing this analog should not be exposed to DTT.

Acknowledgments

I would like to thank members of my laboratory. This work was supported in part by a grant from the Department of Defense (DAMD17-03-1-0625) and Beckman Research Institute excellence award.

References

1. Shapkina, T. G., Dolan, M. A., Babin, P., and Wollenzien, P. (2000) Initiation factor 3-induced structural changes in the 30 S ribosomal subunit and in complexes containing tRNA(f)(Met) and mRNA. *J. Mol. Biol.* **299**, 615–628.

2. Wu, S., and Green, M. R. (1997) Identification of a human protein that recognizes the 3′ splice site during the second step of pre-mRNA splicing. *EMBO J.* **16**, 4421–4432.
3. Meisenheimer, K. M., Meisenheimer, P. L., Willis, M. C., and Koch, T. H. (1996) High yield photocrosslinking of a 5-iodocytidine (IC) substituted RNA to its associated protein. *Nucleic Acids Res.* **24**, 981–982.
4. Willis, M. C., Hicke, B. J., Uhlenbeck, O. C., Cech, T. R., and Koch, T. H. (1993) Photocrosslinking of 5-iodouracil-substituted RNA and DNA to proteins. *Science* **262**, 1255–1257.
5. Wang, Z., and Rana, T. M. (1998) RNA–protein interactions in the Tat-trans-activation response element complex determined by site-specific photo-cross-linking. *Biochemistry* **37**, 4235–4243.
6. Favre, A. (1990) In Morrison, H. (ed.), *Bioorganic Photochemistry*. J. Wiley & Sons, Inc., New York, Vol. 1.
7. Costas, C., Yuriev, E., Meyer, K. L., Guion, T. S., and Hanna, M. M. (2000) RNA–protein crosslinking to AMP residues at internal positions in RNA with a new photocrosslinking ATP analog. *Nucleic Acids Res.* **28**, 1849–1858.
8. Sylvers, L. A., and Wower, J. (1993) Nucleic acid-incorporated azidonucleotides: probes for studying the interaction of RNA or DNA with proteins and other nucleic acids. *Bioconjug. Chem.* **4**, 411–418.
9. Hanna, M. M. (1989) Photoaffinity cross-linking methods for studying RNA–protein interactions. *Methods Enzymol.* **180**, 383–409.
10. Yu, Y. T. (1999) Construction of 4-thiouridine site-specifically substituted RNAs for cross-linking studies. *Methods* **18**, 13–21.
11. Gaur, R. K., and Krupp, G. (1997) Chemical and enzymatic approaches to construct modified RNAs. *Methods Mol. Biol.* **74**, 99–110.
12. Mundus, D., and Wollenzien, P. (2000) Structure determination by directed photo-cross-linking in large RNA molecules with site-specific psoralen. *Methods Enzymol.* **318**, 104–118.
13. Milligan, J. F., Groebe, D. R., Witherell, G. W., and Uhlenbeck, O. C. (1987) Oligoribonucleotide synthesis using T7 RNA polymerase and synthetic DNA templates. *Nucleic Acids Res.* **15**, 8783–8798.
14. Milligan, J. F., and Uhlenbeck, O. C. (1989) Synthesis of small RNAs using T7 RNA polymerase. *Methods Enzymol.* **180**, 51–62.
15. Parang, K., Kohn, J. A., Saldanha, S. A., and Cole, P. A. (2002) Development of photo-crosslinking reagents for protein kinase–substrate interactions. *FEBS Lett.* **520**, 156–160.
16. MacMillan, A. M., Query, C. C., Allerson, C. R., Chen, S., Verdine, G. L., and Sharp, P. A. (1994) Dynamic association of proteins with the pre-mRNA branch region. *Genes Dev.* **8**, 3008–3020.
17. Potter, R. L., and Haley, B. E. (1983) Photoaffinity labeling of nucleotide binding sites with 8-azidopurine analogs: techniques and applications. *Methods Enzymol.* **91**, 613–633.

18. Bowser, C. A., and Hanna, M. M. (1991) Sigma subunit of *Escherichia coli* RNA polymerase loses contacts with the 3′ end of the nascent RNA after synthesis of a tetranucleotide. *J. Mol. Biol.* **220**, 227–239.
19. Gopalakrishna, S., Gusti, V., Nair, S., Sahar, S., and Gaur, R. K. (2004) Template-dependent incorporation of 8-N3AMP into RNA with bacteriophage T7 RNA polymerase. *RNA* **10**, 1820–1830.
20. Kjems, J., Egebjerg, J., and Christiansen, J. (1998) *Analysis of RNA–Protein Complexes In Vitro*. Elsevier, Utrecht, The Netherlands, Vol. 26, pp. 13–53.
21. Krupp, G. (1989) Unusual promoter-independent transcription reactions with bacteriophage RNA polymerases. *Nucleic Acids Res.* **17**, 3023–3036.
22. Pleiss, J. A., Derrick, M. L., and Uhlenbeck, O. C. (1998) T7 RNA polymerase produces 5′ end heterogeneity during in vitro transcription from certain templates. *RNA* **4**, 1313–1317.
23. el-Deiry, W. S., So, A. G., and Downey, K. M. (1988) Mechanisms of error discrimination by *Escherichia coli* DNA polymerase I. *Biochemistry* **27**, 546–553.
24. Conrad, F., Hanne, A., Gaur, R. K., and Krupp, G. (1995) Enzymatic synthesis of 2′-modified nucleic acids: identification of important phosphate and ribose moieties in RNase P substrates. *Nucleic Acids Res.* **23**, 1845–1853.
25. Huang, Y., Beaudry, A., McSwiggen, J., and Sousa, R. (1997) Determinants of ribose specificity in RNA polymerization: effects of Mn^{2+} and deoxynucleoside monophosphate incorporation into transcripts. *Biochemistry* **36**, 13718–13728.
26. Moore, M. J., and Sharp, P. A. (1992) Site-specific modification of pre-mRNA: the 2′-hydroxyl groups at the splice sites. *Science* **256**, 992–997.
27. Query, C. C., Strobel, S. A., and Sharp, P. A. (1996) Three recognition events at the branch-site adenine. *EMBO J.* **15**, 1392–1402.
28. Gozani, O., Patton, J. G., and Reed, R. (1994) A novel set of spliceosome-associated proteins and the essential splicing factor PSF bind stably to pre-mRNA prior to catalytic step II of the splicing reaction. *EMBO J.* **13**, 3356–3367.
29. Gaur, R. K., McLaughlin, L. W., and Green, M. R. (1997) Functional group substitutions of the branchpoint adenosine in a nuclear pre-mRNA and a group II intron. *RNA* **3**, 861–869.

12

A Simple Crosslinking Method, CLAMP, to Map the Sites of RNA-Contacting Domains Within a Protein

Hiren Banerjee and Ravinder Singh

Summary

A large number of proteins contain multiple RNA recognition motifs (RRMs). How multiple RRMs contribute to RNA recognition in solution is, however, poorly understood. Here, we describe a simple biochemical approach called CLAMP (crosslinking and mapping of protein domain) to identify an RRM that is crosslinked to a specific nucleotide in RNA. It involves site-specific incorporation of a chromophore, photochemical RNA–protein crosslinking, and site-specific chemical cleavage of the protein. This technique is suitable for numerous other RNA binding proteins that have multiple RNA binding domains.

Key Words: *N*-Chlorosucccinimide; crosslinking; 5-iodouracil; polypyrimidine tract; Py tract; RNA binding domain; RNA binding proteins; RNA recognition motif; RRM; sex determination; splicing.

1. Introduction

The RNA binding proteins perform many important biological functions by modulating one or more posttranscriptional RNA-processing events (e.g., splicing, polyadenylation, RNA stability, and translation). The *Drosophila melanogaster* protein Sex-lethal (SXL) is one such RNA binding protein that serves to control sex determination ***(1,2)***. It is synthesized in females but not in males and acts as a master sex switch. SXL controls the splicing of *Sxl, transformer (tra)*, and *male-specific lethal-2 (msl2)* pre-mRNAs (messenger RNAs), and the translation of the *msl2* mRNA by binding to uridine-rich sequences or the polypyrimidine tracts (Py tracts) ***(3–6)***. The Py tract adjacent to the 3′ splice sites is an important splicing signal in mammals ***(7,8)***.

SXL contains two RNA recognition motifs (RRMs). The RRM family is the largest family of RNA binding proteins found in all three domains of life

From: *Methods in Molecular Biology, vol. 488: RNA-Protein Interaction Protocols*
Edited by: Ren-Jang Lin © Humana Press Inc, Totowa, NJ

(bacteria, archaea, and eukarya) ***(9,10)***. An RRM motif is made up of 80–90 amino acids and folds into a four-stranded antiparallel β-sheet and two α-helices. Typically, RNA binds to the β-sheet surface of an RRM.

The large majority of RRM proteins contain two to four RRMs. Only a handful of RRM proteins have been characterized for RNA recognition at the structural level. Nonetheless, how multiple RRMs contribute to binding affinity and specificity for cognate RNAs remains poorly understood.

We provide the rationale, salient features, and a detailed biochemical protocol called CLAMP (cross linking and mapping of protein domain) to define RNA recognition by multiple RRMs in solution. As a model for these studies, we have used the SXL protein and its binding site, the non-sex-specific Py tract of *tra* ***(11)***. This approach also complements other methods that are used for studying RNA–protein interactions.

1.1. Rationale and Salient Features

There were two main reasons for developing this biochemical method for the analysis of RNA-protein interactions. First, it is known that SXL, like most Py tract-binding proteins, efficiently but indiscriminantly crosslinks to one or more uridines within the Py tract on exposure to shortwave ultraviolet (254-nm UV) light ***(12–14)***. However, it is difficult to determine which residue in the RNA is crosslinked to the protein. Thus, for a domain-mapping strategy, it is necessary that only a specific RNA residue is crosslinked. Therefore, we chemically synthesized RNAs containing 5-iodouracils (5-IUs) at specific positions ***(15)***. An important advantage of using the 5-IU chromophore is that it can be crosslinked to specific amino acids (tyrosine, phenylalanine, histidine, and methionine) at 325-nm UV light, the wavelength at which other protein or RNA chromophores are not detectably activated, which avoids the risk of photodamage and indiscriminant crosslinking.

Second, once an RNA residue has been crosslinked to protein, it is desirable that the relevant protein domain can be easily identified. Although chemical (e.g., cynogen bromide) and enzymatic (e.g., trypsin) cleavage reagents can provide a protein fingerprint, they cleave proteins frequently and thus typically generate a complex peptide pattern that is usually hard to decipher. Thus, it was necessary to have a simple biochemical approach for identifying the protein domain to which an RNA molecule is crosslinked. It was known that *N*-chlorosuccinimide (NCS) specifically cleaved polypeptides at tryptophan residues ***(16)***, which occurs infrequently in most proteins. SXL lacks a tryptophan.

Thus, we reasoned that by introducing a single tryptophan in SXL, particularly in the linker region between RRM1 and RRM2, we could avoid the above problems, allowing straightforward biochemical identification of relevant RRMs. We chose tyrosine at position 200 (Y200) for site-directed mutagenesis for several reasons ***(17)***: First, based on X-ray studies ***(18,19)***, it is appropriately located in the linker region between RRM1 and RRM2 (**Fig. 1A**). Second, its substitution to tryptophan (W in the figure), another aromatic residue, was expected to have little or no effect on protein function. In fact, our computer modeling suggested that a Y200-to-W substitution would not perturb the structure of SXL (**Fig. 1B**). Finally, it is located away from the RNA binding surface of SXL (**Fig. 1A**) ***(18)***. We have confirmed experimentally that the RNA binding affinity of the Y200W substitution mutant protein was indistinguishable from that of the wild-type protein (**Fig. 2**); the equilibrium dissociation constants K_{ds} for both proteins were approx 3–10 n*M*, consistent with previous studies ***(13,14)***.

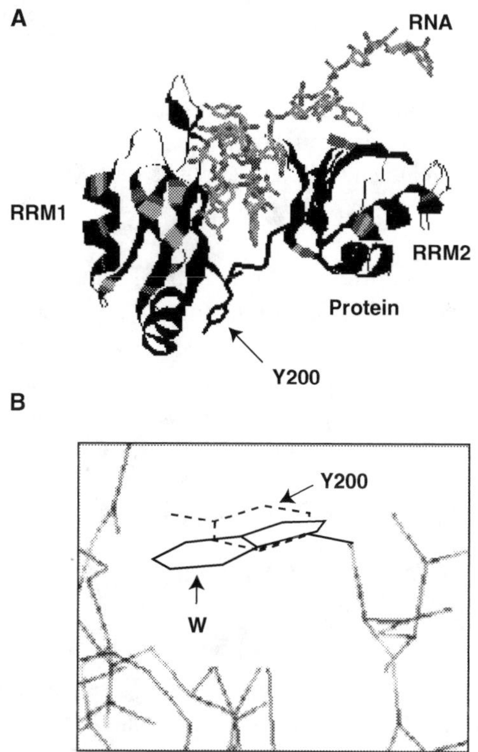

Fig. 1. (**A**) X-ray structure of RNA (non-sex-specific polypyrimidine [Py] tract of *tra*) bound to the RNA binding domain (RNA recognition motif 1 [RRM1] and RRM2) of Sex-lethal (SXL), redrawn with RasMol using coordinates from **ref.** ***18***. An arrow indicates the position of tyrosine 200 (Y200) in the linker region between RRM1 and RRM2. (**B**) Computer modeling shows one of the rotamers of tyrosine (Y200) and tryptophan (W) at position 200.

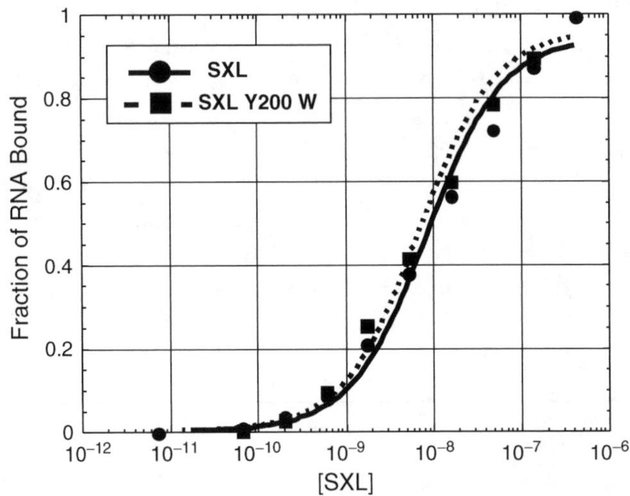

Fig. 2. A nitrocellulose filter-binding assay for the binding of Sex-lethal (SXL) (wild type and Y200W) to the non-sex-specific polypyrimidine [Py] tract of *tra* ***(25)***. Fraction of RNA bound (*y*-axis) and the molar concentrations for SXL (*x*-axis) are shown.

The peptide fragments corresponding to RRM1 and RRM2 are of different sizes (200 vs 154 amino acids) and thus could be easily resolved by electrophoresis on an sodium dodecyl sulfate (SDS) polyacrylamide gel. To confirm the identities of these RRMs, we also generated two deletions outside of the RNA binding domain: deletion of either the carboxyl terminus (SXL WΔC) or both the amino and carboxyl termini (SXL W ΔNΔC). These deletions would reduce in a predicted manner the sizes of relevant RRM fragments to confirm the identities of RRM1 and RRM2 (*see* **Note 1**).

1.2. Potential Site of Photocrosslinking on SXL

As noted, the chemistry of 5-IU crosslinking most likely involves photoelectron transfer mechanism for the crosslinking of π-stacked chromophores (for additional details, see Fig. 2 in **ref.** ***20***). The X-ray structure of SXL bound to the non-sex-specific Py tract shows several pairs of aromatic amino acids properly positioned next to uridines for 5-IU crosslinking ***(18)***. One of them is shown here for illustrative purposes (**Fig. 3**).

1.3. Identification of an RRM Crosslinked to a Specific 5-IU

Appropriate RNAs and proteins were incubated and crosslinked. The crosslinked protein–RNA complexes were cleaved by NCS ***(16)***, and the peptide fragments were resolved by SDS polyacrylamide gel electrophoresis

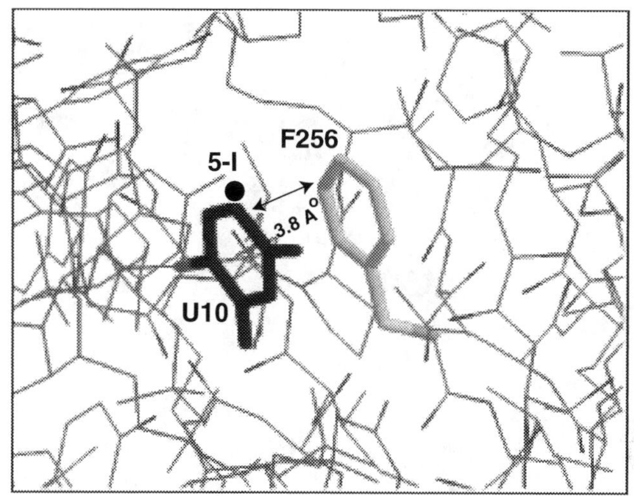

Fig. 3. Arrangement of U10 in RNA (dark) and Phe256 (F256) (light) in Sex-lethal (SXL). The sphere represents the iodine at position 5 of U10. The distance between U10 and F256 is approx 3.8 Å.

(*see* **Subheading 3.3.**). For reference, the position of the tryptophan and the expected sizes of NCS cleavage fragments are schematically shown in **Fig. 4A**. In the absence of 5-IU, no radiolabeled band was observed, indicating that generation of a radiolabeled peptide fragment required 5-IU and did not result from indiscriminate crosslinking typically associated with the exposure of most RRM proteins, in particular the Py tract-binding proteins, to short-wavelength UV light (lanes 1–3). In the absence of NCS cleavage, only full-length proteins of expected sizes were present, indicating that there was no photodamage to SXL under these crosslinking conditions (lanes 4–6). The NCS cleavage of the crosslinked RNA–protein complex, using an RNA with 5-IU at position 16, generated a shorter fragment corresponding to the expected size (200 amino acids) for RRM1 (arrow 1, lane 7). Deletion of only the carboxyl terminus had no effect on the mobility of this band (lane 8). However, deletion of the amino terminus, as expected, reduced the size of this fragment (lane 9), indicating that this fragment is RRM1. Thus, we conclude that 5-IU at position 16 is specifically crosslinked to RRM1.

On the other hand, NCS cleavage of the crosslinked complex (with RNA containing a 5-IU at position 10) generated a band corresponding to the predicted size (154 amino acids) for RRM2 (arrow 2, lane 10). As expected, deletion of only the carboxyl terminus reduced the size (94 amino acids) of this fragment (lane 11), which confirmed the identity of this fragment as RRM2. Additional deletion of the amino terminus had no further reduction in the size

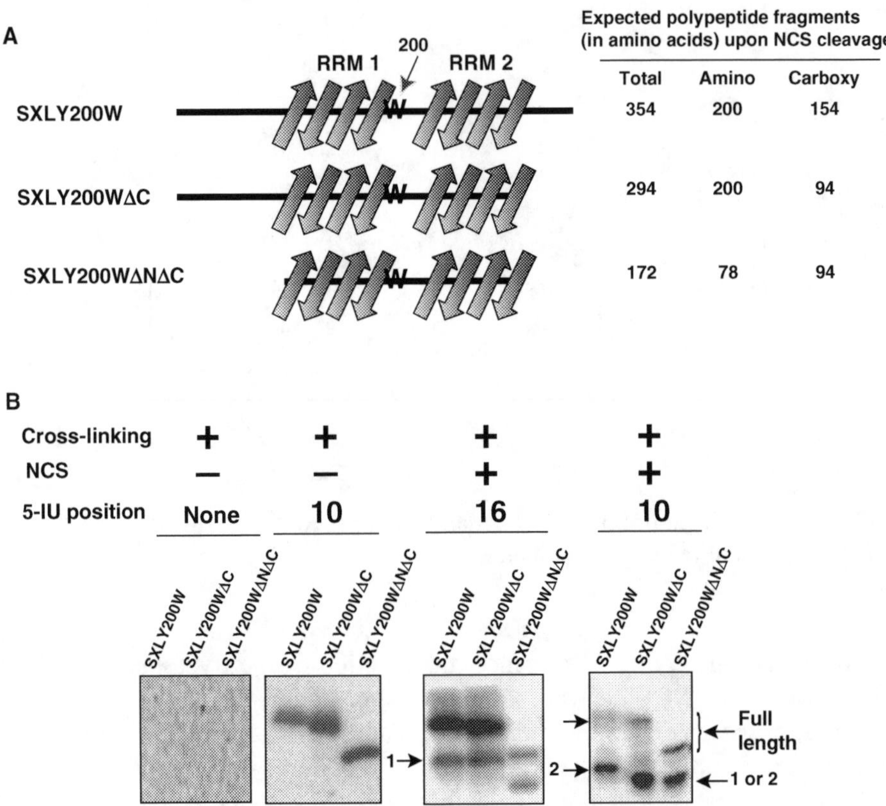

Fig. 4. Identification of an RNA recognition motif (RRM) crosslinked to site-specifically incorporated 5-IU residues. (**A**) A schematic of SXLY200W, SXLY200WΔC, and SXLY200WΔNΔC proteins. Shaded arrows represent the four β-strands of an RRM. For simplicity, α-helices are not shown. The position of Y200 is indicated. The expected sizes of *N*-chlorosuccinimide (NCS) cleavage fragments in amino acids are indicated for each protein. (**B**) Sodium dodecyl sulfate (SDS) polyacrylamide gel analysis of the NCS-cleaved peptides. Proteins are shown at the top of each lane; locations of RRMs are indicated by arrows (1, 2, or 1 or 2). The positions of 5-IU in RNA are shown. At the bottom of the gel (lanes 7–12), the identity of a crosslinked RRM is indicated. The sequence of the RNA used for 5-IU incorporation and the numbering for nucleotides are shown for reference. Lanes 1–3, crosslinking with an RNA lacking 5-IU; lanes 4–6, crosslinking with an RNA containing a 5-IU at position 10 but no NCS cleavage; lanes 7–9, 5-IU at position 16, crosslinking, and NCS cleavage; and lanes 10–12, same as lanes 7–9 except that 5-IU is at position 10.

of this fragment (arrow 1 or 2, lane 12); it was not possible to resolve the RRM1 (78 amino acids) and RRM2 (94 amino acids) fragments of SXL WΔNΔC. We conclude that 5-IU at position 10 is specifically crosslinked to RRM2.

In summary, we conclude that site-specific incorporation of 5-IU and chemical cleavage by NCS provides an easy method to identify a protein domain that is crosslinked to a specific RNA residue. This approach led to the model that $U2AF^{65}$ and SXL recognize the Py tract using multiple modes of binding ***(17)***. This method should be applicable to many other RNA binding proteins that have multiple RRM domains (*see* **Note 2**).

2. Materials

1. RNA: 5-iodouracil (5-IU); DNA synthesizer (Applied Biosystems); γ-^{32}P-ATP (adenosine triphosphate); phenol/chloroform/isoamyl alcohol; ethanol.
2. Protein: Glutathione-Sepharose beads; PreScission protease (Amersham Pharmacia Biotechnology); bovine serum albumin (BSA); glutathione.
3. Crosslinking: semi-UV methacrylate cuvette (Fisher Scientific 14-385-938); NCS (Sigma 10,968-1); HeCd laser source; SDS polyacrylamide gel; X-ray film or phorphorimager screen.
4. Buffer D: 20 m*M* HEPES, pH 8.0, 0.2 m*M* ethylenediaminetetraacetic acid (EDTA), 20% glycerol, 0.05% NP-40, and 1 m*M* dithiothreitol (DTT) ***(21)***.
5. Laemmeli buffer (2X) for SDS gels: 250 m*M* Tris-HCl, pH 6.8, 2% SDS, 10% glycerol, 20 m*M* DTT, and 0.01% bromophenol blue ***(22)***.
6. Binding reaction: 10 m*M* Tris-HCl (pH 7.5), 1 m*M* DTT, 50 m*M* KCl, 0.09 µg/µL acetylated BSA, 0.15 µg/µL transfer RNA, 30 µL protein, 0.5 U/µL RNasin (optional).
7. Tris-base/SDS electrophoresis buffer: 25 m*M* Tris, 250 m*M* glycine, 0.1% SDS ***(22)***.

3. Methods

3.1. RNA

1. Two RNA substrates were chemically synthesized on a DNA synthesizer with a 5-IU at either position 10 or 16 of the non-sex-specific Py tract of *tra*.
2. RNAs were radiolabeled at the 5′ end using polynucleotide kinase and γ-^{32}P-ATP at 37 °C for 30 min and gel purified on a 15% denaturing polyacrylamide gel.
3. The labeled RNAs were eluted from the gel, extracted using phenol/chloroform, and ethanol precipitated with 0.3 *M* sodium acetate and glycogen (*see* **Note 3**).

3.2. Protein

1. Recombinant Gluthione S-transferase (GST) fusion SXL proteins (full length, carboxyl, and carboxyl- and amino-terminal deletions) containing a single tryptophan (SXL W) were expressed in and affinity purified from *Escherichia coli* on glutathione-Sepharose beads ***(17)***.
2. The recombinant SXL protein lacking GST was released from the affinity matrix using PreScission protease (20 U PreScission protease in 2 mL buffer D at 4 °C

overnight), and dialyzed against two changes of buffer D (*see* **Subheading 2., item 4**). It was necessary to prepare SXL without GST because the GST protein has several undesirable tryptophans. Alternatively, one can use a histidine tag or cleave the GST portion by using thrombin or other proteases for which cleavage sites are present in most expression vectors.

3.3. Crosslinking, Cleavage, and Domain Identification

1. Heat denature the labeled RNA at 90 °C for 5 min and chill on ice.
2. Mix the RNA with appropriate dilutions (in buffer D; *see* **Subheading 2., item 4**) of SXL in a 100-μL binding reaction (*see* **Subheading 2., item 6**) ***(13)***. Incubate the binding reaction mix under appropriate conditions; we incubated at room temperature for 20 min for SXL. For these studies, we used an SXL concentration that gave approx 90% binding. Higher crosslinking yields were obtained with saturating protein concentrations.
3. In a semi-UV methacrylate cuvette (*see* **Note 4**), irradiate the sample for 10 min at room temperature using a 325-nm HeCd laser source (300 mW/cm^2) (*see* **Notes 5** and **6**) ***(15)***. The samples were stored in ice after crosslinking.
4. Mix an aliquot (usually 10 μL) of the crosslinked reaction with 4 volumes of 50% acetic acid/8.25 *M* urea containing 10 mg/mL NCS.
5. Mix thoroughly and incubate for 2.5 min at room temperature.
6. Extract the reaction mix twice with chloroform/isoamyl alcohol (24:1).
7. Lyophilize the aqueous phase at 65 °C for 20 min.
8. Dissolve in 2X Laemmeli buffer (*see* **Subheading 2., item 5**) and heat denature at 90 °C for 5 min.
9. Load a portion (~10 μL) of the sample in an SDS polyacrylamide gel of appropriate concentration, depending on the size of the NCS cleavage fragments (15% gel for the SXL fragments) and re-solve by electrophoresis in 1X Tris-glycine/SDS running buffer (*see* **Subheading 2., item 7**) at room temperature.
10. Dry the gel at 65 °C for 45 min.
11. Expose to an X-ray film or a phorphorimager screen for detection and quantitation.

4. Notes

1. A protein with 3 RRMs can also be easily analyzed by introducing one tryptophan between RRM1 and RRM2 (1W23) and another between RRM2 and RRM3 (12W3) in two separate mutants ***(17)***. Although RRM2 is associated with the larger NCS cleavage fragment in both situations, it is possible to unambiguously distinguish crosslinking to each of three RRMs from only two mutants. It is not advisable to construct a double mutant with two tryptophans (1W2W3) to reduce the size of the middle RRM (RRM2) because NCS cleavage is usually partial ***(16)*** and would generate a more complex pattern.
2. It may be necessary to mutate a tryptophan residue if it is present within the relevant portion of a protein.
3. We keep 5-IU RNAs in dark.
4. Although we used the HeCd laser source to obtain 325-nm wavelength of light (courtesy of Dr. Tad Koch), it is possible to explore a standard UV transilluminator

by blocking short-wavelength UV light with a polystyrene Petri dish, provided there is no background crosslinking under these conditions.

5. Crosslinking can be achieved using 5-Br derivatives, which can be excited using a monochromatic 308-nm XeCl excimer laser ***(23)*** or with a 312-nm transilluminator ***(15,24)***.
6. It is advisable to perform a time course for crosslinking.

References

1. Bell, L. R., Maine, E. M., Schedl, P., and Cline, T. W. (1988) Sex-lethal, a *Drosophila* sex determination switch gene, exhibits sex- specific RNA splicing and sequence similarity to RNA binding proteins. *Cell* **55**, 1037–1046.
2. Bell, L. R., Horabin, J. I., Schedl, P., and Cline, T. W. (1991) Positive autoregulation of sex-lethal by alternative splicing maintains the female determined state in *Drosophila. Cell* **65**, 229–239.
3. Singh, R. (2002) RNA–protein interactions that regulate pre-mRNA splicing. *Gene Expression* **10**, 79–92.
4. Smith, C. W., and Valcarcel, J. (2000) Alternative pre-mRNA splicing: the logic of combinatorial control. *Trends Biochem. Sci.* **25**, 381–388.
5. Schutt, C., and Nothiger, R. (2000) Structure, function and evolution of sex-determining systems in Dipteran insects. *Development* **127**, 667–77.
6. Adams, M. D., Rudner, D. Z., and Rio, D. C. (1996) Biochemistry and regulation of pre-mRNA splicing. *Curr. Opin. Cell Biol.* **8**, 331–339.
7. Moore, M. J. (2000) Intron recognition comes of AGe. *Nat. Struct. Biol.* **7**, 14–16.
8. Reed, R. (2000) Mechanisms of fidelity in pre-mRNA splicing. *Curr. Opin. Cell Biol.* **12**, 340–5.
9. Antson, A. A. (2000) Single-stranded-RNA binding proteins. *Curr. Opin. Struct. Biol.* **10**, 87–94.
10. Varani, G., and Nagai, K. (1998) RNA recognition by RNP proteins during RNA processing. *Annu. Rev. Biophys. Biomol. Struct.* **27**, 407–445.
11. Sosnowski, B. A., Belote, J. M., and McKeown, M. (1989) Sex-specific alternative splicing of RNA from the transformer gene results from sequence-dependent splice site blockage. *Cell* **58**, 449–459.
12. Zamore, P. D., and Green, M. R. (1989) Identification, purification, and biochemical characterization of U2 small nuclear ribonucleoprotein auxiliary factor. *Proc. Natl. Acad. Sci. U. S. A.* **86**, 9243–9247.
13. Valcarcel, J., Singh, R., Zamore, P. D., and Green, M. R. (1993) The protein Sex-lethal antagonizes the splicing factor U2AF to regulate alternative splicing of transformer pre-mRNA. *Nature* **362**, 171–175.
14. Singh, R., Valcarcel, J., and Green, M. R. (1995) Distinct binding specificities and functions of higher eukaryotic polypyrimidine tract-binding proteins. *Science* **268**, 1173–1176.
15. Willis, M. C., Hicke, B. J., Uhlenbeck, O. C., Cech, T. R., and Koch, T. H. (1993) Photocrosslinking of 5-iodouracil-substituted RNA and DNA to proteins. *Science* **262**, 1255–1257.

16. Mirfakhrai, M., and Weiner, A. M. (1993) Chemical Cleveland mapping: a rapid technique for characterization of crosslinked nucleic acid-protein complexes. *Nucleic Acids Res.* **21**, 3591–3592.
17. Banerjee, H., Rahn, A., Davis, W., and Singh, R. (2003) Sex lethal and U2 small nuclear ribonucleoprotein auxiliary factor (U2AF65) recognize polypyrimidine tracts using multiple modes of binding. *RNA* **9**, 88–99.
18. Handa, N., Nureki, O., Kurimoto, K., et al. (1999) Structural basis for recognition of the tra mRNA precursor by the Sex-lethal protein. *Nature* **398**, 579–585.
19. Crowder, S. M., Kanaar, R., Rio, D. C., and Alber, T. (1999) Absence of interdomain contacts in the crystal structure of the RNA recognition motifs of Sex-lethal. *Proc. Natl. Acad. Sci. U. S. A.* **96**, 4892–4897.
20. Meisenheimer, K. M., Meisenheimer, P. L., and Koch, T. H. (2000) Nucleoprotein photo-cross-linking using halopyrimidine-substituted RNAs. *Methods Enzymol.* **318**, 88–104.
21. Dignam, J. D., Lebovitz, R. M., and Roeder, R. G. (1983) Accurate transcription initiation by RNA polymerase II in a soluble extract from isolated mammalian nuclei. *Nucleic Acids Res.* **11**, 1475–1489.
22. Sambrook, J., Fritsch, E. F., and Maniatis, T. (1989) *Molecular Biology. A Laboratory Manual*, Cold Spring Harbor Laboratory Press, New York.
23. Gott, J. M., Willis, M. C., Koch, T. H., and Uhlenbeck, O. C. (1991) A specific, UV-induced RNA-protein cross-link using 5-bromouridine-substituted RNA. *Biochemistry* **30**, 6290–6295.
24. Stump, W. T., and Hall, K. B. (1995) Crosslinking of an iodo-uridine-RNA hairpin to a single site on the human U1A N-terminal RNA binding domain. *RNA* **1**, 55–63.
25. Wong, I., and Lohman, T. M. (1993) A double-filter method for nitrocellulose-filter binding: application to protein-nucleic acid interactions. *Proc. Natl. Acad. Sci. U. S. A.* **90**, 5428–5432.

13

Proteins Specifically Modified With a Chemical Nuclease as Probes of RNA–Protein Interaction

Oliver A. Kent and Andrew M. MacMillan

1. Introduction

Hydroxyl radicals generated from probes covalently attached to RNA or protein have been useful for examining both intra- and intermolecular contacts to RNA in both structured RNAs and RNA–protein complexes *(1–3)*. Diffusible radicals produced from a tethered Fe-EDTA (ethylenediaminetetraacetic acid) moiety are excellent probes of local RNA structure since they are only capable of cleaving the phosphodiester backbone within approx 10–20 Å from their site of generation. We describe the preparation and use of a series of proteins chemically modified at their N-terminus with Fe-EDTA to probe the interaction of the N-terminus of an RNA binding protein, the mammalian splicing factor U2AF65 *(4)*, with RNA. These experiments formed part of a larger investigation of the role of U2AF65 in the assembly of the mammalian spliceosome on pre-mRNA (messenger RNA) substrates *(5)*.

Synthesis of the probes used in this study was carried out by chemoselective reaction of an EDTA thioester (EDTA-3MPA [3-mercaptopropionic acid]) with the appropriate U2AF65 derived precursor modified to present an N-terminal cysteine residue following site-specific proteolysis (**Fig. 1A**). Thioesters react specifically with N-terminal Cys since an initial thiotransesterification is followed by rapid rearrangement to yield a stable amide linkage (**Fig. 1B**). Transient modification of internal Cys residues is reversed using an excess of dithiothreitol (DTT) and the pH at which the modification is performed minimizes reaction with Lys side chains. This strategy is based on work by Erlanson and Verdine *(6)*, who extended Kent's peptide ligation chemistry *(7)* to the N-terminal functionalization of proteins with Fe-EDTA; in an elegant study, these

From: *Methods in Molecular Biology, vol. 488: RNA-Protein Interaction Protocols*
Edited by: Ren-Jang Lin © Humana Press Inc, Totowa, NJ

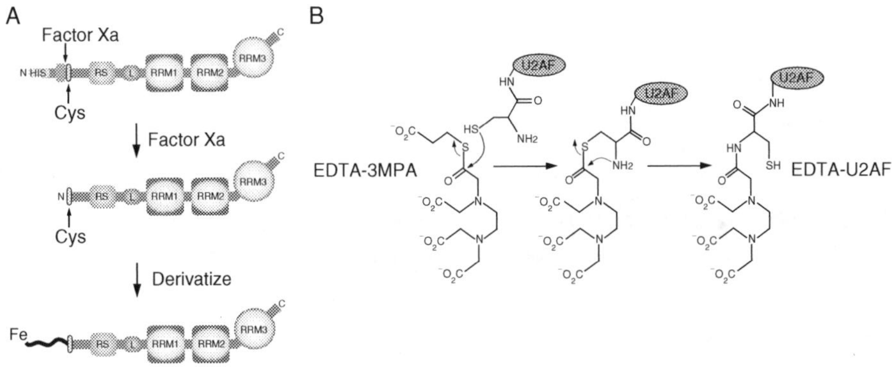

Fig. 1. Chemoselective modification of N-terminal cysteines. (**A**) General strategy for N-terminal functionalization of a recombinant protein (*see* **ref. 7**). (**B**) Chemistry of N-terminal modification. Initial thiotransesterification is followed by rearrangement to yield a stable amide linkage. Modification of basic residues is minimized by carrying out the derivatization at pH 7.8; modification of internal Cys residues is reversed by carrying out the derivatization in the presence of 300 m*M* dithiothreitol (DTT). EDTA, ethylenediaminetetraacetic acid; 3MPA, 3-mercaptopropionic acid; RRM, RNA recognition motif.

workers were able to demonstrate NFAT-mediated modulation of DNA binding by AP1 using patterns of DNA cleavage as a reporter of binding orientation.

We prepared a panel of four modified U2AF65 probes, representing successive N-terminal deletions, each containing an N-terminal His_6 tag and a factor Xa cleavage site followed by an introduced Cys residue. Cleavage of partially purified protein yielded the desired precursor with an N-terminal Cys, which was reacted with EDTA-3MPA to yield the functionalized probe, which could then be activated by the addition of Fe^{2+}, ascorbic acid, and hydrogen peroxide. To confirm that modification of the N-terminal Cys was specific, we performed modification/chase experiments with both EDTA-3MPA and the thioester biotin-3MPA since protein modification with biotin can be monitored by using an avidin-horseradish peroxidase (HRP) conjugate. These experiments showed that modification of U2AF65 was specific to the N-terminal Cys (in the presence of six internal Cys residues; *4*) and was essentially quantitative.

2. Materials

1. Glassware for organic synthesis.
2. Rotary evaporator (Buchi).
3. High-performance liquid chromatograph (HPLC) (Waters).
4. C-18 analytical Sep-Pak (Waters).
5. Thin-layer chromatographic (TLC) plates.

6. Cerium molybdate TLC stain (12g of ammonium molybdate, 0.5g of ceric ammonium molybdate, and 15mL of concentrated sulfuric acid added to 235mL of distilled water).
7. Avidin-HRP conjugate (Bio-Rad).
8. Oligonucleotide primers.
9. pET-30 expression system (Novagen).
10. *Eco*RI and *Hin*dIII restriction enzymes (New England Biolabs).
11. Ni-NTA agarose (Qiagen).
12. Factor Xa protease (Roche).
13. Factor Xa protease removal resin (Qiagen).
14. HPLC buffer A (2% v/v acetonitrile, 0.06% v/v TFA).
15. HPLC buffer B (90% v/v acetonitrile, 0.06% v/v TFA).
16. Buffer D (20m*M* HEPES, pH 7.8, 0.1*M* KCl, 0.5m*M* DTT, 0.2m*M* EDTA, 20% v/v glycerol).
17. Factor Xa buffer (20m*M* HEPES, pH 7.8, 1m*M* $CaCl_2$, 100m*M* KCl, 20% glycerol).
18. T7 transcription buffer (40m*M* Tris, pH 8, 3m*M* nucleotide triphosphates (NTPs), 25m*M* $MgCl_2$).
19. U2AF gel mobility shift buffer (10m*M* HEPES, pH 7.9, 60m*M* KCl, 2m*M* $MgCl_2$, 1µg transfer RNA (tRNA), 0.25m*M* DTT, 0.1m*M* EDTA, and 10% (v/v) glycerol).

3. Methods

3.1. Protein Modifcation Reagents

3.1.1. Synthesis and Purification of EDTA-3MPA

1. 3MPA (1.0mL, 11.7mmol) was added to a stirring suspension of EDTA dianhydride (3.0g, 11.7mmol) in 15mL dimethylformamide (DMF; **Fig. 2A**). The reaction was stirred overnight under a positive pressure of nitrogen.
2. The reaction was heated to 60°C and the DMF removed under pressure with a rotary evaporator. The remaining yellow oil was resuspended in 12mL 1*M* NaOH and stirred at room temperature overnight. The resulting slurry was centrifuged to remove the precipitate, and the supernatant was added to 33mL of methanol and allowed to stand at 4°C for several hours to precipitate EDTA. The reaction was recentrifuged to remove the precipitate and the supernatant lyophilized to a fine yellow powder. The powder, containing EDTA-3MPA, was resuspended in distilled water and the pH adjusted to 2.0 with trifluoroacetic acid (TFA).
3. The crude EDTA-3MPA was purified by C-18 analytical Sep-Pak (Waters). The Sep-Pak was prewashed with 50mL acetonitrile, 20mL buffer A, and 20mL distilled water. The EDTA-3MPA-containing solution was loaded onto the column, washed with distilled water (250mL) and buffer A (250mL) before elution with buffer B. EDTA-3MPA containing fractions were identified by ultraviolet (UV) absorbance (λ_{max} = 236nm), pooled, and lyophilized to a fine powder (1.0g, 33% yield).
4. In preparation for derivatization reactions, EDTA-3MPA was dissolved in 10% (v/v) dimethyl sulfoxide (DMSO) and 100m*M* HEPES, pH 8.5, to a final concentration of 50m*M* (19mg/mL) and stored at −20°C.

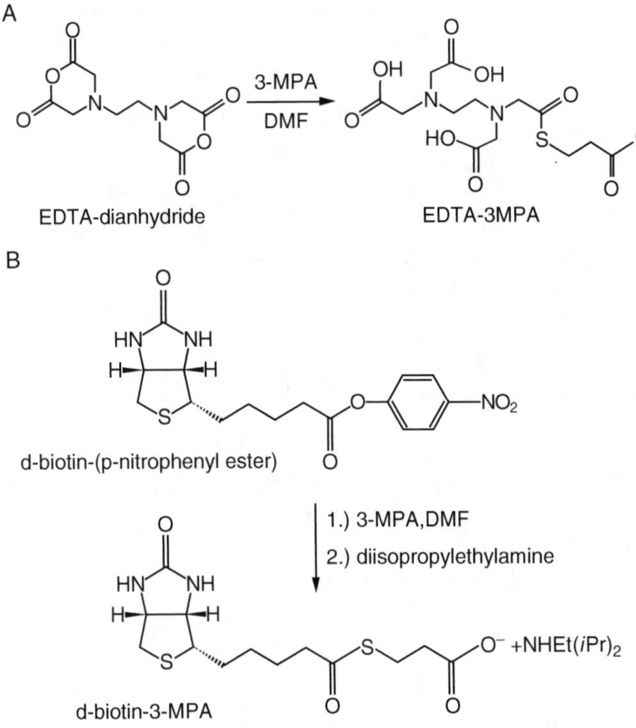

Fig. 2. Preparation of protein modification reagents. (**A**) Synthesis of EDTA-3MPA (ethylenediaminetetraacetic acid-3-mercaptopropionic acid): (i) 3-MPA, dimethylformamide (DMF); (ii) NaOH. (**B**) Synthesis of biotin-3MPA: (i) 3-MPA, diisopropylethylamine, DMF.

3.1.2. Synthesis of Biotin-3MPA

1. 3MPA (48 mL, 0.547 mmol) was added to a stirring suspension of *d*-biotin-(p-nitrophenyl ester) (200 mg, 0.547 mmol) and diisopropylethylamine (0.286 mL, 1.6 mmol) in 0.5 mL DMF (**Fig. 2B**). The reaction was stirred overnight at room temperature.
2. The ammonium salt of biotin-3MPA was precipitated from solution by titrating dropwise the reaction with ethyl acetate. The resulting fine white powder (110 mg, 0.240 mmol, 44% yield) was stored at –20 °C. TLC analysis (5% MeOH/CH_2Cl_2) of biotin-3MPA using cerium molybdate stain: $R_f = 0.3$. High-resolution MS-FAB: MH^+ is 333.0906 ($C_{13}H_{20}O_4N_2S_2$, calculated mass 333.0943). ^1H nuclear magnetic resonance (NMR) (d-DMSO): (ppm) 1.32 (m, 2H, CH_2); 1.44 (m, 2H, CH_2); 1.56 (m, 2H, CH_2); 2.57 (m, 5H, heterocycle CH_2, 2CH_2); 2.86 (d, 2H, CH_2); 2.96 (t, 2H, CH_2); 3.08 (m, 1H, heterocycle CH); 6.38 (s, 1H, NH); 6.46 (s, 1H, NH); 12.25 (broad s, 1H, acid H).

3. In preparation for derivatization reactions, biotin-3MPA was dissolved in 10% (v/v) DMSO and 100 m*M* HEPES, pH 8.5, to a final concentration of 50 m*M* (17 mg/mL) and stored at –20 °C.

3.2. Protein Expression and Modification

3.2.1. Cloning, Protein Expression, and Purification

1. A panel of complementary DNAs representing full-length and deletion mutants of U2AF65, each with a factor Xa site followed by a Cys codon, was prepared by polymerase chain reaction (PCR) and cloned into the pET-30 expression system using *Eco*RI and *Hin*dIII (*see* **Note 1**). Proteins produced from these constructs contain an N-terminal His_6 tag followed by the factor Xa site and introduced Cys residue (**Fig. 3A**).
2. The His_6-tagged proteins were expressed in *Escherichia coli*, partially purified by Ni-NTA chromatography, and dialyzed into factor Xa buffer in preparation for factor Xa proteolysis and derivitization.

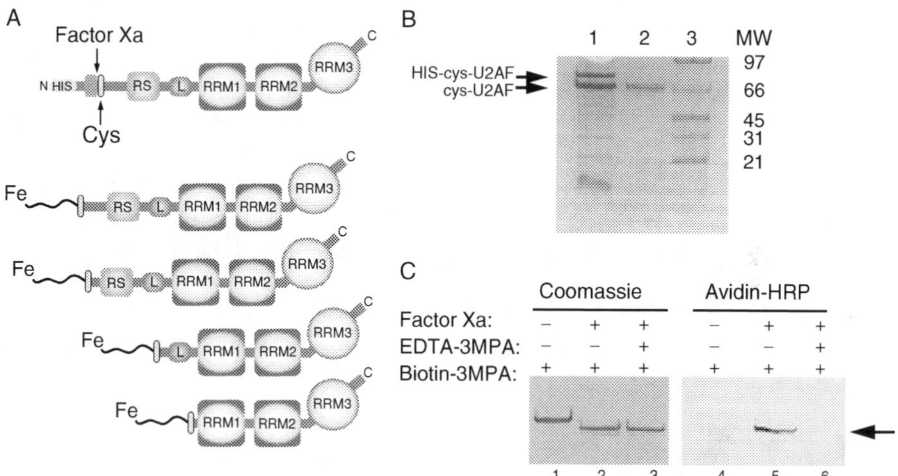

Fig. 3. Preparation of ethylenediaminetetraacetic acid (EDTA)-modified U2AF65. (**A**) Panel of N-terminal EDTA-derivatized U2AF65 deletion constructs. Modified proteins were synthesized in two steps from a His_6-tagged precursor containing a factor Xa site followed by an introduced Cys residue. (**B**) Sodium dodecyl sulfate polyacrylamide gel electrophoresis (SDS-PAGE) analysis of purification of factor Xa cleaved U2AF65. Lane 1, U2AF65 cut with factor Xa. Lane 2, U2AF65 following purification with factor Xa removal resin and Ni-NTA resin. Lane 3, molecular weight markers. (**C**) Analysis of specificity and efficiency of protein modification using biotin-3MPA (3-mercaptopropionic acid). Left, Coomassie stain of SDS-PAGE gel; right, avidin-HRP (horseradish peroxidase) probe of Western blot. Uncut (lanes 1 and 4), cut and reacted with biotin-3MPA (lanes 2 and 5), or cut and reacted with EDTA-3MPA (lanes 3 and 6) and chased with biotin-3MPA and probed with avidin-HRP (lanes 4–6). RRM, RNA recognition motif.

3.2.2. Protein Cleavage and Derivatization

1. Ni-NTA-purified proteins were cleaved using the protease factor Xa (1/20 factor Xa: U2AF g/g, 2–3 h, RT, factor Xa buffer). A proteolysis time course and factor Xa titration were performed on pilot scales to determine the best balance between extent of reaction and specificity of cleavage (*see* **Note 2**). For full-length U2AF65, it was determined that the optimal cleavage time was approx 2 h, at which point the reaction was only 50% complete, but there was minimal second-site cleavage. Preparative-scale reactions were performed on 500 μg of U2AF65 precursor in a reaction volume of 1 mL. Factor Xa, uncleaved U2AF, and cleavage fragments were removed by consecutive batch incubations of the reaction with factor Xa removal resin and Ni-NTA resin (**Fig. 3B**).
2. The purified protein was derivatized in 1-mL reactions containing 300 m*M* DTT, 1 m*M* EDTA-3MPA (or 1 m*M* biotin-3MPA) incubated at 0 °C for 12 h. Following derivatization, products were dialyzed against buffer D for gel shift analysis or buffer D lacking EDTA and glycerol for Fe-EDTA cleavage reactions.
3. To determine whether modification was specific to the N-terminal Cys, derivatization reactions were performed with biotin-3MPA on uncleaved U2AF65 (no N-terminal Cys) and cleaved U2AF65 (N-terminal Cys). These experiments indicated that derivatization is dependent on the presence of an N-terminal Cys and does not occur on any of the six internal Cys residues (**Fig. 3C**). Extent of derivatization was monitored by chase experiments (**Fig. 3C**) in which initial reaction with EDTA-3MPA was followed by incubation with biotin-3MPA, sodium dodecyl sulfate polyacrylamide gel electrophoresis (SDS-PAGE), blotting on nitrocellulose, and analysis with avidin-HRP conjugate (Bio-Rad).

3.3. Mapping Protein–RNA Interactions Using Fe-EDTA-Derivatized Protein

3.3.1. RNA Preparation

1. RNA (5′-GGGCUCGUCUCGAGGGUGCUGACUGGCUUCUUCUCUCUUU UUCCCUCAGGCCUACUCUUCU-3′) was synthesized by T7 transcription from synthetic double-stranded DNA templates (0.4 μ*M* annealed DNA template and 8 μg of T7 RNA polymerase in T7 transcription buffer) containing a T7 promoter sequence attached to the 3′ splice site region of the PIP85.B pre-mRNA *(7)*. Transcriptions were performed at 37 °C for 4 h and then purified by denaturing PAGE (8%, 19:1 acrylamide/bis-acrylamide), visualized by UV, excised, and extracted from the gel.
2. Purified transcriptions were dephosphorylated with calf intestinal alkaline phosphatase (Roche) and 5′ end labeled using γ-^{32}P-ATP (3000 Ci/mmol; PerkinElmer Life Sciences-NEN) and T4 polynucleotide kinase (Roche). Labeled RNAs were gel purified by denaturing PAGE (8%, 19:1 acrylamide/bis-acrylamide), visualized by autoradiography, excised, and extracted from the gel. Gel-purified RNAs were resuspended in double-distilled water and stored at –20 °C.

3.3.2. Gel Mobility Shift

Gel mobility shift assays were used to examine the interaction of functionalized U2AF65 with short RNAs representing the 3′ splice site of mammalian introns (**Fig. 4A**; **5**). This titration is critical to establishing the appropriate conditions for subsequent cleavage experiments, including the amount of probe protein to be used (*see* **Note 3**). Binding reactions were performed at room temperature for 60 min in buffer D and contained 5′ ^{32}P-labeled RNA (50–100 × 10^3 cpm) and 0–100 pmol derivatized U2AF65.

Reactions were immediately analyzed by native PAGE (6%, 89:1 acrylamide/bis-acrylamide) run with 50 m*M* Tris-glycine running buffer at 110 V for 3 h.

3.3.3. RNA Cleavage Reactions

1. We investigated the structure of the 3′ splice site RNA bound to the U2AF65 probes using 5′ end-labeled RNA, initiating Fe-EDTA-mediated cleavage under conditions at which about half of the RNA was bound and analyzing the results by denaturing PAGE. Cleavage experiments were carried out in 25-μL reactions containing 0.6 μ*M* U2AF-EDTA (dialyzed into buffer D lacking EDTA and glycerol) and 5′ ^{32}P-labeled RNA (50–100 × 10^3 cpm; *see* **ref. 5** for details) in buffer D lacking EDTA and glycerol with 1 μg added tRNA. Binding was performed at room temperature for 60 min followed by the addition of 0.5 μ*M* $FeSO_4$ and a further 10-min incubation on ice to allow chelation. Affinity cleavage was initiated by the addition of 0.05% (v/v) H_2O_2 and 5 m*M* ascorbic acid, and the reactions were allowed to proceed on ice for 10 min. Reactions were quenched with 30 m*M* thiourea containing 1% (v/v) glycerol, extracted with phenol/chloroform/isoamyl alcohol, and ethanol precipitated (*see* **Note 4**).
2. Following ethanol precipitation, reactions were subjected to 10% denaturing PAGE (19:1 acrylamide/bis-acrylamide) alongside markers generated by ribonuclease (RNase) A and RNase T1 digestion as well as base hydrolysis of the RNA (*see* **Note 5**). Dried gels were exposed to a Molecular Dynamics phosphor screen and scanned using a Molecular Dynamics Storm 840 PhosphorImager (**Fig. 4B**; *see* **Note 6**). For the experiment described here, regions of significant cleavage can be seen in the vicinity of the RNA substrate branch point region for the WT U2AF65 and the Δ1–14 U2AF65 deletion mutant (**Fig. 4B**). Cleavages were also detected at the 3′ splice site region for Δ1–14, Δ1–64 and near the 3′ splice site with Δ1–94 (**Fig. 4B**). For both Δ1–64 and Δ1–94 constructs, no cleavages were observed in the vicinity of the branch point region. These results demonstrated that association of U2AF65 with its cognate RNA results in a juxtaposition of the branch region and 3′ splice site resulting from pronounced RNA bending (*see* **ref. 5**).

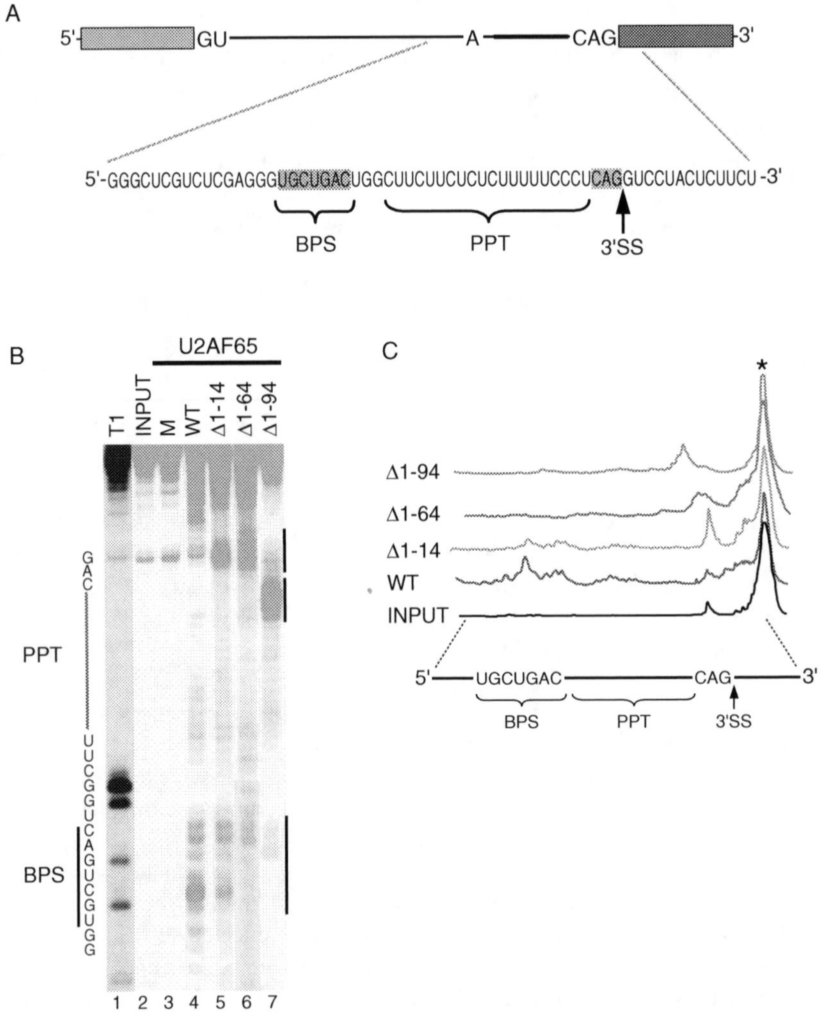

Fig. 4. Directed hydroxyl radical cleavage of RNA by U2AF65-Fe-EDTA (ethylenediaminetetraacetic acid). (**A**) Short RNA derived from PIP85.B pre-mRNA (messenger RNA) *(8)* representing 3′ splice site of a mammalian intron. Sequence includes the branch point sequence (BPS), polypyrimidine tract (PPT), and consensus 3′ splice site (3′SS). (**B**) Cleavage of wild-type RNA with N-terminal deletion mutants of U2AF65. Reactions were incubated with U2AF65-Fe-EDTA (lanes 4–7) under cleavage conditions or with U2AF65-EDTA (lane 3, M) lacking iron. Input (lane 2) and ribonuclease (RNase) T1 sequencing (lane 1) shown. (**C**) Densitometric analysis of cleavage reactions (WT, Δ1–14, Δ1–64, and Δ1–94). Individual graphs were normalized to full-length RNA (asterisk).

4. Notes

1. Amplification of full-length U2AF requires 10% DMSO as a cosolvent in the PCR reaction.
2. The preferred factor Xa cleavage site is C-terminal to the sequence IE/DGR, but nonspecific cleavage after basic residues can occur depending on the sequence and structure context. Full-length U2AF65 is a good model for nonspecific factor Xa cleavage since there are 21 basic (R or K) residues in the N-terminal 63 amino acids.
3. In general, cleavage experiments are performed using probe concentrations corresponding to the experimentally determined K_D for probe–RNA interaction. In practice, a titration of probe concentration within the vicinity of this K_D yields the best cleavage results.
4. It is important for all cleavage reagent stock solutions to be prepared fresh before each use.
5. Ladders were generated by addition of 0.5 ng RNase A (Roche) or 0.3 U RNase T1 (Roche) to reactions containing RNA in buffer D followed by digestion at room temperature for 3 and 1 min, respectively. A nucleotide-resolution ladder was generated by adding 1 µL of 1 *M* NaOH to RNA in 10 µL buffer D with 1 µg added tRNA and allowing cleavage to proceed for 10 s at room temperature. Reactions were quenched with 400 µL 0.3 *M* Na0Ac, pH 5.6, and ethanol precipitated.
6. While significant cleavages can usually be readily identified directly from the PhosphorImager scan, it is helpful to perform a densitometric analysis of individual lanes to normalize individual cleavage patterns against uncleaved input RNA (*see* **Fig. 4C**). This analysis also aids in distinguishing the quite different patterns observed from low-level background degradation and genuine cleavage events (**Fig. 4B,C**). In general, we have been conservative in our identifications and have only commented on the strongest observed cleavages.

Acknowledgments

We would like to thank M. Green for providing the U2AF65 expression plasmid as well as Ayube Reayi and Louise Foong for their contributions to the early part of this work. O.A.K. was supported by a graduate fellowship from the Alberta Heritage Foundation for Medical Research (AHFMR). This work was supported by the Natural Sciences and Engineering Research Council of Canada (NSERC) and the Canadian Institutes for Health Research (CIHR) and by an establishment grant from AHFMR to A.M.M.

References

1. Han, H., and Dervan, P. B. (1994). Visualization of RNA tertiary structure by RNA-EDTA.Fe(II) autocleavage: analysis of tRNA(Phe) with uridine-EDTA. Fe(II) at position 47. *Proc. Natl. Acad. Sci. U. S. A.* **91**, 4955–4959.
2. Kent, O. A., and MacMillan, A. M. (2002). Early organization of pre-mRNA during spliceosome assembly. *Nat. Struct. Biol.* **9**, 576–581.

3. Wilson, K. S., and Noller, H. F. (1998). Mapping the position of translational elongation factor EF-G in the ribosome by directed hydroxyl radical probing. *Cell* **92**, 131–139.
4. Zamore, P. D., Patton, J. G., and Green, M. R. (1992). Cloning and domain structure of the mammalian splicing factor U2AF. *Nature* **355**, 609–614.
5. Kent, O. A., Reayi, A., Foong, L., Chilibeck, K. A., and MacMillan, A. M. (2003). Structuring of the 3′ splice site by U2AF65. *J. Biol. Chem.* **278**, 50572–50577.
6. Erlanson, D. A., Chytil, M., and Verdine, G. L. (1996). The leucine zipper domain controls the orientation of AP-1 in the NFAT.AP-1.DNA complex. *Chem. Biol.* **3**, 981–991.
7. Dawson, P. E., Muir, T. W., Clark-Lewis, I., and Kent, S. B. (1994). Synthesis of proteins by native chemical ligation. *Science* **266**, 776–779.
8. Query, C. C., Moore, M. J., and Sharp, P. A. (1994). Branch nucleophile selection in pre-mRNA splicing: evidence for the bulged duplex model. *Genes Dev.* **8**, 587–597.

14

RNA–Protein Crosslink Mapping Using TEV Protease

Ian A. Turner, Chris M. Norman, Mark J. Churcher, and Andrew J. Newman

Summary

Characterization of novel RNA–protein interactions often demands physical mapping of the RNA binding sites in the protein. This can sometimes be accomplished using radioactively labeled RNA in covalent RNA–protein crosslinking experiments. The position of the radioactive label crosslinked to the protein can then be determined by fragmentation of the protein using a battery of sequence-specific proteolytic enzymes or chemical reagents. However, there are typically many cleavage sites in the natural protein sequence, and for large proteins, particularly when there are multiple sites of RNA–protein interaction, it may be difficult or impossible to determine the sites of crosslink formation unambiguously using this traditional physical mapping approach. We have developed an alternative method for physical mapping of RNA–protein crosslinks based on random insertion into the protein of a short peptide tag that includes the target sequence ENLYFQG (Glu-Asn-Leu-Tyr-Phe-Gln-Gly) for the highly specific TEV protease from tobacco etch virus. Covalent RNA–protein crosslinks can then be physically mapped by TEV protease digestion, fractionation of the proteolytic digestion products by sodium dodecyl sulfate polyacrylamide gel electrophoresis (SDS-PAGE) and visualization of the labeled protein fragments by phosphorimaging.

Key Words: Crosslink mapping; pre-mRNA; RNA–protein interactions; splicing; TEV protease; transposon; yeast.

1. Introduction

RNA–protein crosslink mapping using TEV (tobacco etch virus) protease entails the random insertion of a short peptide tag (including the TEV protease cleavage site) at multiple sites in the protein of interest. This is achieved in vitro using purified Tn5 transposase and a customized transposon that initially includes a kanamycin resistance gene (Kan). After transposition, the Kan gene is removed, leaving the TEV site peptide tag behind in the recipient gene, marked by a unique restriction site. Insertion of short peptide tags will often perturb the

From: *Methods in Molecular Biology, vol. 488: RNA-Protein Interaction Protocols*
Edited by: Ren-Jang Lin © Humana Press Inc, Totowa, NJ

protein's structure or interactions, thereby compromising important functions. Clearly, for crosslink mapping experiments, it is essential to identify and isolate those genes that encode *functional* peptide-tagged copies of the protein. TEV protease site insertions in individual functional genes are then roughly mapped by restriction enzyme cleavage and precisely located by DNA sequencing.

Protocols for RNA–protein crosslinking vary widely according to the components involved. In this chapter, we describe mapping of RNA–protein crosslinks in a protein involved in yeast pre-mRNA (messenger RNA) splicing. Active splicing extracts are made from several yeast strains that express different TEV site-tagged versions of the splicing factor. Spliceosomes are assembled using pre-mRNAs site-specifically modified with a single 4-thiouridine (4-thioU) photocrosslinker and adjacent radiolabeled phosphate. Site-specific RNA–protein crosslinking is induced by irradiation with long-wave ultraviolet (UV), which exclusively activates the 4-thioU to form covalent crosslinks with nearby protein molecules ***(1)***. Spliceosomes are then affinity captured using an N- or C-terminal epitope tag on the splicing factor. After cleavage of the ^{32}P-labeled protein with TEV protease, the RNA is "trimmed" with either ribonuclease (RNase) A or RNase T1, depending on the precise RNA sequence in the vicinity of the photocrosslinker. The radiaoactively labeled TEV-digested polypeptides are fractionated by sodium dodecyl sulfate polyacrylamide gel electrophoresis (SDS-PAGE). Comparison of the ^{32}P-labeling profiles of the protein tagged with TEV sites at different positions allows RNA–protein crosslinks to be assigned to discrete regions of the protein.

2. Materials

1. Oligonucleotides for transposon construction.
2. Restriction enzymes, T7 RNA polymerase, T4 DNA ligase, T4 polynucleotide kinase (New England Biolabs).
3. Qiaex II gel extraction kit (Qiagen).
4. RNasin rribonuclease inhibitor (Promega)
5. Tn5 transposase (Epicentre).
6. High-efficiency electrocompetent *Escherichia coli* cells.
7. 365-nm UV lamp (model B 100 AP, UV Products).
8. TEV protease (Invitrogen).
9. Immunoglobulin G (IgG) Sepharose 6 Fast Flow (Amersham Biosciences).
10. Anti-FLAG M2 monoclonal antibody coupled to Agarose (Sigma). The FLAG epitope is an eight amino acid peptide (AspTyrLysAspAspAspAspLys).
11. RNase T1 (Merck Biosciences).
12. RNase A (Ambion).
13. SDS-PAGE equipment (Invitrogen).
14. 10X EZ::TN buffer: 0.5 *M* Tris-acetate, pH 7.5, 1.5 *M* potassium acetate, 100 m*M* magnesium acetate, 40 mM spermidine.

15. 1% SDS.
16. SOB. Medium: Tryptone 20g/L, Yeast Extract 5g/L, 10 mM NaCl, 2mM KCl.
17. 5X T7 buffer: 200 mM Tris-HCl, pH 8.0, 10 mM spermidine, 50 mM dithiothreitol (DTT), 50 mM $MgCl_2$.
18. 5X nucleotide triphosphate (NTP) mix: 5 mM adenosine triphosphate (ATP), 5 mM cytosine triphosphate (CTP), 5 mM uridine triphosphate (UTP), 2.5 mM guanosine triphosphate (GTP).
19. Phenol buffered with 50 mM sodium acetate, pH 5.3, 10 mM ethylenediaminetetraacetic acid (EDTA), 10 mM 2-mercaptoethanol.
20. Formamide gel loading buffer: 95% Formamide, 10 mM EDTA, 0.1% xylene cyanol, 0.1% bromophenol blue.
21. γ-[^{32}P]-ATP (Amersham).
22. 10X T4 polynucleotide kinase buffer: 700 mM Tris-HCl, pH 7.6, 100 mM $MgCl_2$, 50 mM DTT.
23. Glycogen carrier (Sigma).
24. T4 DNA ligase mix: 2X T4 DNA ligase buffer, 8 U/μL RNasin ribonuclease inhibitor, 400 U/μL T4 DNA ligase.
25. Elution buffer: 0.3 *M* sodium acetate, pH 5.3, 1 mM EDTA, 0.1% SDS.
26. 5X splice cocktail: 10 mM ATP, 12.5 mM $MgCl_2$, 300 mM potassium phosphate, pH 7.0.
27. IP150: 10 mM Tris-HCl, pH 8.0, 150 mM NaCl, 0.1% NP40.
28. mM buffer: 50 mM Tris-HCl, pH 8.0, 0.5 mM EDTA, 1 mM DTT.

3. Methods

RNA–protein crosslink mapping using TEV protease involves the following methods: (1) construction of the TEV protease site insertion transposon; (2) transposition reaction and excision of the Kan marker; (3) isolation of functional genes carrying TEV protease sites; (4) synthesis of pre-mRNAs containing photocrosslinkers at specific positions; (5) RNA–protein crosslinking and crosslink mapping using TEV protease.

3.1. Construction of the TEV Protease Site Insertion Transposon

Construction of the transposon starts with the synthesis of two complementary 73-mer oligonucleotides (AW42 GATCCTGTCTCTTATACACATCTGG CGCGCCGTAGAAAATTTATATTTTCAAGGAGATGTGTATAAGAGACAG and AW43 GATCCTGTCTCTTATACACATCTCCTTGAAAATATAAATTTT CTACGGCGCGCCAGATGTGTATAAGAGACAG), which are 5′-phosphorylated, annealed together, and cloned into the BamHI site of pBluescript KS+ (Stratagene) or a similar cloning vector marked with ampicillin resistance (Amp). This synthetic DNA insert encodes the TEV protease recognition sequence ENLYFQG and a site for the restriction enzyme AscI, flanked by the "outer end" (OE) recognition sequences for the Tn5 transposase *(2)* (*see* **Fig. 1**).

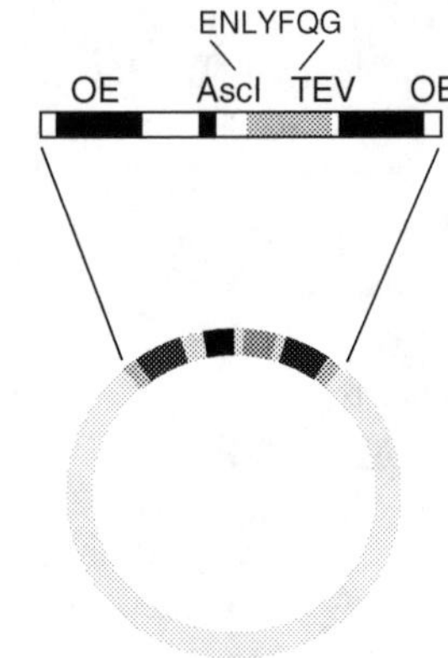

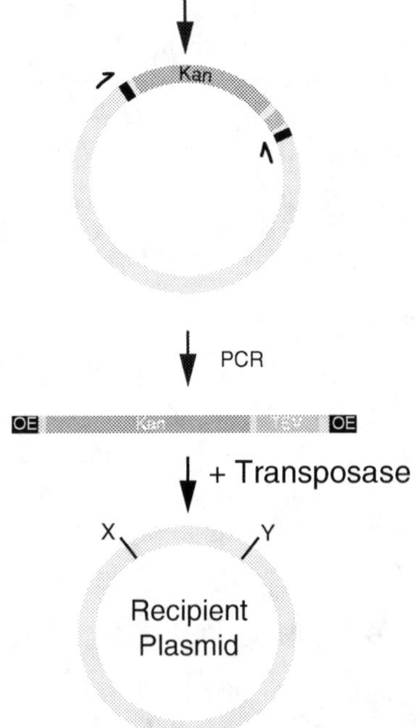

Fig. 1. Schematic of tobacco etch virus (TEV) transposon construction and in vitro transposition. The transposon is built from two synthetic oligonucleotides annealed together and cloned into the BamHI site of pBluescript KS+. A kanamycin resistance (Kan) marker is inserted at the AscI site in the transposon. The transposon is amplified by polymerase chain reaction (PCR) using primers specific for sequences in the polylinker and randomly inserted into the recipient plasmid in vitro using Tn5 transposase (*see* **Subheadings 3.1.** and **3.2.**).

The sequence is designed so that five of the six reading frames are closed by translation stop codons, leaving the frame encoding the TEV protease recognition site as the only open reading frame.

To follow the products of transposition reactions a removable Kan marker is inserted into the TEV transposon sequence. The Kan marker from pTn-Mod-Okm is provided with AscI sites at its ends by polymerase chain reaction (PCR) using primers AW44 GCAGCAGGCGCGCCAAAGCCACG TTGTGTCTCAA and AW45 TGCTGCGGCGCGCCTTAGAAAAACTCA TCGAGCA, digested with AscI and inserted into the AscI ste in the pBluescript KS+ TEV transposon clone. Finally, the Kan transposon is amplified by PCR using primers FTA1 CCCTCGAGGTCGACGGTATCG and RTA1 ATAGGGCGAATTGGAGCTCCA specific for the flanking polylinker sequences in pBluescript KS+. The PCR product is isolated by agarose gel electrophoresis and Qiaex II gel extraction in preparation for the in vitro transposition reaction (**Subheading 3.2.1.**).

3.2. Transposition Reaction and Excision of the Kan Marker

It is advantageous for subsequent steps if the recipient plasmid (which carries an Amp marker) is first modified by site-directed mutagenesis so that unique restriction sites are available at or near the extremities of the protein-coding sequence. After transposon insertion, this will facilitate removal (via a subcloning step) of any transposons lying outside the region of interest.

3.2.1. In Vitro Transposition Reaction

1. A typical 10-μL in vitro transposition reaction is assembled from the following components: 1 μL EZ::TN 10X reaction buffer; 0.5 μg target DNA; molar equivalent of Kan TEV transposon (*see* **Note 1**); distilled water to 9 μL; 1 μL EZ::TN transposase (Epicentre, 1 U/μL).
2. Incubate at 37 °C for 2 h.
3. Stop reaction by addition of 1 μL 10X stop solution (1% SDS) and heating to 70 °C for 10 min.
4. Electroporate 1-μL aliquots into electrocompetent *E. coli* cells.
5. Add 5 volumes SOB plus 2% glucose plus 10 mM $MgCl_2$ and incubate at 37 °C for 2 h.
6. Plate on ampicillin plus kanamycin to select plasmids that have been targeted by the transposon.;
7. Prepare plasmid DNA from pooled Amp Kan cells (*see* **Note 2**).

3.2.2. Removal of Extraneous Transposons by Subcloning

Digest 2 μg of the plasmid DNA from **Subheading 3.2.1.** with the two restriction enzymes (X and Y) that cut at the ends of the protein-coding sequence and gel/Qiaex purify the transposon-containing fragment (*see* **Fig. 2**). Insert into fresh X, Y double-digested vector, again selecting Amp Kan plasmids, and prepare plasmid DNA from the pooled cells. This procedure generates a library in which all the transposons reside within the protein-coding region.

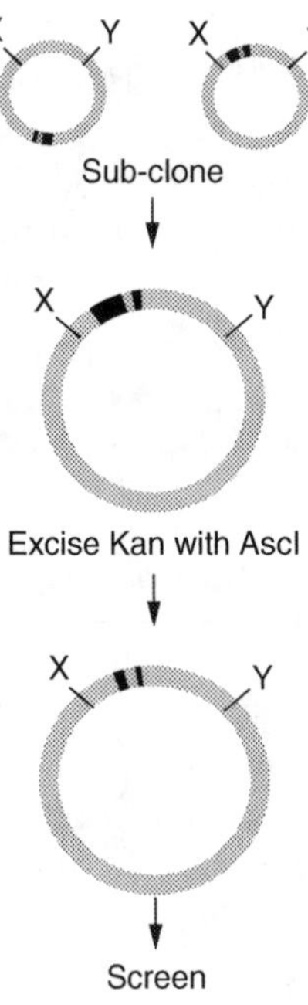

Fig. 2. Schematic showing removal of extraneous transposons by subcloning using restriction enzyme sites X and Y. This produces a library of plasmids carrying transposons in the X-Y region. The Kan marker is removed by cleavage with AscI and recircularization. This produces a library of plasmids carrying tobacco etch virus (TEV) transposon insertions (69 basepairs) in the X-Y region (*see* **Subheadings 3.2** and **3.3.**).

3.2.3. Excision of Kanamycin-Resistance Marker Gene

Digest plasmid DNA from **Subheading 3.2.2.** with AscI to release the Kan marker fragment and gel/Qiaex purify the linearized plasmid (*see* **Fig. 2**). Recircularize by addition of T4 DNA ligase and electroporate into *E. coli* cells as before, this time selecting for Amp. Prepare plasmid DNA from the pooled cells. This generates a library of target genes, each bearing a randomly placed 69-bp insert that includes a unique AscI site. Five of the possible six reading frames for TEV transposon insertion are closed by stop codons in the transposon, and the remaining reading frame introduces a short, internal peptide tag (LSLIHIWRAVENLYFQGDVYKRQ) that includes the recognition site for TEV protease (ENLYFQG). Many positions in the protein will not tolerate insertion of foreign sequences, so it is next necessary to identify genes that remain functional after insertion of the TEV protease target sequence (**Subheading 3.3.**).

3.3. Isolation of Functional Genes Carrying TEV Protease Sites

For genes with a function that is essential for growth, it is possible to screen or select for gene function, typically using a "plasmid shuffle" approach *(3)*. This method requires construction of a haploid yeast strain in which the genomic copy of the gene of interest is deleted, and essential gene functions are provided instead by a copy of the gene on a *URA3*-marked plasmid. The library of candidate genes produced in **Subheading 3.2.3.** is introduced into the gene disruption strain on *TRP1*-marked plasmids. Individual genes are then scored for ability to support growth on plates containing 5-fluoro-orotic acid (5-FOA). 5-FOA is converted to a toxic product (5-fluorouracil) by the action of the decarboxylase encoded by the *URA3* gene. Therefore, only cells that have lost the *URA3*-marked copy and retain a functional copy of the gene on the *TRP1* plasmid can survive 5-FOA counterselection. Plasmids are isolated from the survivors and retested in yeast (*see* **Note 3**). The location of the internal tag in each gene is determined by restriction mapping, using the unique AscI site, followed by DNA sequencing.

Use of proteins containing TEV protease sites in RNA–protein crosslinking experiments typically requires the preparation of extracts from cells expressing the modified protein. A collection of extracts is made from several yeast strains in which the protein of interest carries TEV protease sites at different positions and an N- or C-terminal epitope tag to allow affinity capture of the protein. The example described here (see **Subheading 3.5.**) uses yeast extracts competent for pre-mRNA splicing, made according to a standard protocol ***(4,5)***.

3.4. Synthesis of Pre-mRNAs Carrying Photocrosslinkers at Specific Positions

Site-specific RNA–protein photocrosslinking requires the construction of RNA molecules carrying a single photoactivatable 4-thioU at a specific position, preceded by a single ^{32}P-labeled phosphate. This is achieved by T4 DNA ligase-mediated ligation of two half-molecule RNAs, RNA1 and RNA2 (*see* **Fig. 3**). In the example described here, RNA1 consists of exon1-intron sequences from the yeast *ACT1* gene. RNA2 consists of the first 25 nucleotides of exon 2 from the *ACT1* gene and has a 4-thioU residue at its 5′ end. An *ACT1* pre-mRNA carrying the 4-thioU photocrosslinker immediately downstream of the 3′ splice site is made in three steps: (1) preparative transcription of RNA1 using T7 RNA polymerase; (2) phosphorylation of RNA2; and (3) ligation of the exon1-intron and exon2 RNAs using T4 DNA ligase and a 40-mer antisense DNA bridging oligonucleotide.

3.4.1. Preparative Transcription of RNA1 Using T7 RNA Polymerase

1. Set up a 200-μL transcription reaction from the following components: 40 μL 5X T7 buffer; 40 μL 5X NTP mix; 10 μL T7 actin exon1-intron PCR product (blunt ended; 0.5 μg/μL); 5 μL RNasin ribonuclease inhibitor; 5 μL T7 RNA polymerase (20 U/μL); 100 μL double-distilled water (ddH$_2$O).
2. Incubate at 37 °C for 2–6 h.
3. Add 25 μL 0.5 *M* EDTA to redissolve the magnesium pyrophosphate precipitate.
4. Add 225 μL phenol, vortex 1 min, spin 1 min in a microfuge, and transfer the aqueous phase to a fresh tube.

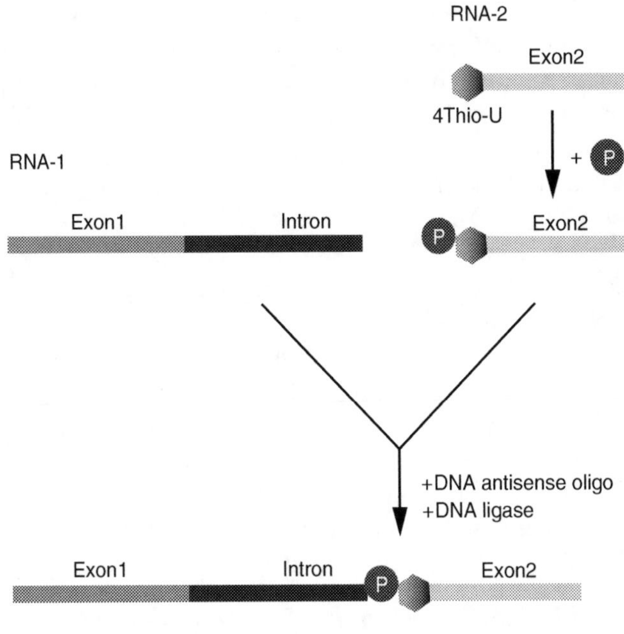

Fig. 3. Schematic showing synthesis of yeast *ACT1* pre-mRNA carrying a single 4-thioU (4-thiouridine) photocrosslinker at the 3′ splice site. RNA1 is made by transcription of a polymerase chain reaction (PCR) product in which the T7 RNA polymerase promoter is fused to exon1-intron sequences from the yeast *ACT1* gene. RNA2 is a synthetic exon2 oligoribonucleotide with a single 4-thioU residue at its 5′ end. A single ^{32}P label is introduced at the 5′ end of RNA2 (using T4 polynucleotide kinase), and RNA1 is ligated to RNA2 using an antisense "splint" oligonucleotide and T4 DNA ligase (*see* **Subheading 3.4.**).

5. Add 225 µL chloforom/isoamyl alcohol (24:1), repeat the extraction, and spin as above.
6. Add 45 µL 5 *M* ammonium acetate and 3 volumes ethanol; mix and chill at –20 °C for 15 min.
7. Spin 2 min in a microfuge and remove the supernatant.
8. Dissolve the RNA pellet in formamide gel loading buffer and denature at 90 °C for 1 min.
9. Purify the RNA by denaturing PAGE (8 *M* urea) followed by passive elution in 300 µL elution buffer for 2–12 h at 37 °C.
10. Precipitate the eluted RNA by addition of 3 volumes of ethanol.
11. Spin 2 min in a microfuge and remove the supernatant.
12. Rinse with 96% ethanol and dissolve the RNA pellet in double-distilled water.
13. Measure the absorbance at 260 nm and calculate the RNA concentration (*see* **Notes 4** and **5**).

3.4.2. Phosphorylation of RNA2

1. Set up a 20-μL phosphorylation reaction from the following components: 1 μL 20 μ*M* RNA2; 12 μL γ-[^{32}P]-ATP (Amersham, 3000 Ci/mmol and 10 μCi/μL); 2 μL 10X T4 polynucleotide kinase buffer; 2 μL ddH_2O; 1 μL RNasin ribonuclease inhibitor (40 U/μL); 2 μL T4 polynucleotide kinase (10 U/μL).
2. Incubate at 37 °C for 60 min.
3. Stop the reaction by addition of 1 μL 0.5 *M* EDTA.
4. Add 90 μL ddH_2O and 100 μL phenol.
5. Vortex 1 min, spin 1 min in a microfuge, and transfer the aqueous phase to a fresh tube.
6. Add 20 μL 5 *M* ammonium acetate, 1 μL glycogen carrier (10 μg/μL), and 3 volumes of ethanol.
7. Mix and chill at –20 °C for 15 min, then spin 2 min in a microfuge.
8. Remove the supernatant and rinse the pellet with cold 96% ethanol.
9. Dissolve the 5′-phosphorylated radiolabeled RNA in 8 μL ddH_2O.

3.4.3. RNA Ligation Using T4 DNA Ligase

Anneal the two RNA species to the complementary bridging DNA oligonucleotide as follows:

1. Transfer the radiolabeled RNA2 (8 μL from **Subheading 3.4.2.**) to a 0.5-mL microfuge tube.
2. Add 1 μL 20 μ*M* RNA1 (from **Subheading 3.4.1.**).
3. Add 1 μL 20 μ*M* antisense bridging oligonucleotide.
4. Cool slowly from 75 °C to 30 °C in a thermal cycler machine.
5. Spin 10 s in a microfuge.
6. Add 10 μL T4 DNA ligase mix.
7. Incubate at 30 °C for 3 h.
8. Add 1 μL 0.5 *M* EDTA, 80 μL ddH_2O, 100 μL phenol.
9. Vortex 1 min and spin in a microfuge for 1 min.
10. Transfer the aqueous phase to a fresh tube.
11. Add 10 μL 3 *M* sodium acetate, pH 5.3, and 3 volumes of ethanol.
12. Mix and chill at –20 °C for 15 min.
13. Dissolve the RNA pellet in formamide gel loading buffer and denature at 90 °C for 1 min.
14. Purify the RNA by denaturing PAGE (8 *M* urea).
15. Visualize the ligated pre-mRNA by autoradiography.
16. Recover the pre-mRNA from the gel slice by passive elution in 300 μL elution buffer for 2–12 h at 37 °C.
17. Precipitate the eluted RNA by addition of 3 volumes of ethanol and 1 μL glycogen carrier (10 μg/μL) and chilling at –20 °C for 15 min.
18. Spin 2 min in a microfuge and remove the supernatant.
19. Rinse with 96% ethanol and dissolve the RNA pellet in 25–50 μL ddH_2O.
20. Quantify the radiolabeled pre-mRNA by Cerenkov counting and adjust to 10,000 cpm/μL (*see* **Note 6**).

3.5. RNA–Protein Crosslinking and Crosslink Mapping Using TEV Protease

1. Set up 200-μL pre-mRNA splicing reactions on ice from the following components: 80 μL pre-mRNA splicing extract; 40 μL 5X splice cocktail; 20 μL 30% (w/v) polyethylene glycol 6000; 40 μL ddH_2O; 20 μL actin pre-mRNA (10,000 cpm/μL, from **Subheading 3.4.3.**).
2. Incubate at 23 °C 5–10 min (*see* **Note 7**).
3. Transfer 50-μL droplets to a chilled (4 °C) Parafilm-covered metal block.
4. UV irradiate (365 nm) for 5 min.
5. Transfer the UV-irradiated reaction to a 1.5-mL microcentrifuge tube.
6. Add 1 mL IP150.
7. Add 20 μL IgG-Sepharose beads (protein A tag) or anti-FLAG M2-agarose beads (FLAG tag).
8. Mix gently at 4 °C for 2 h to capture epitope-tagged proteins/complexes.
9. Spin out the beads (2500 g for 10 s in a microfuge).
10. Wash twice with 1 mL IP150 (ice cold).
11. Wash once with 300 μL 1X TEV buffer.
12. Resuspend the beads in 200 μL 1X TEV buffer.
13. Divide sample into two 100-μL aliquots.
14. Add 1 μL TEV protease (10 U/μL) to one aliquot.
15. Incubate both aliquots 16 h at 18 °C.
16. Add 2 μL RNase T1 (10 U/μL) (*see* **Note 8**).
17. Incubate at 37 °C for 30 min.
18. Wash twice with 1 mL IP150 (ice cold).
19. Elute proteins from beads in 10 μL SDS-PAGE sample buffer at 85 °C for 2 min.
20. Fractionate the ^{32}P-labeled proteins on a 3–8% Tris-acetate gel by electrophoresis at 140 V for 75 min.
21. Fix the gel in 7% acetic acid, 7% methanol (15 min).
22. Dry the gel at 80 °C for 30 min (ThermoSavant Gel Dryer or equivalent).
23. Visualize labeled proteins by phosphorimaging (typically 12- to 48-h exposure).

Using splicing extracts differing in the location of the TEV protease site in the protein of interest, it is straightforward to deduce the approximate position of RNA–protein contact, down to the resolution afforded by the range of TEV protease sites available, simply by comparison of the patterns of ^{32}P-labeled TEV protease digestion products ***(6)***. If the RNA–protein contact has been mapped between two TEV protease sites but greater resolution is required, a secondary screen for additional TEV protease sites can be performed, and this is facilitated if two unique restriction sites are available flanking the relevant area of protein-coding sequence. The procedure then is to repeat the transposition reaction (**Subheading 3.2.1.**) using the small restriction fragment as the target DNA, followed by ligation into a suitable "gapped" plasmid vector and selection for Amp Kan transformants in *E. coli*. Removal of the Kan marker and selection of functional genes are performed as before (**Subheadings 3.2.3.**

and **3.3.**), and the crosslink-mapping experiment **Subheading 3.5.**) is repeated (*see* **Note 9**).

4. Notes

1. The number of micromoles of target and transposon can be calculated using the following formula:

   Micromoles = Micrograms/Size of target of transposon × 660
   For example 0.2 μg of a 6100-bp target = 0.2/6100 × 660 = 0.05×10^{-6} μmol = 0.05 pmol.

2. Good coverage of all possible transposon positions in a target DNA molecule requires 5–10 × 10^3 independent transposon insertions per kilobase of target.
3. Libraries of heterologous or nonessential RNA binding protein genes carrying transposons can be screened for function in yeast using the three-hybrid assay for RNA binding *(7)*.
4. The concentration of the purified transcript can be calculated using the following formula:

   Concentration (μM) = Absorbance at 260 nm × 100/n
   where *n* is the length of the transcript in nucleotides.

5. It is essential to synthesize in parallel a control pre-mRNA carrying an unmodified uridine residue in place of the 4-thioU crosslinker. The two pre-mRNAs can then be used side by side in photocrosslinking experiments to demonstrate that RNA–protein crosslinking is 4-thioU dependent and therefore genuinely site specific.
6. One common problem with RNA transcribed using phage polymerases such as T7 RNA polymerase is that some templates are prone to yield transcripts with length heterogeneity at the 3′ terminus. There can be dramatic reductions in the yield of ligation products from reactions involving heterogeneous T7 transcript 3′ ends. This drawback can be circumvented using a *cis*-acting ribozyme (such as the hammerhead ribozyme) fused to the 3′ end of the transcript *(8)*. Ribozyme cleavage yields clean ends bearing a cyclic 2′,3′-phosphodiester. Such ends can be easily and efficiently decyclized by treatment with T4 polynucleotide kinase to remove the 2′,3′-phosphate in preparation for ligation to a phosphorylated 5′ end. It is also advantageous to gel purify synthetic RNA oligonucleotides by denaturing PAGE to remove shorter RNA contaminants that arise during chemical synthesis.
7. A small sample of the reaction can be withdrawn at this stage, deproteinized by phenol extraction, ethanol precipitated, and fractionated by 8 *M* urea PAGE. Splicing intermediates and products can then be visualized directly by phosphorimaging, allowing splicing activity to be monitored.
8. RNase digestion trims off most of the radioactively labeled RNA molecule, leaving a small oligonucleotide (containing the ^{32}P) covalently crosslinked to the protein, which can then be fractionated normally by SDS-PAGE. The choice of the RNase to be used in any particular experiment (usually we use RNase T1, which cuts after G residues, or RNase A, which cuts after C and U residues) is dictated by the RNA sequence in the vicinity of the 4-thioU crosslinker.

9. An alternative approach is to introduce TEV protease sites by site-directed mutagenesis ***(9)***, targeting regions of poor sequence conservation predicted to be exposed at the surface of the protein.

Acknowledgments

We are grateful to Adam Wilkinson for his contributions to the design and construction of the transposons used in this work. This work was supported by the MRC.

References

1. Sontheimer, E. J. (1994) Site-specific RNA crosslinking with 4-thioUridine. *Mol. Biol. Rep.* **20**, 35–44.
2. Goryshin, I. Y., and Reznikoff W. S. (1998) Tn5 in vitro transposition. *J. Biol. Chem.* **273**, 7367–7374.
3. Sikorski, R. S., and Boeke, J. D. (1991) In vitro mutagenesis and plasmid shuffling: from cloned gene to mutant yeast. *Methods Enzymol.* **194**, 302–318.
4. Lin, R.-J., Newman, A. J., Cheng, S.-C., and Abelson, J. (1985) Yeast mRNA splicing in vitro. *J. Biol. Chem.* **260**, 14780–14792.
5. Stevens, S. W., and Abelson, J. (2002) Yeast pre-mRNA splicing: methods, mechanisms and machinery. *Methods Enzymol.* **351**, 200–220.
6. Turner, I. A., Norman, C. M., Churcher, M. J., and Newman, A. J. (2006) Dissection of Prp8 protein defines multiple interactions with crucial RNA sequences in the catalytic core of the spliceosome. *RNA* **12**, 375–386.
7. SenGupta, D. J., Zhang, B., Kraemer, B., Pochart, P., Fields, S., and Wickens, M. (1996) A three-hybrid system to detect RNA-protein interactions in vitro. *Proc. Natl. Acad. Sci. U. S. A.* **93**, 8496–8501.
8. Price, S. R., Ito, N., Oubridge, C., Avis, J. M., and Nagai, K. (1995) Crystallisation of RNA-protein complexes I. Methods for the large-scale preparation of RNA suitable for crystallographic studies. *J. Mol. Biol.* **249**, 398–408.
9. Kunkel, T. A. (1985) Rapid and efficient site-specific mutagenesis without phenotypic selection. *Proc. Natl. Acad. Sci. U. S. A.* **82**, 488–492.

15

Structural Analysis of Protein–RNA Interactions With Mass Spectrometry

Mamuka Kvaratskhelia and Stuart F.J. Le Grice

Summary

We present a high-resolution mass spectrometric footprinting approach enabling the identification of amino acids in the protein of interest interacting with cognate RNA. This approach is particularly attractive for studying large nucleoprotein complexes that are less amenable to crystallographic or nuclear magnetic resonance analysis. Importantly, our methodology allows examination of protein–RNA interactions under biologically relevant conditions using limited amounts of protein and nucleic acid samples.

Key Words: Footprinting; mass spectrometry; nucleoprotein complex; protein; RNA; structure.

1. Introduction

The structures for many RNA-processing proteins or separate protein subunits are available at atomic resolution ***(1–9)***. However, biologically relevant large protein–RNA structures are often less amenable to crystallographic or nuclear magnetic resonance (NMR) analysis. Therefore, there is a need for devising new and complementary approaches that enable rapid and accurate mapping of protein–RNA contacts. We describe a mass spectrometric (MS) footprinting methodology that allows us to identify amino acids in the protein of interest that interact with cognate RNA ***(10)***.

The experimental strategy is depicted in **Fig. 1**. The method exploits differential accessibility of the primary amine-modifying reagent *N*-hydroxysuccinimide (NHS)-biotin to lysine residues in the free protein vs the protein–RNA complex. Subsequent MS analysis enables accurate identification of these residues. Monitoring lysine accessibility is a logical choice as lysine-phosphate backbone contacts play a key role in formation of many nucleoprotein complexes. Introducing sodium dodecyl sulfate polyacrylamide

From: *Methods in Molecular Biology, vol. 488: RNA-Protein Interaction Protocols*
Edited by: Ren-Jang Lin © Humana Press Inc, Totowa, NJ

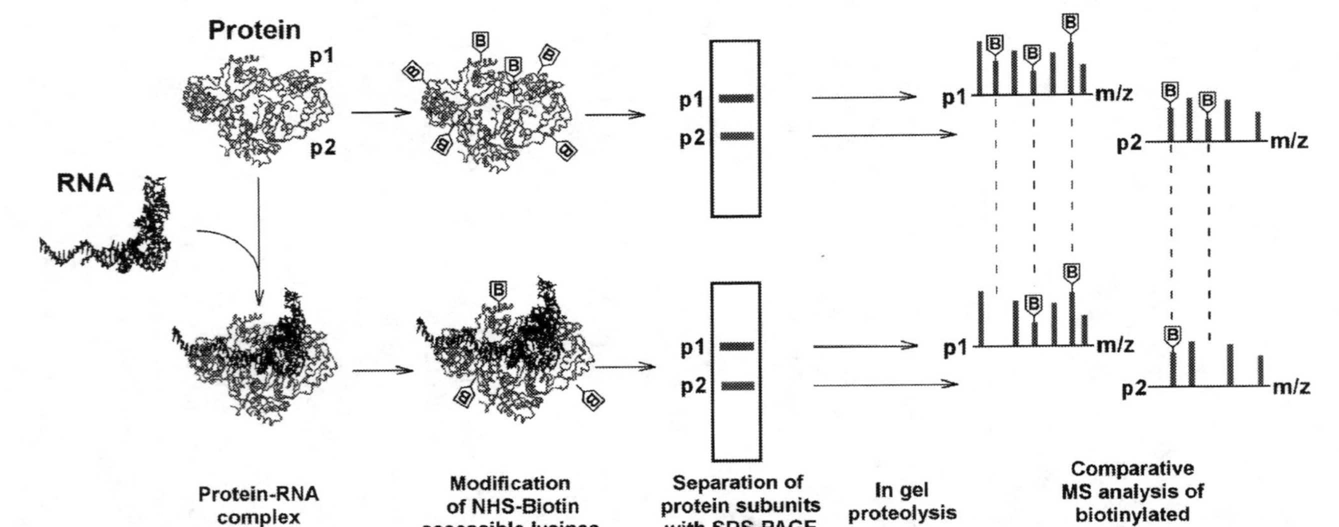

Fig. 1. Protein footprinting strategy. Biotinylation reactions of free protein comprised of separate protein subunits, and the preformed protein–RNA complexes are carried out in parallel. Surface-exposed lysines are modified by NHS-biotin in free protein, while those coordinating RNA become shielded from modification in the nucleoprotein complex. Individual protein subunits are separated by sodium dodecyl sulfate polyacrylamide gel electrophoresis (SDS-PAGE) and then subjected to in-gel proteolysis. Comparative mass spectrometric (MS) analysis of the peptide fragments enables us to identify lysines shielded by RNA contacts from those remaining susceptible to modification in the complex. The experimental scheme is adapted from **ref. *10***.

gel electrophoresis (SDS-PAGE) and in-gel proteolysis prior to MS is important for the following reasons: SDS-PAGE allows separation of individual protein subunits based on their molecular weight differences. Thereafter, contact lysines can be accurately assigned to individual components of a multisubunit complex. For our analysis of HIV-1 reverse transcriptase (RT), this was of particular importance since both subunits are derived from the same gene but differentially processed by the viral protease ***(10)***. Subsequent in-gel proteolysis produces short peptide fragments amenable to MS and MS/MS analysis. The biotinylated peptide peaks can be readily identified from MS data and the modified sites accurately assigned to appropriate lysine residues by MS/MS analysis. Comparative examination reveals lysines readily modified in the free protein but protected in the context of the nucleoprotein complex (**Fig. 2**). The methodology can be expanded to probe other RNA-interacting amino acids such as Arg, Trp, Tyr, His, and Cys using corresponding commercially available reagents ***(11)***.

2. Materials

2.1. Modification of Lysine Residues

1. 50 m*M* HEPES buffer, pH 7.5, 50 m*M* NaCl (*see* **Note 1**).
2. NHS-biotin (Pierce).
3. Quenching solution: 100 m*M* Tris-HCl buffer, pH 8.0, containing 100 m*M* lysine.

2.2. SDS-PAGE and In-Gel Proteolysis

1. Staining solution: 0.5% w/v Comassie brilliant blue (Bio-Rad), 50% methanol, 10% acetic acid, 40% high-performance liquid chromatography (HPLC) pure water.
2. Destain solution: 50% methanol, 10% acetic acid, 40% HPLC pure water.
3. 50 m*M* NH_4HCO_3 buffer, pH 8.0.
4. Acetonitrile (Fisher).
5. SpeedVac (Thermo Savant, Holbrook, NY).
6. Platform shaker (New Brunswick Scientific, Edison, NJ).
7. Scalpel (Fisher).

2.3. Mass Spectrometric Analysis

1. Matrix: 5 mg/mL solution of α-cyano-4-hydroxy-cinnamic acid (Sigma) in 75% acetonitrile/25% water.
2. MALDI-TOF (matrix-assisted laser desorption/ionization time-of-flight) instrument equipped with a curved field reflectron feature (Kratos Analytical Instruments, Manchester, U.K.).
3. Waters Q-ToF-II instrument (Manchester, U.K.) equipped with an electrospray source and a Waters cap-LC (Waters Symmetrie300 precolumn and a Micro-Tech Scientific, Vista, CA, ZC-10-C18SBWX-150 column).

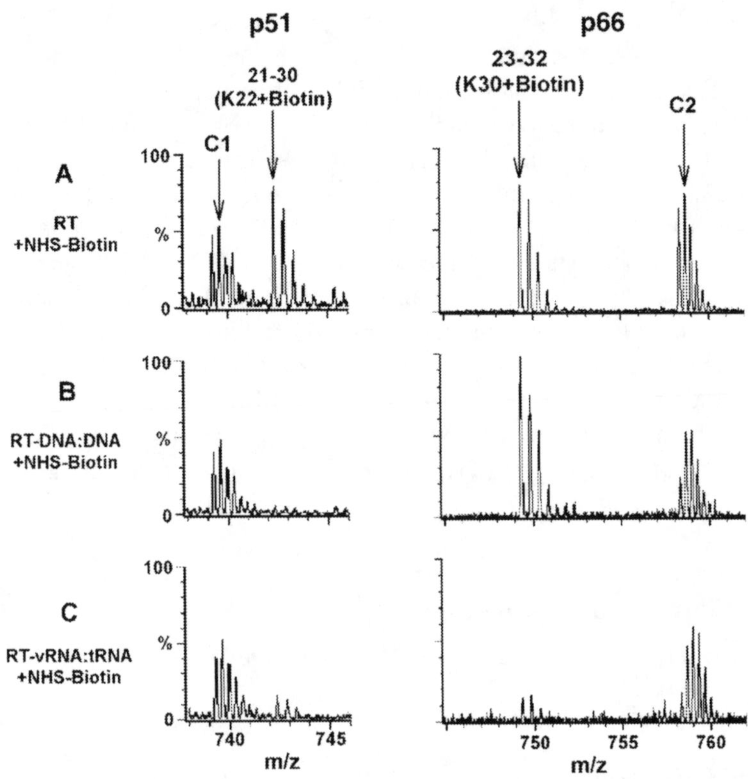

Fig. 2. Mass spectrometric data showing similarity (left panel) and differences (right panel) between HIV-1 reverse transcriptase (RT) when complexed with DNA:DNA and a viral RNA:transfer RNA (tRNA) duplex. RT is comprised of two protein subunits, p66 and p51. Treatment of free RT with *N*-hydroxysuccinimide (NHS)-biotin resulted in modification of K22 of p51 yielding the peak corresponding to the 21- to 30-peptide stretch, plus one biotin molecule (**A**, left panel). In the context of the RT-DNA:DNA (**B**, left panel) or RT-viral RNA:tRNA complex (**C**, left panel), this peak was significantly diminished, most likely due to protection by the cognate nucleic acid contacts. The modification of K30 in p66 is shown in the right panel. The 23–32 (K30 + biotin) peak persisted in RT-DNA:DNA (**B**, right panel) but diminished significantly in the RT-viral RNA:tRNA complex (**C**, right panel). These data indicate that K22 of p51 is coordinating both DNA:DNA and viral RNA:tRNA, while K30 of p66 contacts specifically the viral RNA:tRNA duplex and not DNA:DNA. Unmodified RT peptide peaks C1 and C2 serve as internal controls. Each multiply charged peptide ion resulted in a clearly resolved peak cluster, indicating monoisotopic resolution in our mass spectrometric analysis. Data adapted from **ref. *10***.

3. Methods

3.1. Modification of Lysine Residues

1. Prepare and analyze the following two reaction mixtures in parallel: Free protein at 10–100 ng/μL final concentration, and protein–RNA complex containing the same amount of protein and 2- to 10-fold molar excess of RNA.
2. Add freshly prepared stock solution of NHS-biotin to both reactions to obtain the final concentration of 200–1000 μ*M (see* **Note 2**). Incubate the reactions at room temperature for 30 min.
3. Terminate both reactions by adding the quenching solution to the reaction mixture in 1/10 ratio.
4. If required, concentrate the reaction mixes by vacuum desiccation (SpeedVac) at medium heat (45 °C) until the sample volume is reduced to 20 μL (*see* **Note 3**).

3.2. SDS-PAGE and In-Gel Proteolysis

1. Add SDS-PAGE loading buffer to the sample and fractionate protein subunits via SDS-PAGE.
2. Following electrophoresis, stain the gel with Coomassie blue for 5 min at room temperature.
3. Destain the gel for at least 2 h (or until the protein bands are distinctly visible), exchanging the destaining solution several times.
4. Excise the bands with a clean scalpel (*see* **Note 4**) and slice the protein bands in three to four pieces. Place the sliced gel pieces into 1.5-mL microcentrifuge tubes.
5. Add 1 mL destain solution to each tube and shake overnight at 200 rpm
6. Carefully remove the destain solution using a pipet without touching the gel slice. Add 1 mL fresh destain solution and shake samples for an additional 1 h.
7. Centrifuge samples in a microcentrifuge and carefully remove the destain solution with a pipet.
8. Add 1 mL 50 m*M* NH_4HCO_3 solution and shake the tubes for 15 min. Centrifuge the samples and remove ammonium bicarbonate solution with a pipet. Repeat this step.
9. Add 50% H_20/50% acetonitrile solution to the gel pieces and shake samples for 1 h at room temperature.
10. Remove the solution and add 200 μL of 100% acetonitrile to the gel pieces. Shake samples for 15 min.
11. Remove acetonitrile and desiccate the gel pieces with SpeedVac for 15 min at medium heat (45 °C).
12. Prepare a stock solution (0.2 μg/μL) of trypsin in 10 m*M* HCl (*see* **Note 5**). Immediately prior to use, dilute the stock solution 10-fold with 50 m*M* ammonium bicarbonate buffer.
13. Add 50 μL trypsin solution to each microcentrifuge tube (*see* **Note 6**) and shake at 200 rpm overnight at room temperature.
14. Add 150 μL pure acetonitrile to each digestion and immediately vortex samples.

15. Centrifuge the tubes and carefully pipet out 180 µL supernatant without touching the gel slice (*see* **Note 7**). Transfer the solution into 500-µL microcentrifuge tubes.
16. Dry peptide mixtures completely using vacuum desiccation at medium heat (45 °C).
17. Add 15 µL HPLC-grade water to each sample. At this stage, the samples are ready for MS analysis.

3.3. Mass Spectrometric Analysis

1. Divide each sample in two portions for MALDI-TOF (0.5 to 2 µL) and Q-ToF (6 to 12 µL) analysis.
2. Apply 0.5-µL sample onto the MALDI plate and immediately mix with 0.5 µL matrix solution by pipeting the mix up and down for several times. Air dry samples at room temperature (~15 min). These sample can now be used for MALDI-TOF analysis.
3. Analyze samples in the reflectron mode. Once the peptide peaks are identified, activate the postsource decay feature of the MALDI to determine amino acid sequencing of the peptides.
4. Place 6–12 µL of sample into the vials for Q-ToF analysis. Perform two sequential linear gradients of 5–40% acetonitrile for 35 min and 40–90% acetonitrile for 10 min. Operate the instrument in the MS/MS mode.
5. Analyze the data with the MASCOT engine (http://www.matrixscience.com).
6. For quantitative comparison of lysine modifications in free protein and protein–RNA complexes, use at least two unmodified proteolytic peptide peaks as controls.

4. Notes

1. Use nonamine buffers, pH 7.0–9.0. Should protein or RNA stock solutions contain amines (for example, Tris), dialyze the preparations against nonamine buffers such as phosphate, HEPES, borate, and carbonate.
2. Prior to proceeding to the footprinting experiments, optimize the NHS-biotin concentration and check that the integrity of the protein–RNA complex is fully preserved under the experimental conditions. For this, incubate the protein with increasing concentrations of NHS-biotin and monitor the ability of the protein to bind cognate RNA using conventional assays such as gel retardation or filter-binding analysis. For footprinting experiments, choose the lowest concentration of NHS-biotin at which the ability of the protein to bind cognate RNA is fully impaired. Use this concentration of NHS-biotin to test the integrity of the protein–RNA complex. Preform the protein–RNA complex first and then expose the complex to NHS-biotin. The complex should be stable enough that it does not dissociate on modification with NHS-biotin. For quantitative comparison, examine unmodified protein–RNA complex in parallel experiment.
3. If the sample is completely dried, redissolve the protein by adding 20 µL H_2O.
4. Wear gloves during all experiments. Human keratin is a main contamination observed during MS analysis.

5. Store the remaining stock solution of trypsin at –20 °C. Reuse only once immediately after thawing and discard the remaining stock. Alternatively, store trypsin in aliquots corresponding to those required for proteolysis.
6. Allow the gel slices to fully rehydrate (10–15 min). Prior to overnight hydrolysis, check that the hydrated gel slices are fully submerged in the buffer. If needed, add an additional 10 to 20 µL ammonium bicarbonate buffer to the reaction mix.
7. The peptide extraction step can be repeated to increase the yield by 20–30%. In particular, add 50 µL ammonium bicarbonate buffer to the gel slice and incubate the mix at room temperature for 15 min. Then add 150 µL acetonitrile and vortex the mix immediately. Remove 180 µL supernatant without touching the gel slices and combine with previous extraction of the same sample.

References

1. Ma, J. B., Ye, K., and Patel, D. J. (2004). Structural basis for overhang-specific small interfering RNA recognition by the PAZ domain. *Nature* **429**, 318–322.
2. Opalka, N., Chlenov, M., Chacon, P., Rice, W. J., Wriggers, W., and Darst, S. A. (2003). Structure and function of the transcription elongation factor GreB bound to bacterial RNA polymerase. *Cell* **114**, 335–345.
3. Monzingo, A. F., Gao, J., Qiu, J., Georgiou, G., and Robertus, J. D. (2003). The X-ray structure of *Escherichia coli* RraA (MenG), a protein inhibitor of RNA processing. *J. Mol. Biol.* **332**, 1015–1024.
4. Lau, C. K., Diem, M. D., Dreyfuss, G., and Van Duyne, G. D. (2003). Structure of the Y14-Magoh core of the exon junction complex. *Curr. Biol.* **13**, 933–941.
5. Krasilnikov, A. S., Yang, X., Pan, T., and Mondragon, A. (2003). Crystal structure of the specificity domain of ribonuclease P. *Nature* **421**, 760–764.
6. Augustin, M. A., Reichert, A. S., Betat, H., Huber, R., Morl, M., and Steegborn, C. (2003). Crystal structure of the human CCA-adding enzyme: insights into template-independent polymerization. *J. Mol. Biol.* **328**, 985–994.
7. Calero, G., Wilson, K. F., Ly, T., Rios-Steiner, J. L., Clardy, J. C., and Cerione, R. A. (2002). Structural basis of m7GpppG binding to the nuclear cap-binding protein complex. *Nat. Struct. Biol.* **9**, 912–917.
8. Nagai, K., Muto, Y., Pomeranz Krummel, D. A., et al. (2001). Structure and assembly of the spliceosomal snRNPs. Novartis Medal Lecture. *Biochem. Soc. Trans.* **29**, 15–26.
9. Cramer, P., Bushnell, D. A., and Kornberg, R. D. (2001). Structural basis of transcription: RNA polymerase II at 2.8 angstrom resolution. *Science* **292**, 1863–1876.
10. Kvaratskhelia, M., Miller, J. T., Budihas, S. R., Pannell, L. K., and Le Grice, S. F. (2002). Identification of specific HIV-1 reverse transcriptase contacts to the viral RNA:tRNA complex by mass spectrometry and a primary amine selective reagent. *Proc. Natl. Acad. Sci. U. S. A.* **99**, 15988–15993.
11. Lundblad, R. L. (1995). *Techniques in Protein Modification*, CRC Press, Boca Raton, FL.

16

Analyzing RNA–Protein Crosslinking Sites in Unlabeled Ribonucleoprotein Complexes by Mass Spectrometry

Henning Urlaub, Eva Kühn-Hölsken, and Reinhard Lührmann

Summary

Mass spectrometry is a powerful tool for the analysis of biomolecules, proteins, nucleic acids, carbohydrates, lipids. In combination with genome sequences that are available in the databases, it has proven to be the most straightforward and sensitive technique for the sequence analysis and hence the identification of protein components in the cells, their (post)translational modifications, and their relative and absolute abundance. In addition, mass spectrometric methods are successfully applied for the structural analysis of biomolecules (i.e., deciphering molecule–ligand interactions and spatial quartenary arrangements of molecule complexes). We describe a methodology for the mass spectrometric analysis of protein–RNA contact sites in purified ribonucleoprotein (RNP) particles. The method comprises ultraviolet (UV) crosslinking of proteins to RNA, hydrolysis of the protein and RNA moieties, isolation of cross-linked peptide-RNA oligonucleotides, MALDI (matrix-assisted laser desorption/ionization) mass spectrometry of the isolated conjugates to determine the sequence of the crosslinked peptide and RNA part. The utility of this methodology is demonstrated on crosslinks isolated from UV-irradiated spliceosomal particles; these were [15.5 K-61 K-U4atac] small nuclear ribonucleoprotein (snRNP) particles prepared by reconstitution in vitro and U1 snRNP particles purified from HeLa cells.

Key Words: Crosslinking; mass spectrometry; protein; RNA.

1. Introduction

RNA molecules play a fundamental role in cellular processes such as gene expression (transcription and pre-mRNA [messenger RNA] processing), post-transcriptional control (mRNA stability), RNA export, ribosomal RNA (rRNA) maturation, translation, and translational control. RNA molecules that are involved in these processes are rarely active in the absence of proteins but are found as components of stable ribonucleoprotein (RNP) particles. Since protein–RNA interactions lie at the structural and functional heart of the RNP

From: *Methods in Molecular Biology, vol. 488: RNA-Protein Interaction Protocols*
Edited by: Ren-Jang Lin © Humana Press Inc, Totowa, NJ

particles, much attention is currently being devoted to questions of protein and RNA tertiary structures and to the quaternary arrangements of the individual macromolecules in RNP particles.

We present a mass spectrometric (MS) method that we have established that allows us to characterize sites of direct protein–RNA contact in RNP particles, either native or reconstituted in vitro, after the contacts have been made permanent by ultraviolet (UV) crosslinking ***(1–4)***. This method enables us to identify directly distinct regions of proteins that interact with RNA in RNP particles and thus allows the definition of novel putative RNA binding domains (RBDs) of individual proteins.

Furthermore, in the absence of highly resolved three-dimensional RNP structures, our approach yields information about the orientation and the overall arrangement of proteins and RNA within RNP particles.

In this manner, we have identified the RBDs in spliceosomal RNP particles, for example, the highly conserved Sm proteins in the U1 small nuclear RNP (snRNP) and 25 S [U4/U6.U5] tri-snRNP particles. In this case, we were able to demonstrate that the evolutionarily conserved Sm site of the snRNA is in contact with the inner surface of the heptameric Sm ring ***(2)***. In addition, we found two amino acid residues within the RBD of the U1 70 K protein that are in direct contact with stem I of the U1 snRNA ***(1)***. Comparison with structures of RBD-containing proteins already crystallized (such as U1 A and Sxl; ***5,6***) revealed a similar amino acid–nucleotide interaction and provides further evidence for the highly conserved nature of the RBD–RNA interactions. We demonstrated that the novel U4/U6-specific protein 61 K ***(4)*** binds RNA directly, and that its evolutionarily conserved Nop domain ***(7)*** is in direct contact with the 5′ loop of the 5′ stem-loop of U4 snRNA in the presence of the 15.5 K protein.

In the protocols provided in this chapter, we refer to isolated U snRNP particles from HeLa cells ***(8–10)*** involved in pre-mRNA processing ***(11,12)*** and to [U4atac-15.5 K-61 K] protein–RNA complexes reconstituted in vitro ***(4)***. Importantly, we note that the approach can be regarded as a general one so that the protocols can easily be adapted to investigations of other native RNP particles or of RNP particles reconstituted in vitro.

2. Materials

1. Purified native RNP particles (e.g., U1 snRNP particles, 25 S [U4/U6.U5] tri-snRNP particles).
2. Overexpressed RNA binding protein (e.g., U4/U6-specific 15.5 K protein, U4/U6-specific 61 K protein).
3. RNA prepared by transcription in vitro (e.g., U4atac snRNA).
4. UV crosslinking equipment (**Fig. 1**) or UV Stratalinker 2400 (Stratagene, La Jolla, CA).

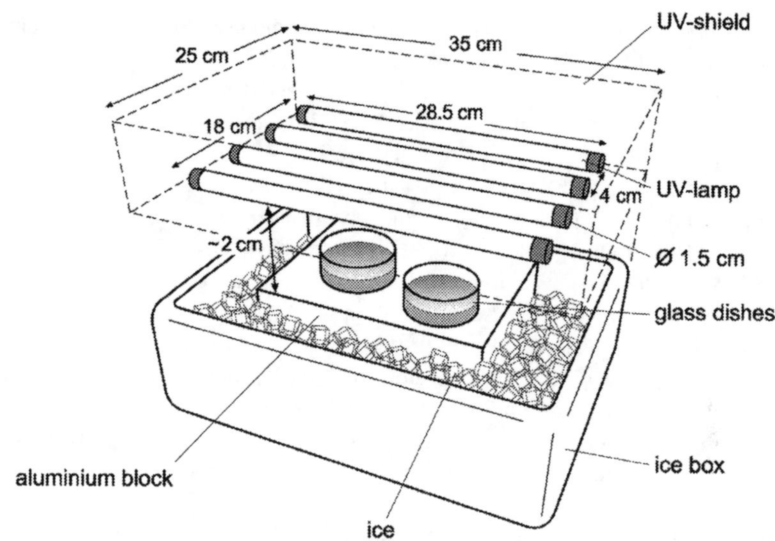

Fig. 1. Apparatus for ultraviolet (UV) crosslinking. The four 8-W germicidal lamps (e.g., G8T5, Herolab, Germany) have a dimension of 1.5 × 28.5 cm each. We used glass dishes with an inner diameter of 3.5 cm. The solution depth should be 1 mm. Dishes of different sizes can be used according to the sample volume. For details, see text.

5. Glass dishes with a planar surface and an inner diameter, for example, of 3.5 or 12.5 cm (according to sample volume).
6. Ethanol p.a. (per analysis) grade (Merck, Darmstadt, Germany).
7. Depending on scale: Standard Eppendorf tubes (Eppendorf, Hamburg, Germany) or Corex centrifugation tubes (30 mL); Sorvall rotor HB-4/HB-6 (Kendro Laboratory Products, Asheville, NC) or equivalent.
8. Depending on scale: Sorvall Evolution RC centrifuge (Kendro Laboratory Products) or cooled tabletop centrifuge (e.g., Heraeus Biofuge fresco; Kendro Laboratory Products) or equivalent.
9. Buffer 1: 6 *M* guanidinium hydrochloride (GHCl), 50 m*M* Tris/HCl, pH 8.0.
10. Buffer 2: 50 m*M* Tris/HCl, pH 7.5, 150 m*M* NaCl, 5 m*M* ethylenediaminetetraacetic acid (EDTA).
11. Endoproteinases trypsin (Promega, Madison, WI), Lys-C (Roche, Mannheim, Germany), Glu-C (Roche), chymotrypsin (Roche).
12. RNasin (Promega).
13. Equipment for size-exclusion (SE) chromatography (e.g., SMART System or FPLC System/Äkta Purifier system, Amersham Biotech/GE Healthcare, Uppsala, Sweden).
14. S75 HR SE column (3.1 × 300 m*M* or 10 × 300 m*M*, Amersham Biotech/GE Healthcare).

15. Sodium dodecyl sulfate polyacrylamide gel electrophoresis (SDS-PAGE) equipment and chemicals for silver staining according to **ref.** ***13***.
16. SDS sample buffer: 60 m*M* Tris-HCl, pH 6.8, 1 m*M* EDTA, 16% glycerol, 2% SDS, 0.1% bromophenol blue (BPB), 50 m*M* dithiothreitol (DTT).
17. Buffer 4: 50 m*M* Tris-HCl, pH 7.5, 2 m*M* EDTA.
18. Ribonucleases A and T1 (Ambion, Austin, TX).
19. Equipment for high-performance liquid chromatography (HPLC; e.g., SMART System; Microgradient system 140C, Applied Biosystems, Foster City, CA).
20. C18 reversed-phase (RP) column (2.1 × 150 m*M*; GraceVydac, Hesperia, CA) or 0.32 × 150 m*M* (Micro-Tech Scientific, Vista, CA).
21. HPLC solvent A: Water containing 0.1% (v/v) trifluoroacetic acid (TFA; Fluka, Buchs, Switzerland). Solvent B: Acetonitrile (ACN, Merck Darmstadt, Germany, LiChrosolv grade) containing 0.085% (v/v) TFA.
22. MALDI (matrix-assisted laser desorption/ionization) MS equipment (e.g., Bruker Reflex series, Ultraflex series, Applied Biosystems Voyager series, API 4700/4800 series).
23. DHB (2,5-dihydroxybenzoic acid; Sigma-Aldrich, Steinheim, Germany).
24. Matrix solution 1: 10 mg/mL solution of DHB in 50% ACN containing 0.1% TFA.
25. THAP (2′,4′,6′,-trihydroxyacetophenone; Fluka, Buchs, Switzerland).
26. Matrix solution 2: 10 mg/mL solution of THAP matrix in 50% ACN containing 0.5% TFA,
27. TFA (Fluka, Buchs, Switzerland).
28. Peptide standard calibration mixture for MALDI MS (e.g., Bruker Daltonics, Bremen, Germany).

3. Methods

The strategy we use for the identification of contact sites in native or in vitro reconstituted protein–RNA complexes is comprised of the following:

1. UV crosslinking of the complexes at 254 nm to generate a covalent bond between the protein and its cognate RNA at their site of interaction.
2. Purification of covalently linked peptide–RNA oligonucleotide heteroconjugates from the UV-irradiated complexes: SE chromatography; RP-HPLC.
3. Analysis of the purified heteroconjugates to identify the crosslinked protein region (including the actual crosslinked amino acid) and the crosslinked RNA region (including the actual crosslinked nucleotide): MALDI time-of-flight (TOF) MS.

3.1. General Remarks

Certain critical points have to be considered when one performs UV crosslinking and the subsequent identification of protein–RNA contact sites at the molecular level.

3.1.1. Amount and Concentration of the Particles

3.1.1.1. Native RNP Particles

Owing to the relatively low yield of UV crosslinking at 254 nm (*see* **Subheading 3.1.3.**, we recommend using a starting amount of RNP particles that is sufficient to allow purification and subsequent analysis of peptide–RNA oligonucleotide heteroconjugates. When one is using a capillary RP-HPLC system (*see* **Subheading 3.4.2., step 5**) with an RP column with an inner diameter of 300 µm for the final step in the purification of covalently linked peptide–RNA oligonucleotides, 10–50 µg of purified or in vitro reconstituted RNP complexes should be used as starting material. The native RNP particles are typically adjusted to a concentration of 0.1 mg/mL before crosslinking.

3.1.1.2. In Vitro Reconstituted Particles

Our purification procedure can also be applied to the analysis of RNP particles assembled in vitro from a known number of defined complexes. RNA can be most easily transcribed in vitro by phage RNA polymerase ***(14,15)*** or by chemical synthesis. Proteins can be produced by recombinant techniques ***(16)*** in *Escherichia coli*, yeast, or insect cells, or they can be purified from any other available biological source. The starting concentration of reconstituted particles is in the same range as for native ones. However, a number of considerations have to be taken into account in work with reconstituted particles. First, reconstitution in vitro can result in artificial crosslinking events when the assembly is incomplete or nonspecific. Incomplete assembly on the RNA can be minimized by using an excess of protein over RNA. In addition, nonspecific crosslinking can be due to the lack of additional specific protein components. An example for this is the crosslinking of the highly conserved Sm proteins, which are assembled in vitro on their cognate U snRNA Sm site ***(17,18)***. UV irradiation studies with radioactively labeled Sm site RNA and only three of the seven proteins (Sm G, E, F) revealed numerous crosslinks between these proteins and the RNA ***(2)***. However, only when all seven Sm proteins are fully assembled on the Sm site RNA is the crosslinking pattern of the proteins similar to that observed in native particles.

3.1.2. Buffer Systems

In general, any buffer systems are suitable, with the proviso that the buffer should not contain substantial concentrations of reagents that are known to scavenge radicals, for example, glycerol with a concentration of >15% (v/v). UV crosslinking is a radical reaction, and radical scavengers therefore drastically reduce its yield.

3.1.3. UV Crosslinking

UV crosslinking at 254 nm is a straightforward technique to detect direct protein–RNA interactions. It generates a covalent bond between an amino acid side chain of the protein and a base of the RNA when both are in a favorable position.

Putative UV-crosslinkable amino acids are tyrosine, histidine, phenylalanine, leucine, and cysteine ***(3)***. The most UV-reactive base is uridine ***(19)***. Note that not all proteins that are tightly associated with RNA can be UV crosslinked. An example of this is the human spliceosomal protein 15.5 K bound to U4/U4atac snRNA ***(20)***. The crystal structure of protein 15.5 K in complex with the 5′ stem-loop of U4 snRNA shows that this protein interacts almost exclusively with a purine-rich internal loop (AAU) within the 5′ stem-loop of U4 snRNA ***(21)***. Although a variety of amino acids within the protein interact through hydrogen bonds and hydrophobic interactions with the bases of the asymmetric internal loop of the RNA, neither of these amino acids contains a crosslinkable side chain. In addition, corresponding regions within the protein and the RNA must have enough flexibility to allow the formation of a new covalent bond between the components. At the RNA level, this is only the case when the base of the RNA is not involved in basepairing. Correspondingly, the most flexible—and thus the most easily crosslinked—areas within proteins are the loop regions ***(2,22)***.

Despite the fact that the yield of UV crosslinking at 254 nm is relatively low when compared with that of the crosslinking of in vitro reconstituted particles bearing an artificially introduced crosslinking label on the RNA, it has the advantage that it can be applied to native purified RNP particles that have been isolated directly from cellular compartments. It further avoids generation of the "false-positive" results that are frequently associated with the heterogeneous populations generated as a result of incomplete assembly in the reconstitution reaction with labeled RNA. Nonetheless, the approach of UV crosslinking at 254 nm followed by identification of the protein–RNA crosslinking sites has been also successfully applied to protein–RNA complexes reconstituted in vitro ***(20)***.

3.1.4. Crosslinking Conditions

UV crosslinking is performed in a flat glass dish under a suitable light source. The result is strongly influenced by the choice of lamp, the distance between the lamp and the sample, and the irradiation time. Our laboratory uses a custom-made UV irradiation device (*see* **Fig. 1** and **Subheading 3.2.**). Alternatively, one can use a commercially available UV irradiation apparatus (e.g., UV Stratalinker 2400). Importantly, the conditions of UV irradiation—in particular the irradiation time and the distance of the sample from the lamp—have to be adjusted. During our studies with snRNP particles, we have observed that 2 min of irradiation is sufficient to achieve a maximum crosslinking yield. In contrast, for more rigid RNP particles such as ribosomes, the irradiation time can be prolonged. The samples are irradiated in custom-made glass dishes with a planar surface. Precooling of the glassware is essential. The depth of the sample solution is ideally 1 mm, and the size of the glass dish or the sample volume should be adjusted accordingly.

3.2. UV Crosslinking Protocol

The first step of purification of covalently linked peptide–RNA oligonucleotides is the UV irradiation of the samples to form a covalent linkage between a protein and the RNA.

1. Starting materials are native purified snRNPs (for example, 17 S U1 snRNP, or 25 S [U4/U6.U5] tri-snRNP) or RNP particles reconstituted in vitro, such as [U4/U6-15.5 K-61 K]. The purification of native U snRNPs and the reconstitution in vitro of [U4atac-15.5 K-61 K] protein–RNA complexes are described in detail elsewhere ***(4,8–10)***. The concentration of the RNP particles is adjusted to approximately 0.05–0.1 mg/mL with the buffer in which they were purified or reconstituted (*see* **Note 1**).
2. The samples are pipeted into precooled glass dishes with a typical diameter of 3.5 cm. The glass dishes must be flat (to provide for a homogeneous layer of liquid), and they should be placed on an aluminum block in ice (**Fig. 1;** *see* **Note 2**).
3. The samples are irradiated for 2 min at a distance of 2 cm from the UV source. We use a custom-made UV irradiation device with four 8-W germicidal lamps (G8T5, Herolab, Germany) mounted in parallel (**Fig. 1**).
4. The samples are pooled and precipitated with 3 volumes of ice-cold ethanol (p.a. grade) in the presence of 1/10 volume of 3 *M* sodium acetate, pH 5.3, for at least 2 h at −20 °C. They are then centrifuged for 30 min at 4 °C in a tabletop centrifuge (13,000 rpm [16,060 *g*]; Eppendorf tubes) or in a HB-4/HB-6 rotor (10,000 rpm [16,340 *g*]; Corex tubes). The pellets are washed with an appropriate volume of ice-cold 80% (v/v) ethanol/water and spun down as before. If in Eppendorf tubes, they are then dried briefly *in vacuo* (for not more than 5 min); if in Corex tubes, they are dried in air on the laboratory bench for 10–20 min (*see* **Note 3**).

3.3. First Separation Step: Size-Exclusion Chromatography

Size-exclusion chromatography ("gel filtration" SEC) is the first step in the purification of crosslinked peptide–RNA oligomers. The principle is illustrated in **Fig. 2**. UV-irradiated RNP complexes are allowed to dissociate, and the protein moiety is hydrolyzed with endoproteinases to yield intact RNA that still carries specific peptides covalently attached to the RNA at the sites of crosslinking. Noncrosslinked RNA and RNA with crosslinked peptides can be separated from the noncrosslinked peptides by SEC. This step is critical since any residual noncrosslinked peptides in the RNA-containing fractions will lead to a

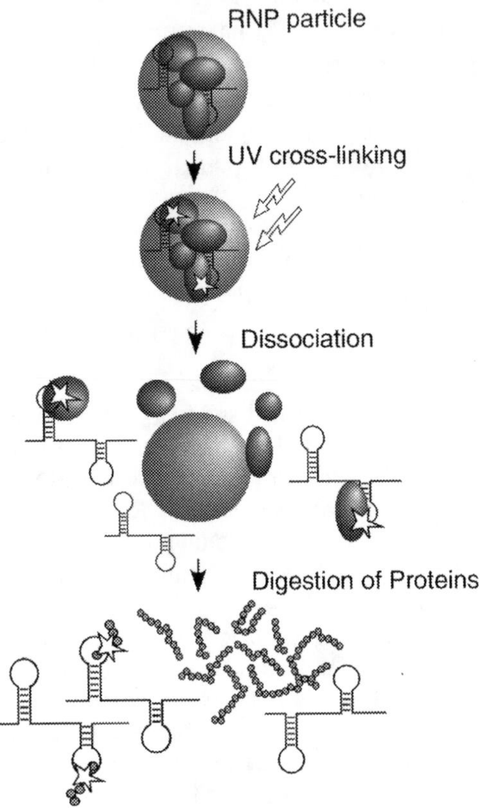

Fig. 2. Schematic representation of the initial step in the purification of crosslinked peptide–RNA oligonucleotides. After ultraviolet (UV) irradiation of ribonucleoprotein (RNP) particles, the particles are dissociated, and the protein moiety is digested with endoproteinases to obtain intact RNA that carries crosslinked peptides (white stars indicate the site of crosslinking) and noncrosslinked peptides. See text for details.

complex elution pattern in the final separation by RP-HLPC (*see* **Subheading 3.4.1.**) and to additional complexity of the mass spectra obtained in the analysis of the crosslinks.

3.3.1. General Remarks

3.3.1.1. Size-Exclusion Columns

Size-exclusion chromatography can only be applied in a first separation step when the RNA is significantly larger than the average size of the peptides generated (10–30 amino acids, i.e., 1000–3000 Da). For example, RNA oligomers (≤30 nucleotides) cannot be separated from peptides. The choice of the SE column is important as separation is influenced by its size (length and diameter) as well as by the matrix. The smallest column for our purpose is a Superose 75 HR column, measuring 3.2 × 300 mm, mounted in a SMART System. The maximum sample volume that can be applied to these is 50 µL. Smaller SE columns will usually have a smaller maximum sample volume and will therefore not be usable as it is difficult to obtain samples below 50 µL by the procedures described above. For large-scale preparations, we use a Superose 74 HR measuring 10 × 300 mm, mounted in a standard FPLC system or in the Äkta Purifier system. The maximum sample volume that can be applied on this column is 200 µL.

3.3.1.2. Dissociation Conditions

Dissociation and the subsequent digestion of the RNP particles with endoproteinases must be complete. Incomplete digestion will result in either larger peptide fragments or in only partially hydrolyzed proteins, which will comigrate with the RNA and will thus interfere with the final detection of peptide–RNA oligonucleotide crosslinks during RP-HPLC (*see* **Subheading 3.4.2.**). Incomplete digestion is a particular danger when proteins are tightly associated with their cognate RNAs (e.g., Sm proteins bound to the Sm site RNA in U snRNAs; ***16,17***).

For dissociation/denaturation and subsequent digestion of the irradiated particles, SDS, urea, or GHCl can be used according to the conditions specified for each endoproteinase. In our hands, best results are obtained by dissociation/denaturation in 6 *M* GHCl at room temperature for reconstituted particles and at elevated temperature (90 °C) for native particles.

SDS has the disadvantage that it prevents endoproteinases, such as chymotrypsin or trypsin, from working properly, even at very low SDS concentrations. For example, when we used trypsin in the presence of 0.1% (w/v) SDS to generate peptides, we noticed that a considerable number of cleavage sites were missed.

The use of 6–8 *M* urea for dissociation/denaturation can cause carbamylation of the lysine residues due to the presence of traces of cyanate in the urea; carbamylated lysine is no longer a substrate for trypsin, and most importantly, carbamylation causes a mass shift of 43.006 (monoisotopic mass) in the mass spectrometer. Such modification has to be considered in calculating the cross-linked peptide and RNA moiety from a measured mass (*see* **Subheading 3.5.3.**).

3.3.1.3. Digestion Conditions

It must be emphasized that the choice of the endoproteinase is also critical for the identification of the cross-linked protein region. If the sequence of the crosslinked protein is known, then one should perform a theoretical digestion of the protein with various endoproteinases and check that reasonable peptides (i.e., 10–20 amino acids) are expected; the endoproteinase should then be chosen accordingly. When one is working with noncharacterized particles, it is difficult to predict which endoproteinase will generate crosslinked peptides of reasonable size. It can happen that a crosslinked peptide cannot be identified because it is too small, is too large, does not elute from the HPLC column, and so on. Thus, a negative result does not necessarily mean that no crosslinking occurred. Changing the endoproteinase can often lead to positive results.

The following endoproteinases are routinely used for digestion of crosslinked proteins: trypsin (sequencing grade, Promega), chymotrypsin (sequencing grade, Roche), Lys-C (Roche), and Glu-C (sequencing grade, Roche). Lys-C has the advantage that it generates larger peptides than does trypsin. Larger, crosslinked peptides elute at a higher percentage of solvent B (water, ACN, 0.085% TFA) from the RP-HPLC column in the final purification step (*see* **Subheading 3.4.2., step 5**), whereas tryptic, crosslinked fragments coelute at lower percentages of buffer B together with larger, noncrosslinked RNA oligomers. As a rule of thumb, we recommend starting the first experiment with trypsin. If no crosslinks can be purified, then one should resort to one of the other endoproteinases.

3.3.1.4. SDS-PAGE Analysis of Crosslinks

We strongly recommend analyzing aliquots of the fractions after SE chromatography by SDS-PAGE *(23)* and visualization of their components by silver staining *(13)*. **Figure 3A** shows a typical SE elution profile of UV-irradiated RNPs treated with endoproteinase (in this case, irradiated U1 snRNPs digested with trypsin in the presence of GHCl; *see* **Subheading 3.3.2., step 1**), and **Fig. 3B** shows the corresponding SDS-PAGE analysis. The RNA-containing fractions contain only U1 snRNA and crosslinked U1 snRNA. The endoproteinase trypsin is detected in fractions that are eluted later, and—importantly—no larger protein fragments due to incomplete digestion are detectable. A similar pattern is observed in reconstituted [15.5 K-61 K-U4atac] RNPs after digestion with trypsin in the presence of GHCl (**Fig. 3C**). In contrast, digestion of the same RNP with chymotrypsin in the presence of 0.1% (v/v) SDS leads to incomplete digestion, as revealed by the presence of numerous larger protein fragments that coelute with U4atac snRNA (**Fig. 3D**).

3.3.2. SE Chromatography Protocol

1. The precipitated UV-irradiated samples are dissolved in an appropriate volume (typically 50 μL) of buffer 1 and heated to 90 °C for 5 min. They are then diluted to give a final concentration of 1 *M* GHCl with buffer 1 without GHCl. Endoproteinase is added to a final concentration of 1:20 (w/w) enzyme/substrate ratio, and the mixtures are incubated overnight at 37 °C in the presence of 40 U RNasin (Promega).

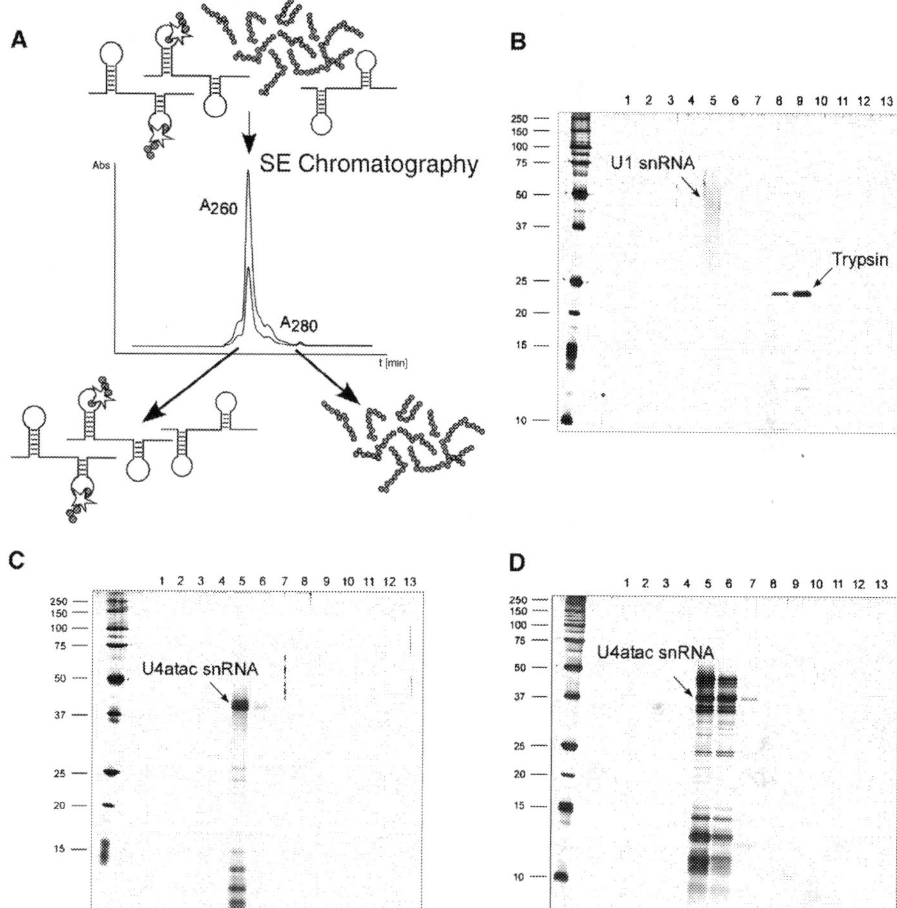

Fig. 3. (**A**) Schematic representation of the first purification step of crosslinked peptide–RNA oligonucleotides by size-exclusion (SE) chromatography. SE chromatography separates RNA molecules—both with and without crosslinked peptides—from the excess of noncrosslinked peptides. (**B**) Sodium dodecyl sulfate polyacrylamide gel electrophoresis (SDS-PAGE) analysis of fractions derived from SE chromatography of ultraviolet (UV)-irradiated U1 small nuclear RNPs (snRNPs) dissociated in the presence of 6*M* guanidinium hydrochloride (GHCl) and digested with trypsin. U1 snRNA and trypsin are visualized by silver staining ***(12)***. Note that except for trypsin no additional proteins (or large protein fragments) are identified. (**C**) SDS-PAGE analysis of fractions derived from SE chromatography of UV-irradiated [U4atac-15.5 K-61 K] RNP dissociated in the presence of 6M GHCl and treated with chymotrypsin. (**D**) SDS-PAGE analysis of fractions derived from SE chromatography of UV-irradiated [U4atac-15.5 K-61 K] RNP dissociated in the presence of 1% SDS and treated with chymotrypsin after adjustment of the SDS concentration to 0.1%. The position of U4atac is indicated. Silver staining of SE fractions shows that the RNA-containing fractions still include numerous protein fragments compared with the products of digestion in the presence of GHCl.

2. The sample is precipitated with ethanol (*see* **Subheading 3.2., step 4**), washed (*see* **Subheading 3.2., step 4**), dried briefly *in vacuo* (e.g., SpeedVac) and redissolved in an appropriate volume of buffer 2 (*see* **Note 4**).
3. The dissolved sample (50 µL) is injected onto a 75 HR column (Amersham Bioscience/GE Healthcare) mounted into a Smart System (Amersham Bioscience/GE Healthcare) running in buffer 2 at a flow rate of 40 µL/min at room temperature. The absorbance is monitored at 260 and 280 nm. Collect 100-µL fractions.
4. Take 10 µL aliquots from each fraction and add 10 µL of SDS-PAGE sample buffer. Samples are heated to 90 °C for 5 min and then spun down and loaded onto a 13% SDS polyacrylamide gel *(23)*. Electrophoresis is performed until the bromophenol blue tracking dye reaches the bottom of the gel. Gels are silver stained according to **ref.** *12*.
5. When several runs on the SE column are performed, the RNA-containing fractions are pooled, and the RNA is precipitated with 3 volumes of ethanol (*see* Subheading 3.2., **step 4**). Proceed with **Subheading 3.4.2., step 1**.
6. When working with such an amount of UV-irradiated RNA particle that only one SE separation was necessary, proceed with **Subheading 3.4.2., step 2**.

3.4 Second Separation Step: Reversed-Phase High-Performance Liquid Chromatography

3.4.1. General Remarks

In the second and final separation step, RP-HPLC is applied to separate crosslinked peptide–RNA oligonucleotides from an excess of noncrosslinked RNA oligonucleotides after the intact RNA with the crosslinked peptides (obtained in **Subheading 3.3.2., step 3**) has been digested with RNases (**Fig. 4**). Under standard RP-HPLC conditions (solvents water/ACN/TFA), RNA oligonucleotides are found in the flowthrough as they are not retained by the alkyl chains of the RP material. Conversely, crosslinked peptide–RNA oligonucleotides are retained because of their peptide moiety and are eluted at a higher percentage of solvent B (80% ACN, 0.085% TFA in water). Peptide–RNA crosslinks are detected by monitoring the absorbance at 220 nm (peptide) and 260 nm (nucleotides). Fractions that show an absorbance at 220 and 260 nm (due to the crosslinked RNA moiety) are collected and are further analyzed by MALDI-TOF MS. **Figure 4** shows a schematic representation of the final purification of peptide–RNA oligonucleotide crosslinks with an RP-HPLC example derived from [U4atac-15.5 K-61 K] RNPs after digestion of the complexes with chymotrypsin and RNases A and T1. Within the chromatogram, two fractions that show an absorbance at 220 and 260 nm could be detected and were thus chosen for further analysis in the MALDI-TOF mass spectrometer (Bruker Daltonics).

The successful detection of peptide–RNA oligonucleotides in RP-HPLC depends not only on the items discussed (i.e., starting material, crosslinking

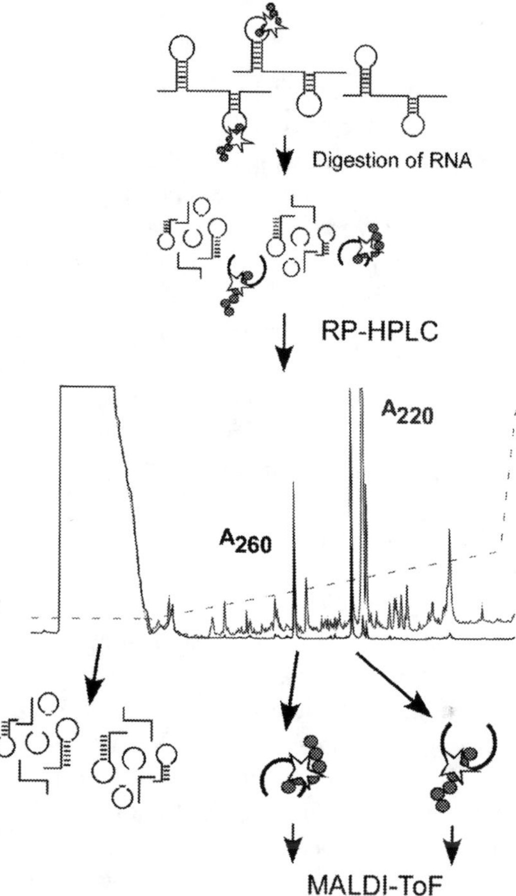

Fig. 4. Schematic representation of the second purification step of crosslinked peptide–RNA oligonucleotides by reversed-phase high-performance liquid chromatography (RP-HPLC). RNA-containing fractions from the size-exclusion (SE) chromatography are pooled, and the RNA is hydrolyzed with ribonucleases (RNase A or RNase T1). The mixture, which consists of noncrosslinked RNA oligonucleotides and crosslinked peptide–RNA oligonucleotides, is injected onto an RP-HPLC column. The noncrosslinked RNA oligonucleotides are found in the flowthrough, while crosslinked peptide–RNA oligonucleotides are retained and are only eluted at a higher concentration of solvent B (dashed line). Fractions that show an absorbance at 220 and 260 nm due to the crosslinked RNA oligonucleotide are analyzed in the MALDI-TOF (matrix-assisted laser desorption/ionization time-of-flight) mass spectrometer.

yield, choice of endoproteinase, etc.) but also on the hardware components used in this final purification step.

In principle, any HPLC system can be used for separation if it is equipped at least with a dual-wavelength recorder (to monitor for peptides at 214–220 nm and RNA at 254–260 nm). The HPLC system should be equipped with a standard peptide C18 column. In our hands, C18 material with a pore size of 300 Å and a particle size of 300 µm works best. The length of the column is not critical. The most important point is the sensitivity of the detection. This depends on the size of the RP column (e.g., for conventional HPLC 4.6-mm inner diameter, for analytical HPLC 2.1-mm inner diameter, for micro-HPLC 0.5- to 1.0-mm inner diameter, for capillary HPLC 0.1- to 0.5-mm inner diameter, and for nano-HPLC ≤ 0.1-mm inner diameter). For example, we have analyzed preparative amounts of UV-irradiated prokaryotic ribosomal subunits with a conventional HPLC ***(24)***, preparative amounts of UV-irradiated U1 snRNP or 25 S [U4/U6.U5] tri-snRNPs on an analytical HPLC ***(1)***, and small amounts (≤20 µg) of U1 snRNP1 and in vitro reconstituted [U4atac-15.5 K-61 K] RNPs on a capillary HPLC.

The purity and homogeneity of the sample also have a strong influence on the final separation and detection of the crosslinked peptide–RNA. Incomplete digestion of the protein moiety, as mentioned, will lead to a crowded chromatogram, in which the crosslinks coelute with noncrosslinked peptides and thus impede detection. Incomplete hydrolysis of the RNA prior to HPLC separation also drastically impairs the detection of crosslinks. Nonhydrolyzed RNA or larger RNA oligomers elute at similar concentration of solvent B (8–15% v/v ACN, 0.08% v/v TFA in water) as crosslinks (12–20% v/v ACN, 0.08% v/v TFA in water). Owing to the very strong absorbance of the RNA at 254 nm in this particular region of the chromatogram, the actual crosslink cannot be detected.

3.4.2. RP-HPLC Protocol

1. When samples have been pooled and precipitated (*see* Subheading 3.3.2., **step 5**), after SE chromatography they are centrifuged, washed, and dried as described in **Subheading 3.2., step 4**. For RNA hydrolysis, these are dissolved in an appropriate volume of buffer 4, and 10 µg of RNase A or 15 µg of RNase T1 (Ambion) are added (*see* **Note 5**).
2. Alternatively, samples can be treated with 1 µg RNase A or 2 µg T1 in buffer 2 (*see* **Subheading 3.3.2., step 1**, and **Note 5**)
3. Incubation is carried out for 2 h at 52 °C.
4. Add 1 µg of endoproteinase (the same as the one used in **Subheading 3.3.2.**), and the mixture is incubated overnight at 37 °C (*see* **Note 6**).
5. The sample is injected onto an RP-HPLC column mounted in a corresponding HPLC system. We use the following systems routinely in our laboratory:

a. For UV-irradiated RNP particles 50 µg or larger: C18 RP column, 150 × 2.1 mm (218TP5215; GraceVydac) mounted in a SMART System (Amersham Biotech/GE Healthcare) equipped with a 10 µL flow cell running at a flow rate of 100 µL/min.
b. For UV-irradiated RNP particles smaller than 50 µg: C18 RP column, 150 × 0.3 mm (Micro-Tech Scientific) mounted in an Applied Biosystems Microgradient system 140C equipped with a 35 nL flow cell running at a flow rate of 2 µL/min.
c. The solvents are as follows: For the SMART System, solvent A is water containing 0.1% (v/v) TFA; solvent B is ACN containing 0.085% (v/v) TFA. For the Cap-LC system, solvent A is water containing 0.1% (v/v) TFA, and solvent B is 80% ACN containing 0.085% (v/v) TFA.
d. The following gradients are applied: For the SMART System, isocratic elution at 5% solvent B until after elution of the injection peak (i.e., until the UV trace returns to the baseline); 5% solvent B to 45% solvent B in 120 min; 45% solvent B to 90% solvent B in 10 min; 10 min at 90% solvent B. It may be necessary to include another isocratic step around 10% solvent B if a constant and strong increase in the absorbance at 260 nm is observed at that percentage (indicating incomplete RNA hydrolysis or larger RNA oligonucleotides). Importantly, it should be borne in mind that this value refers to the theoretical concentration of solvent B in our SMART System, which is equipped with an analytical column. The value might vary in different HPLC systems because of different delay volumes. For the Cap-LC system, isocratic elution at 5% solvent B until after elution of the injection peak (i.e., until the UV trace returns to the baseline); 5% solvent B to 60% solvent B in 60 min, 60% to 90% solvent B in 3 min, 7 min at solvent 90% B.

6. Fractions that show an absorbance at 220 and 260 nm are collected, dried completely *in vacuo*, and stored for further at analysis –20 °C.
7. Proceed with **Subheading 3.5.2., step 1**.

3.5. Mass Spectrometry

3.5.1. General Remarks: MALDI-TOF vs Electrospray Ionization

Mass spectrometry is by now the fastest and most sensitive method for sequence analysis of biomolecules ***(25–28)***. In particular, proteins and peptides including (posttranslational) modifications can be sequenced on an almost-routine basis. In addition, MS has been successfully applied for the sequencing of RNA/DNA molecules ***(29–31)***. MS uses two main methods to ionize biological macromolecules: electrospray ionization (ESI; ***32***) and MALDI ***(33,34)***. The principal difference between ESI and MALDI is that ESI is a continuous ionization method by which multiply charged biomolecules (e.g., $[M+2H]^{2+}$ to $[M+nH]^{n+}$) are produced from a capillary electrode placed at high voltage with respect to a grounded counterelectrode. In contrast, MALDI is a pulsed ionization technique in which the biomolecules in a UV-absorbing matrix (e.g., nicotinic acid) are desorbed from/with the matrix molecules by a laser pulse, finally becoming mainly singly charged (e.g., $[M+H]^+$) by charge transfer from the matrix molecules to the analyte molecules.

The latter has the advantage that the resulting mass spectra are relatively easy to interpret as they do not show several mass peaks from one species. Moreover, in terms of sample preparation and instrumentation, MALDI MS is much easier to handle than the electrospray technique. On the other hand, sequence information, especially from modified peptides/biomolecules, is more reliable when derived from ESI MS combined with collision-induced decay *(35)*. In contrast, sequencing of unknown or modified peptides (i.e., *de novo* sequencing) in a MALDI-TOF mass spectrometer by postsource decay (PSD; *36,37*) has several disadvantages: (1) It requires sufficient amounts of sample, (2) sequencing is often restricted to a certain size of the fragment selected for sequencing (1000–2000 m/z), and (3) much experience in spectrum evaluation and interpretation—or in bioinformatics—is required to deduce the peptide sequence from the mass spectrum.

Notably, MS of peptide–RNA oligonucleotides differs from the routine MS analysis of modified peptides or of pure oligonucleotides as peptides and RNA (oligomers) behave differently in MS. RNA is analyzed under conditions by which it becomes negatively ionized by changing the polarity of the ESI and MALDI mass spectrometer in combination with the use of a neutral, volatile buffer (e.g., ammonium acetate) and an RNA-compatible matrix (e.g., 3-hydroxy-picolinic acid; Sigma-Aldrich), respectively. Conversely, peptides are analyzed under acidic conditions, at which they are positively charged.

Despite these disadvantages, we have chosen MALDI-TOF instead of ESI (on a conventional mass spectrometer with ESI source, as triple-quadrupole instruments, Q-ToF instruments, linear and three-dimensional ion traps) for the analysis of purified peptide–RNA heteroconjugates. This was for the following reasons: (1) MALDI-TOF is highly sensitive; (2) it has an extremely high mass accuracy ($\leq$ 10 ppm, which is important for calculation of possible composition and sequence of the peptide and RNA moieties from the measured mass; *see* **Subheading 3.5.3.**) with external calibration; (3) samples can be easily prepared; and most important, (4) it is much easier to find conditions under which a sufficient number of ionized crosslinked peptide–RNA species travels through the analyzer toward the detector, especially when they have high m/z values.

Note that a successful analysis is only achieved when the samples are measured in the reflectron mode of the MALDI-TOF instrument, as only then can monoisotopic masses be considered. For measuring in the reflectron mode, the use of DHB or THAP as matrices is recommended since α-cyano-4-hydroxycinnamic acid (HCCA) gives poor or no results. Using

HCCA as a matrix for peptide–RNA oligonucleotides allows MS analysis in the instrument's linear mode only, and the mass accuracy in the linear mode is too low to distinguish unambiguously between the bases C and U, which differ by only 1 mass unit.

3.5.2. Preparation of Samples and Acquisition of MS Spectra

1. Dried fractions from the RP-HPLC separation are dissolved in not more than 10 µL 50% ACN (v/v), 0.1% TFA, and sonicated in a sonication bath for 3 min.
2. Follow these procedures next:
   a. 0.5 µL of the sample is mixed on a stainless steel sample plate with the same volume of matrix solution 1. The preparation is air dried and subjected to MS in a Voyager DE-STR (Applied Biosystems) under standard conditions in the reflectron mode. The acceleration voltage is 20 kV, the grid voltage 68%, and the delay time 250 ns. A total of 300 laser shots (N_2-pulsed laser, 20 Hz, 337 nm) are summed.
   b. Alternatively, 0.5 µL of the sample is mixed on a stainless steel sample plate with the same volume of matrix solution 2. The preparation is air dried and subjected to MS in a Voyager DE-STR under standard conditions in the reflectron mode. The acceleration voltage is 20 kV, the grid voltage 68%, and the delay time 250 ns. A total of 300 laser shots (N_2-pulsed laser, 20 Hz, 337 nm) are summed.
   c. We also have good experience with Bruker MALDI-TOF mass spectrometers (Bruker Daltonics). Bruker instruments are equipped with different sample plates, which in turn require a modified sample preparation. When DHB is used as the matrix, 0.5 µL of the sample is mixed on a Bruker Anchor 600 sample plate with the same volume of a 10 mg/mL solution of DHB in 50% ACN or water, respectively. The preparation is air dried and measured in a Reflex IV under standard conditions in the reflectron mode. The acceleration voltage is 20 kV for the IS1 and 16.9 kV for the IS2; the delay time is 400 ns. A total of 300 laser shots (N_2-pulsed laser, 5 Hz, 337 nm) are summed.

3.5.3. Spectrum Evaluation

It is often straightforward to identify the crosslinked peptide and oligonucleotide parts from MALDI MS spectra of purified peaks. This is exemplified in the following case:

Figure 5A shows a MALDI-TOF spectrum from a purified crosslink derived from UV-irradiated [U4atac-15.5 K-61 K] RNPs after digestion with chymotrypsin and RNase T1. Notably, DHB and THAP preparations yielded a resolution higher than 15,000 and mass accuracies of 30 ppm or better by using close external calibration.

To determine the peptide and RNA compositions, we compared all chymotryptic fragments of the protein moiety with all RNase T1 fragments of the RNA moiety. We found that the measured monoisotopic mass ($[M+H]^+ = 2780.818$) matches precisely only a single chymotryptic fragment of the 61 K protein, encompassing protein positions 263–273 (SSTSVLPHTGY, $[M+H]^+_{cal} = 1148.558$) and to

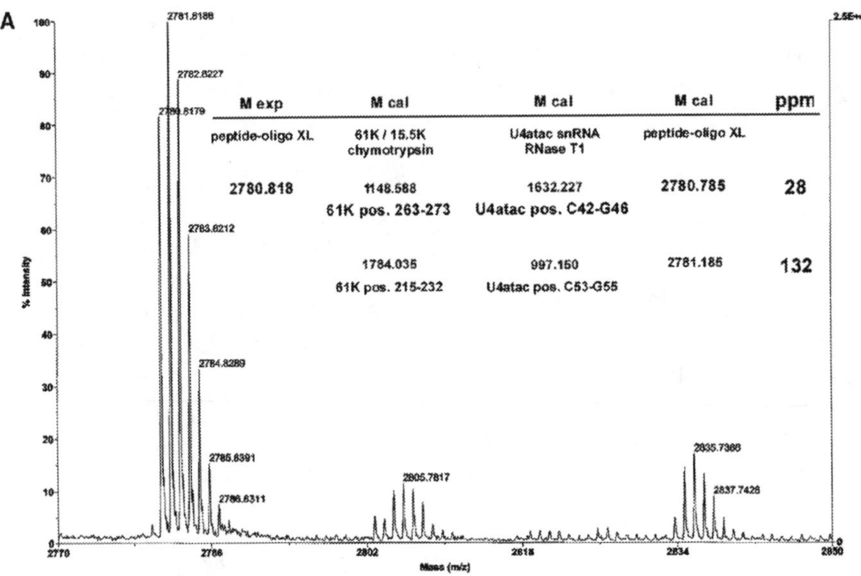

Fig. 5. MALDI-ToF (matrix-assisted laser desorption/ionization time-of-flight) mass spectra from purified peptide–RNA oligonucleotide crosslinks. (**A**) MALDI mass spectrometric (MS) spectrum of a purified crosslink derived from ultraviolet (UV)-irradiated [U4atac-15.5 K-61 K] RNP (ribonucleoprotein) treated with endoproteinase chymotrypsin and ribonuclease (RNase) T1. The spectrum was recorded on a Voyager DE-STR MALDI mass spectrometer (Applied Biosystems, Foster City, CA) under standard conditions. The measured monoisotopic mass (M_{exp}) of 2780.818 matches precisely a peptide of protein 61 K encompassing positions 263–273 (SSTSVLPHTGY, $[M+H]^+_{cal}$ = 1148.558) and to an RNase T1 fragment of U4atac RNA between nucleotide positions C42 and G46 (5′-CAUAG-3′, M_{cal} = 1632.227), M_{cal} = 1148.558 + 1632.227 = 2780.785. Notably, the mass deviation is less than 20 ppm ($M_{exp} - M_{cal}$ = 2780.818 – 2780.785 = 0.033 amu). The next-best hit from our database search was a chymotryptic fragment of protein 61 K from positions 215–232 crosslinked to a U4atac snRNA T1 fragment from positions C53–G55. Since the mass deviation between experimental and calculated mass is more than 130 ppm, this was not considered to be the actual crosslinked peptide–RNA oligonucleotide. See text for further details.

an RNase T1 fragment of U4atac RNA between nucleotide positions C42 and G46 (5′-CAUAG-3′, M_{cal} = 1632.227), 1148.558 + 1632.227 = 2780.785. The mass accuracy was better than 20 ppm (i.e., ± 0.033 amu [absolute mass units]). The next best "hit" from this calculation (61 K positions 215–232, $[M+H]^+_{cal}$ = 1784.035 and U4atac nucleotide positions C53–G55, M_{cal} = 997.150) had a significantly higher mass deviation (130 ppm) and was thus not considered to represent the crosslinked complex. From this highly accurate

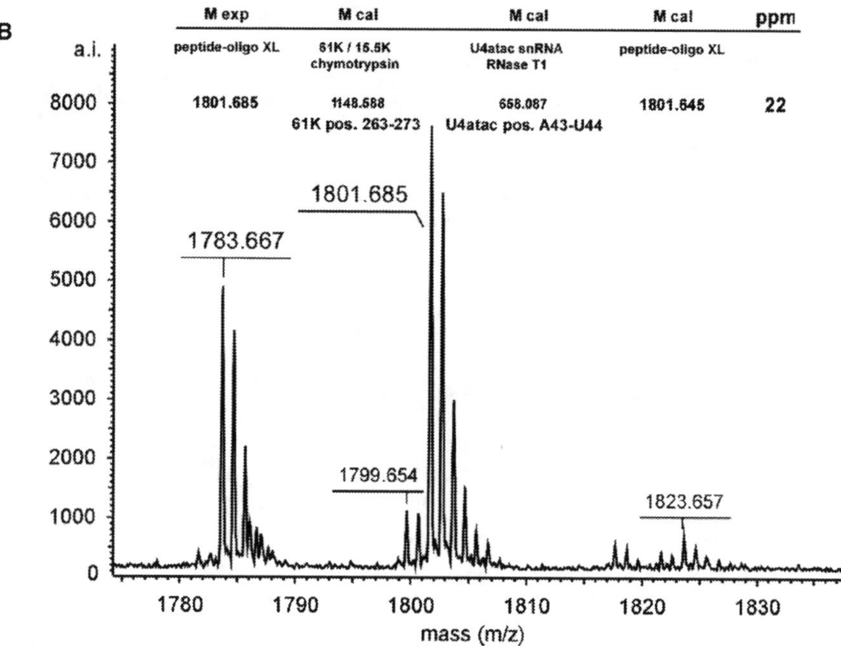

Fig. 5. (**B**) (continued) MALDI MS spectrum of a purified crosslink derived from UV-irradiated [U4atac-15.5 K-61 K] RNP treated with endoproteinase chymotrypsin and RNases A and T1. The spectrum was recorded on a Reflex IV MALDI-TOF mass spectrometer (Bruker Daltonics, Bremen, Germany) under standard conditions. The measured monoisotopic mass (M_{exp}) of 1801.685 matches precisely a peptide of protein 61 K encompassing positions 263–273 crosslinked to an RNase A and T1 fragment comprising positions A43 and U44 in the 5′ stem-loop of U4atac snRNA (M_{cal} = 1801.645). As in **Fig. 4A**, the mass deviation is less than 30 ppm.

mass analysis, we conclude that the 61 K protein between residues 263 and 273 must be crosslinked to U4atac between nucleotide positions C42 and G46.

This is strongly supported by the MALDI analysis of a second crosslink derived from the same particles after digestion with chymotrypsin and RNases A and T1. Our highly accurate MALDI analysis revealed a mass peak of $[M+H]^+$ = 1801.685 (**Fig. 5B**), corresponding exactly to a peptide of protein 61 K, comprising positions 263–273 crosslinked to an AU dinucleotide. Notably, AU (A43 and U44, 5′-AU-3′, M_{cal} = 658.087) is located within the previously identified T1 fragment of U4atac.

Figure 5C gives a third example of the determination of the crosslinked peptide and RNA moieties from a single MS spectrum. The spectrum shows the MALDI MS analysis of a crosslink derived from native UV-irradiated U1

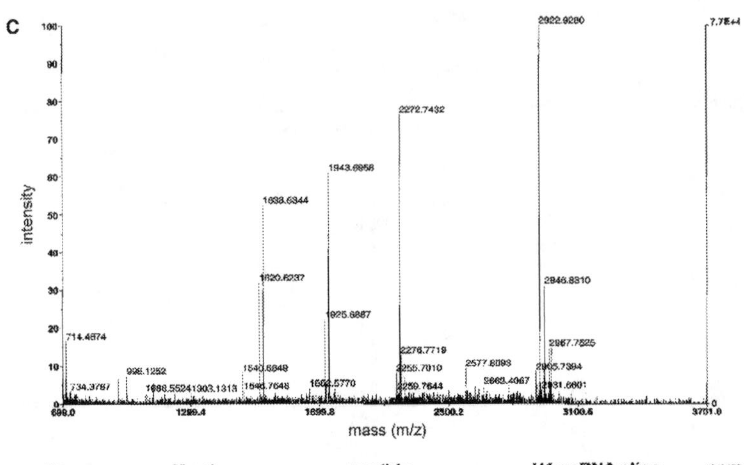

M exp	M cal	peptide	U1 snRNA oligo	ppm
2922.9280	2922.8469	U1 70K $_{173}$RVLVDVER$_{180}$	5′ $_{29}$AUCACG$_{34}$ 3′	**28**
2577.8093	2577.7999	U1 70K $_{173}$RVLVDVER$_{180}$	5′ $_{29}$AUCAC$_{33}$ 3′	**4**
2272.7432	2272.7589	U1 70K $_{173}$RVLVDVER$_{180}$	5′ $_{29}$AUCA$_{32}$ 3′	**7**
1943.6958	1943.7069	U1 70K $_{173}$RVLVDVER$_{180}$	5′ $_{29}$AUC$_{31}$ 3′	**6**
1638.6344	1638.6659	U1 70K $_{173}$RVLVDVER$_{180}$	5′ $_{29}$AU$_{30}$ 3′	**19**

Fig. 5. (**C**) (continued) MALDI MS spectrum recorded under our highly accurate conditions from purified peptide–RNA oligonucleotide crosslinks derived from UV-irradiated U1 snRNP hydrolyzed with endoproteinase trypsin and RNase T1. The spectrum was recorded on a Voyager DE-STR MALDI mass spectrometer under standard conditions. The measured monoisotopic masses (M_{exp} = 2922.928, 2577.809, 2272.743, 1943.696, 1638.634) match precisely a tryptic fragment of the U1 protein 70K from positions 173–180 crosslinked to a U1 snRNA T1 fragment from positions 29–34 and to its 3′-hydrolysis products, respectively. Importantly, the mass deviation over the entire spectrum was less than 30 ppm.

snRNPs treated with trypsin and RNase T1. By comparing all possible tryptic fragments from the 10 U1 proteins (70K, U1A, U1C, and the Sm proteins B/B′, D1–D3, E, F, and G) with all possible U1 snRNA RNase T1 fragments, we found that the mass peaks correspond to a U1 70K tryptic fragment from positions 173–180 (RVLVDVER) crosslinked to a U1 snRNA T1 fragment encompassing positions 29–34 (5′-AUCACG-3′) and to the 3′ hydrolysis products of the latter. Notably, the mass accuracy over the entire spectrum is better than 30 ppm using close external calibration.

Our preparations (DHB and THAP) also showed excellent properties in the PSD analysis of selected peptide–oligonucleotide mass peaks, revealing structural information about the crosslinked peptide moiety. An example is given in **Fig. 5D**, which shows a PSD sequence analysis under standard conditions in a

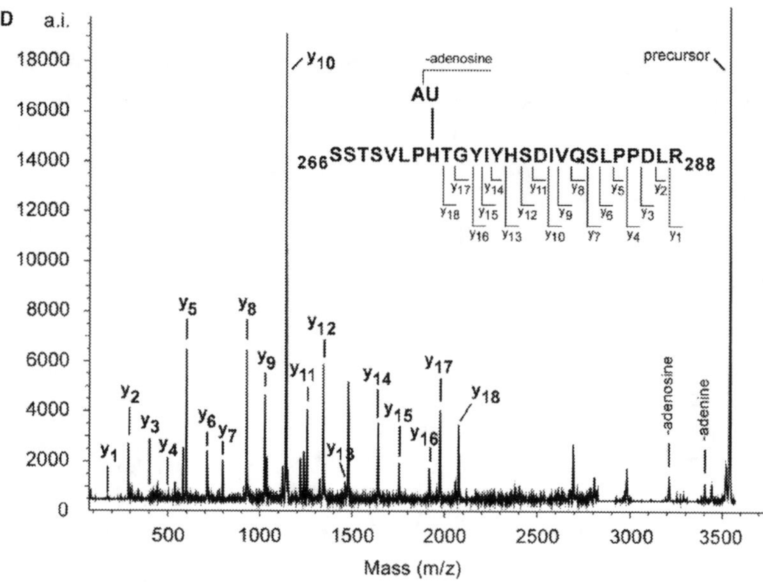

Fig. 5. (**D**) (continued) Postsource decay (PSD) spectrum of a selected crosslink (precursor mass m/z = 3535.489) derived from [U4atac-15.5 K-61 K] RNP treated with trypsin and RNases A and T1. The spectrum was recorded on a Reflex IV MALDI mass spectrometer. The sequence with the experimentally found y-type ion series is listed within the spectrum. Importantly, PSD sequence analysis revealed that His270 is the potentially crosslinked amino acid, and U44 rather than A43 within the identified T1/A fragment (5′-AU-3′) is the actual crosslinked nucleotide.

Reflex IV MALDI-TOF instrument (Bruker Daltonics). The species analyzed is a 61 K–U4atac crosslink (selected precursor mass m/z = 3535.489) derived from the UV-irradiated [U4atac-15.5 K-61 K] RNPs after digestion with trypsin and RNases A and T1. Importantly, the measured precursor mass of m/z = 3535.489 did not yield a specific tryptic 61 K fragment crosslinked to an RNase A/T1 fragment of U4atac snRNA in our database search. The PSD spectrum shows a complete series of C-terminal fragment ions (so-called y-type ions) up to y18, revealing the sequence TGYIYHSDIVQSPPDLR (**Fig. 5D**). In combination with the measured precursor mass, we thus identified the crosslink as a fragment of protein 61 K encompassing positions 263–288 (SSTSVLPHTGYIYHSDIVQSP PDLR) crosslinked to an AU dinucleotide (A43 and U44, 5′-AU-3′). The missing y19 (H) and y20 (P) indicate that His270 is the actual crosslinked amino acid. The fragments of a peptide derived from the U4/U6-specific protein 61 K encompassing positions 263–288 (SSTSVLPHTGYIYHSDIVQSPPDLR) up to y18 (residues 270–288). The observed losses of adenine and adenosine, on the other hand, suggest that the RNA is crosslinked via a uracil rather than an adenine base.

Acknowledgments

We thank Thomas Conrad, Gabi Heyne, Peter Kemkes, and Hossein Kohansal for their excellent technical assistance in preparation of snRNPs from HeLa cells; Uwe Plessmann and Monika Raabe for their help in HPLC; and Christof Lenz from Applied Biosystems Europe for his support in MALDI MS on the Voyager DE-STR. This work is supported by a YIP grant from the EURASNET (within the sixth EU framework) to H.U. and by a BMBF grant (031U215B) to R.L.

4. Notes

1. The buffer contains 20 m*M* HEPES KOH (pH 7.5), 250 m*M* NaCl, 1.5 m*M* $MgCl_2$, and 0.5 m*M* DTT in the case of U1 snRNPs, in 20 m*M* HEPES KOH (pH 7.9), 1.5 m*M* $MgCl_2$, 250 m*M* NaCl, 0.5 m*M* DTT, and 0.2 m*M* EDTA in the case of 25 S[U4/U6.U5] tri-snRNPs, or in 20 m*M* HEPES KOH (pH 7.5), 150 m*M* NaCl, 1.5 m*M* $MgCl_2$, 0.1% Triton X-100, 0.2 m*M* EDTA in the case of reconstituted [U4atac-15.5 K-61 K] protein–RNA complexes. The ideal sample volume is 1 mL.
2. If the starting sample volume is smaller, it is adjusted with the respective buffer to yield a solution depth of 1 mm. However, the sample concentration should not be below 0.05 mg/mL. If the starting sample volume is larger, glass dishes with a greater diameter are used, if available; alternatively, the sample can be irradiated in successive batches in the same dish. The latter procedure has the advantage that the sample recovery rate is higher than when glass dishes with a larger diameter are used. In particular, protein–RNA particles that contain numerous relatively large proteins and relatively small RNAs tend to stick to the glass surface; saturation of the glass surface (with sample) minimizes this effect. An example of such a complex is the 25 S [U4/U6.U5] tri-snRNP, which contains large proteins such as the U5-specific proteins 220 K (human Prp8) and 200 K, the U4/U6-specific proteins 90 K and 60 K, and snRNAs of length 116, 145, and 106 nt, respectively ***(12)***.
3. For sample volumes of 1 mL or less, the sample is distributed into three (or fewer) 1.5-mL Eppendorf tubes and precipitated as above. For larger sample volumes, we recommend using Corex glass tubes (Kendro Laboratory Products) for the precipitation. Note that the Corex tubes must be completely free of RNases; our laboratory uses a set of Corex tubes reserved exclusively for this purpose.
4. The sample volume depends on the size of the SE column. We usually dissolve the sample in 50 μL; however, large-scale preparations require volumes of up to 200 μL or more. As a rule of thumb, 100 pmol to 1 nmol of RNA in a sample volume of 50 μL leads to excellent separation on a 75 HR column (3.1 × 300 mm, Amersham Biotech/GE Healthcare) in the Smart System (Amersham Bioscience/GE Healthcare). For larger volumes, we perform several runs on small SE columns instead of using a larger column on a different system. Larger columns generate larger volumes of eluted sample, and the detector systems of preparative

chromatography systems are generally not as sensitive as those in semimicro- or microsystems.

5. The RNA hydrolysis must be complete. Therefore, we recommend using a sufficient amount of RNases. However, when RNase A and T1 are used together, the amount of RNase A should be kept as low as possible as RNase A is readily cleaved by endoproteinases in a second digestion (*see* **Subheading 3.4.2., step 4**), and its fragments interfere with the detection of the crosslinks during HPLC. In contrast, RNase T1 remains almost intact and elutes as a single peak from the RP column around 30% solvent B (*see* **Subheading 3.4.1., Fig. 4**).
6. This second endoproteinase treatment is necessary as it substantially improves the yield of crosslinked peptide–RNA oligonucleotides in the RP-HPLC. We conclude that the first digestion (**Subheading 3.3.2., step 1**) generates mainly larger peptide fragments because of steric hindrance of the endoproteinase caused by the large intact RNA.

References

1. Urlaub, H., Hartmuth, K., Kostka, S., Grelle, G., and Lührmann, R. (2000) A general approach for identification of RNA–protein cross-linking sites within native human spliceosomal small nuclear ribonucleoproteins (snRNPs). Analysis of RNA–protein contacts in native U1 and U4/U6.U5 snRNPs. *J. Biol. Chem.* **275**, 41458–41468.
2. Urlaub, H., Raker, V. A., Kostka S., and Lührmann, R. (2001) Sm protein–Sm site RNA interactions within the inner ring of the spliceosomal snRNP core structure. *EMBO J.* **20**, 187–196.
3. Urlaub, H., Hartmuth, K., and Lührmann, R. (2002) A two-tracked approach to analyze RNA–protein crosslinking sites in native, nonlabeled small nuclear ribonucleoprotein particles. *Methods* **26**, 170–181.
4. Nottrott, S., Urlaub, H., and Lührmann, R. (2002) Hierarchical, clustered protein interactions with U4/U6 snRNA: a biochemical role for U4/U6 proteins. *EMBO J.* **21**, 5527–5538.
5. Oubridge, C., Ito, N., Evans, P. R., Teo, C. H., and Nagai, K. (1994) Crystal structure at 1.92 Å resolution of the RNA-binding domain of the U1A spliceosomal protein complexed with an RNA hairpin. *Nature* **372**, 432–438.
6. Handa, N., Nureki, O., Kurimoto, K., et al. (1999) Structural basis for recognition of the tra mRNA precursor by the Sex-lethal protein. *Nature* **398**, 579–585.
7. Gautier, T., Berges, T., Tollervey, D., and Hurt, E. (1997) Nucleolar KKE/D repeat proteins Nop56p and Nop58p interact with Nop1p and are required for ribosome biogenesis. *Mol. Cell. Biol.* **17**, 7088–7098.
8. Will, C. L., Kastner, B., and Lührmann, R. (1994) Analysis of ribonucleoprotein interactions in *RNA processing*. In: *A Practical Approach*, Vol. 1 (Higgins, S. J., and Hames, B. D., eds.), IRL Press, Oxford, U.K., pp. 141–177.
9. Kastner, B. (1998) Purification and electron microscopy of spliceosomal snRNPs. In: *RNP Particles, Splicing and Autoimmune Diseases* (Schenkel, J., ed.), Springer-Verlag, Berlin, pp. 95–140.

10. Kastner, B., and Lührmann, R. (1999) Purification of U small nuclear ribonucleoprotein particles. In: *RNA–Protein Interaction Protocols, Methods in Molecular Biology*, Vol. 118 (Haynes, S. R., ed.), Humana Press, Totowa, NJ, pp. 289–298.
11. Burge, C. B., Tuschl, T., and Sharp, P. A. (1999) Splicing of precursors to mRNA by the spliceosome. In: *The RNA World* (Gesteland, R. F., Cech, T. R., and Atkins, J. F., eds.), Cold Spring Harbor Laboratory Press, Cold Spring Harbor, NY, pp. 525–560.
12. Will, C. L., and Lührmann, R. (2001) Spliceosomal UsnRNP biogenesis, structure and function. *Curr. Opin. Cell Biol.* **13**, 290–301.
13. Blum, H., Beier, H., and Gross, H. J. (1987), Improved silver staining of plant proteins, RNA and DNA polyacrylamid gels. *Electrophoresis* **8**, 93–99.
14. Yisraeli, J. K., and Melton, D. A. (1989) Synthesis of long, capped transcripts in vitro by SP6 and T7 RNA polymerases. *Methods Enzymol.* **180**, 42–50.
15. Milligan, J. F., and Uhlenbeck, O. C. (1989) Synthesis of small RNAs using T7 RNA polymerase. *Methods Enzymol.* **180**, 51–62.
16. Sambrook, J., Fritsch, E. F., and Maniatis, T. (1989) *Molecular Cloning, a Laboratory Manual*, 2nd ed., Cold Spring Harbor Laboratory Press, Cold Spring Harbor, NY.
17. Raker, V. A., Hartmuth, K., Kastner, B., and Lührmann, R. (1999) Spliceosomal U snRNP core assembly: Sm proteins assemble onto an Sm site RNA nonanucleotide in a specific and thermodynamically stable manner. *Mol. Cell. Biol.* **19**, 6554–6565.
18. Kambach, C., Walke, S., Young, R., et al. (1999) Crystal structures of two Sm protein complexes and their implications for the assembly of the spliceosomal snRNPs. *Cell* **96**, 375–387.
19. Meisenheimer, K. M., Meisenheimer, P. L., and Koch, T. H. (2000) Nucleoprotein photo-cross-linking using halopyrimidine-substituted RNAs. *Methods Enzymol.* **18**, 88–104.
20. Nottrott, S., Hartmuth, K., Fabrizio, P., et al. (1999) Functional interaction of a novel 15.5 kD [U4/U6.U5] tri-snRNP protein with the 5′ stem-loop of U4 snRNA. *EMBO J.* **18**, 6119–6133.
21. Vidovic, I., Nottrott, S., Hartmuth, K., Lührmann, R., and Ficner, R. (2000) Crystal structure of the spliceosomal 15.5 kD protein bound to a U4 snRNA fragment. *Mol. Cell* **6**, 1331–1342.
22. Urlaub, H., Kruft, V., Bischof, O., Müller E. C., and Wittmann-Liebold, B. (1995) Protein-rRNA binding features and their structural and functional implications in ribosomes as determined by cross-linking studies. *EMBO J.* 14, 4578–4588.
23. Laemmli, U. K. (1970) Cleavage of structural proteins during the assembly of the head of bacteriophage T4. *Nature* **227**, 680–685.
24. Urlaub, H., Thiede, B., Müller, E. C., and Wittmann-Liebold, B. (1997) Contact sites of peptide-oligoribonucleotide cross-links identified by a combination of peptide and nucleotide sequencing with MALDI MS. *J. Protein Chem.* **16**, 375–383.
25. Yates, J. R., 3rd. (2004) Mass spectral analysis in proteomics. *Annu. Rev. Biophys. Biomol. Struct.* **33**, 297–316.

26. Mann, M., Hendrickson, R. C., and Pandey, A. (2001) Analysis of proteins and proteomes by mass spectrometry. *Annu. Rev. Biochem.* **70**, 437–473.
27. Wang, R., and Chait, B. T. (1994) High-accuracy mass measurement as a tool for studying proteins. *Curr. Opin. Biotechnol.* **5**, 77–84.
28. McLachlin, D. T., and Chait, B. T. (2001) Analysis of phosphorylated proteins and peptides by mass spectrometry. *Curr. Opin. Chem. Biol.* **5**, 591–602.
29. Berkenkamp, S., Kirpekar, F., and Hillenkamp, F. (1998) Infrared MALDI mass spectrometry of large nucleic acids. *Science* **281**, 260–262.
30. Kirpekar, F., and Krogh, T. N. (2001) RNA fragmentation studied in a matrix-assisted laser desorption/ionisation tandem quadrupole/orthogonal time-of-flight mass spectrometer. *Rapid Commun. Mass Spectrom.* **15**, 8–14.
31. McCloskey, J. A., Whitehill, A. B., Rozenski, J., Qiu, F., and Crain, P. F. (1999) New techniques for the rapid characterization of oligonucleotides by mass spectrometry. *Nucleosides Nucleotides* **18**, 1549–1553.
32. Fenn, J. B., Mann, M., Meng C. K., Wong, S. F., and Whitehouse, C. M. (1989) Electrospray ionisation for mass spectrometry of large biomolecules. *Science* **246**, 64–71.
33. Karas, M., and Hillenkamp, F. (1988), Laser desorption ionization of proteins with molecular masses exceeding 10,000 daltons. *Anal. Chem.* **60**, 2299–2301.
34. Koy, C., Mikkat, S., Raptakis, E., et al. (2003) Matrix-assisted laser desorption/ionization-quadrupole ion trap-time of flight mass spectrometry sequencing resolves structures of unidentified peptides obtained by in-gel tryptic digestion of haptoglobin derivatives from human plasma proteomes. *Proteomics* **3**, 851–858.
35. Tanaka, K., Waki, H., Ido, Y., Akita, S., Yoshida, Y., and Yoshida, T. (1988) Protein and polymer analyses up to m/z 100 000 by laser ionization time-of-flight mass spectrometry. *Rapid Commun. Mass Spectrom.* **2**, 151–153.
36. Steen, H., and Jensen, O. N. (2002) Analysis of protein–nucleic acid interactions by photochemical cross-linking and mass spectrometry. *Mass Spectrom. Rev.* **21**, 163–182.
37. Kaufmann, R., Spengler, B., and Lützenkirchen, F. (1993) Mass spectrometric sequencing of linear peptides by product-ion analysis in a reflectron time-of-flight mass spectrometer using matrix-assisted laser desorption ionization. *Rapid Commun. Mass Spectrom.* **7**, 902–910.

17

In Vitro Selection of Random RNA Fragments to Identify Protein-Binding Sites Within Large RNAs

Ulrich Stelzl and Knud H. Nierhaus

Summary

In vitro selection experiments have various goals depending on the composition of the initial pool and the selection method applied. We developed an in vitro selection variant that is useful for the identification of minimal RNA binding sites for proteins within large RNAs. A pool of randomly fragmented RNA is constructed from a large RNA, which is the natural binding partner for a protein. Such a pool contains all the potential binding sites and is therefore used as starting material for affinity selection. A successful in vitro selection with the purified protein will identify the protein's natural RNA target site. The method has been developed for ribosomal systems and is a general approach providing a basis for the functional and structural characterization of large ribonucleoprotein particles.

Key Words: In vitro selection; ribosome; RNA binding site; RNA–protein interaction; SELEX; SERF.

1. Introduction

An important step for the elucidation of the function of an RNA–protein complex is a characterization of the corresponding RNA–protein interactions. But, even if the RNA interactions are known at atomic level as in the case of ribosomes, one still has a problem: Ribosomal proteins usually have multiple interactions with ribosomal RNA (rRNA) sequences, and an extreme case such as L22 even contacts all six domains of the 23S rRNA *(1)*. In this case, the primary binding site, which binds with high affinity the ribosomal protein first to trigger the assembly process, is of high interest but cannot be derived in a simple way from the final ribosome structure. Identification of the primary binding site can be achieved with a modification of an in vitro selection method (SELEX, systematic evolution of ligands by exponential enrichment; *2*), the

From: *Methods in Molecular Biology, vol. 488: RNA-Protein Interaction Protocols*
Edited by: Ren-Jang Lin © Humana Press Inc, Totowa, NJ

so-called SERF method (selection of random RNA fragments; ***3,4***). In this approach a pool of random fragments from large, naturally occurring RNAs is used for affinity selection using a purified protein in vitro. This method has been used successfully for the identification of the primary binding site of ribosomal proteins from an operon of over 5000 nt ***(5)***. We have successfully applied this method for the identification of rRNA binding sites of ribosomal proteins ***(3,6)***, but the method is equally applicable for other large ribonucleoprotein particles ***(7)*** as well as noncoding RNAs ***(8)***.

SERF deviates from the usual SELEX protocols by the use of a pool of randomly fragmented RNAs. **Figure 1** outlines the experimental strategy of SERF. DNA encoding the noncoding RNA is digested with deoxyribonuclease (DNase) I in a manganese (Mn^{2+})-dependent manner ***(9)***, producing random DNA fragments. DNase I cuts double-stranded DNA (dsDNA) in the presence of Mn^{2+}, producing blunt-ended DNA or overhangs with 1–2 nt. The replacement of Mg^{2+} with Mn^{2+} also strongly slows the DNA digestion, so that the required fragment size can be conveniently steered and determined. T4 DNA

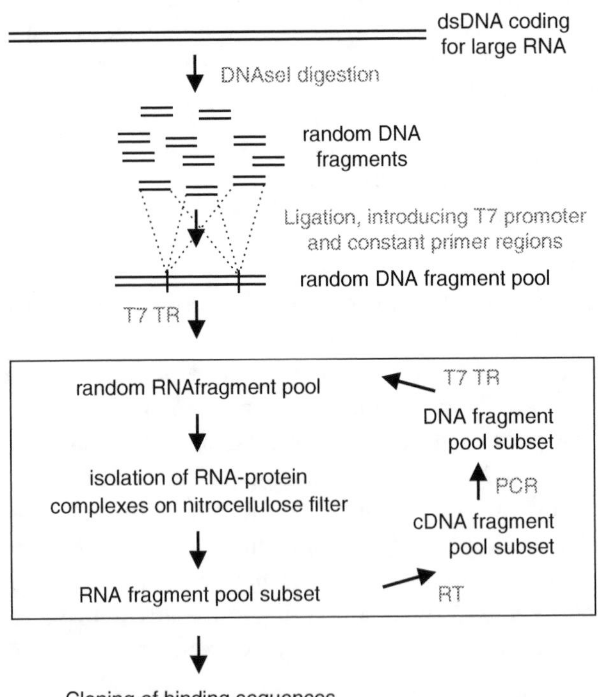

Fig. 1. Outline of the SERF (selection of random RNA fragments) procedure. cDNA, complementary DNA; dsDNA, double-stranded DNA; PCR, polymerase chain reaction; RT, reverse transcriptase.

polymerase is then used to make the fragments blunt ended. After cloning into a vector and creating a DNA fragment pool via polymerase chain reaction (PCR), a pool of randomly fragmented RNA is prepared by T7 transcription. RNA–protein complexes are formed and collected on nitrocellulose filters, thus separating the nonbound RNA. The level of RNA background binding to nitrocellulose and nonspecific binding of RNA by the protein are factors that essentially determine the success of a specific selection ***(10)***. Bound RNA is recovered from the filter, reverse transcribed to complementary DNA (cDNA), and amplified in a PCR reaction. The PCR products provide the template for repeated T7 in vitro transcription to produce RNA fragments for the next round of selection. Binding of the RNA to the target protein is measured after every round by filter-binding assays. If binding increases, typically after four rounds of selection ***(4)***, the fragments are cloned and sequenced. As the initial pool is derived from a large RNA, which is the natural binding partner for the protein, the random RNA fragment pool contains all the potential binding sites. Therefore, the winner sequence family reveals the native primary RNA binding site to which the protein binds with the highest affinity.

2. Materials

1. Purified target protein.
2. DNA encoding large RNAs.
3. Ribonuclease (RNase)-free water (e.g., from milliQ purification systems).
4. pGem-3Zf(–) (Promega, Madison, WI).
5. DNase I, RNase free (Roche Diagnostics).
6. DNase I digestion buffer: 50 m*M* Tris-HCl (pH 8.0 at 25 °C), 0.01 m*M* $MnCl_2$ (freshly prepared).
7. T4 DNA polymerase (New England Biolabs).
8. Polymerization buffer: 10 m*M* Tris-HCl (pH 7.9 at 25 °C), 10 m*M* $MgCl_2$, 50 m*M* NaCl, 1 m*M* dithiothreitol (DTT), 0.1 mg/mL bovine serum albumin (BSA).
9. Klenow fragment DNA polymerase (New England Biolabs).
10. Restriction enzymes: Sma I, EcoRI, BamHI (New England Biolabs).
11. CIP (calf intestine alkaline phosphatase; Roche Diagnostics).
12. T4 DNA ligase (New England Biolabs).
13. Ligation buffer: 50 m*M* Tris-HCl (pH 7.5 at 25 °C), 10 m*M* $MgCl_2$, 10 m*M* DTT, 10 m*M* adenosine triphosphate (ATP), 25 mg/mL BSA.
14. AMV reverse transcriptase (RT; Roche Diagnostics).
15. RT buffer: 50 m*M* Tris-HCl (pH 8.5 at 25 °C), 8 m*M* $MgCl_2$, 30 m*M* KCl, 1 m*M* DTT.
16. T7–5 plus single-stranded (ssDNA) primer: 5′-TAATACGACTCACTATAGGGC GAATTCGAGCTCG-3′.
17. 3 minus ssDNA primer: 5′-GTCGACTCTAGAGGATCC-3′.
18. PCR thermocycler.
19. PCR buffer: 10 m*M* Tris-HCl (pH 8.3 at 25 °C), 1.5 m*M* $MgCl_2$, 50 m*M* KCl.

20. QIAquick PCR purification kit (Qiagen, Hilden, Germany).
21. T7 RNA polymerase in vitro transcription equipment.
22. Transcription buffer: 40 m*M* Tris-HCl (pH 8.0 at 25 °C), 22 m*M* $MgCl_2$, 5 m*M* DTT, 1 m*M* spermidine, 100 µg/mL BSA, 1 U/µL RNasin, 5 U/mL inorganic pyrophosphatase.
23. RNA extraction buffer: 10 m*M* Tris-HCl (pH 7.8 at 25 °C), 100 m*M* NaCl, 1% sodium dodecyl sulfate (SDS), 1 m*M* DTE.
24. µCi [α^{32}P]UTP (uridine triphosphate; Amersham, Braunschweig, Germany).
25. Agarose gel electrophoresis.
26. PPA (urea) gel electrophoresis.
27. Phenol, pH 8.1, Tris-HCl buffered (Roth, Karlsruhe, Germany).
28. Binding buffer: 20 m*M* HEPES KOH (pH 7.5 at 4 °C), 4 m*M* $MgCl_2$, 400 m*M* NH_4Cl, and 6 m*M* β-mercaptoethanol.
29. Nitrocellulose filter, 0.45 µm (Schleicher and Schüll).
30. RNase-free glycogen (Roche Diagnostics).

3. Methods

3.1. Construction of a Random RNA Fragment Pool

An RNA pool is constructed that consists of a statistical distribution of RNA fragments of certain lengths. This pool contains the RNA sites to which the proteins bind in vivo provided that the length of the RNA fragments covers the corresponding binding site. The pool construction starts from a plasmid that codes for the large target RNA (*see* **Note 1**) and is digested with DNase I. T4 DNA polymerase is used to make the fragments completely blunt ended. The fragments are ligated into the Sma I site of the vector pGem-3Zf(−), thereby introducing the 5′ and 3′ constant regions for PCR amplification.

1. DNase I digestion (**Fig. 2**, left): Treat 35 ng/µL DNA with DNase I (RNase free, 0.3 µg per µg DNA) in DNase I digestion buffer at 16 °C for 15 to 50 min. Stop the reaction by the addition of ethylenediaminetetraacetic acid (EDTA; 0.1 m*M*). Extract twice with phenol and concentrate the DNA fragments via precipitation with three volumes EtOH in the presence of 0.3 *M* NaAc (pH 6.0) (*see* **Note 2**).
2. Blunt ending of the fragments: Resuspend the DNA fragments in water and incubate with T4 DNA polymerase (2.5 U per µg DNA) in polymerization buffer and 100 µ*M* of each deoxyribonucleotide 5′-triphosphates (dNTPs) at 25 °C for 15 min. Add 1 U Klenow fragment per microgram DNA and further incubate at 25 °C for 15 min.
3. Ligation of the fragments into the digested and dephosphorylated vector with T4 DNA ligase: Ligation is performed in ligation buffer at 16 °C for 5–7 h: 0.01–1 µg DNA (vector plus fragment) per microliter, 0.2 U T4 DNA ligase per microliter (*see* **Note 3**). The multiple cloning sites on the vector, flanking the insert, were used as constant regions for PCR amplification.

3.2. Cycles of Selection

SELEX is a widely used and well-established method ***(11)*** and detailed descriptions of the experimental procedures are available ***(12)***. Nevertheless, the selective step of each SELEX experiment needs to be optimized to be efficient and to avoid artifacts. The specialty of the SERF variant presented here is that the RNA pool consists of sequences of different sizes derived from a large RNA. During the various cycles of selection, a pressure favoring shorter fragments exists (*see* **Note 4**). Therefore, it is important to be specific in the reverse transcription reaction (see below; a reaction time not longer than 30 min; stoichiometry of 3′(–) primer: template ≤ 2:1) and not to perform too many PCR cycles. Furthermore, a control without RNA during the whole procedure is required.

1. The PCR amplification of the random fragment pools: Reactions are carried out in 150-μL batches: PCR buffer, 0.5 m*M* of each of the T7–5′(+) and 3′(–) primers, 2 m*M* of each of the dNTPs, and 20 U TAQ polymerase/milliliter. To amplify the initial pool, use up to 10 vol% of the ligation reaction as a template. During

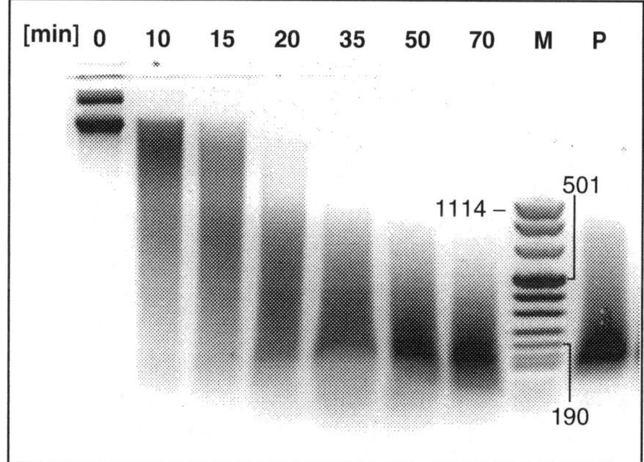

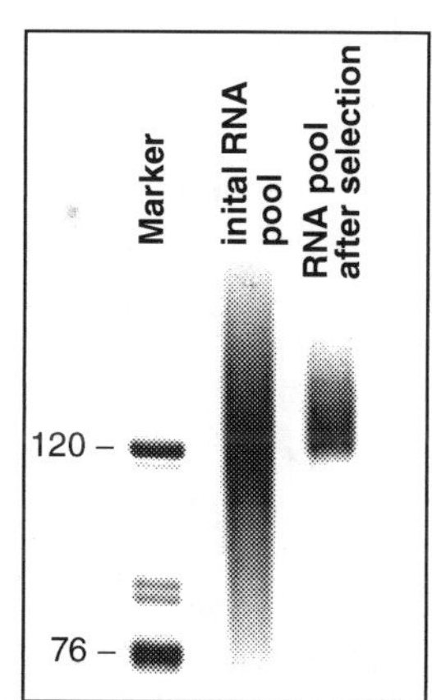

Fig. 2. Size distributions of random fragment pools. Left: Size distribution of random DNA fragments on a 2% ethidium bromide-stained agarose gel. The 0- to 70-min time course of the Mn^{2+}-dependent (0.01 m*M*) deoxyribonuclease (DNase) I digestion of the plasmid ptac-1 (10.6-kbp) carrying ribosomal DNA (rrnB operon). M, marker (1114 to 124 bp); P, polymerase chain reaction (PCR)-amplified DNA fragment pool ready for in vitro T7 transcription. Right: Size distribution of a random ribosomal RNA (rRNA) fragment pools from a SERF (selection of random RNA fragments) experiment on a 7% PAA/urea gel. Binding of the initial pool to ribosomal protein L11 was 1.5%, and it increased after four rounds of selection to 12% of RNA retained on nitrocellulose filter ***(4)***. Lane 1, marker: 5S RNA (120 nt) and transfer RNA (tRNA) bulk (76 nt); and tRNAs with a long extra arm (two bands above 76 nt); lane 2, initial RNA starting pool; lane 3, RNA after four rounds of selection.

the subsequent selection rounds, use up to 10 vol% of the reverse transcription reaction mixture. After 2 min at 92 °C, perform only 10–20 cycles of amplification, 92 °C for 30 s, 56 °C for 30 s, and 70 °C for 45 s. Purify the product with the QIAquick PCR purification kit according to the user's manual and thus reduce the volume by one-half. Analyze the PCR fragments (7 µL of PCR reaction) on a 2% agarose gel (**Fig. 2**, left, lane P).

2. Preparation of the RNA random fragment pool by T7 in vitro transcription (*see* **Note 5**): Incubate a 100-µL reaction contained 30–50 µL template (50–150 pmol/mL), 3.75 m*M* of each NTP, and 40 µg/mL of purified T7 RNA polymerase in transcription buffer at 37 °C for 5 h or overnight. For ^{32}P labeling of the RNA, add 2–5 µCi [α^{32}P]UTP to the reaction mixture. Add DNase I (RNase free, 10 µg/mL) after the reaction has finished and keep at 37 °C for a further 10 min. Dilute to 200 µL with water, extract with phenol, and precipitate with EtOH (*see* **Note 6**).
3. Gel purification of the RNA (*see* **Note 7**): The RNA fragments are purified via denaturing urea PAA–gel electrophoresis ***(13)*** and are seen as a smear (ultraviolet [UV] shadowing) on an acrylamide gel (**Fig. 2**, right). Cut out RNA of the desired size (100–200 nt, including constant primer regions), smash the gel into pieces by pressing through a syringe, and extract in RNA extraction buffer and 25% phenol overnight. Remove phenol and gel pieces by centrifugation. Extract the aqueous phase with chloroform and precipitate the RNA with EtOH twice. A 100-µL T7 transcription reaction purified this way yields about 2000 pmol of RNA fragments (*see* **Note 8**).
4. Formation of RNA–protein complexes: Incubate RNA fragments for 3 min at 70 °C and slowly cool to 37 °C in binding buffer. Incubate the proteins (*see* **Note 9**) for 15 min at 37 °C in binding buffer before mixing with the RNA fragments. Incubate the mixture for 10 min at 37 °C and 10 min at 0 °C (ice bath) to form RNA--protein complexes. Apply the mixture on a 0.45-µm nitrocellulose filter (prewetted and degassed in an evacuator for 30 min in binding buffer) under mild suction (water pump). Wash the filter with 500 µL of ice-cold binding buffer. Place the filter on a clean glass plate and cut it in pieces. Extract with 200 µL phenol (Tris-HCl, pH 7.8) and 400 µL 7 *M* urea (freshly prepared) for 60 min at room temperature. Add 200 µL chloroform/isoamyl alcohol (24:1) to facilitate phase separation by centrifugation. Treat the aqueous phase with chloroform/isoamyl alcohol, add 1 µg/mL RNase-free glycogen and recover the RNA from the aqueous phase by EtOH precipitation.
5. Reverse transcription of the RNA: Dissolve the selected RNA in 10 µL water containing 1.5 pmol/µL 3′(–) primer. For annealing, heat the RNA primer mixture at 70 °C for 5 min and cool to room temperature during 5 min. Add AMV RT in 30 µL RT buffer containing 500 µ*M* of each dNTP for 30 min at 42 °C. Use 10 µL of the reaction directly in a 150 µL PCR reaction.

3.3. Monitoring Progress of Selection

Progress in the selection procedure can be detected by measuring the binding of the newly synthesized RNA to the protein after each round of selection. This is most easily done with a filter-binding assay, in which radiolabeled RNA is incubated with excess protein and filtered through nitrocellulose ***(14,15)***. Another widely used method is a band-shift assay (***6;*** *see also* Chapter 7 of this book), by which the protein–RNA complex is separated from free RNA on a native PAA gel. Since the random RNA fragment pool contains RNAs of different length, only the filter-binding assay is used for measuring the progress of selection. However, the binding of cloned fragments (i.e., winner sequences that have a defined length) can be measured conveniently by band-shift assays and characterized by RNA-probing methods ***(3,16)***.

1. Nitrocellulose filter-binding assay: Incubate RNA fragments (70 n*M* f.c.) for 3 min at 70 °C in binding buffer and then cool to 37 °C. Incubate the protein for 15 min at 37 °C in binding buffer, mix with the RNA fragments, incubate 10 min at 37 °C, and further incubate 10 min on ice. RNA and protein should add up to a volume of 300 µL. Assay several increasing protein concentrations (starting from, e.g., 33 n*M* f.c.) while keeping RNA concentrations constant (*see* **Note 10**). Apply the protein complexes directly onto a 0.45-µm nitrocellulose filter (prewetted and degassed for 30 min in binding buffer) under suction. Determine the radioactivity retained on the filter by liquid scintillation counting or phosphorimaging.
2. Cloning of sequences that bind to the protein: Once an increase in binding between the initial pool and pools after selection is detected, digest the corresponding PCR product with BamHI and EcoRI and ligate into pGem-3Zf(−). Pick up to 100 clones for sequencing and identify overlapping sequences that indicate a potential high-affinity binding site. Test individual fragments for binding.

4. Notes

1. The RNA sequences can also be obtained as PCR products (e.g., amplified from libraries) and need not be present as cloned fragments in a vector. However, vector sequences need not be excluded from the pool preparation as the sequence variability of random fragment pools is low ***(4)***, and the vector sequences do not interfere with the selection.
2. It is difficult to remove DNase I from the mixture completely. Since the rate of DNase I digestion decreases when the fragments reach a size around 100–250, the remaining activity does not interfere with further procedures (e.g., blunt ending with T4 DNA polymerase).
3. Best results are obtained with an excess of vector over ribosomal DNA fragments in the ligation reaction. The cut as well as the religated vector alone do not yield any PCR product.
4. The process might favor shorter fragments due to possible processivity problems of the RT and a biased amplification of the cDNAs during PCR.
5. For optimized T7 in vitro transcription protocols, see also **refs.** ***17*** and ***18*** and Chapter 9 of this volume.
6. A white pellet is obtained that has to be re-solved completely with a pipet before further purification of the RNA.
7. The RNA fragments are seen as a smear on a denaturing acrylamide gel as the pool contains fragments of different sizes. In addition to the DNase I digestion

step that roughly sets the size of the fragments, in this step the size distribution of the random RNA fragment pool can be further defined. Please take into account that the fragments contain 5′ and 3′ constant regions.

8. The assumption is that 1 A_{260} of a approx 140-nt fragment corresponds to 800 pmol.
9. Proteins have to be essentially pure to rule out that RNA fragments are selected that interact with contaminating proteins. However, most important, the proteins must not show any RNase activity. This has to be tested by 15 min, 37 °C incubation of the purified proteins with 5′-labeled radioactive RNA in binding buffer and must not produce cleavage products.
10. Binding levels strongly depend on the affinity of the proteins to RNA; thus the protein concentrations have to be adjusted accordingly. As a rule, nonspecific binding of RNA fragments to a protein of the initial pool should not exceed 5% of the input RNA after correcting for background binding. Selection will become more stringent by increasing the RNA-to-protein ratio. Background binding (i.e., the amount of RNA retained on the nitrocellulose filter without protein), usually amounts to 6–10% and can be reduced by washing once with 500 µL binding buffer to 3–5%.

References

1. Klein, D. J., Moore, P. B., and Steitz, T. A. (2004) The roles of ribosomal proteins in the structure assembly, and evolution of the large ribosomal subunit. *J. Mol. Biol.* **340**, 141–177.
2. Bartel, D. P., and Szostak, J. W. In: *RNA–Protein Interactions* (Nagai, K., and Mattaj, I. W., eds.), Oxford University Press, New York, 1994, pp. 248–268.
3. Stelzl, U., Spahn, C. M. T., and Nierhaus, K. H. (2000) Selecting rRNA binding sites for the ribosomal proteins L4 and L6 from randomly fragmented rRNA: application of a method called SERF. *Proc. Natl. Acad. Sci. U. S. A.* **97**, 4597–4602.
4. Stelzl, U., Spahn, C. M. T., and Nierhaus, K. H. (2000) RNA–protein interactions in ribosomes: in vitro selection from randomly fragmented rRNA. *Methods Enzymol.* **318**, 251–268.
5. Nierhaus, K. H. (1991) The assembly of prokaryotic ribosomes. *Biochimie* **73**, 739–755.
6. Stelzl, U., and Nierhaus, K. H., (2001) A short fragment of 23S rRNA containing the binding sites for two ribosomal proteins, L24 and L4, is a key element for rRNA folding during early assembly. *RNA* **7**, 598–609.
7. Nilsen, T. W. (2003) The spliceosome: the most complex macromolecular machine in the cell? *Bioessays* **25**, 1147–1149.v
8. Eddy, S. R. (2001) Non-coding RNA genes and the modern RNA world. *Nat. Rev. Genet.* **2**, 919–929.
9. Melgar, E., and Goldthwait, D. A. (1968) Desoxyribonucleic acid nucleases: II. The effects of metals on the mechanism of action of deoxyribonuclease I. *J. Biol. Chem.* **243**, 4409–4416.
10. Irvine, D., Tuerk, C., and Gold, L. (1991) SELEXION systematic evolution of ligands by exponential enrichment with integrated optimization by non-linear analysis. *J. Mol. Biol.* **222**, 739–761.

11. Gold, L., Polisky, B., Uhlenbeck, O., and Yarus, M. (1995) Diversity of oligonucleotide functions. *Annu. Rev. Biochem.* **64**, 763–797.
12. Marshall, K. A., and Ellington, A. D. (2000) In vitro selection of RNA aptamers. *Methods Enzymol.* **318**, 193–214.
13. Doudna, J. A. (1997) Preparation of homogenous ribozyme RNA for crystallization. *Methods Mol. Biol.* **74**, 365–370.
14. Carey, J., Cameron, V., deHaseth, P. L., and Uhlenbeck, O. C. (1983) Sequence-specific interaction of R17 coat protein with its ribonucleic acid binding site. *Biochemistry* **22**, 2601–2610.
15. Draper, D. E., Deckman, I. C., and Vartikar, J. V. (1988) Physical studies of ribosomal protein–RNA interactions. *Methods Enzymol.* **164**, 203–220.
16. U. Stelzl, et al. (2003) RNA-structural mimicry in *E. coli* ribosomal protein L4-dependent regulation of the S10 operon. *J. Biol. Chem.* **278**, 28237–28245.
17. Milligan, F., and Uhlenbeck, O. C. (1989) Synthesis of small RNAs using T7 RNA polymerase. *Methods Enzymol.* **180**, 51–62.
18. Triana-Alonso, F. J., Dabrowski, M., Wadzack, J., and Nierhaus, K. H. (1995) Self-coded 3′-extension of run-off transcripts produces aberrant products during in vitro transcription with T7 RNA polymerase. *J. Biol. Chem.* **270**, 6298–6307.

18

Immunoprecipitation Analysis to Study RNA–Protein Interactions in *Xenopus* Oocytes

Naoto Mabuchi, Kaoru Masuyama, and Mutsuhito Ohno

Summary

Results obtained from in vitro experiments often need to be confirmed by in vivo experiments. The study of RNA–protein interactions is no exception. Information on RNA–protein complex formation in the cell is important for understanding the mechanisms of cellular RNA metabolism such as RNA processing and transport. For such purposes, *Xenopus* oocytes are extremely useful cells thanks to their large size. Interactions of microinjected proteins and RNAs with their binding partners can be examined easily by immunoprecipitation experiments with nuclear or cytoplasmic fractions from microinjected *Xenopus* oocytes. We describe a method to study how RNAs that have been microinjected into the nucleus of *Xenopus* oocytes are assembled into complexes with specific endogenous proteins.

Key Words: Immunoprecipitation; microinjection; RNA export; RNA processing; RNA–protein interaction; *Xenopus* oocytes.

1. Introduction

In eukaryotic cells, RNA molecules that have been transcribed from the chromatin template are subjected to specific processing and transport. Such processes are largely mediated by protein factors that associate with RNA directly or indirectly. To study such RNA–protein associations in vivo, *Xenopus* oocytes are extremely useful cells. Thanks to their large size, recombinant proteins or in vitro-transcribed RNAs can be introduced easily into the nucleus or cytoplasm by means of microinjection. In many cases, the microinjected proteins and RNAs can precisely recapitulate the events in the cells ***(1–7)***. Moreover, the interactions of injected proteins and RNAs with their binding partners can be detected easily by immunoprecipitation assays ***(8–14)***. This chapter describes immunoprecipitation of RNA–protein complexes after microinjection of in vitro-transcribed^{32} P-labeled RNAs into the nucleus of *Xenopus* oocytes.

From: *Methods in Molecular Biology, vol. 488: RNA-Protein Interaction Protocols*
Edited by: Ren-Jang Lin © Humana Press Inc, Totowa, NJ

2. Materials

1. SP6, T3, or T7 bacteriophage RNA polymerase (10–20 U/∝L) (Promega, Madison, WI). Optimized 5X buffer and 100 m*M* dithiothreitol (DTT) are supplied with the enzymes.
2. NTP mix: 5 m*M* adenosine triphosphate (ATP), cytosine triphosphate (CTP), 1 m*M* guanosine triphosphate (GTP), uridine triphosphate (UTP); or 5 m*M* ATP, CTP, GTP, 1 m*M* UTP for the synthesis of m^7G (7-methyl guanosine)-capped or uncapped RNAs, respectively.
3. m^7G cap analog: $m^7G(5')ppp(5')G$ (New England BioLabs, Beverly, MA). Prepare as a 10 m*M* stock solution in water.
4. Ribonuclease inhibitor (RNasin; 10–20 U/μL) (Promega) or equivalent.
5. [α-^{32}P]UTP (800 Ci/mmol; 20 mCi/mL) of in vitro transcription quality.
6. Phenol/chloroform/isoamyl alcohol (25:24:1) (PCI).
7. Quick spin column (G-50) for radiolabeled RNA purification (Roche, Indianapolis, IN) or equivalent.
8. Micromanipulator: Mk1 Micromanipulator (Singer Instrument, Somerset, England) or equivalent.
9. Microinjector: IM 300 Microinjector (Narishige, Tokyo, Japan) or equivalent.
10. Stereomicroscope: Zeiss Stemi 2000-C (Carl Zeiss, Ismaning, Germany) or Leica MZ12 (Leica Microsystems, Heerbrugg, Switzerland) or equivalent.
11. Illuminator KL 2500 LCD (Carl Zeiss) or equivalent.
12. Micropipette Grinder EG-44 (Narishige) or equivalent (optional).
13. Glass capillary: Microcaps 6.66 μL (Drummond Scientific Company, Broomall, PA).
14. 60-well microtestplate (Greiner Bio-One, Kremsmuenster, Austria).
15. Blue dextran (MW 2 million; Sigma-Aldrich, St. Louis, MO): 30 mg/mL in H_2O, filter sterilized.
16. 5X Barth solution with Ca^{2+}: 440 m*M* NaCl, 5 m*M* KCl, 12 m*M* $NaHCO_3$, 50 m*M* HEPES, 4 m*M* $MgSO_4$, 1.7 m*M* $Ca(NO_3)_2$, 2 m*M* $CaCl_2$. Adjust to pH 7.6 with NaOH. Store at 4 °C.
17. Antibiotics stock: 10 mg/mL ampicillin and 10 mg/mL streptomycin. Store at –20 °C.
18. 1X Barth solution with Ca^{2+} containing antibiotics: Add 1/1000 volume of antibiotics stock to 1X Barth solution with Ca^{2+} and store at 19 °C. This solution can be used for 3–5 d.
19. 10X J-buffer salts: 700 m*M* NH_4Cl, 70 m*M* $MgCl_2$, 1 m*M* ethylenediaminetetraacetic acid (EDTA), 100 m*M* HEPES (pH 7.6).
20. J-buffer: 1/10 volume of 10X J-buffer salts and glycerol (final 10% v/v). Store at 19 °C.
21. Homomix: 50 m*M* Tris-HCl (pH 7.4), 5 m*M* EDTA, 1.5% sodium dodecyl sulfate (SDS), 300 m*M* NaCl, 1.5 mg/mL proteinase K. Store at –20 °C.
22. RNA loading dye: 80% formamide (deionized), 0.05% SDS, 0.05% xylene cyanol FF, 0.05% bromophenol blue. Store at –20 °C.
23. Denaturing polyacrylamide gel stock solution: 0.5X TBE, 7 *M* urea, and appropriate concentration (% w/v) of acrylamide/bis mix (37.5:1). Filtrate and store in a dark

bottle. The gel is polymerized by the addition of 1/100 volume of 10% ammonium persulfate and 1/2000 volume of TEMED.

24. Protein A-Sepharose Fast Flow (Amersham Biosciences, Piscataway, NJ).
25. PBS (phosphate-buffered saline): 10 m*M* phosphate buffer (pH 7.4), 137 m*M* NaCl, 2.7 m*M* KCl.
26. RSB100N: 10 m*M* Tris-HCl (pH 7.4), 100 m*M* NaCl, 2.5 m*M* $MgCl_2$, 0.1% Nonidet P-40.
27. Protease inhibitor cocktail: Complete, EDTA free (Roche, Indianapolis, IN) 50X stock solution is prepared by dissolving 1 tablet in 500 µL H_2O and stored at –20 °C.
28. Intensifying screen: Biomax MS screen (Kodak, Rochester, NY) or equivalent (optional).

3. Methods

The methods described are (1) in vitro transcription of ^{32}P-labeled RNA substrates for microinjection, (2) microinjection into the nucleus of *Xenopus* oocytes, (3) transport analysis of the microinjected RNAs in oocytes, and (4) immunoprecipitation analysis.

3.1. In Vitro Transcription of ^{32}P-Labeled RNA Substrates for Microinjection

3.1.1. Preparation of Template DNA

A DNA fragment of interest is cloned into a plasmid vector suitable for in vitro transcription with bacteriophage SP6, T7, or T3 polymerase. The resultant plasmid is linearized with an appropriate restriction enzyme with a cleavage site at or near the 3′ end of the insert. Alternatively, template DNA can be prepared by polymerase chain reaction (PCR) with primers complementary to the 5′ and 3′ ends of the insert. The 5′ region of the 5′ PCR primer must have an appropriate bacteriophage promoter sequence. More complete descriptions of in vitro transcription systems have been published elsewhere ***(15,16)***. The linear template DNA is extracted twice with PCI, ethanol precipitated, and dissolved in water at 1 mg/mL.

3.1.2. In Vitro Transcription

1. A reaction mixture (in 10 µL) containing 1 µL template DNA (1 mg/mL), 2 µL 5X transcription buffer, 1 µL 100 m*M* DTT, 1 µL NTP mix (5 m*M* ATP, CTP, 1 m*M* GTP, UTP), 1 µL m^7G cap analog (10 m*M*), 1 µL bacteriophage RNA polymerase (10–20 U/µL), 0.3 µL ribonuclease (RNase) inhibitor (40 U/µL), 1 µL [α-^{32}P]UTP, and 1.7 µL H_2O is incubated at 37 °C for 30 min (*see* **Note 1**).
2. PCI extraction is carried out twice after the addition of 90 µL H_2O to the reaction mixture. Unincorporated nucleotides in the recovered aqueous phase are removed by a Quick spin column G50.

3. Check 1 μL of eluate by denaturing polyacrylamide gel electrophoresis (PAGE) followed by autoradiography to ensure that the synthesized RNA is of the expected size. The standard procedure for denaturing PAGE is described elsewhere *(17)*. If prematurely terminated RNA products are present at significant levels, the full-length transcript should be gel purified.
4. Several RNAs of different sizes can be mixed (*see* **Note 2**) in an appropriate ratio so that each RNA has similar radioactivity (as estimated by the band intensity of the autoradiogram in **step 3**). The RNA mixture is ethanol precipitated and dissolved in an appropriate amount of water (*see* **Note 3**).

3.2. Microinjection Into the Nucleus of Xenopus Oocytes

This section focuses on the RNA microinjection assay into the nucleus of *Xenopus* oocytes. Detailed descriptions of how to keep *Xenopus* frogs and how to isolate oocytes from them can be found elsewhere *(18)*.

3.2.1. Microinjection Equipment

The basic microinjection system consists of a microinjector, a micromanipulator, and a stereomicroscope. Although several different microinjection systems are commercially available, our laboratory mainly uses the combination of Narishige microinjector, Singer micromanipulator, and Zeiss stereomicroscope (*see* **Heading 2.**). Microinjection needles are prepared by pulling glass capillaries using a micropipet puller. To make the needle sharp, the tip is broken with forceps or cut with a knife (*see* **Note 4**).

3.2.2. Preparation of the Microinjection Mixture

To make the microinjection mixture, 1 μL of the RNA mixture prepared in **Subheading 3.1.2.** is mixed with 1 μL filter-sterilized blue dextran (30 mg/mL in H_2O), 0.2 μL RNasin, and H_2O to 4 μL. Blue dextran is used to mark the injected nuclei. A successful nuclear injection should result in a blue nucleus. Recombinant proteins or inhibitors, if necessary, can be added at the expense of water. The injection mixture may be subjected to a brief centrifugation (5 min at room temperature in a microfuge) to remove any particulates that may clog the needle.

3.2.3. Microinjection

The injection needle is marked with a thin marker pen at 0.8-mm intervals. One mark should correspond to a volume of 66 nL. The needle is then filled with the microinjection mixture, and the injection time is calibrated so that one injection blow (one activation of the foot switch) corresponds to the half-way mark (33 nL). Healthy stage V and stage VI oocytes are put in a 60-well microtestplate that has been filled with 1X Barth solution with Ca^{2+} containing antibiotics. The needle filled with the injection mixture is then used to pierce the top of the animal hemisphere of the oocyte to a depth of approximately one-quarter of the oocyte diameter, and an injection blow is carried out using the foot switch.

3.3. Transport Analysis of Microinjected RNAs in Oocytes

RNAs injected in the nucleus are often processed and transported to the cytoplasm. In addition, injected RNAs are frequently unstable and degraded. It is important to check the behavior of injected RNAs by performing a microinjection test experiment prior to the immunoprecipitation experiments.

To analyze RNA export to the cytoplasm, the nuclear-injected oocytes, after appropriate incubation time at 19 °C, are manually dissected with sharp forceps in J-buffer into nuclear and cytoplasmic fractions. To do so, the injected oocyte is stabbed using the sharp forceps near the top of the animal hemisphere to make a small hole. The oocyte is then gently squeezed so that the nucleus comes out of the hole. It is important not to damage the nucleus with the sharp forceps. The oocyte from which the nucleus has been removed is considered to be the "cytoplasm." Only the oocytes with a blue nucleus are selected, and RNA is extracted from each fraction. The extracted RNA is analyzed by denaturing PAGE (**Fig. 1**).

1. Five dissected nuclei and cytoplasms are resuspended in two separate tubes, each containing 200 μL homomix, by continuous vigorous voltexing (*see* **Note 5**). The resuspended samples are incubated at 50 °C for 30 min.
2. Extract the samples once with 200 μL PCI and recover only 120 μL of the aqueous phase.
3. RNA in the samples is precipitated by adding 375 μL ethanol (no addition of salt).

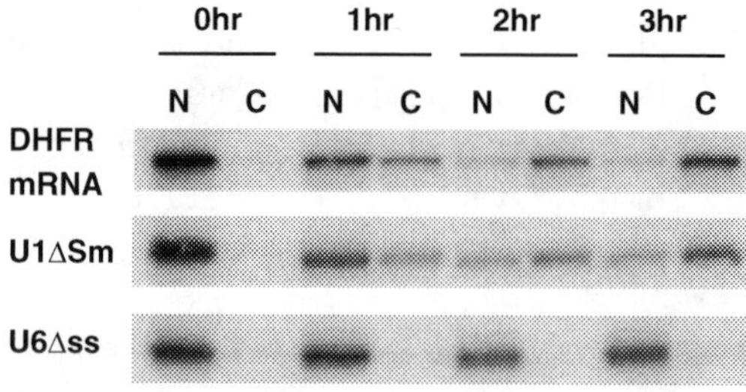

Fig. 1. Export of microinjected RNAs in *Xenopus* oocytes. A mixture of in vitro-transcribed ^{32}P-labeled DHFR (dihydrofolate reductase) messenger RNA, U1ΔSm, and U6Δss RNAs was injected into the nucleus of *Xenopus* oocytes. The injected oocytes were incubated at 19 °C for 0, 1, 2, or 3 h and dissected into the nucleus and cytoplasm. RNA extracted from the nuclear (N) and cytoplasmic (C) fractions was analyzed by denaturing 8% polyacrylamide gel electrophoresis (PAGE).

4. After centrifugation for 15 min in a microfuge, the RNA pellet is rinsed with 70% ethanol and dissolved in 15 µL RNA loading dye (equivalent to 5 µL per nucleus or cytoplasm).
5. Heat the sample to 80 °C for 2 min and chill immediately on ice.
6. Analyze 5 µL per lane by denaturing PAGE.
7. After electrophoresis, urea is removed by soaking the gel in water for 10 min. The gel is then dried and exposed to X-ray film. Intensifying screens may be used to enhance the signals.

3.4. Immunoprecipitaion Analysis

The association of microinjected RNAs with specific endogenous proteins can be studied by the analysis of RNA that has been coprecipitated by antibodies against specific proteins (**Fig. 2**).

3.4.1. Preparation of Antibody-Conjugated Protein A-Sepharose Beads

An appropriate amount of antibody (*see* **Note 6**) is mixed with 15 µL of preswollen protein A-Sepharose beads in 500 µL PBS containing 0.1% Nonidet P-40, and the mixture is incubated with constant rotation for at least 1 h at 4 °C. The beads are then washed twice with RSB100N.

3.4.2. Preparation of Nuclear and Cytoplasmic Extracts for Immunoprecipitation

After an appropriate period of incubation (*see* **Note 7**), the microinjected oocytes are dissected into nuclei and cytoplasms. The dissected nuclei and cytoplasms are collected separately in RSB100N containing 1 U/µL RNasin and 1X protease inhibitor cocktail. The nuclear and cytoplasmic fractions from 4 oocytes (in 100 µL buffer) are often used per immunoprecipitation. The sample is homogenized by vigorous pipeting with a yellow tip or by passing through a 25-gauge needle five times. The mixture is centrifuged at full speed in a microfuge at 4 °C for 15 min, and the supernatant is taken as a nuclear or cytoplasmic extract. In the case of the cytoplasmic extract, care should be taken not to take the yellowish lipid layer on top of the solution.

3.4.3. Immunoprecipitation

1. Mix 100 µL nuclear or cytoplasmic extract with 15 µL antibody-bound protein A-Sepharose beads, as prepared in **Subheading 3.4.1.**, and incubate the mixture with constant rotation for 1 h at 4 °C.
2. Spin down the beads at 500 *g* for 30 s and collect the supernatant. Mix 10 µL supernatant with 90 µL homomix and keep this as the "unbound" sample.
3. Wash the beads four times with 1 mL RSB100N.
4. Add 1 mL RSB100N to the bead pellet and transfer it to a new tube. This step is important to reduce background due to nonspecific binding of RNA to the tube.

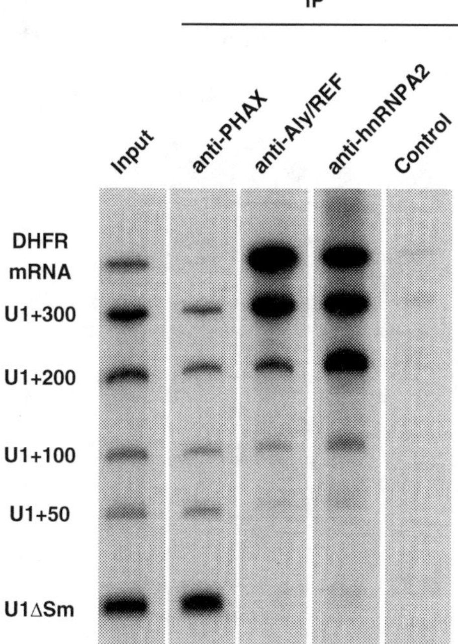

Fig. 2. Immunoprecipitation assay after microinjection of various RNAs into *Xenopus* oocytes. A mixture of in vitro-transcribed ^{32}P-labeled U1 RNA derivatives, in which U1 RNA was elongated by the insertion of DHFR (dihydrofolate reductase) messenger RNA (mRNA) sequences of various lengths (50, 100, 200, or 300 nt) together with ^{32}P-labeled DHFR mRNA and U1ΔSm RNA, was injected into the nucleus of *Xenopus* oocytes. The injected oocytes were incubated at 19 °C for 1 h. Immunoprecipitation was carried out from the nuclear fraction as described in the text, with anti-PHAX, anti-Aly/REF, anti-hnRNPA2, and control (anti-Myc tag) antibodies.

5. Spin down the beads and remove the supernatant thoroughly.
6. Add 85 μL homomix to the beads and mix.

3.4.4. Extraction of RNA and Denaturing PAGE

1. Incubate the above samples for 30 min at 50 °C.
2. Add 100 μL PCI and recover 70 μL of the aqueous phase.
3. Add 7 μL 3 *M* NaOAc, 0.5 μL glycogen (20 mg/mL), and 175 μL ethanol to the aqueous phase and spin the mixture at full speed in a microfuge for 15 min.
4. Rinse the pellet with 70% ethanol and dry.
5. Dissolve the pellet in 10 μL RNA loading dye.
6. Heat the sample to 80 °C for 2 min and chill on ice.
7. Fractionate 5-μL sample by denaturing PAGE.

8. After electrophoresis, urea is removed by soaking the gel in water for 10 min. The gel is then dried and exposed to an X-ray film. Intensifying screens may be used to enhance the signals.

4. Notes

1. All RNA polymerase II transcripts naturally receive a m^7G cap structure. The m^7G cap plays important roles in stability, processing, transport, and translation of RNA. To add an m^7G cap to the RNA substrates, the in vitro transcription reaction is carried out in the presence of a 10-fold molar excess of m^7G cap analog over GTP. For the synthesis of uncapped RNAs such as transfer RNAs, the reaction is carried out in the absence of m^7G cap analog with the alternative NTP mix (5 m*M* ATP, CTP, GTP, 1 m*M* UTP).
2. Several control RNAs are routinely coinjected. In the microinjection assay, for instance, U6Δss RNA is a useful marker for accurate nuclear injections since U6Δss RNA is neither exported nor imported ***(1)***.
3. The amount of added water determines the amount of RNA microinjected into one oocyte. The proper amount of RNA microinjected into one oocyte should be determined experimentally by considering the transport capacity of the oocytes. Less than 10 fmol of RNA molecules per nucleus should not saturate the RNA export pathways in our experimental conditions.
4. The needle may be sharpened using a micropipet grinder.
5. Five uninjected whole oocytes are added to the nuclear fraction to make the total amount of RNA in the nuclear fraction similar to that in the cytoplasmic fraction. This is important to achieve quantitative RNA recovery through the RNA extraction procedures.
6. Since different antibodies have different titers and affinity, the amount of antibody should be determined experimentally. In addition, control antibodies such as the preimmune serum should be included as the specificity control of immunoprecipitation (anti-Myc tag is used in **Fig. 2**).
7. Appropriate incubation time should be determined experimentally since the state of RNA processing or transport is dependent on the incubation time. The time frame should be 30 min to 3 h after injection. We frequently use 1 h.

References

1. Jarmolowski, A., Boelens, W. C., Izaurralde, E., and Mattaj, I. W. (1994) Nuclear export of different classes of RNA is mediated by specific factors. *J. Cell Biol.* **124**, 627–635.
2. Fornerd, M., Ohno, M., Yoshida, M., and Mattaj, I. W. (1997) CRM1 is an export receptor for Leucine-rich nuclear export signals. *Cell* **90**, 1051–1060.
3. Ohno, M., Segref, A., Bachi, A., Wilm, M., and Mattaj, I. W. (2000) PHAX, a mediator of U snRNA nuclear export whose activity is regulated by phosphorylation. *Cell* **101**, 187–198.
4. Lund, E., and Dahlberg, J. E. (1998) Proofreading and aminoacylation of tRNAs before export from the nucleus. *Science* **282**, 2082–2085.

5. Lund, E., Guttinger, S., Calado, A., Dahlberg, J. E., and Kutay, U. (2004) Nuclear export of microRNA precursors. *Science* **303**, 95–98.
6. Green, M. R., Maniatis, T., and Melton, D. A. (1983) Human beta-globin pre-mRNA synthesized in vitro is accurately spliced in *Xenopus* oocyte nuclei. *Cell* **32**, 681–694.
7. Yu, Y.-T., Shu, M.-D., and Steitz, J. (1998) Modifications of U2 snRNA are required for snRNP assembly and pre-mRNA splicing. *EMBO J.* **17**, 5783–5795.
8. Ohno, M., Segref, A., Kuersten, S., and Mattaj, I. W. (2002) Identity elements used in export of mRNAs. *Mol. Cell* **9**, 659–671.
9. Kataoka, N., Yong, J., Kim, V. N., et al. (2000) Pre-mRNA splicing imprints mRNA in the nucleus with a novel RNA-binding protein that persists in the cytoplasm. *Mol. Cell* **6**, 673–682.
10. Kim, V. N., Kataoka, N., and Dreyfuss, G. (2001) Role of the nonsense-mediated decay factor hUpf3 in the splicing-dependent exon–exon junction complex. *Science* **293**, 1832–1836.
11. Le Hir, H., Gatfield, D., Izaurralde, E., and Moore, M. J. (2001) The exon–exon junction complex provides a binding platform for factors involved in mRNA export and nonsense-mediated mRNA decay. *EMBO J.* **20**, 4987–4997.
12. Masuyama, K., Taniguchi, I., Kataoka, N., and Ohno, M. (2004) RNA length defines RNA export pathway. *Genes Dev.* **18**, 2074–2085.
13. Masuyama, K., Taniguchi, I., Kataoka, N., and Ohno, M. (2004) SR proteins preferentially associate with mRNAs in the nucleus and facilitate their export to the cytoplasm. *Genes Cells* **9**, 959–965.
14. Masuyama, K., Taniguchi, I., Okawa, K., and Ohno, M. (2007) Factors associated with a purine-rich exonic splicing enhancer sequence in *Xenopus* oocyte nucleus. *Biochem. Biophys. Res. Commun.* **359**, 580–585.
15. Yisraeli, J. K., and Melton, D. A. (1989) Synthesis of long, capped transcripts in vitro by SP6 and T7 RNA polymerases. *Methods Enzymol.* **180**, 42–50.
16. Clarke, P. A. (1999) Labeling and purification of RNA synthesized by in vitro transcription. *Methods Mol. Biol.* **118**, 1–10.
17. Sambrook, J., Fritsch, E. F., and Maniatis, T., eds. (1989) *Molecular Cloning: A Laboratory Manual*, 2nd ed., Cold Spring Harbor Laboratory Press, Cold Spring Harbor, NY.
18. Terns, M. P., and Goldfarb, D. S. (1998) Nuclear transport of RNAs in microinjected *Xenopus* oocytes. *Methods Cell Biol.* **53**, 559–589.

19

Mapping the Regions of RNase P Catalytic RNA That Are Potentially in Close Contact With Its Protein Cofactor

Phong Trang and Fenyong Liu

Summary

Ribonuclease P (RNase P) from *Escherichia coli* is a transfer RNA (tRNA)-processing enzyme and consists of a catalytic RNA subunit (M1 RNA) and a protein component (C5 protein). M1GS, a gene-targeting ribozyme derived from M1 RNA, can cleave a target messenger RNA (mRNA) efficiently in vitro and inhibit its expression effectively in cultured cells. It has been shown that C5 protein can significantly increase the activities of M1 ribozyme and M1GS RNA in cleaving a natural tRNA substrate and a target mRNA, respectively. Understanding how C5 binds to M1GS RNA and affects the specific interactions between the ribozyme and its target mRNA substrates may facilitate the development of gene-targeting ribozymes that function effectively in vivo in the presence of cellular proteins. We describe the methods to determine the regions of a M1GS ribozyme that are potentially in close proximity to C5 protein. Specifically, methods are described in detail in using Fe(II)-ethylenediaminetetraacetic acid (EDTA) cleavage and nuclease footprint analyses to map the regions of the ribozyme in the absence and presence of C5 protein. These methods intend to provide experimental protocols for studying the regions of RNase P ribozyme that are in close contact with C5 protein.

Key Words: Fe(II)-EDTA; nuclease footprint; M1 RNA; ribozyme; RNA–protein interaction; RNase P; structural analysis.

1. Introduction

Ribonuclease (RNase) P is a ribonucleoprotein complex responsible for the 5′ maturation of transfer RNAs (tRNAs) ***(1,2)***. It catalyzes a hydrolysis reaction to remove a 5′ leader sequence from tRNA precursors (pre-tRNA) and several small RNAs, including 4.5S RNA, 10Sa RNA, small-model substrates, and plant virus RNA ***(3–6)***. In *Escherichia coli*, RNase P consists of a catalytic RNA subunit (M1 RNA) of 377 nucleotides and a protein subunit (C5 protein) of 119 amino acids ***(1,2)***. In vitro, M1 RNA can cleave its pre-tRNA substrates

From: *Methods in Molecular Biology, vol. 488: RNA-Protein Interaction Protocols*
Edited by: Ren-Jang Lin © Humana Press Inc, Totowa, NJ

at high divalent ion concentration (e.g., 100 m*M* Mg^{2+}) in the absence of C5 protein ***(7)***. The addition of C5 protein dramatically increases the rate of cleavage by M1 RNA under low concentration of Mg^{+2} in vitro and is required for RNase P activity and cell viability in vivo ***(1,2)***.

Studies on substrate recognition by M1 RNA and RNase P have led to the development of a general strategy in which M1 RNA and RNase P can be used as gene-targeting tools to cleave any specific messenger RNA (mRNA) sequences. RNase P and M1 RNA can cleave an mRNA sequence efficiently if an additional small RNA (called an external guide sequence [EGS]), which contains a sequence complementary to the mRNA substrate and a 3´ proximal CCA, is present ***(8)***. EGSs are antisense oligoribonucleotides that have been used to diminish gene expression in mammalian cells ***(9,10)*** with the aid of either RNase P or its catalytic subunit (e.g., M1 RNA). The EGS-based technology takes advantage of RNase P or M1 RNA to cleave a targeted mRNA when the EGS hybridizes to the target RNA and forms a structure resembling a portion of the natural tRNA substrates of the enzymes ***(8,10)***. Recent studies have shown that expression of EGSs in tissue culture inhibits the gene expression of herpes simplex virus, human cytomegalovirus (HCMV), and influenza virus and furthermore abolishes the replication of HCMV and influenza virus ***(11–13)***. To increase the targeting efficiency, the EGS can be covalently linked to M1 RNA (e.g., to the 3´ end) to generate a sequence-specific ribozyme, M1GS RNA ***(9,14)***. Thus, in principle, any RNA could be targeted by a custom-designed M1GS for specific cleavage.

The roles of M1 RNA have been well documented, with extensive phylogenetic and biochemical analyses establishing models for the secondary structure ***(15–17)*** and the three-dimensional structure of the ribozyme ***(18–20)***. However, our understanding of the function of C5 protein remains relatively limited, although the three-dimensional structures of C5 protein and its homolog in *Bacillus subtilis* have recently been determined, and models of the interactions between M1 RNA and C5 protein have been proposed ***(19,21–23)***. It has been shown that C5 protein, an extremely basic protein, stabilizes the active conformation of M1 RNA, enhances the interactions between the enzyme and the pre-tRNA substrate, reduces the deleterious effects of mutations in M1 RNA, and broadens substrate specificity of the enzyme ***(2,24–28)***. C5 protein also significantly enhances the activity of M1GS RNA to cleave a target mRNA. Furthermore, it is believed that, when expressed in human cells, M1GS RNA may interact with cellular proteins, including those associated with human RNase P activity ***(9,29)***. The interactions among M1GS RNA and these human cellular proteins may stabilize the ribozyme and enhance its activity in cleaving an mRNA substrate and inhibiting gene expression in cells ***(9,29,30)***. Several reviews covering the recent progress on the function of C5 protein and its potential interaction with M1 RNA to achieve cleavage of a natural tRNA substrate have recently been published ***(2,27,28)***. This chapter focuses on using RNase mapping and Fe(II)-EDTA (ethylenediaminetetraacetic acid) footprinting analyses to study the RNA–protein interaction between M1 RNA and C5 protein to understand how C5 protein increases the activity of M1GS RNA in cleaving an mRNA substrate.

2. Materials

2.1. Chemicals and Solutions

1. 1 *M* Tris-HCl, pH 8.0.
2. 1 *M* Tris-HCl, pH 7.5.
3. 1 *M* Tris-HCl, pH 7.0.
4. 0.5 *M* EDTA..
5. 5 *M* NaCl.
6. 1 *M* $MgCl_{2.}$
7. 5 *M* NaOH.
8. 8 *M* urea.
9. 10X TBE: 0.89 *M* Tris-borate, 10 m*M* EDTA.
10. DEPC-treated H_2O: Double-distilled water is mixed with 0.1% diethylpyrocarbonate (DEPC) and stirred overnight. The DEPC is inactivated by autoclaving for 20 min.
11. Phenol.
12. Chloroform/isoamyl alcohol (24:1).
13. 1 *M* dithiothreitol (DTT).
14. Thiourea (Sigma).
15. [^{32}P]-labeled nucleotides (Amersham).
16. [^{32}P]pCp (Amersham).
17. 5% nondenaturing polyacrylamide gels in 1X TBE.
18. 8% denaturing gels that contain 7 *M* urea in 1X TBE.
19. 4% denaturing gels that contain 7 *M* urea in 1X TBE.

2.2. Solutions and Buffers

1. Buffer A: 50 m*M* Tris-HCl, pH 7.5, 100 m*M* NH_4Cl, 100 m*M* $MgCl_2$, 4% polyethylene glycol (PEG).
2. Buffer B: 50 m*M* Tris-HCl, pH 7.5, 100 m*M* NH_4Cl, 10 m*M* $MgCl_2$, 0.5–1 μ*M* C5 protein.
3. 10X folding buffer C: 500 m*M* Tris-HCl, pH 7.5, 1000 m*M* NH_4Cl, and 100 m*M* $MgCl_2$.
4. 2X RNA dye solution: 8 *M* urea, 20 m*M* EDTA, 0.25 mg/mL bromophenol blue (BPB), 0.25 mg/mL xylene cyanol FF (XCFF).
5. Alkaline lysis buffer: 50 m*M* Na_2CO_3, pH 9.2, 1 m*M* EDTA.

2.3. Enzymes and Reagents

1. T4 polynucleotide kinase and 10X kinase buffer (New England Biolabs).
2. T4 RNA ligase and ligase buffer (New England Biolabs).

3. S1 nuclease (Promega, Madison, WI).
4. RNase V1 (Pierce).
5. RNase T1 (Industrial Research).
6. T7 in vitro transcription system, 5X transcription buffer (Promega).
7. Polymerase chain reaction (PCR) system including 10X PCR buffer, 25 m*M* $MgCl_2$, four 10 m*M* dNTP (deoxyribonucleotide 5´-triphosphate), Taq DNA polymerase (Promega).
8. RNasin RNase inhibitor (Promega).
9. $(NH_4)_2Fe(SO_4)_2$ (Sigma).
10. Calf intestine alkaline phosphatase (CIAP) and 10X CIAP buffer (New England Biolabs).

2.4. C5 protein and Plasmids

1. M1 RNA clone (plasmid FL117).
2. C5 protein (kindly provided by Dr. Venkat Gopalan of Ohio State University).

2.5. Miscellaneous Material

Also needed is a phosphorimager screen.

2.6. Equipment

1. Automated thermal cycler.
2. Molecular Dynamics PhosphorImager.
3. Oligonucleotide synthesis facility.

3. Methods

3.1. Generating M1 Ribozyme

3.1.1. DNA Template

The DNA template for generation of M1 ribozyme is constructed using PCR with plasmid FL117 (wild-type M1 cloned into pUC19 vector) ***(31)*** as the DNA template and the following oligonucleotides as the primers:

1. 5´ primer (OliT7) 5´ TAATACGACTCACTATAG 3´ (containing the T7 promoter sequence).
2. 3´ primer (M1EI12) 5´ TGGTGCAACGAGAACCCTGTGGAATTG 3´.

Add the following sequentially: 10 µL 10X PCR buffer 10X (Mg^{2+} free); 2 µL 10 m*M* deoxyadenosine 5´-triphosphate (dATP); 2 µL 10 m*M* deoxythymidine 5´-triphosphate (dTTP); 2 µL 10 m*M* deoxyguanosine 5´-triphosphate (dGTP); 2 µL 10 m*M* deoxycytidine 5´-triphosphate (dCTP); 8 µL $MgCl_2$ (25 m*M*); 100 pmol 5´ primer (OliT7); 100 pmol 3´ primer (M1IE12); 1 µg PvuII-digested FL117 plasmid; 1 µL Taq DNA polymerase; and PCR-grade water to a final reaction volume of 100 µL.

PCR amplification of the DNA template is performed using the following amplification program: Denaturing for 2 min at 94 °C; 30 cycles of denaturing for 2 min at 94 °C, annealing for 1 min at 47 °C, extension for 1 min at 72 °C, and final extension for 10 min at 72 °C.

The PCR DNA products are separated in 5% polyacrylamide gels under nondenaturing conditions and are purified and used as the template for the in vitro transcription synthesis of the ribozymes.

3.1.2. Synthesis of M1-IE Ribozyme

The M1-IE ribozyme is synthesized by an in vitro transcription.

1. 4 μL (2 μg) M1 DNA from PCR reaction.; 8 μL 5X transcription buffer; 4 μL 100 m*M* DTT; 4 μL 10 m*M* adenosine 5′-triphosphate (ATP); 4 μL 10 m*M* guanosine 5′-triphosphate (GTP); 4 μL 10 m*M* cytosine 5′-triphosphate (CTP); 4 μL 10 m*M* uridine 5′-triphosphate (UTP); 4 μL H_2O; 1 μL RNasin (5 U/μL); 2 μL T7 RNA polymerase. Incubate the reaction mixture at 37 °C for 4 h or overnight.
2. Add an equal volume of 2X RNA dye solution and load onto 8% polyacrylamide-7 *M* urea gels.
3. Place the gel on a thin-layer chromatographic (TLC) plate (silica gel UV_{256}). Visualize RNA bands by briefly shadowing with a shortwave ultraviolet (UV) lamp. Extract RNA from the excised gel slice by the crush-soak method using DEPC-treated water (*see* **Note 1**).

3.2. In Vitro Cleavage of an mRNA Substrate by M1GS RNA in the Absence and Presence of C5 Protein

3.2.1. M1-IE Ribozyme Cleavage of Substrate ie37 In Vitro in the Absence of C5

For M1-IE ribozyme cleavage of substrate ie37 in vitro in the absence of C5 ***(31)***:

1. Mix 10 n*M* of M1-IE that has been allowed to fold properly (*see* **Note 2**) and 10 n*M* of ^{32}P-labeled ie37 mRNA, 1 μL of 10X buffer A, and water in a final volume of 10 μL.
2. Incubate the reaction mixture at 37 °C to 50 °C for 30 min.
3. Terminate the reaction by adding 10 μL of 2X RNA dye solution and 1 μL of phenol. Incubate at 90 °C for 2 min. Load onto an 8% polyacrylamide-7 *M* urea gel and electrophorese for 1–2 h.
4. Expose the gel to a phosphorscreen and scan with a PhosphorImager.

3.2.2. M1-IE Ribozyme Cleavage of Substrate ie37 In Vitro in the Presence of C5

1. Mix 1 n*M* of M1-IE that has been allowed to fold properly (*see* **Note 2**) and 20 n*M* of C5 protein and incubate the mixture at 37 °C for 5 mins to allow binding. Then

add to the mixture 10 n*M* of ^{32}P-labeled ie37 mRNA, 1 μL of 10X buffer B, and water in a final volume of 10 μL.

2. Incubate the reaction mixture at 37 °C to 50 °C for 30 min.
3. Terminate the reaction by adding 10 μL of 2X RNA dye solution and 1 μL of phenol. Incubate at 90 °C for 2 mins. Load onto an 8% polyacrylamide-7 *M* urea gel.
4. Expose the gel to a phosphorscreen and scan with a PhosphorImager.

3.3. Nuclease Cleavage Footprint Analysis to Determine the Regions of M1 RNA That Are Potentially in Close Contact With C5 Protein

To determine the regions of the ribozymes that are protected by C5 protein, M1 ribozyme is incubated either alone or with C5 protein to allow binding and then subjected to digestion by nucleases. The ribozyme RNA is either radiolabeled with ^{32}P at the 5′ terminus by kinase or the 3′ end with ligase. Three different nucleases are used: RNase T1, which recognizes single-stranded RNA regions and only cleaves at guanosine positions; nuclease S1, which only cleaves at single-stranded regions; and RNase V (cobra venom RNase), which only reacts toward basepaired regions. The reaction mixtures that contain ribozymes in the presence and absence of C5 protein are subjected to digestion by different nucleases of various concentrations to accurately determine the susceptibility of the ribozyme regions to digestion. The cleavage products are then separated in denaturing polyacrylamide gels and quantitated. The amounts of the cleavage products at a particular position correlate with the susceptibility of this position to degradation by nucleases. If a position is located at a single-stranded region and directly interacts with C5 protein, the nucleotide at this position is now expected to be less susceptible to cleavage by RNase T1 and nuclease S1, which cleave at single-stranded RNA regions. Meanwhile, the susceptibility of a position to cleavage by RNase V may be reduced if this position is bound by C5 protein and located at a double-stranded region or involved in tertiary interactions. The levels of protection by C5 protein are calculated by obtaining the ratio of the amounts of cleavage products in the absence of C5 protein over those in the presence of the protein cofactor. The ribozyme RNA is either radiolabeled with ^{32}P at the 5′ terminus by kinase or the 3′ end with ligase.

3.3.1. Dephosphorylation of M1 Ribozyme

Dephosphorylate M1 ribozyme by adding the following (final reaction volume is 30 μL):

1. 3 μL of 10X CIAP buffer.
2. 27 μL of M1 RNA and water (at least 20 pmol).
3. 0.5 μL of RNasin.
4. 1 μL of CIAP.

Incubate at 37 °C for 1 h. Bring the volume to 100 μL. The RNA samples are extracted using phenol and chloroform, followed by ethanol precipitation.

3.3.2. 5′ End Labeling With γ-[^{32}P]ATP

Perform 5′-end labeling with γ-[^{32}P]ATP by adding the following:

1. 2 μL of 10X T4 polynucleotide kinase buffer.
2. 10 μL of RNA (at least 10 pmol and water).
3. 3 μL of γ-[^{32}P]ATP (at least 10 pmol).
4. 1 μL of RNasin.
5. 1 μL of T4 polynucleotide kinase.
6. 5 μL of DEPC H_2O.

Incubate the mixture at 37 °C for 1 h. Add 20 μL of 2X RNA dye, load on 4% denaturing polyacrylamide gel, and extract the labeled M1 RNA as above (*see* **Notes 1** and **2**).

3.3.3. 3′ End Labeling With [^{32}P]-pCp

Perform 3′-end labeling with [^{32}P]-pCp (reaction volume of 15 μL) by adding the following:

1. 5 μL 3X T4 RNA ligase buffer.
2. 1 μL of 100 m*M* ATP.
3. 5 μL of RNA (at least 10 pmol).
4. [^{32}P]-pCp.
5. 1 μL of T4 RNA ligase.

Incubate the reaction at 4 °C for 16 h. Add 30 μL of 2X RNA dye, load on 4% denaturing polyacrylamide gel, and purify as above (*see* **Note 2**).

3.3.4. RNase Mapping

RNase mapping is carried out in the following steps:

1. Either 5′ or 3′ end-labeled M1 ribozyme (about 10,000 cpm per sample) is incubated in a volume of 10–20 μL (*see* **Note 3**) in buffer A (50 m*M* Tris-HCl, pH 7.5, 100 m*M* NH_4Cl, 100 m*M* $MgCl_2$, 4% PEG) (without C5 protein) and in buffer B (50 m*M* Tris-HCl, pH 7.5, 100 m*M* NH_4Cl, 10 m*M* $MgCl_2$, 0.5 μ*M* C5 protein) (with C5 protein, ribozyme, and C5 ratio of 1:20) for 30 min at 37 °C.
2. Diluted RNases T1, V, and S1 (0.5 μL) are added to the incubated mixture (*see* **Note 4**), and the reactions are incubated for various length of time at 37 °C: T1 for 10 min; V for 14 min; S1 for 40 min.
3. The reactions are stopped by adding phenol (50 μL) and chloroform (50 μL).
4. Precipitate the RNA using ethanol and resuspend it in 6 μL of DEPC H_2O and 6 μL of 2X RNA dye. Run the samples on 8% denaturing acrylamide gels along with T1 and A control (*see* below). Dry gel and expose it to a phosphorscreen and scan with a PhosphorImager.

3.3.5. T1 Control

For T1 control, a complete digestion of RNA sample with RNase T1 under denaturing conditions is performed: Add labeled M1 RNA substrate (about 50,000 cpm) to 20 μL of denaturing buffer (0.5X RNA dye buffer). Add 2 μL of a 1:20 dilution of RNase T1 to the solution, incubate for 5–10 min at 50 °C, and put on ice immediately (*see* **Note 5**). The sample is now ready for gel electrophoresis.

3.3.6. Alkaline Hydrolysis Control

For the alkaline hydrolysis control (A control): Make a nucleotide size ladder by alkaline hydrolysis of the RNA substrate.

1. Dry 1 μL of each 5′ and 3′ end-labeled M1 RNA (about 100,000 cpm each) in a SpeedVac.
2. Resuspend dried sample into 20 μL of alkaline lysis buffer. Resuspend well and transfer to another Eppendorf tube (*see* **Note 6**).
3. Incubate at 95 °C to 100 °C for 7 min.
4. The samples are precipitated in ethanol and resuspended in 20 μL of 1X RNA dye.

3.4. Fe(II)-EDTA-Mediated Mapping of the Regions of M1 RNA That Are Potentially in Close Contact With C5 Protein

The Fe(II)-EDTA mapping technique has been used extensively to study the tertiary structures of group I intron ribozyme and RNase P RNA and M1 RNA–tRNA interactions ***(32,33)***. The cleavage of nucleic acids by this method is largely independent of base identity or secondary structure. Therefore, Fe(II)-EDTA mapping can be used to probe the tertiary interaction of M1 RNA and C5 protein ***(24,33)***.

1. Either 5′ or 3′ end-labeled M1 ribozyme is incubated in a volume of 10–20 μL in the absence and presence of C5 protein (ribozyme and C5 ratio of 1:20) (buffer A or B) for 30 min at 37 °C.
2. The mixture is then supplemented with 2.5 m*M* DTT, 2 m*M* $(NH_4)_2Fe(SO_4)_2$, and 4 m*M* EDTA using 10X stock solutions (25 m*M* DTT, 20 m*M* $(NH_4)_2Fe(SO_4)_2$, 40 m*M* EDTA).
3. The entire mixture is incubated for 90 min at 37 °C.
4. 1 μL of thiourea (100 m*M*) is added to the mixture to stop the reaction.
5. The samples are ethanol precipitated, resuspended in 10 μL DEPC H_2O, and 10 μL of 2X RNA dye.
6. Load the samples along with T1 and A control on 8% denaturing acrylamide gel.
7. Dry the gel and expose it to a phosphorscreen and scan with a PhosphorImager.

Acknowledgments

We thank Dr. Venkat Gopalan of Ohio State University for providing the purified C5 protein and Kihoon Kim for helpful discussions. P.T. was a recipient of the

American Heart Association Predoctoral Fellowship (Western States Affiliate). F.L. was a Scholar of the Lymphoma and Leukemia Society of America and an Established Investigator of the American Heart Association. The research has been supported by the National Institutes of Health.

4. Notes

1. This gel-running step should be completely RNase free since RNA is extremely sensitive to degradation. It is recommended that DEPC-treated water is used for the entire gel-running process from making the gel to running buffer. The ribozymes that are in vitro synthesized by T7 RNA polymerase should be purified using denaturing gels containing 8 *M* urea. The gel slices that contain the ribozyme fractions are crushed in an Eppendorf tube with a minihomogenizer bar and then soaked in RNase-free water for 20 min in ice. Precautions should be taken to avoid degradation of RNA as a result of RNase contamination. Since most of RNases do not require divalent cations, it is not recommended to soak the gel crush with TE buffer. The ribozyme fractions in the soaked tube are separated from the crushed gel slices by microcentrifugation. The ribozyme-containing supernatants are extracted with phenol/chloroform solutions twice, and then the ribozymes are precipitated in the presence of ethanol.
2. Before proceeding to the in vitro cleavage assays, the RNase mapping, and Fe(II)-EDTA footprinting analyses, the ribozyme is resuspended in 1X folding buffer C, incubated at 65 °C for 5 min, and then allowed to fold with gradually lowering the temperature to 37 °C. This treatment will allow most ribozyme to fold into active conformations.
3. Since M1 ribozyme is 377 nucleotides long, both the 5′ and 3′ end-labeled ribozymes are needed in the analysis to cover the full-length M1 RNA sequence. Furthermore, 4–10% polyacrylamide gels containing 8 *M* urea should be used in the analysis.
4. RNase T1, V, and S1 can be diluted in the buffers used for RNase digestion (buffers A and B) to 1:20, 1:200, and 1:2000 for use in the reaction. The RNases can also be diluted into DEPC-treated water prior to their use. The RNase dilution varies to correspond to the incubation time of the user preference.
5. Placing the reaction in ice immediately is necessary to stop the RNase T1 digestion. This digestion sample can be reused as long as it is stored at 4 °C.
6. This step is necessary to remove excess salt that may precipitate in the original tube.

References

1. Altman, S., and Kirsebom, L. A. (1999) In: *The RNA World* (Gesteland, R. F., Cech T. R., and Atkins J. F., eds.), Cold Spring Harbor Laboratory Press, Cold Spring Harbor, NY, pp. 351–380.
2. Frank, D. N., and Pace, N. R. (1998) Ribonuclease P: unity and diversity in a tRNA processing ribozyme. *Annu. Rev. Biochem.*, **67**, 153–180.
3. Liu, F., and Altman, S. (1994) Differential evolution of substrates for an RNA enzyme in the presence and absence of its protein cofactor. *Cell* **77**, 1093–1100.

4. Komine, Y., Kitabatake, M., Yokogawa, T., Nishikawa, K., and Inokuchi, H. (1994) A tRNA-like structure is present in 10Sa RNA, a small stable RNA from *Escherichia coli. Proc. Natl. Acad. Sci. U. S. A.* **91**, 9223–9227.
5. Liu, F., and Altman, S. (1996) Requirements for cleavage by a modified RNase P of a small model substrate. *Nucleic Acids Res.* **24**, 2690–2696.
6. Guerrier-Takada, C., van Belkum, A., Pleij, C. W., and Altman, S. (1988) Novel reactions of RNase P with a tRNA-like structure in turnip yellow mosaic virus RNA. *Cell* **53**, 267–272.
7. Guerrier-Takada, C., Gardiner, K., Marsh, T., Pace, N., and Altman, S. (1983) The RNA moiety of ribonuclease P is the catalytic subunit of the enzyme. *Cell* **35**, 849–857.
8. Forster, A. C., and Altman, S. (1990) External guide sequences for an RNA enzyme. *Science* **249**, 783–786.
9. Liu, F., and Altman, S. (1995) Inhibition of viral gene expression by the catalytic RNA subunit of RNase P from *Escherichia coli. Genes Dev.* **9**, 471–480.
10. Yuan, Y., Hwang, E. S., and Altman, S. (1992) Targeted cleavage of mRNA by human RNase P. *Proc. Natl. Acad. Sci. U. S. A.* **89**, 8006–8010.
11. Plehn-Dujowich, D., and Altman, S. (1998) Effective inhibition of influenza virus production in cultured cells by external guide sequences and ribonuclease P. *Proc. Natl. Acad. Sci. U. S. A.* **95**, 7327–7332.
12. Kawa, D., Wang, J., Yuan, Y., and Liu, F. (1998) Inhibition of viral gene expression by human ribonuclease P. *RNA* **4**, 1397–1406.
13. Dunn, W., Trang, P., Khan, U., Zhu, J., and Liu, F. (2001) RNase P-mediated inhibition of cytomegalovirus protease expression and viral DNA encapsidation by oligonucleotide external guide sequences. *Proc. Natl. Acad. Sci. U. S. A.* **98**, 14831–14836.
14. Frank, D. N., Harris, M. E., and Pace, N. R. (1994) Rational design of self-cleaving pre-tRNA-ribonuclease P RNA conjugates. *Biochemistry* **33**, 10800–10808.
15. Haas, E. S., Brown, J. W., Pitulle, C., and Pace, N. R. (1994) Further perspective on the catalytic core and secondary structure of ribonuclease P RNA. *Proc. Natl. Acad. Sci. U. S. A.* **91**, 2527–2531.
16. Haas, E. S., Armbruster, D. W., Vucson, B. M., Daniels, C. J., and Brown, J. W. (1996) Comparative analysis of ribonuclease P RNA structure in Archaea. *Nucleic Acids Res.* **24**, 1252–1259.
17. Haas, E. S., and Brown, J. W. (1998) Evolutionary variation in bacterial RNase P RNAs. *Nucleic Acids Res.* **26**, 4093–4099.
18. Chen, J. L., Nolan, J. M., Harris, M. E., and Pace, N. R. (1998) Comparative photocross-linking analysis of the tertiary structures of *Escherichia coli* and *Bacillus subtilis* RNase P RNAs. *EMBO J.* **17**, 1515–1525.
19. Massire, C., Jaeger, L., and Westhof, E. (1998) Derivation of the three-dimensional architecture of bacterial ribonuclease P RNAs from comparative sequence analysis. *J. Mol. Biol.* **279**, 773–793.
20. Krasilnikov, A. S., Yang, X., Pan, T., and Mondragon, A. (2003) Crystal structure of the specificity domain of ribonuclease P. *Nature* **421**, 760–764.

21. Stams, T., Niranjanakumari, S., Fierke, C. A., and Christianson, D. W. (1998) Ribonuclease P protein structure: evolutionary origins in the translational apparatus. *Science* **280**, 752–755.
22. Kazantsev, A. V., Krivenko, A. A., Harrington, D. J., et al. (2003) High-resolution structure of RNase P protein from *Thermotoga maritima. Proc. Natl. Acad. Sci. U. S. A.* **100**, 7497–7502.
23. Tsai, H. Y., Masquida, B., Biswas, R., Westhof, E., and Gopalan, V. (2003) Molecular modeling of the three-dimensional structure of the bacterial RNase P holoenzyme. *J. Mol. Biol.* **325**, 661–675.
24. Kim, J. J., Kilani, A. F., Zhan, X., Altman, S., and Liu, F. (1997) The protein cofactor allows the sequence of an RNase P ribozyme to diversify by maintaining the catalytically active structure of the enzyme. *RNA* **3**, 613–623.
25. Niranjanakumari, S., Stams, T., Crary, S. M., Christianson, D. W., and Fierke, C. A. (1998) Protein component of the ribozyme ribonuclease P alters substrate recognition by directly contacting precursor tRNA. *Proc. Natl. Acad. Sci. U. S. A.* **95**, 15212–15217.
26. Reich, C., Olsen, G. J., Pace, B., and Pace, N. R. (1988) Role of the protein moiety of ribonuclease P, a ribonucleoprotein enzyme. *Science* **239**, 178–181.
27. Hsieh, J., Andrews, A. J., and Fierke, C. A. (2004) Roles of protein subunits in RNA–protein complexes: lessons from ribonuclease P. *Biopolymers* **73**, 79–89.
28. Gopalan, V., Vioque, A., and Altman, S. (2002) RNase P: variations and uses. *J. Biol. Chem.* **277**, 6759–6762.
29. Mann, H., Ben-Asouli, Y., Schein, A., Moussa, S., and Jarrous, N. (2003) Eukaryotic RNase P: role of RNA and protein subunits of a primordial catalytic ribonucleoprotein in RNA-based catalysis. *Mol. Cell* **12**, 925–935.
30. Hsu, A. W., Kilani, A. F., Liou, K., Lee, J., and Liu, F. (2000) Differential effects of the protein cofactor on the interactions between an RNase P ribozyme and its target mRNA substrate. *Nucleic Acids Res.* **28**, 3105–3116.
31. Trang, P., Lee, M., Nepomuceno, E., Kim, J., Zhu, H., and Liu, F. (2000) Effective inhibition of human cytomegalovirus gene expression and replication by a ribozyme derived from the catalytic RNA subunit of RNase P from Escherichia coli. *Proc. Natl. Acad. Sci. U. S. A.* **97**, 5812–5817.
32. Celander, D. W., and Cech, T. R. (1991) Visualizing the higher order folding of a catalytic RNA molecule. *Science* **251**, 401–407.
33. Westhof, E., Wesolowski, D., and Altman, S. (1996) Mapping in three dimensions of regions in a catalytic RNA protected from attack by an Fe(II)-EDTA reagent. *J. Mol. Biol.* **258**, 600–613.

20

Quantification of MicroRNAs, Splicing Isoforms, and Homologous mRNAs With the Invader Assay

Peggy S. Eis and Mariano A. Garcia-Blanco

Summary

The understanding of physiology and pathology requires accurate quantification of intracellular concentrations of important molecules such as unique RNA species. Accurate quantification of highly homologous messenger RNAs (mRNAs) ***(1–3)***, alternatively spliced mRNAs ***(4)***, and the short microRNAs (miRNAs) ***(5,6)*** has been successfully achieved using the Invader assay. This method directly detects specific RNA molecules in preparations of pure total cellular RNA (1–100 ng) or in crude cell lysate (10^3–10^4 cells) samples using an isothermal signal amplification process with a fluorescence resonance energy transfer (FRET)-based fluorescence readout. Features of the Invader assay include the ability to detect 1–10 RNA molecules per cell, to discriminate between RNAs that differ by a single base, and to precisely measure 1.2-fold changes in RNA expression. Further, an isothermal format and the ability to detect two different RNA molecules with a biplex format make the Invader assay suitable for high-throughput screening applications.

Key Words: Alternative splicing; Cleavase enzyme; FRET; gene expression; high-throughput screening; HTS; Invader assay; invasive cleavage; microRNA; miRNA; mRNA; RNA quantification/quantitation; splice variant.

1. Introduction

Biomedical research has greatly benefited from the the Human Genome Project, and comparative analyses of DNA sequences from other mammalian genomes will likely accelerate our understanding of disease and the development of new treatments. The expanding role of RNA molecules in cellular biology has been highlighted by the discovery of the approx 22-nt microRNAs (miRNAs) ***(7–9)*** as regulators of gene expression and by the realization of the very high prevalence of alternatively spliced messenger RNAs (mRNAs) for a given gene ***(10)***. Precise quantification of mRNA isoforms and miRNAs will be essential in understanding the proper regulation of gene expression. For

From: *Methods in Molecular Biology, vol. 488: RNA-Protein Interaction Protocols*
Edited by: Ren-Jang Lin © Humana Press Inc, Totowa, NJ

example, the role of splicing factors can be investigated by measuring the level and type of spliced mRNAs that result when a splicing factor is excluded or altered. Similarly, miRNA biogenesis, which requires processing by Drosha and Dicer *(8)*, can be investigated by quantifying the level of unprocessed primary miRNA transcript vs the final approx 22-nt miRNA product. Finally, quantification of both protein and RNA expression levels may provide insight on RNA–protein interactions from a stoichiometric perspective in our efforts to understand cellular chemistry.

Accurate quantification of RNAs of any type (herein the target), including the short miRNAs, can be achieved using the Invader® assay (*see* **Note 1**) as this method does not rely on a reverse transcription step followed by target amplification but instead directly quantifies a specific RNA in the sample. Invader assays (*see* **Note 2**) can be used to detect DNA or RNA and have been successfully used for genotyping and quantitative detection of DNA and RNA ***(1–6,11–16)***. While several detection formats have been used to detect Invader assay cleavage products ***(13,16,17)***, the method for RNA quantification described here ***(1)*** is a detection format based on fluorescence resonance energy transfer (FRET) that is performed isothermally in 96-well microplates. Either total RNA or cell lysate samples can be used, two different RNA molecules can be detected in the same reaction (biplex format; *see* **Note 3**), and the assay can be implemented in a high-throughput screening (HTS) format ***(15,18)*** (*see* **Note 4**).

One of the hallmarks of the Invader assay is its ability to discriminate single-base changes in nucleic acids ***(13,14)***. **Figure 1**, which illustrates the mechanics of this RNA detection method, shows an application of the assay for the detection of homologous mRNAs (human cytochrome P450 [CYP] 3A4 and 3A7). The assay uses target-specific oligonucleotides (oligos) and the 5′ nuclease activity of engineered variants ***(1,19)*** of *Thermus thermophilus* RNA polymerase, called Cleavase® enzymes (*see* **Note 1**), to generate signal without target amplification. The basic principles of this method (*see* **Fig. 1**) are as follows:

1. *Formation of an overlap structure and Probe cleavage*. Two oligos, the Invader oligo and Probe, are designed to adjacently bind to the RNA target being detected (*see* **Fig. 1A**). The Probe comprises two regions, a target-specific region (TSR) and a noncomplementary 5′ arm. The Invader oligo is fully complementary to the target except for its 3′ end (*see* **Note 5**) and must extend at least 1 bp into the TSR, forming an overlap structure consisting of the 3′ end of the Invader oligo, the 5′ end of the TSR, and the corresponding nucleotide in the RNA target. A Cleavase enzyme ***(19,20)*** then cleaves the 5′ arm of the Probe on the 3′ side of the "invaded" base, also known as the cleavage site (*see* **Note 6**).
2. *Specificity of 5 arm cleavage*. While general quantification of nucleic acids simply requires formation of an overlap structure on any portion of the DNA or RNA, single-base discrimination by the Invader assay requires that the overlap structure be formed at the polymorphic nucleotide. In the example shown in **Fig. 1A**, specific

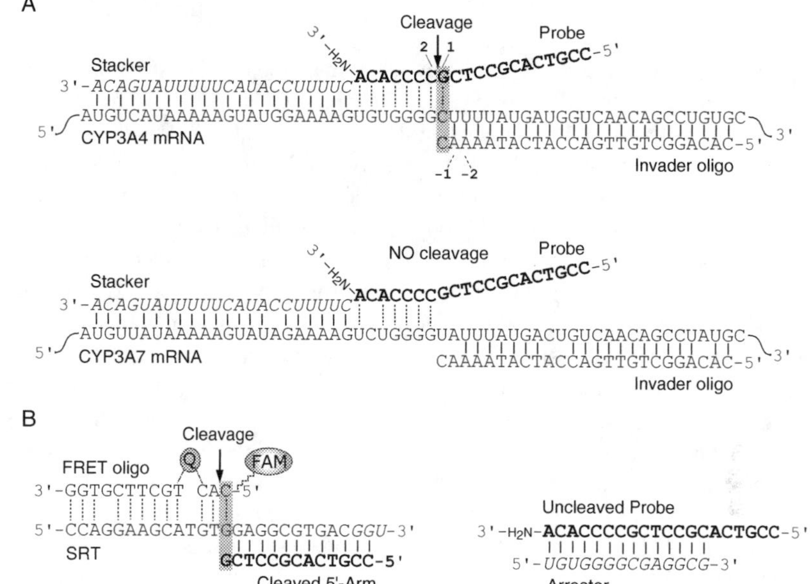

Fig. 1. Invader assay design for specific detection of human CYP3A4 messenger RNA (mRNA). *See* Heading 1. and **Subheading 3.1.** for additional discussion of assay principles. The Probe sequence is shown in **bold**, and *italic* sequences are 2´-*O*-methylated nucleotides (Stacker, Arrestor, and last three nucleotides of the secondary reaction template [SRT]). Solid vertical lines indicate stable basepairing; dotted vertical lines indicate cycling of the oligo (Probe or FRET oligo) on and off the target at the optimal assay temperature. Vertical arrows indicate the site of cleavage by the Cleavase enzyme, and shaded boxes highlight the overlap structures. (**A**) Primary reaction for detection of CYP3A4 mRNA (*see* **Note 7**). The CYP3A4 Invader probe set (Invader oligo, Probe, and Stacker) is shown bound to the CYP3A4 mRNA (top sequences) and the homologous CYP3A7 mRNA (bottom sequences; mismatches are indicated by the absence of a vertical line). Numbering above the Probe (2 1) and below the Invader oligo (–1 –2) identifies positions in the oligos with respect to the overlap structure. Position 1 in the Probe is the first nucleotide of its target-specific region (TSR). The top sequences illustrate the formation of a 1-bp overlap structure comprising position 1 in the Probe, the 3´ end of the Invader oligo, and the corresponding position in the CYP3A4 mRNA. Probe cleavage releases its 5´ arm (noncomplementary region plus 1 nt of the TSR). The 3´ end of the Invader oligo is designed to form a mismatch with the mRNA (*see* **Note 5**). Stackers, which are optional, are used to enhance assay performance (*see* **Note 10**). In the bottom sequences, cleavage cannot occur when the CYP3A4 Invader probe set binds the CYP3A7 mRNA as the two adjacent bases (U and A) at the overlap site result in a 2-nt gap structure. (**B**) Secondary reaction for fluorescence resonance energy transfer (FRET) detection of primary reaction products. Cleaved 5´ arms from the primary reaction (panel **A**) function as Invader oligos, forming an overlap structure with a generic sequence FRET oligo and SRT. The Cleavase enzyme cleaves between the FAM fluorophore and quencher dye (Q), resulting in increased fluorescence signal, which is proportional to the number of cleaved 5´ arms generated in the primary reaction. Use of an Arrestor enhances fluorescence signal by sequestering excess, uncleaved Probe (*see* **Note 8**).

detection of CYP3A4 mRNA results by location of the overlap structure at one of two consecutive nucleotide differences in the target (CU in CYP3A4 mRNA vs UA in CYP3A7 mRNA). A 1-bp overlap structure forms with CYP3A4 mRNA, but a 2-nt gap structure forms with CYP3A7 mRNA (*see* **Note 7**). The precise sequence-based alignment of the Invader probe set enables formation of the overlap structure, which leads to cleavage of the 5´ arm. Thus, specificity of the Invader assay is conferred on two levels, sequence-specific binding of the Invader probe set and structure-specific recognition of the overlap substrate by the Cleavase enzyme.

3. *Amplification of cleaved 5´ arms.* The molecule being detected in the assay is the cleaved 5´ arm that results from Invader oligo-directed cleavage of the Probe. Thus, a means for amplifying the number of cleaved 5´ arms is needed to generate enough molecules for detection with currently available instrumentation. This is accomplished by performing the assay at an isothermal temperature near the melting temperature T_m of the Probe/target duplex, which enables the Probe to cycle on and off the target ***(13)***. A high concentration of Probe (~1 µ*M*) is used to enhance Probe cycling ***(21)***. The Invader oligo is designed to remain bound to the target at this temperature. In the presence of Cleavase enzyme and the Invader oligo/target duplex, each Probe binding event results in formation of an overlap structure and cleavage of its 5´ arm. Thus, multiple cleaved 5´ arms are generated per RNA molecule, which are the amplified signal.
4. *FRET-based detection of cleaved 5´ arms.* Signal is further amplified and converted into a fluorescence signal by using the cleaved 5´ arms as Invader oligos in a second Invader reaction (*see* **Fig. 1B**). An overlap structure is formed with the cleaved 5´ arm, a secondary reaction template (SRT), and the FRET oligo. The Cleavase enzyme that was used in the primary reaction now cleaves between the fluorophore and quencher dye on the FRET oligo, resulting in enhanced fluorescence. As for the primary reactions, the secondary Invader reaction is run near the melting temperature of the FRET oligo, enabling multiple cleavage events (second round of signal amplification) per 5´ arm/SRT duplex. Fluorescence is further increased by inclusion of an Arrestor oligo (*see* **Note 8**) in the secondary reaction. Since the cleaved 5´ arms from the primary reaction are not target specific, the FRET detection oligos (FRET oligo and SRT) can be nearly any sequence (*see* **Note 9**).

As the Invader probe sets directly bind the RNA being detected, assay performance is impacted by the secondary and tertiary structure of the RNA. Probes are most efficiently cleaved when they have ready access to unstructured regions in the RNA. Sampling various regions in the RNA is one way of optimizing assay performance. A second method uses a Stacker oligo as part of the Invader probe set (*see* **Note 10**). Stackers bind immediately downstream of the Probe and coaxially stack ***(22)*** with its 3´ end. Use of both a stably bound Invader oligo and Stacker can help reduce RNA structure. Stackers can be synthesized with natural nucleotides or with 2´-*O*-methylated nucleotides, which have increased binding stability with RNA ***(23)***.

Other applications for the Invader assay include detection of alternatively spliced mRNAs ***(4)*** and miRNAs ***(5,6)***. **Figures 2** and **3** show examples of Invader probe

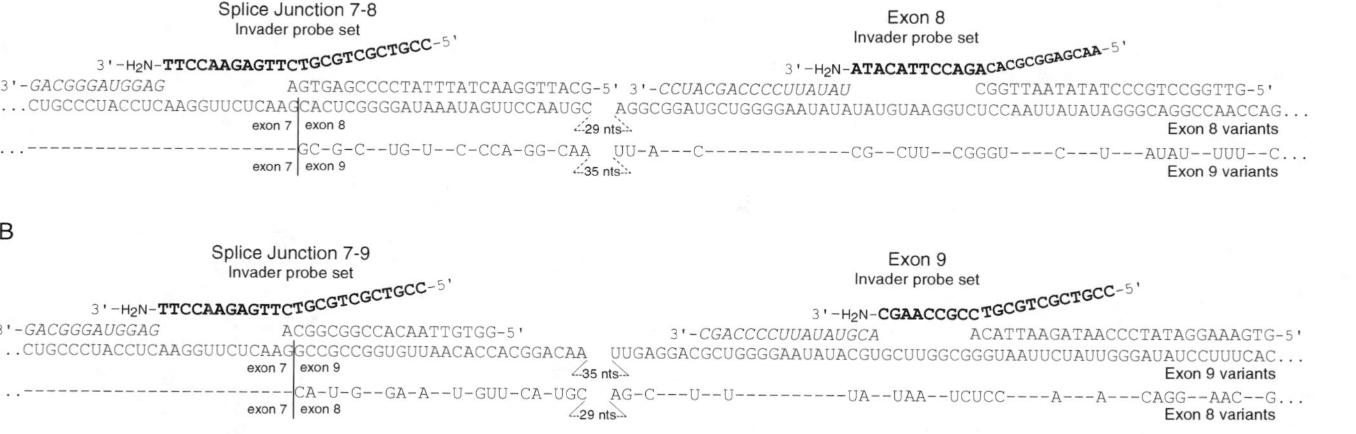

Fig. 2. Invader assays for detection of human FGFR2 messenger RNA (mRNA) splice variants containing either exon 8 or exon 9 (*see* **Note 16**). Splice junctions in the mRNA sequences are indicated by a vertical line. Each Invader probe set consists of a Stacker, Probe, and Invader oligo (displayed left to right on the mRNA sequence). For simplicity, basepairing between the Invader probe sets and the mRNAs is implicit, and secondary reaction oligos (*see* **Fig. 1B**) are not shown. Probe sequences are shown in **bold**, and *italic* sequences are 2′-*O*-methylated nucleotides (Stackers). Assay oligos are fully complementary to the mRNAs except for the 3′ ends of the Invader oligos and the noncomplementary 5′ arms (angled sequence) of the Probes. (**A**) Invader probe sets, splice junction 7–8 or exon 8, for detection of exon 8 splice variants. A partial mRNA sequence for exons 7 and 8 is shown with the exon 9 sequence aligned below (consensus sequence indicated by a dash). (**B**) Invader probe sets, splice junction 7–9 or exon 9, for detection of exon 9 splice variants. A partial mRNA sequence for exons 7 and 9 is shown with the exon 8 sequence aligned below (consensus sequence indicated by a dash).

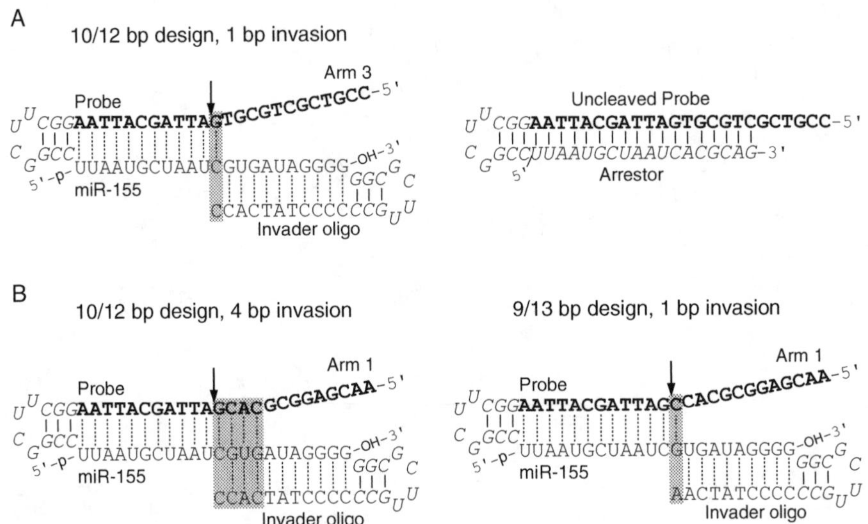

Fig. 3. MicroRNA (miRNA) Invader assays for detection of human miR-155. The Invader oligos and Probes (shown in **bold**) contain generic sequence 2´-*O*-methylated hairpins (*italic* sequence) at their 5´ and 3´ ends, respectively, which coaxially stack with the ends of the miRNA and enhance binding. Dotted vertical lines indicate cycling of the Probe or Invader oligo (*see* **Note 25**) on and off the miRNA target at the optimal assay temperature. Vertical arrows indicate the site of cleavage by the Cleavase enzyme, and shaded boxes highlight the overlap structures. (**A**) 10-/12-bp design: 10-bp duplex between the Invader oligo and miRNA and 12-bp duplex between the Probe and miRNA. Arm 3 (*see* **Table 2**) is used to form a 1-bp overlap. Cleaved 5´ arms are detected in a fluorescence resonance energy transfer (FRET)-based secondary reaction (*see* **Fig. 1B**), which uses the Arrestor sequence shown on the right (solid lines indicate stable basepairing). (**B**) Example of a 4-bp overlap that inadvertently results from the use of arm 1 in a 10/12 design. A 9/13 design is shown on the right using arm 1, which yields a 1-bp invasion (overlap structure).

sets for these applications, which are discussed in greater detail in **Subheadings 3.1.** and **3.2.** While locating Invader probe sets at splice junctions, in most cases, ensures detection of RNA only, assay conditions have been optimized to preclude unintended detection of DNA ***(1,24)*** (*see* **Note 11**). This enables greater flexibility in assay design, which can be particularly relevant when detecting mRNAs with only a single-nucleotide change, such as in edited RNA ***(25,26)***.

2. Materials

2.1 Reagents

Items 11–20 are used for polyacrylamide gel electrophoresis (PAGE) (*see* **Subheading 3.3.**).

1. Specific oligos: Invader oligo, Probe, Stacker (optional), and Arrestor. Store at –20 °C.
2. Generic oligos: FRET oligo and SRT. Store at –20 °C.
3. Invader RNA Assay Generic Reagents Kit (Third Wave Technologies, Madison, WI). Two generic kits are currently available, one for mRNA detection and one for miRNA detection; consult with the manufacturer for the appropriate kit (*see* **Note 12**). Store at –20 °C.
4. Negative control: Transfer RNA (tRNA) Carrier, 20 ng/µL (Sigma, St. Louis, MO, cat. no. R5636). Store at –20 °C.
5. Positive control: In vitro transcript or synthetic RNA (*see* **Note 13**). Store at –20 °C or –70 °C for long-term storage.
6. T7 RNA polymerase kit (e.g., T7-MEGAshortscript kit, Ambion, Austin, TX, or Ampliscribe kit, Epicentre Technologies, Madison, WI) for generation of positive controls. Store at –20 °C.
7. Total RNA sample preparation reagent/kit (e.g., TRIzol® reagent, Invitrogen, Carlsbad, CA, or RNeasy® kit, Qiagen, Valencia, CA) or 10X cell lysis buffer (*see* **Note 14**) if performing the cell lysate format.
8. Phosphate-buffered saline (PBS), no $MgCl_2$/no $CaCl_2$ (for cell lysate method only).
9. $TE_{0.1}$ buffer: 10 m*M* Tris-HCl, pH 8.0, 0.1 m*M* ethylenediaminetetraacetic acid (EDTA).
10. Ribonuclease (RNase)-free (diethylpyrocarbonate [DEPC]-treated) H_2O.
11. Distilled, deionized water (ddH_2O), 0.22 µ*M* filtered.
12. 40% acrylamide/bis (19:1). *Caution*: Highly toxic. Store at 4 °C.
13. Denaturing acrylamide solution: Variable percentage acrylamide, 7 *M* urea, 0.5X TBE. Store at 4 °C.
14. 10% ammonium persulfate (APS). *Caution*: Corrosive. Store at 4 °C.
15. TEMED (*N,N,N ,N* -tetramethylethylenediamine). *Caution*: Corrosive. Store at 4 °C.
16. Urea.
17. 0.5X TBE buffer: 44.5 m*M* Tris-borate, 1 m*M* EDTA.
18. Gel loading solution (for PAGE): 95% deionized formamide, 10 m*M* EDTA, 0.025% sodium dodecyl sulfate (SDS), 0.025% bromophenol blue (BB), 0.025% xylene cyanol FF (XCFF). Store at 4 °C.
19. 0.5 µg/mL EtBr (ethidium bromide) in water. *Caution*: Mutagen.
20. SYBR Gold stain (Molecular Probes, Eugene, OR, cat. no. S-11494).

2.2. Equipment

1. Fluorescence plate reader equipped with filters for detection of fluorescein and red dye (optional): 485/20 nm and 560/20 nm excitation; 530/25 nm and 620/40 nm emission.
2. Thermal cycler, gradient capability recommended (*see* **Note 15**).
3. Gel electrophoresis equipment (for PAGE only; *see* **Subheading 3.3.**).
4. P2, P200, and P1000 pipets or LTS version (e.g., Rainin, Woburn, MA, cat. no. L20, L200, L1000).
5. Electronic repeat pipet, 20–200 µL (e.g., Rainin, Woburn, MA, cat. no. E3-200).
6. Multichannel pipet, 8 channel, 2–20 µL (e.g., Rainin, Woburn, MA, cat. no. L8–20).
7. Rack, 96-well format for 0.2-mL tubes.

2.3. Consumables

Items 6 and **7** are used for PAGE (*see* **Subheading 3.3.**).

1. Aerosol-barrier pipet tips.
2. 1.5- and 0.2-mL polypropylene tubes.
3. Skirted 0.2-mL polypropylene 96-well microplate (MJ Research, Waltham, MA, cat. no. MSP-9601; ISC Bioexpress, Kaysville, UT, cat. no. T-3084-1).
4. Clear Chill-out liquid wax (MJ Research) or polymerase chain reaction (PCR)-grade mineral oil.
5. Microplate sealing film (ISC Bioexpress, cat. no. T-3021-7).
6. NAP-10 columns (Amersham Biosciences, Piscataway, NJ, cat. no. 17-0854-02).
7. Glass thin-layer chromatographic (TLC) plates, C18 silica with fluorescence indicator (Aldrich, Milwaukee, WI, cat. no. Z292923).

3. Methods

The Invader assay methods for mRNA or miRNA detection described in this section encompass (1) assay design strategies and sequence analysis, (2) oligo design and modifications, (3) PAGE procedures for oligos and RNA controls, (4) positive control generation and sample preparation, (5) Invader assay optimization and standard protocol, and (6) troubleshooting and oligo redesign strategies for improving assay performance.

3.1. Design of mRNA Detection Assays

The single-base discrimination capability of the Invader assay is best achieved at the site of overlap between the Invader oligo and the Probe (*see* Notes 6 and 7 and Fig. 1). While the assay performs best when the Probe TSR is fully complementary to the RNA target, mismatches can be tolerated in the Invader oligo and Stacker at positions away from the cleavage site (oligos are usually lengthened to maintain the stability of the interaction with the mRNA; *see* Subheading 3.1.3.). This feature, combined with the relatively small binding footprint for the Invader probe set (~50 nt), enables flexibility in mRNA assay design.

3.1.1. Design Strategies

The following is a series of questions to consider before designing Invader assays for mRNA detection:

1. Is the gene part of a highly homologous gene family (e.g., CYP genes)? Should the assay detect all gene family members (inclusive detection), or should separate assays be designed to specifically detect each individual family member (exclusive detection)?
2. Is the gene evolutionarily conserved? If so, a single assay can often be designed to detect the same gene in multiple species (e.g., human, mouse, and rat).

3. Is the pre-mRNA for the gene alternatively spliced? Should the assay inclusively detect all known splice variants, or should separate assays be developed? If specific detection is desired, the assays can be designed to detect splice junctions, exons, or both in an effort to understand the alternative splicing profile for a given gene (*see* **Fig. 2**).
4. Will the assay be run in a biplex detection format? For example, two genes of interest or the gene of interest and a housekeeping gene can be assayed in a single reaction (*see* **Note 3** and **Subheading 3.7.2.**).

3.1.2. Sequence Download and Analysis

Download all sequences for the gene of interest, particularly in the context of the detection goals for the system being studied (*see* **Subheading 3.1.1.**). Several public databases contain sequence information for mRNAs. A few of the commonly used Web sites are http://www.ncbi.nlm.nih.gov/, http://www.ensembl.org/, http://genome.ucsc.edu/.

Many public databases are linked, and it is often useful to look up the gene of interest at multiple sites to find the most current and complete sequence information, particularly when the goal is detection of alternatively spliced mRNAs.

Perform a multiple sequence alignment on the downloaded sequences using a sequence analysis software package such as LaserGene (DNAStar, Madison,WI), the GCG Wisconsin Package (Accelrys, San Diego, CA), or shareware (e.g., typing in "multiple sequence alignment" in a Web browser identifies a number of Web sites with sequence alignment algorithms). For discrimination of closely related sequences, scan through the alignment for positions in the sequence that are variable, preferably two adjacent differences, and position the cleavage site for the Invader probe set at one of these unique positions (*see* **Fig. 1A**). Good discrimination can also be achieved by a single-base difference at position 1 in the Probe and a second position toward the middle of the Probe. For detection of alternatively spliced mRNAs, identify all known splice variants, perform a multiple sequence alignment, and map the splice junctions (i.e., exon–intron structure). Position the cleavage site at the splice junctions or within exons, depending on the detection goals (*see* **Subheading 3.1.1.**). Examples of Invader probe sets for alternatively spliced mRNAs are shown in **Fig. 2**.

The last step in the sequence analysis is to BLAST (http://www.ncbi.nlm.nih.gov/BLAST/) the candidate sites for Invader probe sets to ensure that only the sequence of interest is detected. Analysis of approx 60 nt of the mRNA sequence corresponding to approx 30 nt on both sides of the candidate cleavage site using the "Search for short, nearly exact matches" module of BLAST will identify if there are other sequences in the genome that may result in crossreactive detection of another mRNA. If the search yields hits in other species, this information can be used to develop an Invader probe set that detects the mRNA in multiple species. If an unrelated mRNA sequence is identified in the search, examine the sequence with respect to the Invader probe set to determine if crossreactive signal would be generated. A couple of mismatches near the cleavage site position will generally not be a problem. Often, the cleavage site can be repositioned a few nucleotides in either direction to preclude crossreactive detection of unintended mRNAs.

3.1.3. Invader Probe Set Design

Assuming crossreactivity with other mRNAs is not an issue (*see* **Subheading 3.1.1.**), one can theoretically design an Invader probe set to almost any position in the mRNA. To focus design efforts, it is often useful to target splice junctions. Design Invader probe sets to two or three different splice junctions that are relatively heterogeneous in sequence (i.e., not excessively AT or GC rich). If the mRNA is expected to be expressed at low levels (<less than>10 molecules per cell), it may be more efficient to design three or four probe sets to ensure that at least one of them yields sufficient assay sensitivity to detect the mRNA in approx 100 ng total RNA sample or a cell lysate sample equivalent to approx 5000 cells per Invader reaction (*see* **Note 17**).

As discussed in **Heading 1**, the best-performing Invader assays are those in which the Invader probe set binds to an unstructured region in the mRNA. Thus, an alternative design method is to use an RNA folding program such as Mfold *(27)* to guide Invader probe set design (*see* **Note 18**). Fold the entire RNA sequence, or portions of it, to identify putative unstructured regions. Preferentially position the 3´ end of the Probe in an unstructured region (e.g., loop region of a hairpin) provided that the detection goals of the assay are not sacrificed (*see* **Subheading 3.1.1.**).

Guidelines for Invader probe set oligo length and modification are provided in **Table 1**. Commonly used noncomplementary 5´ arms for Probes and generic sequence FRET oligos and SRTs are listed in **Table 2**. Select an appropriate 5´ arm sequence for the Probe. Arms 1–4 work well for detecting many different mRNAs. An alternative 5´ arm is selected if a portion of the 5´ arm is complementary to the target mRNA, thus forming a 2-bp or greater invasion (*see* **Fig. 3B** and **Note 20**). A primer/probe analysis program can be used to test for unintended 5´ arm–TSR interactions in the Probe or other potentially problematic structures (*see* **Note 21**). Arrestors are designed to be fully complementary to the Probe TSR plus 6 nt of the 5´ arm (*see* **Fig. 1B**).

For mRNA detection, Invader assay primary reactions are designed to perform optimally at 60 °C (*see* **Note 22**). As Probe stability determines the primary reaction temperature (along with the Stacker, if used), prediction of the Probe

Table 1.
Guidelines for Invader Probe Set Oligonucleotide Design and Modifications (*see* Note 19)

Oligonucleotide	Length (nt)	Modification
Invader oligo	20–25	3′ end:target sequence: A:A, C:C, A:G, C:U[a]
Probe	10–12 (TSR)[b]	3′-amine
Stacker	15–17	All 2′-*O*-methylated nucleotides
Arrestor	16–18[c]	All 2′-*O*-methylated nucleotides

TSR, target-specific region.

[a]Mismatches commonly used between the 3′ end of the Invader oligo and the corresponding position in the RNA target sequence (*see* **Note 5**).

[b]Length corresponds to the Probe TSR and does not include the noncomplementary 5′ arm region (*see* **Table 2** for 5′ arm sequences).

[c]Fully complementary to the Probe TSR plus 6nt of the Probe of the Probe 5′ arm.

Table 2.
5′ Arm and Fluorescence Resonance Energy Transfer (FRET) Detection Oligonucleotide Sequences and Modifications (*see* Note 19)

5′ Arm/oligo[a]	Sequence (5′ to 3′)[b]	Modification
Arm 1	aacgaggcgcac	None
Arm 2	ccgtcacgcctc	None
Arm 3	ccgtcgctgcgt	None
Arm 4	ccgccgagatcac	None
FAM FRET oligo	(FAM)-cac-(EQ)-tgcttcgtgg	5′ FAM, position 4 EQ
Red FRET oligo	(Red)-ctc-(EQ)-ttctcagtgcg	5′ Red, position 4 EQ
FAM/arm 1 SRT	ccaggaagcatgtggtgcgcctcgUUU	Last 3 nucleotides 2′-*O*-methylated
FAM/arm 2 SRT	ccaggaagcatgtggaggcgtgacGGU	Last 3 nucleotides 2′-*O*-methylated
FAM/arm 3 SRT	ccaggaagcatgtgacgcagcgacGGU	Last 3 nucleotides 2′-*O*-methylated
FAM/arm 4 SRT	ccaggaagcatgtggtgatctcggCGG	Last 3 nucleotides 2′-*O*-methylated
Red/arm 1 SRT	cgcagtgagaatgaggtgcgcctcgUUU	Last 3 nucleotides 2′-*O*-methylated
Red/arm 2 SRT	cgcagtgagaatgaggaggcgtgacGGU	Last 3 nucleotides 2′-*O*-methylated
Red/arm 3 SRT	cgcagtgagaatgagacgcagcgacGGU	Last 3 nucleotides 2′-*O*-methylated
Red/arm 4 SRT	cgcagtgagaatgaggtgatctcggcGGU	Last 3 nucleotides 2′-*O*-methylated

SRT, secondary reaction template.

[a]FAM is a fluorescein derivative. Red is Redmond Red™ dye, and EQ is Eclipse Quencher™ dye (Epoch BioSciences, Bothell, WA).

[b]Capitalized nucleotides are 2′-*O*-methylated.

melting temperature is the most important factor in designing assays for a particular temperature. While software (InvaderCreator™ software, Third Wave Technologies) exists for automated design of Invader probe sets *(24)*, a melting temperature analysis for Invader probe set oligos (*see* **Table 3**) can be performed using the two-state hybridization module of the Mfold Web server *(27)* as follows:

Table 3.
Melting Temperature T_m Analysis for Invader Probe Sets

Invader probe set[a]	Oligo[b]	Length[c] (nt)	mFold Input[d]	T_m (°C)
CYP3A4	I	24	cacaggctgttgaccatcataaaa; ttttatgatggtcaacagcctgtg	81.5
	P	8	gccccaca;tgtggggc	54.3
	P +1		gccccacac;gtgtggggc	60.3
	S	22	cttttccatactttttatgaca; tgtcataaaaagtatggaaaag	70.1
FGFR2, sj 7–8	I	26	gcattggaactatttatccccgagtg; cactcggggataaatagttccaatgc	85.1
	P	12	cttgagaacctt;aagguucucaag	55.4
	P +1		cttgagaaccttg;caagguucucaag	59.2
	S	14	gagguagggcag;cugcccuaccuc	69.7
FGFR2, sj 7–9	I	18	ggtgttaacaccggcggc;gccgccggtgttaacacc	85.5
	P	12	same as sj 7-8	—
	P +1		same as sj 7-8	—
	S	14	same as sj 7-8	—
FGFR2, e8	I	23	gttggcctgccctatataattgg; ccaattatatagggcaggccaac	84.2
	P	13	agaccttacatat;atatgtaaggtct	55.7
	P +1		agaccttacatata;tatatgtaaggtct	57.7
	S	16	uauauuccccagcauc; gatgctggggaatata	70.8
FGFR2, e9	I	26	gtgaaaggatatcccaatagaattac; gtaattctattgggatatcctttcac	79.0
	P	9	ccgccaagc;gcttggcgg	58.4
	P +1		ccgccaagca;tgcttggcgg	62.6
	S	16	acgtatattccccagc; gctggggaatatacgt	74.3
miR-155, 10/12	I	10	cccctatcaca;cgugauagggg	60.6
	P	12	gattagcattaa;uuaaugcuaauc	45.6

CYP, cytochrome P450; FGFR2, fibroblast growth factor receptor 2.
[a]Invader probe sets correspond to the sequences shown in **Figs. 1–3.** e, exon; sj, splice junction.
[b]I, Invader oligo; P, Probe; P + 1, Probe + 1 bp of Stacker; S, Stacker.
[c]Length of oligo corresponding to the target-specific region (TSR) only.
[d]Mfold sequence input for two-state hybridization server (*see* **Subheading 3.1.3.**).

1. At the main Mfold Web site (http://www.bioinfo.rpi.edu/applications/mfold/), select the "2-state hybridization server" application. Use the default parameters of Hybridization temperature = 55 °C, $[Na^+] = 1.0\,M$, $[Mg^{++}] = 0.0\,M$, correction type oligomer. Change the total nucleic acid concentration to 0.000001 *M*.
2. Enter the sequence for melting temperature prediction (*see* **Note 23**), a semicolon, and the complement of the sequence (*see* **Table 3** for examples). Select "Hybridize RNA" at the bottom of the page to calculate the predicted melting temperature for the entered sequence. Use the "Back" button on the Web browser to return to the previous page for the next sequence entry.
3. For Probes, enter the TSR only. For assays using a Stacker, enter the Probe TSR plus the next nucleotide at its 3´ end to approximate the contribution of the stacking interaction *(22)*. For the Invader oligo, exclude the 3´ nucleotide as it is designed to form a mismatch with the RNA target (*see* **Table 1**). The following are suggested melting temperature ranges for Invader probe sets (*see* **Note 24**): Invader oligos, 75–85 °C; Probes plus 1 nt, 57–63 °C; Stackers, 65–75 °C.

3.2. Design of miRNA Detection Assays

Sequence information for miRNAs can be obtained from the miRNA Registry *(28)* database (http://www.sanger.ac.uk/Software/Rfam/mirna/index.shtml). Design the Invader oligo and Probe to bind to half of the miRNA (*see* **Fig. 3**), with the Probe getting the extra nucleotide if the miRNA target is 23 nt *(5)*. As described for mRNA detection assays (*see* **Subheading 3.1.3.**), melting temperature analysis of the Invader oligo and Probe can be performed to guide assay design (*see* **Table 3**). If the Probe is AT rich, at least two Invader probe set designs are recommended to ensure the best-performing design is used for sample measurements (*see* **Note 25**). For example, two Invader oligo/Probe TSR lengths (10/12 and 9/13) were tested for the 22-nt miR-155 miRNA (*see* **Fig. 3**). Both assays performed optimally at approx 42 °C, but the 10/12 design yielded a better signal/background ratio (P. S. Eis and J. E. Dahlberg, personal communication).

The universal FRET detection oligos used for mRNA detection (*see* **Table 2**) can also be used for miRNA detection. In most cases, arm 1 (*see* **Table 2**) can be used in the Probe unless unintended complementary with the miRNA results in <greater than> 1-bp invasion, which may excessively destabilize the Invader oligo *(5)*. For these cases (e.g., the 10/12 design for miR-155; *see* **Fig. 3B**), use arm 3 (*see* **Fig. 3A**) along with the corresponding SRT. To increase the stability of the Invader probe set with the miRNA target, a 2´-*O*-methylated stem-loop structure (5´-*GGCUUCGGCC*-3´) is added to the 5´ end of the Invader oligo and the 3´ end of the Probe *(5)*. The Arrestor (synthesized with 2´-O-methyl nucleotides) is designed as for mRNA detection assays (*see* **Table 1** and **Fig. 3A**).

3.3. PAGE Analysis and Purification Procedures

High-purity Probes and FRET oligos are essential for minimizing background signal in the assay, whereas Invader oligos, Stackers, Arrestors, and SRTs can simply be desalted prior to use (*see* **Note 26**). Oligo purification can be performed using reverse-phase or anion exchange high-performance liquid chromatographic (HPLC) methods or with preparative PAGE. Many oligo synthesis vendors offer purification services and most deliver oligos desalted and dried. Analytical PAGE is used to assess the quality of positive controls (e.g., in vitro transcripts) and total RNA samples. Use denaturing gels containing 7 *M* urea with variable percentage of acrylamide (19:1) depending on the size of DNA or RNA: 20% for <less than> 40 nt, 8–15% for 40–500 nt, and 5% for <greater than> 500 nt. Migration of DNA and RNA is monitored relative to the gel loading solution marker dye XCFF (BB migrates faster and usually runs off the gel during the run), which migrates coincident with approx 22 nt in 20%, approx 75 nt in 8%, and approx 130 nt in 5%. Both analytical and preparative PAGE methods are described with variations for oligos and RNA.

Prepare the appropriate concentration acrylamide solution using 40% acrylamide stock, distilled and deionized water, solid urea to a final concentration of 7 *M*, and 10X TBE buffer to a final concentration of 0.5X TBE. *Caution*: Wear eye protection and gloves when preparing acrylamide solutions and pouring gels because of toxicity. Heat can be used during stirring to help solubilize the urea but only in a fume hood or well-ventilated area to prevent accumulation of toxic fumes. Do *not* leave the acrylamide solution unattended during this step. Denaturing acrylamide stocks can be stored for several weeks at 4 °C.

3.3.1. Preparative PAGE

The following procedure is for purification of Invader assay oligos (*see* **Note 27**):

1. Clamp together 10 × 10 cm gel plates using 1.5-mm spacers. Have ready a comb with approx 2-cm wells.
2. Mix 15 mL 20% denaturing acrylamide solution with 150 µL 10% APS and 10 µL TEMED. Immediately pour the gel, insert the comb, and let it polymerize for 20–30 min.
3. Set up the gel with 0.5X TBE buffer and prerun for approx 15 min at 15–30 mA (400-V limit).
4. Prepare oligo sample by mixing 50 µL of 4 µg/µL crude oligo with 50 µL gel loading solution. Denature the oligo sample 1–2 min at 95 °C. Flush the wells in the gel with 0.5X TBE buffer and load the oligo samples (200 µg per 2-cm lane).
5. Run the gel at 15–30 mA (400-V limit) until the XCFF dye migrates three-quarters into the gel.
6. Remove the gel from the plates and sandwich it in plastic wrap. Place the plastic-wrapped gel onto a silica gel TLC plate containing fluorescent indicator. Visualize the

oligo with a ultraviolet (UV) lamp (*caution*: use UV eye protection) and use a marker to outline the main band in each lane. Turn off the UV light, flip the plastic-wrapped gel, and cut open the top layer of wrap. Using the marker outline of the oligo band as a guide, excise the main band with a razorblade and then cut it into approx 1-mm wide strips. Transfer the gel strips into a 1.5-mL tube (1 lane of strips per tube).

7. Elute the oligo with 1 mL of $TE_{0.1}$ buffer per 1.5-mL tube at 37 °C for 2–3 h. Shaking or rotation of the tube is recommended to maximize the purification yield (typically 30–50%).
8. Prepare the eluted oligo for desalting by transferring the approx 1-mL volume to a fresh 1.5-mL tube. Rinse the leftover gel slices by adding 200 µL $TE_{0.1}$ buffer, vortexing the tube, and combining the rinse solution with the eluted oligo solution. The total volume of eluted oligo should be 1 mL.

Desalt the eluted oligo using a NAP-10 column and RNase-free water as follows:

1. Pour off the NAP-10 column storage buffer and place the column in a clamp over a beaker.
2. Fill the column with RNase-free water and let it drain into the beaker.
3. Repeat **step 2** twice more.
4. Using a P1000 pipet, add the 1 mL solution of eluted oligo and let it enter the column.
5. Have ready two 1.5-mL tubes to collect the effluent (total elution volume is 1.5 mL, collect ~ 750 µL per 1.5-mL tube). Add 1.5 mL RNase-free water to the column and immediately collect the effluent into the 1.5-mL tubes.
6. Reduce the desalted oligo volume to approx 50 µL in a vacuum evaporator (e.g., SpeedVac) and then combine the two fractions into one tube (or dry completely and resuspend in 100 µL $TE_{0.1}$ buffer). Quantify the oligo using the procedure described in **Subheading 3.5.**

3.3.2. Analytical PAGE

Next is a procedure that can be used for analysis of oligos, in vitro transcript reactions, or assessing the integrity of total RNA samples:

1. Clamp together 10 × 10 cm gel plates using 0.75-mm spacers. Have ready a 10- or 16-well comb.
2. Mix 10 mL 8% denaturing acrylamide solution with 100 µL 10% APS and 10 µL TEMED. Immediately pour the gel, insert the comb, and let it polymerize for 20–30 min.
3. Set up the gel with 0.5X TBE buffer and prerun for approx 15 min at 15–30 mA (400-V limit).
4. For in vitro transcription reaction analysis, load 4 µL of a 1:10 dilution of the transcription reaction (*see* **Subheading 3.4.1.**). For total RNA samples, mix 1.5 µL of a 50–200 ng/µL total RNA sample with 4.5 µL gel loading solution. Denature the samples 1–2 min at 95 °C. Flush the wells in the gel with 0.5X TBE buffer and load 4 µL of sample per lane (avoid using the outside lanes).

5. Run the gel at 15–30 mA (400-V limit) until the XCFF dye migrates to the bottom of the gel (or three-quarters into the gel if the sample is a 50- to 100-nt transcript).
6. Visualize the RNA in the gel by staining 15 min in 0.5 µg/mL EtBr, destaining 15 min in water, and placing the gel on a UV light box (*caution*: wear UV eye protection); or by staining 5 min in a 1:10,000 dilution of SYBR Gold stain and scanning the gel on a fluoroimager with fluorescein settings (excitation ~ 485 nm, emission ~ 520 nm).

3.4. Positive Control Preparation

Positive controls are not required for relative quantification experiments, but they are helpful for assay optimization, particularly when developing assays for low-expression RNAs. For absolute quantification of mRNA or miRNA, it is critical to purify and quantify the RNA controls in a consistent manner to ensure accurate measurements. Always wear gloves and use additional precautions (*see* **Note 28**) when preparing control RNA to prevent degradation by RNases. Also, use care in the preparation and handling of RNA controls to prevent contamination of work areas that are used for assay setup (*see* **Note 29**).

3.4.1. mRNA Controls: In Vitro Transcript Generation

In vitro transcripts for mRNA detection assays can be prepared from plasmid templates, PCR-generated templates, or synthetic oligos that comprise the Invader probe set binding region (called minitranscripts; *see* **Notes 13** and **30**). A commonly used transcription enzyme is T7 RNA polymerase. An example of oligos for use in a transcription reaction and the resulting minitranscript are as follows (5′ to 3′, T7 promoter sequence in lower case):

Universal T7 promoter oligo: aatttaatacgactcactatagg

Human FGFR2 (fibroblast growth factor receptor 2) splice junction 7–8 oligo for T7 transcription reaction:

CTTCTGCATTGGAACTATTTATCCCCGAGTGCTTGAGAACCTTGAGGUA GGGCAGCCCtatagtgagtcgtattaaatt

Human FGFR2 splice junction 7–8 minitranscript (*see* **Fig. 2A**):

GGGCUGCCCUACCUCAAGGUUCUCUUGCACUCGGGGAUAAAUAGUU CCAAUGCAGAAG

Use a T7 transcription kit (*see* **Subheading 2.1.**) to perform a 20-µL transcription reaction for the type of template used. After the transcription reaction is completed, add 2 U of RNase-free deoxyribonuclease (DNase) and incubate for 15 min at 37 °C. Terminate the reaction by addition of 20 µL gel loading solution (reactions can be stored at –20 °C at this point). Remove a 2-µL aliquot, mix with 8 µL gel loading solution (final dilution of transcription reaction is 1:10), and reserve for analytical PAGE. Assess the quality of the RNA generated in the in vitro transcription reactions using the analytical PAGE procedure in **Subheading 3.3.2.** A 12% gel is recommended for <less than> 100 nt, 8% for 100- to 500-nt transcripts, and 5% for transcripts <greater than>500 nt.

Purify the in vitro transcript using the PAGE and desalting procedures in **Subheading 3.3.1.** with the following modifications: At **step 2**, use the appropriate percentage gel for the size of transcript purified; at **step 4**, denature and load the 40-μL transcription reaction (already mixed with gel loading solution) into one approx 2-cm lane. After the desalting step (*see* **Note 31**), combine the two 750-μL fractions into one tube and measure the concentration using the procedure in **Subheading 3.4.3.**

3.4.2. miRNA Controls

Order the synthetic RNA corresponding to the miRNA sequence of interest with a 5′ phosphate (0.05-μ*M* scale). If the RNA was ordered protected, follow the manufacturer's instructions for 2′ deprotection and reduce the volume in a vacuum evaporator to approx 40 μL. Purify and desalt the RNA using the procedure in **Subheading 3.3.1.** with the following modifications: At **step 4**, add 40 μL gel loading solution to the approx 40 μL volume of RNA, denature, and load into two lanes (~40 μL per lane); at **step 5**, run the XCFF marker dye three-quarter distance into the gel (XCFF migrates at approximately the same distance as 22-nt miRNAs). After the desalting step (*see* **Note 31**), combine the four 750-μL fractions into one tube and measure the concentration using the procedure in **Subheading 3.4.3.**

3.4.3. RNA Control Quantification and Preparation of Working Stocks

Quantify the purified RNA controls by absorbance measurement at 260 nm (to ensure accuracy, measurements should be in the range 0.1–1.0 $A_{260\,\text{nm}}$ units). A microcuvette (e.g., 50-μL volume) can be used to conserve on the amount of RNA needed for the measurement (*see* **Note 32**). Use $TE_{0.1}$ buffer to dilute the RNA, if needed (*see* **Note 31**). To calculate the RNA concentration in nanograms and femtomoles per microliter, use the conversion factor and molecular weight calculation method for single-stranded RNA:

$$1A_{260\text{nm}}\ \text{unit} = 40\text{ng}/\mu\text{L}$$

$$\text{MW}=(\#\text{nt}^{*}320.5) + 159$$

For example, a 1000-nt in vitro transcript diluted 1:20 gives an $A_{260\,\text{nm}}$ measurement of 0.200. The RNA concentration is calculated as follows:

$$0.200*20*(40\text{ng}/\mu\text{L}) = 160\text{ng}/\mu\text{L}$$

$$\text{MW} = (1000 * 320.5) + 159 = 320{,}659$$

$$160\text{ng}/\mu\text{L}*(1\ \text{nmol}/320{,}659\text{ng}) * (10^6\ \text{fmol}/\mu\text{L}$$

For synthetic miRNA controls, use the extinction coefficient (*see* **Note 33**) and the Beer-Lambert law to calculate the RNA concentration. The Beer-Lambert law equation is

$$A = \varepsilon bc$$

where A is the absorbance, ε is the molar extinction coefficient (L/mol·cm), b is the path length of the cuvette in centimeters (usually 1 cm), and c = concentration. Using miR-155 (*see* **Fig. 3**) as an example ($\varepsilon_{260\,\text{nm}} = 229{,}800$), the concentration of a solution diluted 1:20 with an $A_{260\,\text{nm}}$ measurement of 0.200 would be

$$((0.200 * 20)/229{,}800) * (10^6\ \mu M/\text{mol}) = 17.4\ \mu M = 17.4\ \text{pmol}/\ \mu L$$

Store concentrated stocks of RNA controls (in vitro transcript or synthetic miRNA) in multiple aliquots at −70 °C to ensure the availability of an intact control in case a working stock becomes degraded. Working stocks of 1 n*M* (1 fmol/μL) or lower can be stored at −20 °C. Use tRNA carrier to dilute RNA controls below the 1-n*M* level (*see* **Note 34**).

3.5. Oligo Quantification and Preparation of Invader Assay Working Stocks

After purification or desalting of the Invader assay oligos (*see* **Subheading 3.3.1.**), dilute the oligo solution 1:50 with $\text{TE}_{0.1}$ buffer and measure the $A_{260\,\text{nm}}$ value. Calculate the oligo concentration from the $A_{260\,\text{nm}}$ value and its $\varepsilon_{260\,\text{nm}}$ value (*see* **Note 33**) using the calculation example shown for synthetic miRNA in **Subheading 3.4.3.** Oligo stock concentrations of 100–400 μ*M* are recommended; store at −20 °C.

Invader probe set working stocks consist of the Invader oligo (I), Probe (P), and Stacker (S) and are referred to as IPS mixes (for miRNAs, IP mixes). The FRET oligo and SRT are combined into one stock, and the Arrestor is diluted as a separate working stock. Recommended working stock concentrations for both the mRNA and miRNA Invader assays are summarized as follows (*see* **Note 35**):

mRNA IPS oligos: 20 μ*M* I, 32 μ*M* P, 12 μ*M* S
miRNA IP oligos: 40 μ*M* I, 40 μ*M* P
FRET/SRT detection oligos: 13.3 μ*M* FRET oligo, 2 μ*M* SRT
Arrestor: 43 μ*M* for mRNA assays, 54 μ*M* for miRNA assays

3.6. Sample Preparation

Use RNase precautions when preparing samples (*see* **Note 28**). Total RNA or cell lysate samples can be tested in the Invader assay.

3.6.1. Total RNA Method

Prepare total RNA from tissue or cells using an RNA preparation reagent or kit (*see* **Note 36**). Resuspend total RNA in RNase-free water and determine the concentration by absorbance measurement at 260 nm (*see* **Note 32**). Calculate the RNA concentration as follows (for 1-cm path length cuvettes):

$$A_{260nm} \text{ value} * \text{Diluton factor} * 40 \text{ng}/\mu\text{L} = \text{RNA concentration in ng}/\mu\text{L}$$

An assessment of sample quality is recommended. Use the analytical PAGE procedure in **Subheading 3.3.2.** to determine if samples are intact or degraded (*see* **Note 37**). Keep samples at –20 °C for short-term storage and at –70 °C for long-term storage.

3.6.2. Cell Lysate Method

For adherent cells (*see* **Note 38**) cultured in a 96-well plate (10,000–40,000 cells per well):

1. Dilute 10X cell lysis buffer to 1X with RNase-free water (~5 mL for a 96-well plate) and transfer to a buffer reservoir used for multichannel pipet addition. Set up a second buffer reservoir with approx 20 mL PBS (no $MgCl_2$/no $CaCl_2$).
2. Using a multichannel pipet, carefully remove the tissue culture media from the cells without disturbing the cell monolayer. Wash the cells one time with 200 μL PBS per well. Empty the PBS buffer from the wells by inverting the plate and gently blotting it onto paper towels to remove residual buffer (can be inhibitory to the assay).
3. Using a multichannel pipet, add 40 μL of 1X cell lysis buffer per well. Let plate sit at room temperature for 3–5 min to lyse cells.
4. Transfer 25–30 μL of lysates to a 0.2-mL polypropylene microplate. Overlay the lysates with 10 μL Chill-out liquid wax or PCR-grade mineral oil. Cover the plate with well tape (optional) and immediately heat inactivate cellular nucleases by incubating the microplate at 75–80 °C for 15 min.
5. Immediately add heat-treated lysates to the Invader reaction or store the microplate in a –70 °C freezer for testing at a later time (long-term stability has not been established for this step). If high expression levels are anticipated, lysate samples can be diluted with 1X cell lysis buffer before aliquoting 5 μL into the Invader reaction.

3.7. Invader Assay Optimization

3.7.1. Primary Reaction Temperature Optimization

The principal assay optimization protocol involves testing Invader probe sets at variable primary reaction temperatures to determine their optimal assay temperature and relative performance. A gradient thermal cycler works best for this step as several temperatures can be tested using a single instrument (*see* **Note 15**). Test a 10 °C temperature range for Invader probe sets designed for mRNA detection. For example, if the Invader probe sets were designed for performance at 60 °C, a temperature range of 55–65 °C in 2–3 °C increments is recommended. For screening miRNA detection Invader probe sets, a 15–20 °C temperature range (e.g., 42–58 °C) is recommended (*see* **Note 25**). A negative control (tRNA carrier) is tested to assess background signal in the assay, and a positive control (in vitro transcript or synthetic miRNA) is tested to determine relative assay performance for the candidate Invader probe sets. Use 1 amol RNA control per reaction for screening mRNA detection assays and 10 amol per reaction for miRNA detection assays (*see* **Note 39**).

Figure 4 shows 96-well plate layouts for three different Invader probe set screening strategies (*see* **Note 40**). Use **Table 3** to prepare primary and secondary reaction master mixes for the **Fig. 4A** format (adjust accordingly or use **Table 4** for **Fig. 4B,C** formats). The Invader assay temperature optimization protocol is as follows:

1. Program the thermal cycler for the desired gradient temperature range (or single temperature if using **Fig. 4B** format) for a 90-min incubation. Add a second step for a 60-min incubation at 60 °C.
2. Put on lab gloves (*see* **Note 28**). Thaw the generic Invader reagents and oligos and vortex to mix. Vortex the Cleavase enzyme and keep it on ice during reaction setup. Thaw the tRNA carrier and RNA control (or total RNA sample) at room temperature and then keep them on ice during setup. Dilute the RNA control or sample to the appropriate concentration (*see* **Note 41**).
3. Determine the total number of reactions needed and multiply by 1.25 (e.g., the **Fig. 4A** format requires 20 reactions as follows: 12 primary reactions with negative control and 12 primary reactions with positive control, which corresponds to 25 secondary reactions). Prepare the primary reaction master mixes and vortex after the last reagent addition.
4. Using an electronic repeat pipet (*see* **Note 42**), aliquot 10 µL per well according to the plate layout. Next, add 10 µL clear Chill-out wax (or PCR-grade mineral oil) to each well. Cover the plate with well tape (optional) and incubate it using the thermal cycler program from **step 1**.
5. During the primary reaction incubation (**step 4**), prepare the secondary reaction master mix. Subaliquot the mix into 0.2-mL tubes (**Fig. 4A** format uses two sets of four tubes, ~ 30 µL secondary reaction mix per tube) and place them in a 96-well 0.2-mL tube rack to enable multichannel pipet addition (**step 6**).
6. Using a multichannel pipet, add 5 µL secondary reaction mix per well, pipeting up and down twice to mix. The microplate can remain in the thermal cycler during this step or be placed on the benchtop.
7. Incubate the microplate at 60 °C for 60 min (*see* **Note 43**) and measure the fluorescence in a plate reader with filters having the following wavelength/bandpass specifications (*see* **Note 44**):

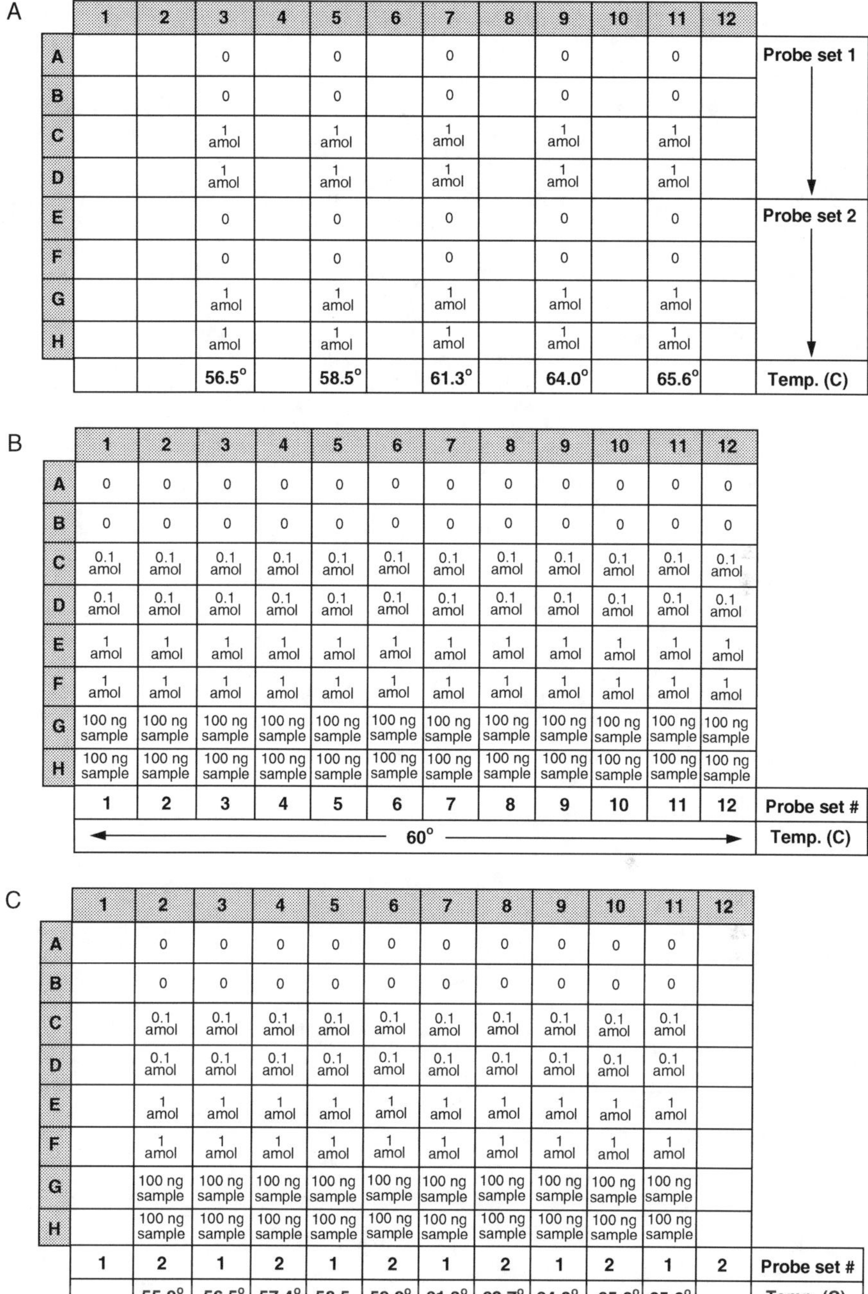

A

	1	2	3	4	5	6	7	8	9	10	11	12	
A			0		0		0		0		0		Probe set 1
B			0		0		0		0		0		
C			1 amol		1 amol		1 amol		1 amol		1 amol		
D			1 amol		1 amol		1 amol		1 amol		1 amol		
E			0		0		0		0		0		Probe set 2
F			0		0		0		0		0		
G			1 amol		1 amol		1 amol		1 amol		1 amol		
H			1 amol		1 amol		1 amol		1 amol		1 amol		
			56.5°		58.5°		61.3°		64.0°		65.6°		Temp. (C)

B

	1	2	3	4	5	6	7	8	9	10	11	12	
A	0	0	0	0	0	0	0	0	0	0	0	0	
B	0	0	0	0	0	0	0	0	0	0	0	0	
C	0.1 amol	0.1 amol	0.1 amol	0.1 amol	0.1 amol	0.1 amol	0.1 amol	0.1 amol	0.1 amol	0.1 amol	0.1 amol	0.1 amol	
D	0.1 amol	0.1 amol	0.1 amol	0.1 amol	0.1 amol	0.1 amol	0.1 amol	0.1 amol	0.1 amol	0.1 amol	0.1 amol	0.1 amol	
E	1 amol	1 amol	1 amol	1 amol	1 amol	1 amol	1 amol	1 amol	1 amol	1 amol	1 amol	1 amol	
F	1 amol	1 amol	1 amol	1 amol	1 amol	1 amol	1 amol	1 amol	1 amol	1 amol	1 amol	1 amol	
G	100 ng sample	100 ng sample	100 ng sample	100 ng sample	100 ng sample	100 ng sample	100 ng sample	100 ng sample	100 ng sample	100 ng sample	100 ng sample	100 ng sample	
H	100 ng sample	100 ng sample	100 ng sample	100 ng sample	100 ng sample	100 ng sample	100 ng sample	100 ng sample	100 ng sample	100 ng sample	100 ng sample	100 ng sample	
	1	2	3	4	5	6	7	8	9	10	11	12	Probe set #
	← 60° →												Temp. (C)

C

	1	2	3	4	5	6	7	8	9	10	11	12	
A		0	0	0	0	0	0	0	0	0	0		
B		0	0	0	0	0	0	0	0	0	0		
C		0.1 amol	0.1 amol	0.1 amol	0.1 amol	0.1 amol	0.1 amol	0.1 amol	0.1 amol	0.1 amol	0.1 amol		
D		0.1 amol	0.1 amol	0.1 amol	0.1 amol	0.1 amol	0.1 amol	0.1 amol	0.1 amol	0.1 amol	0.1 amol		
E		1 amol	1 amol	1 amol	1 amol	1 amol	1 amol	1 amol	1 amol	1 amol	1 amol		
F		1 amol	1 amol	1 amol	1 amol	1 amol	1 amol	1 amol	1 amol	1 amol	1 amol		
G		100 ng sample	100 ng sample	100 ng sample	100 ng sample	100 ng sample	100 ng sample	100 ng sample	100 ng sample	100 ng sample	100 ng sample		
H		100 ng sample	100 ng sample	100 ng sample	100 ng sample	100 ng sample	100 ng sample	100 ng sample	100 ng sample	100 ng sample	100 ng sample		
	1	2	1	2	1	2	1	2	1	2	1	2	Probe set #
		55.9°	56.5°	57.4°	58.5	59.9°	61.3°	62.7°	64.0°	65.0°	65.6°		Temp. (C)

Fig. 4. Microplate layouts for Invader assay temperature/performance optimization experiments (*see* **Subheading 3.7.** and **Note 40**). (**A**) Basic format for optimizing the primary reaction temperature of two Invader probe sets per microplate. A gradient thermal cycler (*see* **Note 15**) is used to test a range of temperatures across the plate (the bottom row indicates the temperature for each column). (**B**) Format for preliminary assessment of 12 Invader probe sets at a single primary reaction temperature (60°C) on both control RNA and a sample. (**C**) Format for assessment of primary reaction temperature and performance on both control RNA and a sample. Two Invader probe sets are alternated in the columns to take advantage of the gradient temperature across the microplate.

Table 4.
Master Mix Preparation for Primary Reaction Temperature Optimization of the Invader Assay

		tRNA negative control:	RNA or sample positive control:
Primary reaction components	1X volume	____ × Volume (No. of reactions × 1.25)	____ × Volume (No. of reactions × 1.25)
RNA primary buffer	4.0 μL		
IP or IPS oligos	0.25 μL		
$TE_{0.1}$ buffer	0.25 μL		
tRNA carrier or positive control	5.0 μL		
Cleavase enzyme	0.5 μL		
Total mix volume	10.0 μL		
Secondary reaction components	1X volume	____ × Volume (No. of reactions × 1.25)	
FRET/SRT detection oligos	0.75 μL		
Arrestor oligo	0.75 μL		
$TE_{0.1}$ buffer	1.5 μL		
RNA secondary buffer[a]	2.0 μL		
Total mix volume	5.0 μL		

FRET, fluorescence resonance energy transfer; IP, Invader oligo and Probe mix; IPS, Invader oligo, Probe, and Stacker mix; SRT, secondary reaction template; tRNA, transfer RNA.

[a]$TE_{0.1}$ Buffer for miRNA Invader assays (combine last two reagent additions into one 3.5 μL addition)

	Excitation	Emission
FAM (carboxyfluorescein) detection	485/20 nm	530/25 nm
Redmond Red detection	560/20 nm	620/40 nm

Process the data as described in **Subheading 3.9.**

3.7.2. Additional Optimization: Specificity, Sample Levels, Biplex Compatibility

Additional optimization procedures include specificity testing for Invader probe sets designed to detect closely related RNAs (*see* **Subheading 3.1.1.**), titration of total RNA or cell lysate sample levels to determine the optimal amount to use in the assay, and biplex compatibility testing to determine if two Invader assays can be performed in the same reaction well. These optimizations can be set up using **Table 5** or **6** and the standard Invader assay protocol in **Subheading 3.8**.

If homologous RNA sequences are being detected (e.g., *see* **Fig. 1A**), cross-reactivity experiments (*see* **Note 29**) will determine if the Invader probe sets are specific for the RNA they are designed to detect ***(1,15)***. Test each Invader probe set with its own in vitro transcript at levels corresponding to the linear detection range and homologous in vitro transcripts at significantly higher levels (e.g., 1, 10, 100, 1,000, and 10,000 amol per reaction). The discrimination level ***(1,15)*** is calculated by dividing the highest level of homologous transcript at which no crossreactive signal is detected by the lowest level of transcript detected for a given assay. For example, in **Fig. 1A**, if the CYP3A4 Invader assay has a limit of detection (LOD) of 0.01 amol (*see* **Note 10**) when tested with the CYP3A4 transcript and CYP3A7 transcript is not detected at 1000 amol but crossreactive

Table 5.
Master Mix Preparation for Single-Assay Detection Format (mRNA or miRNA) of the Invader Assay

Primary reaction components	1X volume	____ × Volume (No. of reactions × 1.25)
RNA primary buffer	4.0 µL	
IP or IPS oligos	0.25 µL	
$TE_{0.1}$ buffer	0.25 µL	
Cleavase enzyme	0.5 µL	
Total mix volume	5.0 µL	
Secondary reaction components	**1X volume**	**____ × Volume (No. of reactions × 1.25)**
FRET/SRT detection oligos	0.75 µL	
Arrestor oligo	0.75 µL	
$TE_{0.1}$ buffer	1.5 µL	
RNA secondary buffer[a]	2.0 µL	
Total mix volume	5.0 µL	

FRET, fluorescence resonance energy transfer; IP, Invader oligo and Probe mix; IPS, Invader oligo, Probe, and Stacker mix; miRNA, microRNA; mRNA, messenger RNA; SRT, secondary reaction template.

[a]$TE_{0.1}$ buffer for miRNA Invader assays (combine last two reagent additions into one 3.5-µL addition).

Table 6.
Master Mix Preparation for Biplex Assay Detection Format (mRNA or miRNA) of the Invader Assay

Primary reaction components	1X volume	____ × Volume (No. of reactions × 1.25)
RNA primary buffer	4.0 μL	
IP or IPS oligos (RNA 1)	0.25 μL	
IP or IPS oligos (RNA 2)	0.25 μL	
Cleavase enzyme	0.5 μL	
Total mix volume	5.0 μL	
Secondary reaction components	**1X volume**	**____ × Volume (No. of reactions × 1.25)**
FRET/SRT detection oligos (RNA 1)	0.75 μL	
FRET/SRT detection oligos (RNA 2)	0.75 μL	
Arrestor oligo (RNA 1)	0.75 μL	
Arrestor oligo (RNA 2)	0.75 μL	
RNA secondary buffer[a]	2.0 μL	
Total mix volume	5.0 μL	

FRET, fluorescence resonance energy transfer; IP, Invader oligo and Probe mix; IPS, Invader oligo, Probe, and Stacker mix; miRNA, microRNA; mRNA, messenger RNA; SRT, secondary reaction template.

[a]$TE_{0.1}$ Buffer for miRNA Invader assays.

signal is detected at 10,000 amol, the discrimination level is at least 1 in 100,000 (1000 amol CYP3A7 transcript/0.01 amol CYP3A4 transcript).

Optimization of total RNA or cell lysate levels for a newly developed Invader assay ensures that samples are quantified in the linear detection range (*see* **Notes 43** and **45**). This is best accomplished using two samples (e.g., control and treated) that are anticipated to encompass the full mRNA or miRNA expression range. For total RNA, test two or three levels in the 1- to 100-ng range (e.g., 20 and 80 ng) for both samples and measure the fluorescence at secondary reaction times of 15, 30, and 60 min. For cell lysates, a test range equivalent to 500–5000 cells per reaction is recommended.

If two RNAs are assayed in the same reaction, a biplex compatibility test is recommended to ensure performance of either Invader probe set is not impaired. Use **Tables 4** and **5** to prepare the primary and secondary master mixes. Test the Invader assays individually and together in the biplex format (an alternate primary reaction temperature may be required if the two assays have different optimums). If the background signal is substantially increased in either assay, try swapping the FRET detection systems (*see* **Table 2**) or redesign one or both assays (*see* **Subheading 3.11.**). Often, both background and sample signal are lower in the biplex format, but assay performance is not compromised if the signal/background values are comparable between the single and biplex assay formats. If signal/background values are reduced for one or both assays, try testing a twofold lower concentration of IPS (or IP) oligos for one or both Invader probe sets.

3.8. Invader Assay Procedure

To maximize the quantitative performance of the assay, *see* **Notes 33** and **42–47**. The assay procedure described in detail under **Subheading 3.7.1.** is used except the primary and secondary reaction master mixes are prepared using **Table 5** or **6**. For clarity, a condensed version of the steps follows:

1. Put on lab gloves and thaw the generic Invader reagents, oligos, RNA control, and samples.
2. Prepare a dilution series of the RNA control and dilute the total RNA samples to the appropriate test level. If cell lysates are being tested, dilute the RNA control with 1X cell lysis buffer and prepare the lysates as indicated under **Subheading 3.6.2.**
3. Determine the total number of reactions needed and multiply by 1.25. Prepare the primary reaction master mix and vortex after the last reagent addition to mix. Using an electronic repeat pipet, aliquot 5 µL per well according to the plate layout.
4. Add 5 µL of total RNA or cell lysate sample (can be added with a multichannel pipet) and pipet up and down twice to mix.
5. Overlay reactions with 10 µL clear Chill-out wax (or PCR-grade mineral oil) to each well. Cover the plate with well tape (optional) and incubate it at the appropriate primary reaction temperature for 90 min.
6. During the primary reaction incubation, prepare the secondary reaction master mix. Subaliquot the mix into 0.2-mL tubes (~70 µL per tube for a full row of 12 reactions) and place them in a 96-well format rack to enable multichannel pipet addition.
7. Using a multichannel pipet, add 5 µL secondary reaction mix per well and pipet up and down twice to mix.
8. Incubate the microplate at 60 °C and measure the fluorescence signal in a plate reader at 15, 30, or 60 min at the appropriate wavelengths:

	Excitation	Emission
FAM detection	485/20 nm	530/25 nm
Redmond Red detection	560/20 nm	620/40 nm

3.9. Data Analysis

Data analysis can be performed multiple ways depending on the experimental design. For relative quantification, a standard curve is not needed, and samples of interest are compared to a control sample after subtraction of the tRNA carrier background signal (*see* **Note 46**). For absolute quantification, a standard curve is generated using an RNA control (*see* **Note 34**), which is used to calculate the expression level in the samples. In either case, it is helpful to remember that signal generation is linear with respect to both time and target level as modulation of either variable can be used to extend the dynamic range of the assay (*see* **Notes 43, 46**, and **48**). The following are a set of steps for calculating the RNA expression level; use **steps 1–4** for relative quantification and **steps 1–6** for absolute quantification. Data analysis templates can be set up using software such as Excel (Microsoft, Redmond, WA). An example calculation using miR-155 data is shown in **Fig. 5**.

A

	miR-155 control RNA (amoles/reaction)					Total RNA samples	
	0	1.5	3	12	24	#1	#2
Raw signal (RFU)	2,294	3,112	4,061	9,756	16,649	3,110	12,838
	2,209	3,296	4,046	9,141	17,206	3,064	12,775
	2,227	3,048	3,889	9,145	17,150	3,249	13,189
Average	2,243	3,152	3,999	9,347	17,002	3,141	12,934
SD	44.8	128.7	95.3	353.9	306.7	96.3	223.1
CV	2.0%	4.1%	2.4%	3.8%	1.8%	3.1%	1.7%
Signal/background		1.4	1.8	4.2	7.6	1.4	5.8
Net signal (RFU)		**909**	**1,755**	**7,104**	**14,758**	**898**	**10,691**
Expression level							
(amoles per reaction)						1.624	17.53
(molecules per cell)						**978**	**10,558**

B

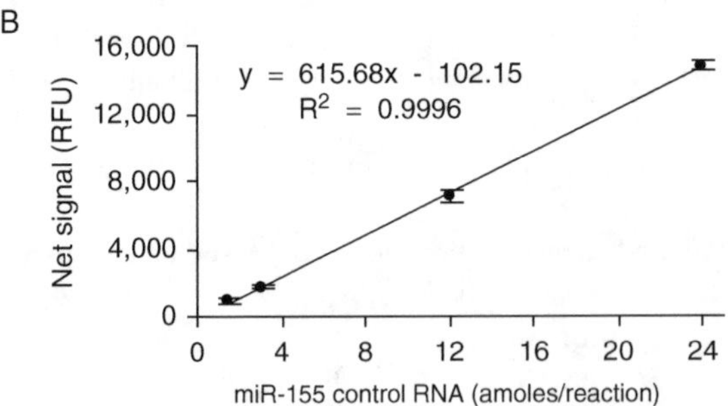

Fig. 5. Invader assay data analysis example (*see* **Subheading 3.9.**): human miR-155 detection. (**A**) Table of raw signal values (measured in a fluorescence plate reader, 30-min secondary reaction) and calculated values for control RNA (synthetic 22-nt miR-155) and two total RNA samples (20 ng total RNA per reaction). (**B**) Standard curve generated with the miR-155 control RNA net signal values (panel **A**). A linear fit encompassing the signal range exhibited by the samples is used to calculate the miRNA expression level, which is reported as attomoles per reaction and molecules per cell (assuming 20 pg of total RNA per cell; *see* **Note 17**).

1. Average the replicates for each control and sample.
2. Calculate the standard deviation (SD) for the replicates and the percent coefficient of variation (%CV; *see* **Note 45**) using the following equation:

$$\%CV = (SD/Average) * 100$$

3. Divide the average for each RNA control and sample by the average tRNA carrier value (0 value in **Fig. 5**) to determine the signal/background values (*see* **Note 46**).
4. Subtract the average tRNA carrier value (0 value in **Fig. 5**) from the average for each RNA control and sample to determine the net signal values.
5. For absolute quantification, generate an RNA control standard curve by plotting net signal vs RNA control level tested. Use a linear fit to generate a standard curve equation (*see* **Note 48**).
6. Convert net signal values for the samples to expression levels (*see* **Note 48**) using the standard curve equation determined in **step 5**.

3.10. Troubleshooting

Assay performance problems generally fall into three categories: (1) no signal, (2) high background signal, and (3) large variation in the replicate reactions. Many problems can be corrected as described next, but in some cases redesign of the assay may be required (*see* **Subheading 3.11.**).

If the positive control or sample signal is very low or indistinguishable from the negative control signal (tRNA carrier), potential causes are
The master mixes (*see* **Tables 4–6**) were improperly prepared or an error was made in the procedure (*see* **Subheading 3.8**).

The Invader oligo and Probe do not form an overlap structure (i.e., oligos were not designed or synthesized correctly).

The wrong oligos were used to prepare the IP or IPS mix.

The wrong FRET oligo, SRT, or Arrestor was used in the secondary reaction.

Improper dilution of the control or sample (recheck the dilution calculations or $A_{260\,nm}$ measurements) (*see* **Subheadings 3.4.3** and **3.6**).

The wrong incubation temperature was used for the primary and/or secondary reactions.

Incorrect fluorescence plate reader settings were used (*see* **Note 44**).

The sample contains an inhibitor; test by running increasing amounts of the sample (*see* **Subheading 3.7.2**); signal will decrease with increasing sample level if an inhibitor is present.

Genomic DNA contamination of the RNA sample is present (*see* **Note 11**).

Samples and/or positive controls are degraded; check RNA quality on a gel (*see* **Subheading 3.3.2**).

Nuclease contamination (RNase) of the reagents occurred; repeat the assay with fresh reagents or newly prepared samples (*see* **Note 28**).

If the negative control (tRNA carrier) signal is very high, potential causes are as follows:

Nuclease contamination (DNase) of the reagents and/or samples occurred; repeat the assay with fresh reagents or newly prepared samples.

Low-purity Probe or FRET oligo was used (*see* **Subheading 3.3**).

If there is high variation in the replicate measurements (*see* **Notes 42** and **45**), potential causes are

There was improper mixing of reagents after thawing and preparation of the primary and secondary reaction master mixes (*see* **Tables 4–6**).

There was primary reaction master mix pipeting error; visually inspect the microplate after dispensing the primary reaction mix for uniform volumes in the wells.

There was secondary reaction master mix pipeting error; visually inspect the multichannel pipet after drawing up the secondary reaction mix to ensure an equal volume will be dispensed to all reaction wells.

Control/sample addition pipeting error occurred; visually inspect the pipet before adding the control or sample to the reaction to ensure a 5-μL volume is dispensed.

Check that pipets are properly calibrated or use pipets recommended in **Subheading 2.2.**

3.11. Invader Assay Redesign

As with other nucleic acid detection methods, such as PCR, it may be necessary to redesign the Invader assay to achieve the desired detection goals. There are three principal reasons for redesign: (1) high background, (2) poor sensitivity, and (3) crossreactivity with a similar sequence RNA. This section describes redesign strategies for improving the performance of the assay. These methods are described in the context of mRNA detection assays, with the final section pertaining specifically to miRNA detection assays. See **Subheading 3.10.** for troubleshooting assay problems not related to Invader probe set design.

3.11.1. High Background

The easiest way to diagnose high background is to perform the Invader assay in a consistent manner and to measure the signal with the same fluorescence plate reader and settings for every assay. This will enable comparison of background signals between different Invader probe sets and make it easier to recognize when the background is unusually high. The background signal is the fluorescence generated when adding the tRNA carrier only to the Invader reaction. Most background problems are caused by deleterious interactions between the primary reaction oligos and often involve the Probe. A thorough investigation of the cause can include performing a time course of the primary reaction (e.g., 30-, 60-, and 90-min primary reaction times and a constant secondary reaction time) and oligo swap-out experiments (e.g., Probe-only and Invader oligo-only reactions). In a primary reaction time-course, a good-performing Invader probe set will exhibit minimal increase in background signal over time (*see* **Note 49**).

While it may be necessary to rigorously identify the cause of high background for an Invader probe set (e.g., when there is almost no flexibility in the location of the cleavage site, such as when detecting a particular mRNA splice junction or edited RNA), it is generally more efficient to redesign and test a new Invader probe set. Commonly used redesign methods include (1) changing the 5′ arm (*see* **Table 2**), (2) moving the cleavage site 1–2 nt in either direction, (3) shortening or lengthening the Probe 1 nucleotide (also need to order a 1-nt longer or shorter Stacker if one is being used), and (4) testing the Invader probe set with the other FRET oligo (*see* **Table 2**).

3.11.2. Poor Sensitivity

The sensitivity of the assay is mainly dependent on the degree of secondary/tertiary structure present in the Invader probe set binding region. Thus, an effective way to increase sensitivity is to test Invader probe sets that contact other sites in the mRNA (screening at least three Invader probe sets is recommended if high sensitivity is needed). However, if the detection goal is <less than> 5000 RNA molecules per reaction (i.e., ~ 1 copy per cell; *see* **Note 17**), it may be more efficient to systematically determine accessible regions in the RNA (*see* **Note 50**) *(29)*. If the detection goal for the assay (e.g., a particular splice junction) does not enable moving the Invader probe set to other regions in the mRNA, the following redesign strategies can improve sensitivity: (1) move the cleavage site 1–2 nt in either direction, (2) extend the Invader oligo's 5′ end or the Stacker's 3′ end in 2- to 3-nt increments, and (3) lengthen the Probe (also requires a new Stacker) to run the assay at a higher primary reaction temperature (*see* **Note 22**).

3.11.3. Crossreactivity With Similar Sequences

An Invader assay exhibiting crossreactivity (*see* **Subheading 3.7.2.**) can be redesigned as follows: (1) test Invader probe sets corresponding to other discrimination sites and (2) test Invader probe sets corresponding to all design permutations for a discrimination site (*see* **Note 7**). For example, a single-base discrimination site can be tested with a −1 or +1 Invader probe set, whereas a dual-base site can be tested with three Invader probe set designs (−2/−1, −1/+1, or +1/+2). Finally, if it is not possible to design an Invader probe set without crossreactivity in its linear detection range, the assay can be performed with sample levels that generate specific signal below the threshold of the crossreactive signal.

3.11.4. Redesign for miRNA Detection Assays

Since miRNAs are only approx 22 nt long, there are fewer redesign strategies that can be applied for this class of Invader assays. All three types of assay problems—high background, low sensitivity, and insufficient specificity—can potentially be solved using a subset of the redesign strategies listed. These include (1) change the 5′ arm, (2) move the cleavage site (can only be moved 1–2 nt in either direction), and (3) test the other FRET detection system (*see* **Table 2**).

Acknowledgments

We acknowledge support from National Institutes of Health grants GM 63090 (to M.A. G.-B.) and GM 30220 (to J.E. Dahlberg). We thank James Dahlberg for permission to use the miR-155 microRNA data and for critical reading of the manuscript.

4. Notes

1. Invader and Cleavase are registered trademarks of Third Wave Technologies, Madison, Wisconsin.
2. The Invader assay, also known as the invasive cleavage assay or invasive signal amplification reaction, uses at least two oligos (the Invader oligo and Probe) to detect DNA or RNA. Some RNA Invader assays include a third oligo in the primary reaction (Stacker). Different names have been used in the literature for the oligos and regions within the Probe. For clarification: invasive oligonucleotide = Invader oligo, stacking oligonucleotide = Stacker, 5′ flap = 5′ arm, and analyte-specific region = target-specific region (TSR). The term *Invader probe set* refers to all the primary reaction oligos—Invader oligo, Probe, and Stacker (optional; *see* **Introduction** and **Note 10**)—for a given assay. Invader probe sets bound to the target are depicted in the 3′-to-5′ orientation, which enables the display of target RNA sequences in the 5′-to-3′ direction to be consistent with public databases.
3. The biplex format of the Invader assay enables the detection of two different RNAs in the same reaction (*see* **Subheading 3.7.**). The second RNA could be the mRNA for a housekeeping gene (a gene with invariant expression levels in the samples being tested, e.g., GAPDH) or a second mRNA or miRNA of interest. A unique sequence 5′ arm is required for each Probe, two different FRET oligos labeled with spectrally distinct fluorophores, and the corresponding SRTs (*see* **Table 2**). For optimal performance in the biplex format, the two Invader probe sets are designed to have comparable primary reaction temperatures. An alternative multiplex format for detection of 5–20 mRNAs per Invader reaction ***(17)*** uses capillary electrophoresis for readout.
4. Detection of RNA in an HTS format ***(15)*** with the Invader assay can be performed in both 96- or 384-well microplates. Since the assay does not require thermal cycling, the isothermal incubation steps can be performed in an oven, which can hold many microplates. Detection is also simpler as a low-cost fluorescence plate reader can be used for endpoint measurements as opposed real-time monitoring during the entire reaction.

5. Invader oligos can be fully complementary or designed with a noncomplementary base at their 3′ end ***(13,20)***. For RNA detection, Invader oligos with 3 mismatches often perform better than the fully complementary version (World Intellectual Property Organization, publication number WO01/90337). Design rules for Invader oligo/target mismatch preferences are listed in **Table 1**.
6. *Cleavage site* refers to the position in the Probe that is cleaved as directed by the 3 end of the Invader oligo. In a 1-bp overlap structure (*see* **Fig. 1A**), cleavage occurs between positions 1 and 2 in the Probe. For overlap structures <greater than> 1 bp, the cleavage site shifts downstream in the Probe. For example, in **Fig. 3B**, the 4-bp overlap structure would result in cleavage of the Probe between positions 4 and 5. The term cleavage site can also be used to identify the position in the RNA that basepairs with the Probe position that is cleaved. This enables a convenient means for describing the location of Invader probe sets on the RNA by referencing positions within the sequence for a given accession number. For example, the Invader probe sets shown in **Fig. 2A** (splice junction 7–8 and exon 8) are located at cleavage sites 1531 and 1619, respectively, for human FGFR2 accession number NM_022969.1 (transcript variant 2).
7. If the Invader oligo 3′ end is mismatched with the mRNA (*see* **Fig. 1A**), a single-nucleotide difference in the mRNA at position 1 in the Probe would yield a 1-nt gap structure between the annealed Invader oligo and Probe. If the Invader oligo is fully complementary to the target, a mismatch between position 1 of the Probe and the altered nucleotide in the mRNA would yield a nick structure (i.e., contiguous binding of the Probe and Invader oligo on the target without overlap). Neither structure results in efficient Probe cleavage by the Cleavase enzyme ***(13,21)***. While not as effective ***(21)***, discrimination of a single-nucleotide difference can also be achieved by positioning the polymorphic site in the mRNA at the −1 position in the Invader oligo. In the **Fig. 1A** example, discrimination of the 2-nt difference (at the site of overlap) between CYP3A4 and CYP3A7 mRNAs can be accomplished by three different locations of the overlap structure: Probe positions 1 and 2, Probe position 1 and Invader oligo position −1, or Invader oligo positions −1 and −2.
8. Arrestor oligos enhance fluorescence signal in the secondary Invader reaction by preventing uncleaved Probes leftover from the primary reaction (only a few percent of Probes are cleaved in the primary reaction) from competing with the cleaved 5′ arms for the binding site on the SRT (**Fig. 1B**). The Arrestor is added in fourfold excess over the Probe and is fully complementary to the Probe TSR but extends only halfway into the 5′ arm sequence.
9. The ability to use a generic—or universal—sequence for the FRET oligo and SRT simplifies assay design and reduces assay costs. The FAM-labeled FRET oligo sequence listed in **Table 2** has been used to detect over 100 different mRNAs ***(1)***.
10. The LOD (i.e., assay sensitivity) for mRNA Invader assays that do not use a Stacker are commonly approx 0.1 amol per reaction. With Stackers, the sensitivity is often improved to approx 0.01 amol (6000 molecules) per reaction. Use of Stackers also reduces the risk of assay background as shorter Probes can be used, which minimizes nonspecific Probe interactions (*see* **Note 21**). The optimal primary reaction temperature is increased 5–10 °C when a Stacker is included in the Invader reaction.

11. Detection of DNA with Invader probe sets designed for RNA detection is generally not observed due to a combination of three factors: careful sample preparation, whether using total RNA or cell lysates (*see* **Subheading 3.6.**), minimizes presence of DNA in the sample; RNA samples do not require a denaturation step (nondenatured DNA exhibits ~ 10-fold lower signal than denatured DNA) ***(1)***; and Invader probe sets have different temperature optimums when bound to RNA vs DNA ***(24)***. In fact, presence of genomic DNA in the sample, which shows up as a high molecular weight band on a gel, can be inhibitory to the Invader reaction. Be sure to carefully follow the sample preparation procedure or try a different total RNA preparation method if the inhibition problem continues.
12. The Invader RNA Assay Generic Reagents Kit used for mRNA detection includes the following enzyme and reaction buffers ***(24)***: Cleavase IX enzyme (40 ng/μL), RNA primary buffer (25 m*M* 4-morpholinepropanesulfonic acid [MOPS], pH 7.5, 250 m*M* KCl, 0.125% Tween-20, 0.125% Nonidet P40, 31.3 m*M* $MgSO_4$, 10% polyethylene glycol [PEG]), and RNA secondary buffer (87.5 m*M* $MgSO_4$). The Generic Reagents Kit used for miRNA detection contains an alternate Cleavase enzyme and reaction buffers: Cleavase XII enzyme (60 ng/μL), RNA primary buffer (25 m*M* MOPS, pH 7.5, 62.5 m*M* KCl, 0.125% Tween-20, 0.125% Nonidet P40, 62.5 m*M* $MgSO_4$, 5% PEG), and $TE_{0.1}$ buffer (used instead of RNA secondary buffer). Consult with the manufacturer (Third Wave Technologies) for updates in kit components used for mRNA and miRNA detection.
13. The positive control for mRNA assays is an in vitro transcript comprising the Invader probe set binding region only (minitranscript of ~ 60 nt) or a nearly full-length transcript (*see* **Note 30** and **Subheading 3.4.1.**). For miRNA detection assays, a synthetic RNA corresponding to the actual miRNA sequence, including a 5′ phosphate, is used.
14. Two compositions have been described for 10X cell lysis buffer, one for mRNA assays (200 m*M* Tris-HCl, pH 8.0, 50 m*M* $MgCl_2$, 5% Nonidet P40, 200 μg/mL tRNA) ***(1,24)*** and one for miRNA assays (200 m*M* Tris-HCl, pH 8.5, 5% Nonidet P40, 200 μg/mL tRNA) ***(5)***. Consult with the manufacturer (Third Wave Technologies) for the recommended formulation.
15. A thermal cycler with gradient capability enables easy optimization of the primary reaction temperature (*see* **Subheading 3.7.1.**). The outside columns of the thermal cycler are often not used as the temperature difference between columns 1 and 2 and 11 and 12 is small and provides no additional information on assay performance. Assays for mRNA detection can be optimized to run isothermally at 60 °C for both the primary and secondary reactions. This enables use of an oven for the incubation steps (*see* **Note 4**).
16. Thirteen human FGFR2 splice variants have been thus far characterized and annotated in GenBank (http://www.ncbi.nlm.nih.gov/, accessed on July 15, 2004). The Invader probe sets shown in **Fig. 2** are designed to detect the following splice variants: for splice junction 7–8 and exon 8 assays, NM023030.1, NM023031.1, NM022969.1, NM022975.2, NM022974.1, NM022976.1; for the splice junction 7–9 assay, NM_023029.1, NM_000141.2, NM_022973.1 (NM_022972.1 and

NM_023028.1 would require a different Invader oligo); for the exon 9 assay, NM_023029.1, NM_000141.2, NM_022973.1, NM_022972.1, NM_023028.1. Splice variants NM_022970.1 and NM_022971.1 would not be detected by any of the Invader probe sets in **Fig. 2** as they lack exons 8 and 9. The splice junction Invader probe sets (7–8 and 7–9) use the same Probe and Stacker (only the Invader oligos need to be unique to discriminate these two splice junctions), which conserves on the total number of assay oligos needed. All four Invader probe sets shown in **Fig. 2** are design examples only; they have not been tested in actual assays.

17. Assuming 20 pg total RNA per cell ***(30)***, 1 molecule of RNA per cell could be detected using 100 ng of total RNA or a cell lysate sample corresponding to 5000 cells in an Invader assay with a 5000-molecule (0.008-amol) detection limit.
18. The Mfold program can be accessed at http://www.bioinfo.rpi.edu/applications/mfold/. Another program that incorporates the Mfold algorithm but has additional features is RNAStructure ***(31)***, which can be accessed at http://rna.chem.rochester.edu/RNAstructure.html. Use of this algorithm for Invader assay design has been described by Olson et al. ***(24)***. The success of RNA structure prediction approaches has not been rigorously examined for the application of Invader assay design. While potentially more costly, it may be more efficient to design and test multiple Invader probe sets for a given RNA without regard for predicted RNA structures.
19. IDT (Coralville, IA) is a convenient source for oligo synthesis as nearly all modifications (e.g., amino modifier C7 for Probe 3′ ends and 2′-*O*-methylated nucleotides) and purification methods are available. FRET oligos are available from Third Wave Technologies or Eurogentec (San Diego, CA) or can be synthesized using dye phosphoramidites from Glen Research (Sterling, VA).
20. A 1-bp invasion of the Probe by the Invader oligo results in a more precise cleavage site reaction ***(13)***, which ensures more optimal cleavage of the FRET oligo in the secondary reaction. However, 2-bp invasions can perform well ***(4)*** and are sometimes used to avoid design of new 5′ arms (and SRTs) when there is no choice in the location of the Invader probe set on the mRNA.
21. Assay performance can be impaired if the Invader probe set oligos, particularly the Probe, form hairpin structures or homodimers or heterodimers with other oligos in the assay. Stable, alternative structures involving the Probe can decrease assay performance by impairing its ability to bind to the intended RNA target (i.e., the effective concentration of the Probe is decreased) or by generating a cleaved 5′ arm in a target-independent manner through a background cleavage event (the Cleavase enzyme can cleave duplex DNA structures). High signal in a no target control reaction (Invader assay run with tRNA carrier) is often indicative of background cleavage of the Probe (*see* **Subheading 3.10.**). While Probe hairpin and homodimer analysis is recommended, comprehensive oligo interaction analyses can be time consuming, and the output may not reflect actual assay performance. Thus, it may be more efficient to simply test candidate Invader probe sets with tRNA carrier and control RNA (or a total RNA sample) and redesign accordingly if problems result (*see* **Subheading 3.10.**).

22. While mRNA detection assays can perform well at a range of primary reaction temperatures, it is convenient to standardize the assays to perform at 60 °C. For example, if the biplex format is used, any two mRNA assays could be combined and run at a single primary reaction temperature (*see* **Note 3**). For HTS experiments (see **Note 4**), an oven can be set at 60 °C and used for both the primary and secondary reaction incubations as the FRET oligos (**Table 2**) also perform optimally at 60 °C. Assays with primary reaction temperatures 2–3 °C different from 60 °C can usually be run at the standard 60 °C with minimal sacrifice in performance. When high sensitivity is required and there is little flexibility in assay design due to specificity requirements, it may be beneficial to run a higher primary reaction temperature (e.g., 65–70 °C) to reduce RNA structure and thus improve Probe binding.
23. Since a graphic of the hybridized sequences is not displayed, be sure to correctly enter the sequence input being analyzed. Results can be conveniently tabulated and compared (*see* **Table 3**) using a spreadsheet program such as Excel.
24. The two-state hybridization algorithm *(27)* does not account for the additional stability of 2´-*O*-methylated oligos when bound to RNA *(23)*. Thus, for Stackers, the actual melting temperature for the RNA target will be higher than the 65–70 °C range. When possible, use G:C basepairs at the ends of oligos.
25. The melting temperature prediction results for miRNA Invader assay oligos should only be used as a general guideline for assay design. Unlike mRNA detection assays, in which Probe stability principally determines assay temperature as the Invader oligo is designed to remain stably bound to the mRNA, the influence of Invader oligo stability on assay performance and temperature with respect to Probe stability is not presently known. For this reason, a wider temperature range is recommended for optimizing miRNA Invader assays. Further, if the melting temperature for the Probe TSR is <45 °C, it may be helpful to test a longer Probe TSR along with a correspondingly shorter Invader oligo *(5)*.
26. Invader oligos, Stackers, and SRTs typically do not require purification prior to the desalt step. However, if very high sensitivity is needed (i.e., for low-expression mRNAs or miRNAs) or a poor-quality oligo synthesis is suspected, HPLC or PAGE purification should be performed as described for Probes and FRET oligos (*see* **Subheading 3.3.**).
27. A variety of gel plates, spacers, and comb sizes can be used for both preparative and analytical gels; gel reagents and sample preparation should be adjusted accordingly if different from the procedures listed in **Subheading 3.3.** Alternate gel-loading solutions can be used, including a formamide-EDTA solution containing the positively charged dye crystal violet, which is used to visualize sample loading but migrates out of the well when electrophoresis starts.
28. Since RNases are readily present on work surfaces and hands, a set of precautions is used at all times to prevent degradation of RNA controls and samples. The following precautions should be used when handling RNA and during the Invader assay setup:
    a. Wear gloves at all times (including when handling equipment such as pipets).
    b. Use RNase-free water (DEPC-treated) for preparation and dilution of RNA con-

trols, RNA samples, and buffers used to dilute oligos used in the Invader assay.

c. Thaw RNA controls and samples at room temperature and then immediately place on ice during dilution and assay setup procedures.

d. Use RNase-/DNase-free consumables such as pipet tips (aerosol-barrier tips are recommended), tubes, and microplates.

Reagent stocks suspected of RNase contamination should be thrown out. If reagents are routinely shared, subaliquoting (e.g., 1-mL aliquots) commonly used dilution reagents such as RNase-free water, $TE_{0.1}$ buffer, and tRNA carrier is recommended.

29. Unlike PCR-based detection methods, the Invader assay is not subject to contamination problems because the target is not amplified. However, contamination prevention measures should be used when handling highly concentrated RNA control stocks. These include preparation and dilution of <greater than> 10 n*M* RNA stocks in a separate area from the Invader assay setup area, changing gloves frequently (particularly after diluting RNA controls prior to setting up an experiment), extra caution when opening tubes, and use of aerosol-barrier pipet tips. If crossreactivity experiments are to be performed for Invader assays that detect closely related sequences (*see* **Subheadings 3.1.1.** and **3.11.3.**), control RNA preparation for each assay should be performed on separate days or in separate areas to prevent cross-contamination of the main stocks.
30. Since Invader assay performance is dependent on the degree of structure present in the RNA where the Invader probe set binds, there can be differences in assay performance with minitranscripts (~60 nt) vs nearly full-length transcripts for a given mRNA. Comparable performance between minitranscripts and longer transcripts has been reported ***(4)***, while in some cases twofold higher signal was observed for a minitranscript vs a 1400-nt transcript (P. S. Eis and J. E. Dahlberg, personal communication), presumably due to decreased structure in the shorter transcript.
31. Since RNA controls are substantially diluted for use in the Invader assay, it is usually not necessary to concentrate the in vitro transcript after the desalting step. A portion of the desalted RNA can be quantified directly by $A_{260\,nm}$ measurement.
32. Microcuvettes (e.g., 50 µL) are available for some spectrophotometers; consult with the manufacturer for availability (the accuracy of the microcuvette measurement can be verified by comparison with a standard cuvette measurement using the same nucleic acid dilution). Alternatively, RNA controls and samples can be quantified using RiboGreen (Molecular Probes). If absolute quantification of the mRNA or mRNA is the goal, *see* **Note 34**.
33. The extinction coefficients for synthetic DNA and RNA are often provided by the manufacturer. IDT provides several nucleic acid analysis tools on their Website: http://www.idtdna.com/SciTools/SciTools.aspx. Select "Oligo Analyzer," paste the DNA or RNA sequence in the field, select target type (DNA or RNA), and select "Analyze." The extinction coefficient is listed in the "Results" section.
34. Accurate determination of absolute levels (e.g., molecules per cell) of an mRNA or miRNA in a sample is dependent on the accuracy of the standard curve generated with the RNA control. Thus, it is critical to use nondegraded RNA control

stocks that have been accurately quantified. Use of high-quality pipets and appropriate dilution volumes (e.g., 10 µL + 90?µL, as opposed to 1 µL + 10?µL) helps ensure the accuracy of the standard curve. A good practice is to develop and consistently use a protocol for preparing, quantitating, and diluting RNA controls and samples to be used in Invader assay experiments. It is also useful to routinely monitor Invader assay background signal (using tRNA carrier) and RNA control signal to identify assay performance problems (*see* **Subheading 3.10.**).

35. The recommended working stock concentrations for IPS and IP mixes have been used for many different Invader assays. While Invader oligo and Stacker concentrations typically do not have a significant impact on assay performance, adjustment of the Probe concentration can sometimes result in better performance. For mRNA detection, 8 and 10 pmol Probe per reaction have been used ***(1,4)***; and for miRNA detection, 5 and 10 pmol of both the Probe and Invader oligo have been used (P. S. Eis and J. E. Dahlberg, personal communication). If an alternative Probe concentration is used, the Arrestor concentration should be adjusted accordingly to maintain a fourfold excess of Arrestor. For complete flexibility in the use of FRET oligo/SRT detection systems, the Arrestor, which is assay specific, is added separately (*see* **Tables 4–6**). However, for some applications, such as HTS experiments (*see* **Note 4**), it may be more convenient to include the Arrestor with the FRET oligo/SRT working stock.
36. Several reagents and kits are available for preparing total RNA from tissue or cells. Commonly used products include TRIzol reagent (Invitrogen) and the RNeasy kit (Qiagen). While mRNA preparations can be used in the Invader assay, they do not necessarily result in enhanced detection capability.
37. Due to the small binding site size (~50 nt) on the RNA by the Invader probe set, the assay can detect RNA in a partially degraded sample. However, for accurate comparison between samples (e.g., a control sample vs treated sample), it is critical to use samples in which the RNA is intact.
38. The cell lysate preparation method can also be used with suspension cell lines ***(5)***; consult with Third Wave Technologies for recommendations.
39. Invader probe sets can be screened with total RNA samples (20–100 ng per reaction); however, this is not recommended if low expression is expected for the mRNA or miRNA being detected.
40. **Figure 4A** shows a commonly used experimental setup to determine the optimal primary reaction temperature for two Invader probe sets. The setup in **Fig. 4B** enables prescreening of up to 12 Invader probe sets at a single primary reaction temperature. The optimal primary reaction temperature for the best-performing Invader probe sets can then be determined in a follow-up experiment. The advantages of the **Fig. 4B** setup for the initial optimization are only 8 reactions are run per Invader probe set instead of 20, which conserves on reagent usage; as many as 12 different Invader probe sets can be screening instead of 2; and performance information is obtained for both the positive control and a sample (preliminary assessment of RNA expression level). The setup in **Fig. 4C** is a combination of **Fig. 4A,B** setups, which would yield the most information in the shortest amount of time per Invader probe set.

41. Samples/controls are added in a 5-µL aliquot. For example, to add a 50-ng total RNA sample to the Invader reaction, dilute the sample to a 10-ng/µL concentration.
42. Careful pipeting at all three steps in the Invader assay minimizes variation (use high-quality pipets such as the LTS type by Rainin; *see* **Subheading 2.2.**). The following pipeting techniques are recommended:
    a. Primary reaction mix addition: Use an electronic repeat pipet and place the pipet tip in the bottom of the microplate well, dispense the aliquot, touch the pipet tip to one side of the well, then lift out the pipet tip and repeat in next well.
    b. Control/sample addition: Use a single-channel pipet (a multichannel pipet can be used if many samples are being tested) and place the pipet tip in the bottom of the microplate well, pipet up and down two times to both dispense the sample and to mix the reaction, keep the pipet in the dispense position (i.e., plunger down) while touching to one side of the microplate well before withdrawing the tip.
    c. Secondary reaction mix addition: Perform the same as control/sample addition procedure except with a multichannel pipet.
43. After addition of the secondary reaction mix, the microplate can be read in the fluorescence plate reader at any time and then incubated further if there is insufficient signal/background (*see* **Subheading 3.9.**). A series of readings at 15, 30, and 60 min (or real-time monitoring of the reaction after addition of the secondary reaction mix) provides a means for extending the dynamic range of the assay (*see* **Note 48**). For example, high-expression samples can be analyzed using the 15-min data and low -expression samples with the 60-min data. Expression measurements from high-expression samples are simply multiplied by 4 to adjust the values to the 60-min level used for the low-expression samples.
44. Some variation in the filter specifications can be tolerated; consult with the fluorescence plate reader manufacturer for recommendations. Some plate readers have monochromators that enable the user to set the wavelength and bandwidth. Gain settings are arbitrary, although it is recommended to find a setting that maximizes signal over background (*see* **Subheading 3.9.**) and minimizes well-to-well variation. Be sure to read the plate from the top (some readers allow both top and bottom reads) and to define a new plate map for the microplates if required. Use of standard plate reader settings enables interassay data comparisons and can facilitate troubleshooting assay performance problems such as degraded RNA controls.
45. The CV is a measure of assay variation. Typical intraassay CVs for the Invader assay range from 0% to 10% but often average 3% within an experiment. The pipeting steps in the assay protocol are a primary source of error (*see* **Note 42**).
46. The optimal total RNA or cell lysate test level (i.e., the sample amount that produces a signal level within the linear quantification range) depends on the mRNA/miRNA expression level in the sample and the detection capability of the assay. A typical test range for total RNA samples is 10–100 ng per reaction; for cell lysates, 1000–5000 cells per reaction. For either sample type, fluorescence signal generation is typically linear at signal/background values <less than> 10. This can be confirmed by comparing the net signal values (*see* **Subheading 3.9.**) obtained from a sample optimization experiment (see **Subheading 3.7.2.**).

A signal/background value of approx 1.2 usually indicates the RNA of interest is present in the sample. Additional statistical analyses (e.g., a *t* test between the tRNA carrier replicates and the sample replicates) can be performed to verify the sample signal is significant. Sample inhibition is frequently observed at 100 ng or more total RNA and high cell number lysate samples. Different sample levels can be used to extend the dynamic range of the assay (*see* **Note 48**).

47. The accuracy of the standard curve is dependent on the RNA control being used (*see* **Notes 30** and **34**) and the dilution procedure. A twofold dilution series is recommended to minimize pipeting errors; however, not every dilution needs testing (e.g., seven two-fold dilutions encompassing 0.02–1.28 amol can be prepared, but only the 0.02-, 0.08-, 0.32-, and 1.28-amol levels tested). To minimize loss of RNA on the inside of the pipet tips, particularly when preparing dilutions at levels below 1 amol/μL, the fresh pipet tip can be prerinsed with tRNA carrier before making the dilution.
48. The linear dynamic range for the standard Invader assay format is two orders of magnitude (i.e., a 100-fold range). However, since fluorescence signal is generated linearly with both time and target level, the dynamic range can be extended by continuous monitoring of the secondary reaction (*see* **Note 43**) or modulating the sample level (*see* **Note 46**). For example, high-expression samples could be tested at the 1- or 10-ng level and low-expression samples at 100 ng. After converting the sample signal to attomoles per reaction (*see* **Subheading 3.9.**), divide by the total RNA level tested to yield the expression level per nanogram of total RNA. Use the expression units of choice, such as attomoles/nanogram, molecules/nanogram (using Avogadro's number of 6.023×10^{23}, 1 amol = 602,300 molecules), or molecules/cell (if cell number per Invader reaction is not known, an estimated value of 20 pg total RNA per cell can be used; ***30***). Alternative curve fits (instead of the standard linear fit) can also be used to extend the dynamic range of the assay, particularly with complex samples such as cell lysates ***(24)*** or when testing more than 100 ng total RNA.
49. Background signal in the Invader assay is mainly generated in the secondary reaction, which results from nonspecific cleavage of the FRET oligo by the Cleavase enzyme.
50. The Reverse Transcriptase-Random Oligonucleotide Libraries (RT-ROL) method described by Allawi et al. ***(29)*** identifies unstructured regions in the RNA, and Invader assays designed for these regions have increased sensitivity. Description of this method is beyond the scope of this chapter as the majority of mRNAs and miRNAs can be detected with the design methods described here (*see* **Subheadings 3.1.** and **3.11.2.**).

References

1. Eis, P. S., Olson, M. C., Takova, T., et al. (2001) An invasive cleavage assay for direct quantitation of specific RNAs. *Nat. Biotechnol.* **19**, 673–676.
2. Burczynski, M. E., McMillian, M., Parker, J. B., et al. (2001) Cytochrome p450 induction in rat hepatocytes assessed by quantitative real-time reverse-transcription polymerase chain reaction and the RNA invasive cleavage assay. *Drug Metab. Dispos.* **29**, 1243–1250.

3. Mills, J. B., Rose, K. A., Sadagopan, N., Sahi, J., and de Morais, S. M. (2004) Induction of drug metabolism enzymes and MDR1 using a novel human hepatocyte cell line. *J. Pharmacol. Exp. Ther.* **309**, 303–309.
4. Wagner, E. J., Curtis, M. L., Robson, N. D., Baraniak, A. P., Eis, P. S., and Garcia-Blanco, M. A. (2003) Quantification of alternatively spliced FGFR2 RNAs using the RNA invasive cleavage assay. *RNA* **9**, 1552–1561.
5. Allawi, H. T., Dahlberg, J. E., Olson, S., et al. (2004) Quantitation of microRNAs using a modified Invader assay. *RNA* **10**, 1153–1161.
6. Eis, P. S., Tam, W., Sun, L., et al. (2005) Accumulation of miR-155 and BIC RNA in human B cell lymphomas. *Proc. Natl. Acad. Sci. U. S. A.* **102**, 3627–3632.
7. Lee, R. C., Feinbaum, R. L., and Ambros, V. (1993) The *C. elegans* heterochronic gene lin-4 encodes small RNAs with antisense complementarity to lin-14. *Cell* **75**, 843–854.
8. Bartel, D. P. (2004) MicroRNAs: genomics, biogenesis, mechanism, and function. *Cell* **116**, 281–297.
9. Bartel, D. P., and Chen, C. Z. (2004) Micromanagers of gene expression: the potentially widespread influence of metazoan microRNAs. *Nat. Rev. Genet.* **5**, 396–400.
10. Johnson, J. M., Castle, J., Garrett-Engele, P., et al. (2003) Genome-wide survey of human alternative pre-mRNA splicing with exon junction microarrays. *Science* **302**, 2141–2144.
11. Neville, M., Seltzer, R., Aizenstein, B., et al. (2002) Characterization of cytochrome P450 2D6 alleles using the Invader system. *Biotechniques* **32**, S34–S43.
12. Ohnishi, Y., Tanaka, T., Ozaki, K., Yamada, R., Suzuki, H., and Nakamura, Y. (2001) A high-throughput SNP typing system for genome-wide association studies. *J. Hum. Genet.* **46**, 471–477.
13. Lyamichev, V., Mast, A. L., Hall, J. G., et al. (1999) Polymorphism identification and quantitative detection of genomic DNA by invasive cleavage of oligonucleotide probes. *Nat. Biotechnol.* **17**, 292–296.
14. Hall, J. G., Eis, P. S., Law, S. M., et al. (2000) Sensitive detection of DNA polymorphisms by the serial invasive signal amplification reaction. *Proc. Natl. Acad. Sci. U. S. A.* **97**, 8272–8277.
15. de Arruda, M., Lyamichev, V. I., Eis, P. S., et al. (2002) Invader technology for DNA and RNA analysis: principles and applications. *Expert Rev. Mol. Diagn.* **2**, 487–496.
16. Berggren, W. T., Takova, T., Olson, M. C., Eis, P. S., Kwiatkowski, R. W., and Smith, L. M. (2002) Multiplexed gene expression analysis using the Invader RNA assay with MALDI-TOF mass spectrometry detection. *Anal. Chem.* **74**, 1745–1750.
17. Chan-Hui, P. Y., Stephens, K., Warnock, R. A., and Singh, S. (2004) Applications of eTag assay platform to systems biology approaches in molecular oncology and toxicology studies. *Clin. Immunol.* **111**, 162–174.
18. Nagano, M., Yamashita, S., Hirano, K., et al. (2002) Two novel missense mutations in the CETP gene in Japanese hyperalphalipoproteinemic subjects: high-throughput assay by Invader assay. *J. Lipid Res.* **43**, 1011–1018.

19. Ma, W. P., Kaiser, M. W., Lyamicheva, N., et al. (2000) RNA template-dependent 5 nuclease activity of *Thermus aquaticus* and *Thermus thermophilus* DNA polymerases. *J. Biol. Chem.* **275**, 24693–24700.
20. Kaiser, M. W., Lyamicheva, N., Ma, W., et al. (1999) A comparison of eubacterial and archaeal structure-specific 5 - exonucleases. *J. Biol. Chem.* **274**, 21387–21394.
21. Lyamichev, V. I., Kaiser, M. W., Lyamicheva, N. E., et al. (2000) Experimental and theoretical analysis of the invasive signal amplification reaction. *Biochemistry* **39**, 9523–9532.
22. Lane, M. J., Paner, T., Kashin, I., et al. (1997) The thermodynamic advantage of DNA oligonucleotide "stacking hybridization" reactions: energetics of a DNA nick. *Nucleic Acids Res.* **25**, 611–617.
23. Majlessi, M., Nelson, N. C., and Becker, M. M. (1998) Advantages of 2 -*O*-methyl oligoribonucleotide probes for detecting RNA targets. *Nucleic Acids Res.* **26**, 2224–2229.
24. Olson, M. C., Takova, T., Chehak, L., et al. (2004) *Invader Assay for RNA Quantitation*, Humana Press, Totowa, NJ.
25. Levanon, E. Y., Eisenberg, E., Yelin, R., et al. (2004) Systematic identification of abundant A-to-I editing sites in the human transcriptome. *Nat. Biotechnol.* **22**, 1001–1005.
26. Bass, B. L. (2002) RNA editing by adenosine deaminases that act on RNA. *Annu. Rev. Biochem.* **71**, 817–846.
27. Zuker, M. (2003) Mfold Web server for nucleic acid folding and hybridization prediction. *Nucleic Acids Res.* **31**, 3406–3415.
28. Griffiths-Jones, S. (2004) The microRNA Registry. *Nucleic Acids Res.* **32**, D109–D111.
29. Allawi, H. T., Dong, F., Ip, H. S., Neri, B. P., and Lyamichev, V. I. (2001) Mapping of RNA accessible sites by extension of random oligonucleotide libraries with reverse transcriptase. *RNA* **7**, 314–327.
30. Alberts, B., Bray, D., Lewis, J., Raff, M., Roberts, K., and Watson, J. D. (1994) *Molecular Biology of the Cell*, 3rd ed., Garland, New York.
31. Mathews, D. H., Burkard, M. E., Freier, S. M., Wyatt, J. R., and Turner, D. H. (1999) Predicting oligonucleotide affinity to nucleic acid targets. *RNA* **5**, 1458–1469.

21

Analysis of RNA Structure and RNA–Protein Interactions in Mammalian Cells by Use of Terminal Transferase-Dependent PCR

Hsiu-Hua Chen, Jeanne LeBon, and Arthur D. Riggs

Summary

Terminal transferase-dependent polymerase chain reaction (TDPCR) can be used after reverse transcription (RT) to analyze RNA. This method (RT-TDPCR) has been used for study of RNA structure and RNA–protein interactions at nucleotide-level resolution. A detailed protocol of RT-TDPCR is presented here with examples of its use with ribonuclease T1 in mammalian cells for detecting (1) RNA structure and protein footprints of the human ferritin heavy-chain messenger RNA and (2) in vivo structure of exon 4 of human XIST (X chromosome inactivation-specific transcript) RNA.

Key Words: Footprinting; mRNA; protein–RNA interaction; RNA structure; RNase T1; RT-TDPCR.

1. Introduction

Terminal transferase-dependent polymerase chain reaction (TDPCR) is a method developed for study of DNA footprints and chromatin structure ***(1–5)***. However, we found that by adding a reverse transcription (RT) step, TDPCR also provides an extremely sensitive, versatile, and quantitative method for RNA analysis with nucleotide-level resolution. This procedure, RT-TDPCR, has adequate specificity and sensitivity for mammalian cells and has already been used for in vivo detection of messenger RNA (mRNA) lesions, stem-loop structures, splicing sites, protein footprints, and transcriptional start sites ***(6–9)***. Yeast ribozyme cleavage intermediates also have been studied both in vitro and in vivo by RT-TDPCR ***(6–11)***.

As illustrated in **Fig. 1**, RT-TDPCR includes the following steps: (1) First-strand complementary DNA (cDNA) synthesis using a gene-specific, biotinylated

From: *Methods in Molecular Biology, vol. 488: RNA-Protein Interaction Protocols*
Edited by: Ren-Jang Lin © Humana Press Inc, Totowa, NJ

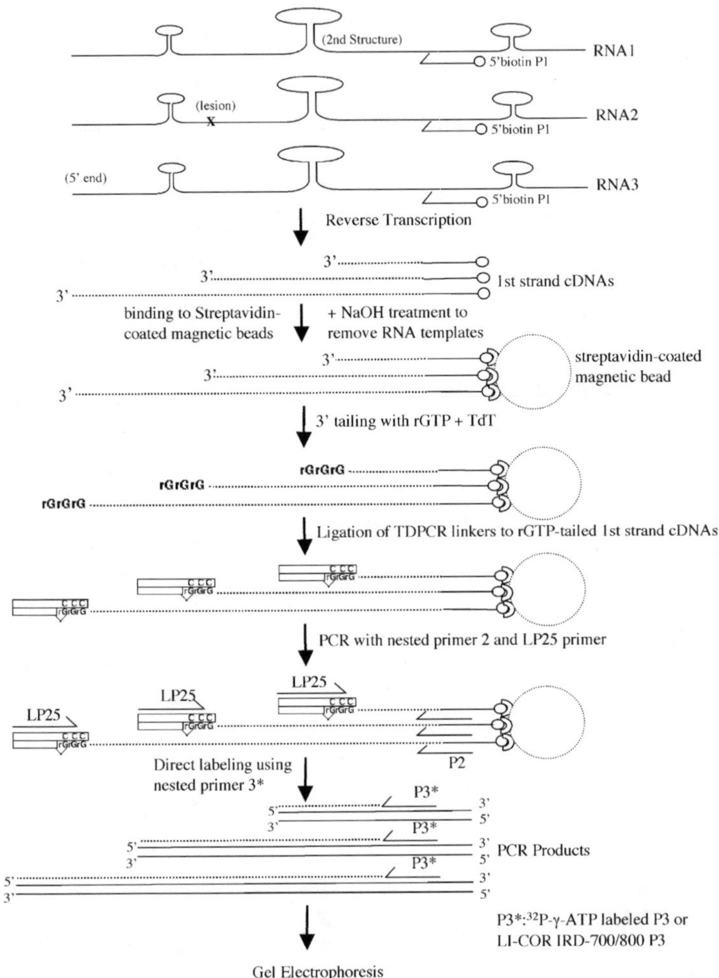

Fig. 1. Schematic outline of the reverse transcriptase terminal transferase-dependent polymerase chain reaction (RT-TDPCR) procedure. (1) RNA is reverse transcribed using a biotin-labeled, gene-specific primer (P1); (2) the newly synthesized complementary DNA (cDNA) strand is captured by streptavidin-coated magnetic beads and is (3) ribotailed using terminal deoxynucleotidyl transferase (TdT) and riboGTP (guanosine 5′-triphosphate). (4) An oligonucleotide linker with a blocked 3′ terminus is ligated to the tailed, 3′ end of the cDNA strand. The cDNA molecules, now having a defined sequence on both the 5′ and 3′ ends, are (5) PCR amplified using a nested gene-specific primer (P2) and a linker-specific primer (LP25). (6) The amplified DNA fragments are used as templates for primer extension using a third gene-specific primer (P3) that is either radioactive (^{32}P) or fluorescent-dye (LI-COR IRD-700/800) labeled at its 5′ end. (7) The labeled products are separated by use of a DNA sequencing gel and visualized by autoradiography or phosphorimaging (^{32}P) or by direct readout from a LI-COR sequencer.

first primer and reverse transcriptase; (2) capture and enrichment of the biotinylated strand cDNA using magnetic streptavidin beads; (3) riboG-tailing of cDNA on the beads using terminal deoxynucleotidyl transferase (TdT), a procedure that adds on average three riboGs; (4) ligation of the riboG-tailed cDNA on the beads to a double-stranded DNA linker; (5) PCR amplification using a nested second primer and a universal linker primer; (6) labeling of PCR products using a ^{32}P-γ-ATP- (adenosine 5′-triphosphate)- or fluorescent-dye-labeled nested-third primer; and (7) sequencing gel electrophoresis to resolve the labeled products. Gel bands are seen where primer extension by RT reaches the end of an RNA molecule or is terminated (or strongly paused) by lesions in the RNA or by stable secondary structure. Lesions or breaks in the RNA can be introduced by various agents, for instance, by treatment of permeabilized cells with a ribonuclease *(6,8)*. RNA footprint studies using RNase T1 have identified specific in vivo protein-binding sites *(6,8)*. **Figures 2** and **3** show the type of results that can be obtained using the RT-TDPCR procedure for analysis of RNA structure and RNA–protein interaction in mammalian cells.

Figure 2 depicts RNase T1 footprinting of human ferritin H-chain mRNA using ^{32}P-γ-ATP labeling and a standard laboratory gel sequencing apparatus *(6)*. Ferritin is an intracellular iron storage protein whose biosynthesis is upregulated by iron only at the posttranscriptional level *(12)*. A stem-loop structure bearing an iron-responsive element (IRE) near the 5′ end of the ferritin H-chain mRNA has been identified and a model proposed *(13)*. In the absence of iron, the IRE-binding protein (IRE-BP) is thought to bind to the IRE stem-loop structure and inhibit translation. Addition of iron eliminates IRE-BP binding; thus, ferritin mRNA is translated *(13)*.

We applied RT-TDPCR to see if the proposed RNA structure and IRE-BP binding could be confirmed, reasoning that protein binding should protect from ribonuclease cleavage. RNase T1 specifically cleaves bonds between the 3′ phosphate group of a guanine ribonucleotide (rG) and the 5′ hydroxyl of the adjacent nucleotide, preferentially in single-stranded RNA *(14,15)*. There are two rGs separated by a single U in the IRE loop region *(6,13)*. We cultured human hepatoma cells (Hep G2) in the presence and absence of iron, treated the cells with ribonuclease (RNase) T1, and isolated total cytosolic RNA *(6,15)*.

Our first RT-TDPCR experiments were done using Moloney murine leukemia virus (M-MLV) reverse transcriptase and primers specific for ferritin H-chain mRNA. We found that full-length fragments were difficult to obtain when the RT step was done with this enzyme at 45 °C, most likely due to the high G+C content (73%) of the region (data not shown). Substituting ThermoScript reverse transcriptase for M-MLV RT and using 55 °C for the RT step gave gel bands indicative of full-length cDNA transcripts, yet the signal strength was relatively low (data not shown). The problem was mainly with

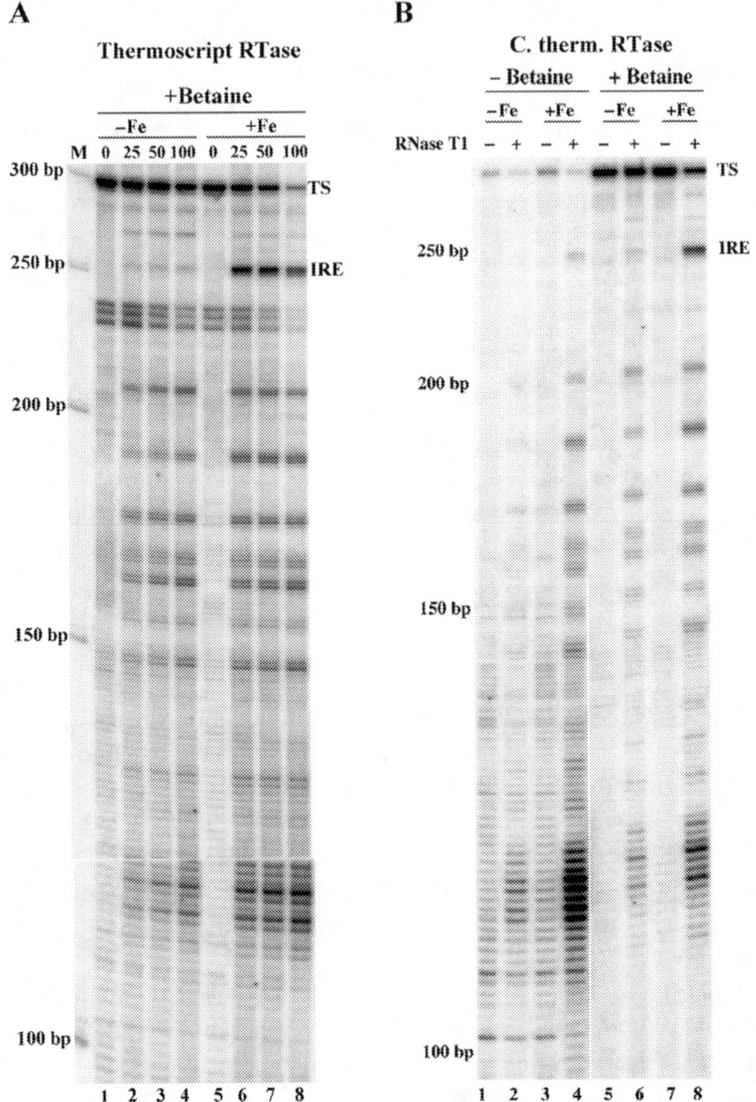

Fig. 2. Ribonuclease (RNase) T1 footprinting by reverse transcriptase terminal transferase-dependent polymerase chain reaction (RT-TDPCR) of human ferritin H-chain messenger RNA (mRNA). Hepatoma cells (Hep G2) were cultured in the presence of either 100 μ*M* desferrioxamine (–Fe) or 100 μ*M* hemin (+Fe) and then treated with RNase T1. Cytosolic mRNA was isolated after RNase T1 treatments as described in **Subheading 3.3.** For each reaction, 250 ng of each RNA sample was used. (**A**) ThermoScript (55 °C for 1 h with hot start; Bio-HFRP1) was used for the reverse transcription (RT) step and AmpliTaq at the polymerase chain reaction (PCR) step with 1.5 *M* betaine. Cells had been treated with 0 U (lanes 1 and 5), 25 U (lanes 2 and 6), 50 U (lanes 3 and 7), and 100 U (lanes 4 and 8) of RNase T1. M: ^{32}P-α-dCTP-labeled 50-bp DNA ladder. (**B**) RT-TDPCR was done using C. therm. DNA polymerase at 70 °C (Bio-HFRP1A) for 30 min for the RT step and AmpliTaq for the PCR step either with (lanes 5–8) or without (lanes 1–4) 1.5 *M* betaine. Cells had been treated with either 0 U (lanes 1, 3, 5, and 7) or 50 U (lanes 2, 4, 6, and 8) of RNase T1. TS denotes the transcription start site, and IRE denotes the iron response element. RTase, reverse transcriptase.

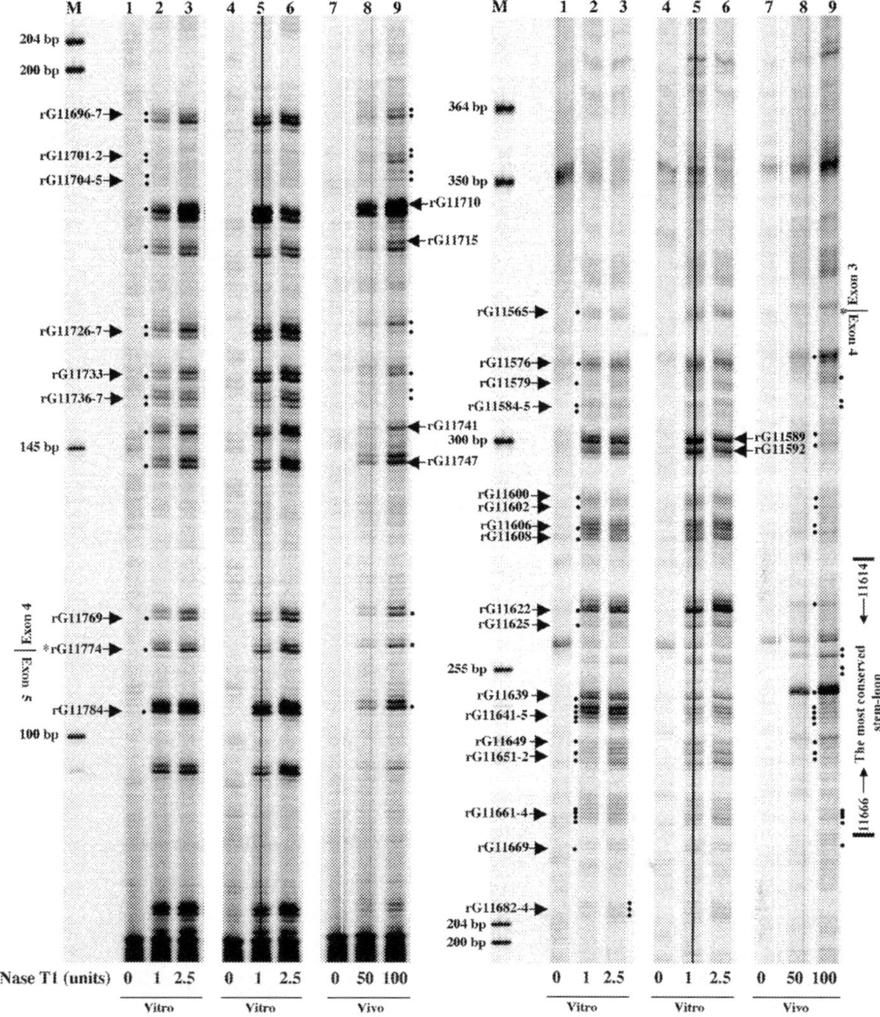

Fig. 3. Ribonuclease (RNase) T1 analysis of exon 4 of human X chromosome inactivation-specific transcript (XIST) RNA. X8 cells were cultured, treated with RNase T1, and RNA prepared as described in **Subheading 3.3.** Reverse transcriptase terminal transferase-dependent polymerase chain reaction (RT-TDPCR) was done using C. therm. at 70°C for 30 min for the reverse transcription (RT) step and the Expand Long system (10X buffer 3 was used) without betaine for the PCR step. The human XIST exon 4-specific primer set used was primer 1, 5′-biotin-TTCAGGTGGGAAGGCTGACTTCCTTCAGTGTG-3′; primer 2, 5′-GCTGACTTCCTTCAGTGTGTTCAAATTCT-3′; primer 3, 5′-IRD-700-GTGTTCAAATTTCTTGGACCTGCTGAAGACTC-3′. For each reaction, 2 μg of in vitro RNase T1-treated RNA was used; this RNA had been treated with 0, 1, and 2.5 U of RNase T1 in buffer with either 0 *M* (lanes 1–3) or 0.3 *M* added NaCl (lanes 4–6). In each reaction, 1 μg of the in vivo RNase T1-treated RNA was used; cells had been treated with 0, 50, and 100 U of RNase T1 (lanes 7–9). Lane M: IRD-700-labeled 50- to 700-bp Sizing Standard from LI-COR. The 5′ end of the third primer used here was at nucleotide 11864; the linker is 25 nt, and tailing adds 2–3 nt. Thus, one can calculate base position ± 1 nt. Faint bands are usually seen at every position, facilitating base counting and confirming that most strong bands are at guanine ribonucleotides.

the PCR step, which used AmpliTaq. We found that addition of 1.5 *M* betaine, which is known to aid PCR amplification of difficult sequences *(16)*, to our PCR mixtures resulted in mostly full-length molecules in samples without RNase T1 treatment (**Fig. 2A**, lanes 1 and 5). The band representing full-length ferritin H-chain cDNAs is labeled TS and is seen in each lane of **Fig. 2**, illustrating that RT-TDPCR can be used to identify transcription start sites. Some less than full-length bands are present in the controls not treated with RNase T1; most of these are due to secondary structure in the RNA, which causes RT pause sites. Treatment with RNase T1 produces faint bands at every position, with much stronger bands at some positions. One can observe in **Figs. 2** and **3** that most strong bands resulting from RNase T1 treatment are doublets or triplets because the ribonucleotide tailing step usually adds either two or three bases but seldom more.

Figure 2 illustrates the use of RT-TDPCR to study protein–RNA binding. In the absence of iron, the ferritin IRE is relatively resistant to cleavage by RNase T1. Addition of iron to the cells results in a dramatic increase in cleavage at the IRE (**Fig. 2A**, compare lanes 2–4 to 6–8), completely consistent with the model described for regulation of ferritin levels by protein binding to the IRE. **Figure 2B** shows results previously published *(6)* using the thermostable reverse transcriptase C. therm.™. Again, iron dramatically increases cleavage at the IRE by RNase T1. The detailed cleavage pattern is also consistent with the previously proposed stem-loop structure of the IRE *(6,13)*. It is also apparent that other changes take place in the RNA when the cells are treated with iron, with numerous RNase T1-generated bands increasing in intensity.

The reverse transcriptase C. therm. may be somewhat better than ThermoScript for RT-TDPCR of high G+C mRNAs or RNAs with stable structures. Unfortunately, this enzyme is not currently available in the right formulation. A formulation change by Roche to a double-enzyme cocktail of C. therm. plus Taq has led to the single enzyme no longer being sold, and the double-enzyme cocktail does not work well for RT-TDPCR. ThermoScript is readily available, and we have recently confirmed that it works well, as should other thermostable reverse transcriptases.

Figure 3 shows application of RT-TDPCR to analysis of human X chromosome inactivation-specific transcript (XIST) RNA. In mammals, one of the two X chromosomes in females is silenced, so only one X is active, the same as in XY males *(17,18)*. In females only, the large noncoding RNA XIST *(19,20)* is produced. The XIST gene is X linked but silent on the active X (the Xa); it is transcribed exclusively from the inactive X chromosome (the Xi), in contrast to most other genes, which are transcriptionally silent on the Xi. XIST remains in the nucleus and coats the Xi in *cis*, an early event in embryogenesis that may be necessary for heterochromatin formation and subsequent silencing of almost all other genes located *cis* to XIST. This silent heterochromatin structure is mitotically stable and can be propagated through all subsequent cell divisions ***(21,22)***.

Since the discovery of XIST in humans and its homolog *Xist* in mouse, efforts have been focused on determining specific elements within the sequence that are essential for coating and silencing, such as the repeat units in the transcript ***(19,20)***. Since only low homology between the primary sequences is found among mammalian species, speculation has been that the secondary structures of the specific elements rather than the primary sequence are important for XIST/*Xist* function ***(23)***. The most highly conserved sequence between mouse *Xist* and human XIST is found in exon 4 ***(24)***, and a stable stem-loop structure has been predicted ***(22,23)***. Evolutionary conservation implies that this structure might have an important role in X-chromosome inactivation, although mouse exon 4 knockout studies have not fully supported this notion ***(24)***.

We treated permeabilized cells with RNase T1, isolated nuclear RNA, and performed RT-TDPCR. For in vitro controls, we isolated total RNA and did RNase T1 treatments in the presence of either no extra NaCl or 0.3 *M* NaCl. For these experiments (**Fig. 3**), a fluorescent dye (IRD-700)-labeled third primer and a LI-COR IR^2 Long Readir DNA Sequencer 4200 system from LI-COR were used. The gene-specific primers used for these experiments start from exon 5 with primer extension going through exon 4 and into exon 3. Exon 4 starts at nucleotide number 11566 and ends at nucleotide 11774 of the sequence reported by Brown et al. ***(19)***. The most highly conserved region in exon 4 is also indicated in **Fig. 3**, with the predicted stem-loop structure ***(24)*** shown in **Fig. 4**. The in vitro studies (**Fig. 3**, lanes 1–6) showed that there is little difference in the cleavage pattern between 10 m*M* Tris 7.5 buffer and buffer plus 0.3 *M* NaCl. There are 49 rGs in exon 4 of XIST, and somewhat more than half (~27) are relatively resistant to RNase T1, suggesting protection by basepairing or steric hindrance due to higher structure.

Importantly, there are major differences between the in vitro and in vivo patterns. For example, focusing on the most conserved region, nucleotide 11639 is much more sensitive to cleavage in vivo than in vitro, whereas nucleotide 11622 becomes very resistant to cleavage in vivo. One interesting observation is that, whereas the accessibility of nucleotide 11639 is in agreement with the predicted structure shown in **Fig. 4**, the other rGs in the proposed loop are not sensitive to cleavage. Also in conflict with the proposed structure is the finding that rG nucleotide 11669 is not sensitive to cleavage, although it is shown by computer modeling to be in a large bulge. At this stage of analysis, it seems that the in vivo structure of the region analyzed is not compatible with those suggested by computer modeling. One is tempted to propose that protein interactions dictate a different structure. Additional studies, of course, will be needed

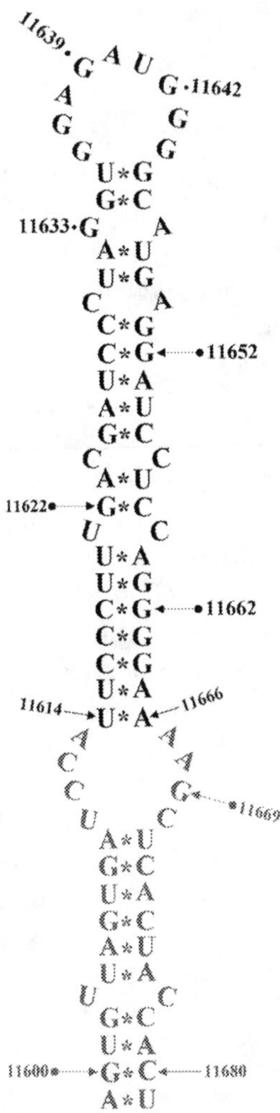

Fig. 4. Predicted stem-loop structure of the most highly conserved region of human exon 4. The structure was obtained using the Mfold server at http://www.bioinfo.rpi.edu/~zukerm/nph-home.cgi. Shown in **bold** is the stretch of 53 bases with 100% homology between human X chromosome inactivation-specific transcript (XIST) (nucleotides 11614 to 11666) and mouse *Xist*.

to fully understand the cleavage pattern. Other cleavage agents certainly can be employed; in theory, any agent that directly or indirectly cleaves the RNA or any agent that stops or pauses RT can be used. The point to be made is that RT-TDPCR provides a very sensitive, informative assay readily applicable to mammalian cells or cell extracts.

An eightfold increase of intensity was observed in vivo compared to in vitro at nucleotide 11639 when the data were quantitated using Gene ImagIR from Scanalytics (not shown). Quantitation studies for transferrin mRNA cleavage have previously been published ***(6)***. Quantification is much more easily done with the LI-COR system than the conventional sequencing gels because of the even band spacing (compare **Fig. 2** to **Fig. 3**). The LI-COR system is now routinely used in our laboratory due to the advantages of nonradioactive detection, evenly spaced bands, and digitized data suitable for quantification. However, the LI-COR instrument is expensive and may not be available, so the conventional, radioactive detection system is also described.

2. Materials

2.1. Chemicals, Reagents, Buffers, and Solutions

1. ^{32}P-γ-ATP and ^{32}P-α-dCTP (deoxycytidine 5′-triphosphate).
2. Hemin (Sigma).
3. Desferrioxamine (Sigma).
4. Dynabeads M-280 streptavidin (Dynal).
5. SequaGel solutions (National Diagnostics).
6. 100 m*M* GTP (guanosine 5′-triphosphate), 100 m*M* ATP, and 100 m*M* dNTPs (deoxynucleotide 5′-triphosphates) (Roche).
7. 1 *M* dithiothreitol (DTT; Sigma).
8. 5 *M* betaine (Sigma).
9. DMEM (Dulbecco's modified Eagle's medium) (1X) liquid (Invitrogen).
10. 1XD-PBS (Dulbecco's phosphate-buffered saline [1X] liquid, contains no calcium or magnesium) (Invitrogen).
11. Physiological buffer: 11 m*M* KH_2PO_4/K_2HPO_4, pH 7.4, 108 m*M* KCl, 22 m*M* NaCl, 1 m*M* $MgCl_2$, 1 m*M* DTT, 1 m*M* ATP.
12. Cytosolic RNA extraction buffer: 10 m*M* Tris-HCl, pH 7.5, 150 m*M* NaCl, 5 m*M* ethylenediaminetetraacetic acid (EDTA), 1% sodium dodecyl sulfate (SDS).
13. RNA STAT-60 (Tel-Test).
14. AG^R 501-X8 (D) resin (20–50 mesh) (Bio-Rad Bioscience).
15. 2X BW buffer: 10 m*M* Tris-HCl, pH 7.5, 1 m*M* EDTA, 2 *M* NaCl.
16. 2X formamide loading dye (2X FD): 98% formamide, 20 m*M* EDTA, pH 8.0, 0.05% xylene cyanol, 0.05% bromophenol blue.
17. 5X Taq buffer: 200 m*M* NaCl, 50 m*M* Tris-HCl, pH 8.9, 0.05% (w/v) gelatin.
18. 3 *M* sodium acetate, pH 5.2.
19. 0.5 *M* EDTA, pH 8.0: Dissolve 93. 06 g of ((ethylenedinitrilo)tetraacetic acid, disodium salt, dihydrate) (FW 372.24) in 450 mL ddH_2O (double-distilled water), adjust to pH 8.0 with 10 *N* NaOH, and adjust volume to 500 mL total with double-distilled water.
20. TE, pH 8.0: 10 m*M* Tris-HCl, pH 8.0, 1 m*M* EDTA.
21. TE, pH 7.5: 10 m*M* Tris-HCl, pH 7.5, 1 m*M* EDTA.

22. 0.1X TE, pH 8.0: 1 m*M* Tris-HCl, pH 8.0, 0.1 m*M* EDTA.
23. 0.1X TE, pH 7.5: 1 m*M* Tris-HCl, pH 7.5, 0.1 m*M* EDTA.

2.2. Enzymes

1. AMV reverse transcriptase (20 U/μL; Life Sciences).
2. M-MLV reverse transcriptase (200 U/μL; Invitrogen).
3. ThermoScript reverse transcriptase (15 U/μL; Invitrogen).
4. Two-step RT-PCR kit using C. therm. DNA polymerase (4 U/μL; Roche),
5. RNasin (20 U/μL; Promega).
6. TdT (15 U/μL; Invitrogen).
7. T4 DNA ligase (3 U/μL; Promega).
8. AmpliTaq polymerase (5 U/μL; Applied Science/PerkinElmer).
9. T4 polynucleotide kinase (10 U/μL; New England Biolabs).
10. *Escherichia coli* exonuclease I (Exo I) (10 U/μL; Amersham/USB).
11. T4 DNA polymerase (3 U/μL; New England Biolabs).
12. Expand™ Long Template PCR system (3.5 U/μL; Roche).
13. RNase T1 (1000 U/μL; Pharmacia/Amersham or Ambion).

2.3. Equipment

1. A thermocycler such as a PTC-100 programmable Thermal Controller (MJ Research) with hot bonnet.
2. PhosphorImager 425 S (Molecular Dynamics).
3. Electrophoresis model SA system (Invitrogen).
4. 6000-V power supply.
5. LI-COR IR^2 Long Readir DNA Sequencer 4200 system (LI-COR).

2.4. Primers and TDPCR Linker

1. Gene-specific primers: Either the Oligo 4™ or the Oligo 5.1™ (National Biosciences) program is used in our laboratory to design the primers. Melting temperatures T_ms are obtained with the following settings: Salt concentration is at 50 m*M*, nucleic acid concentration is at 250 p*M*, and temperature for ΔG calculation is at 25 °C. The first gene-specific primer (Bio-P1) is 5′-biotinylated and designed to have a melting temperature of around 50 °C for use with AMV or M-MLV reverse transcriptase or around 70 °C for use with ThermoScript or C. therm. polymerase. Since the second primer (P2) is used together with LP25 (the linker primer) at the PCR step, it is designed to have a melting temperature of about 63 °C, while the melting temperature of the third primer (P3) is around 65 °C.
2. Universal linker primer (LP25) (25mer): 5′-GCGGTGACCCGGGAGATCTGAA TTC-3′.
3. TDPCR linker primers: Upper primer (LP27) 5′-GCGGTGACCCGGGAGATCTGAAT TCCC-3′ (27mer); lower primer (LP24*C_5) 5′-AATTCAGATCTCCCGGGTCACCGC-pentylNH2–3′ (24mer). Note that this primer is made with an aminopentyl blocking group at its 3′ terminus ***(6)***.

3. Methods

3.1. Preparation of the TDPCR Linkers

1. Kinasing the lower primer: Prepare the kinase mix in a 1.7-mL microcentrifuge tube. (Usually, 10 tubes are done at the same time.) Incubate at 37 °C for 1.5–2 h followed by 20 min at 65 °C *(6)*. Chill on ice (*see* **Table 1**).
2. Annealing with the upper primer: Add 44 µL of 200 µ*M* upper primer directly to the kinased mix (the final concentration of the linkers is 20 µ*M*), mix well, quick spin, add a lid lock to each tube, and then denature at 95 °C for 5 min using a heat block. Turn off the power but leave tubes in the block and allow gradual cooling to room temperature. Leave at 4 °C overnight and store at –20 °C.
3. Always thaw the linkers on ice before preparing the ligation mix.

3.2. Labeling DNA Molecular Weight Standards

1. Label 5 µg of 25- or 50-bp DNA ladder (Invitrogen) with ^{32}P-α-dCTP using T4 DNA polymerase:
   a. Make a 20-µL exonuclease mix in a 0.65-mL tube containing 7 µL of sterile double-distilled water, 4 µL of 5X T4 DNA polymerase buffer (supplied from the manufacturer, NEB), 5 µL of 50-bp ladder (1 µg/µL, NEB), and 4 µL of T4 DNA polymerase (5 U/µL, NEB); mix well, do a quick spin, and incubate at room temperature or 25 °C for 2–4 min. Chill on ice.
   b. Add 30 µL of fill-in reaction mixture containing 18 µL sterile ddH$_2$O, 6 µL of 5X T4 DNA polymerase buffer, 5 µL of 2 m*M* dATG mix (mix 1 µL of each of 100 m*M* dATP [deoxyadenosine 5′-triphosphate], dTTP [deoxythymidine 5′-triphosphate], and dGTP [deoxyguanosine 5′-triphosphate]with 47 µL sterile ddH$_2$O), and 1 µl of ^{32}P-α-dCTP. Mix well (total volume is 50 µL), do a quick spin, and incubate at 37 °C for 2 min.
   c. Add 5 µL of 2 m*M* dCTP (mix 1 µL of 100 m*M* dCTP with 49 µL sterile ddH$_2$O) and incubate at 37 °C for 2 min.
   d. Stop the reaction by the addition of 2.5 µL of 0.5 *M* EDTA, pH 8.0.

Table 1.
Terminal Transferase-Dependent Polymerase Chain Reaction (TDPCR) Linker Lower Primer Kinasing Mix

Component	1x ′µL
Water	292
10X kinase buffer (NEB)	40
100 m*M* adenosine 5′-triphosphate (ATP) (BMB)	4
200 µ*M* lower primer	44
T4 DNA kinase (NEB)[a]	20
Total	400

[a]*See* **Note 6**.

2. Purify the labeled DNA ladder using a G-25 spin column suitable for a sample volume of 50–100 µL (follow the instructions from the supplier).
3. Add an equal volume of 2X FD and store at –20 °C. Markers can be used for up to 3 mo, compensating for radioactive decay by increasing the amount loaded (count an aliquot to help track the radioactive decay).
4. When the ^{32}P markers are fresh, 1.0 µL is sufficient for an overnight exposure. Adjust the volume with 1X FD to approx 10 µL and denature for 2 min at 95 °C before loading alongside RT-TDPCR samples on a sequencing gel.

3.3. In Vivo and In Vitro RNase T1 Treatments

1. RNase T1 treatment of Hep G2 cells and the isolation of cytosolic RNA: Human Hep G2 cells are cultured, treated, and used for RNA preparation as described by Bertrand et al. ***(15)***. In brief, cells are first grown in 10-cm plates in DMEM medium plus 10% calf serum to near 80% confluence. The cells are washed with 1X DPBS and then incubated overnight in serum-free DMEM in the presence of either 100 µ*M* hemin (+Fe) or desferrioxamine (–Fe). The cells are trypsinized, and after addition of serum-containing medium to stop trypsin, the cells are washed once with 1X DPBS, washed once with 1 mL ice-cold physiological buffer, resuspended in 1 mL ice-cold physiological buffer, and counted. For RNase T1 treatment, 1×10^6 cells in 100 µL of ice-cold physiological buffer were mixed with 100 µL of ice-cold physiological buffer containing 0.2% Nonidet P-40 and various amounts (0–200 U) of RNase T1. After incubation at 4 °C for 3 min and centrifugation at 5500 *g* for 15 s at 4 °C, the supernatants are transferred to new Eppendorf tubes containing 200 µL of cytosolic RNA extraction buffer and 400 µL of phenol/isoamyl alcohol (30:1 v/v) were added. The mixtures are extracted and centrifuged at 12,000 *g* in a microcentrifuge for 15 min at 4 °C; the aqueous phases are then transferred to new tubes and extracted two more times with phenol/isoamyl alcohol (30:1 v/v). Finally, RNA is precipitated with 2.5 volumes of 100% ethanol, washed with 1 mL 75% ethanol, and dried briefly. The pellets are dissolved in 100 µL of DEPC (diethylpyrocarbonate) water, and aliquots are taken for $OD_{260/280}$ readings and concentration determination. The extent of RNase T1 digestion is checked on a 2% agarose gel.
2. RNase T1 treatment of X8 cells and isolation of nuclear RNA: X8 cells ***(25,26)*** are grown in 15-cm plates in DMEM medium plus 10% fetal calf serum to near 90% confluence, trypsinized, recovered in serum-containing medium, washed once with 1X DPBS, washed once with 1 mL ice-cold physiological buffer, resuspended in 1 mL ice-cold physiological buffer, and counted. Then, 1×10^6 cells in 100 µL ice-cold physiological buffer ***(15)*** are mixed with 100 µL of ice-cold physiological buffer containing 0.2% Nonidet P-40 and various amounts (0–200 U) of RNase T1. After incubation on ice for 3 min followed by centrifugation at 5500 *g* for 15 s at 4 °C, the supernatant is removed. Cells and nuclei in the pellets are lysed with 200 µL of RNA STAT-60 by pipeting up and down. Add 40 µL of chloroform per sample, extract for 15 s with vigorous shaking, and incubate at room temperature for 5 min. After centrifugation at 12,000 *g* for 15 min in a microcentrifuge at 4 °C,

the aqueous phase is transferred to a new tube, and 100 µL of isopropanol are added. RNA pellets are obtained after centrifugation at 12,000 *g* for 15 min at 4 °C, followed by a 1 mL 75% ethanol wash, spin at 7500 *g* for 5 min, and dry briefly. Each dried pellet is resuspended in 50 µL of DEPC H_2O. One-tenth of each sample is taken for $OD_{260/280}$ readings, concentration is calculated, and approx 1 µg each is used for RT-TDPCR (*see* **Note 12**).

3. RNase T1 treatment of purified total cellular RNA isolated from X8 cells: X8 cells are grown in 15-cm plates in DMEM medium plus 10% fetal calf serum to near 90% confluence. The medium is removed, and 10 mL of RNA STAT-60 is added per plate. Cellular total RNA is isolated according to the manufacturer's instructions. The cells are lysed well by pipeting up and down and incubate at room temperature for 5 min. The lysed mixture is transferred from the plate to a sterile 50-mL centrifuge tube, 2 mL of chloroform is added, and the mixture is extracted for 15 s with vigorous shaking followed by incubation at room temperature for 5–10 min. The aqueous phase is transferred to a new sterile 50-mL tube after spinning at 12,000 *g* for 15 min at 4 °C. Then, add 5 mL of 2-propanol and mix. After incubation at room temperature for 3 min, cellular RNA is precipitated by centrifugation at 12,000 *g* for 15 min at 4 °C. The pellet is washed with 10 mL of 75% ethanol and spun at 7500 *g* for 10 min at 4 °C. Add 0.5 mL 75% ethanol twice sequentially to transfer the RNA pellet from a 50-mL tube to a 1.5-mL microcentrifuge tube, and RNA is reprecipitated by centrifugation at 12,000 *g* for 5 min at 4 °C. After brief drying, the pellet is resuspended in 250 µL of DEPC-treated sterile water. After determining the total RNA concentration, an RNase T1 titration curve is generated as follows: 20–30 µg of total RNA are incubated with various amounts of RNase T1 (0–25 U) in 200 µL of 10 m*M* Tris-HCl, pH 7.5, plus and minus 0.3 *M* NaCl, for 5 min at room temperature. Add 20 µL of 3 *M* sodium acetate, pH 5.2, per reaction, and precipitate RNA with 2.5 volumes of 100% ethanol (*see* **Note 15**). After washing with 1 mL 75% ethanol, each dried pellet is dissolved in 40–60 µL of DEPC H_2O (final RNA concentration is ~ 2 µg/µL). The degree of digestion is checked by running 1 µL of each sample on a 2% agarose gel, and 2 µg of each sample is used for RT-TDPCR (*see* **Note 12**).

3.4. RT-TDPCR Procedure

3.4.1. RT Step

1. First-strand cDNA synthesis using C. therm.: Usually, 0–2 µg of RNA in a volume of 5 µL is used per reaction (*see* **Note 12**). Prepare C. therm. RT mix, 15 µL per sample. Prepare enough for total number of samples plus two extra (*see* **Table 2**). Add 15 µL RT mix to each 5-µL RNA sample. Mix well and do a quick spin. Add 10–20 µL mineral oil on top (omit if the thermocycler has a hot bonnet). Primer extension is carried out at 70 °C or at the melting temperature of Bio-P1 for 30 min. Chill on ice.
2. First-strand cDNA synthesis using ThermoScript: Prepare a 10-µL pre-RT mix containing 0–2 µg of RNA (*see* **Note 12**) and 10 pmol of Bio-P1 in DEPC H_2O and

Table 2.
Reverse Transcription (RT) Mixes

C. therm.	
Component	*1x* µL
DEPC H_2O	5.76
5X RT buffer	4.0
100 m*M* DTT	1.0
25 m*M* dNTP	0.64
RNasin	0.5
DMSO	0.6
20 µ*M* Bio-P1	1.0
C. therm. polymerase[a]	1.5
Total	15

ThermoScript	
Component	1x µL
DEPC H_2O	1.5
5X complementary DNA synthesis buffer	4.0
100 m*M* DTT	1.0
10 mM dNTP	2.0
RNasin	0.5
Total	9

DEPC H_2O, water treated with diethylpyrocarbonate; DMSO, dimethyl sulfoxide; dNTP, deoxynucleotide 5′-triphosphate; DTT, dithiothreitol.

[a]*See* **Notes 1, 6,** and **7**.

add 20 µL of mineral oil on top. Heat the pre-RT mix at 80 °C for 5 min. Quickly chill on ice. Prepare RT mix, 9 µL per sample. Prepare enough for total number of samples plus two extra (*see* **Table 2**). Add 9 µL of RT mix to each sample, mix, and spin. Incubate reaction 2 min at 55 °C, then add 1 µL of ThermoScript, and mix with a Pipetman. Primer extension is carried out at 50–65 °C, depending on the melting temperature of Bio-P1 (usually melting temperature plus 5 °C to increase the specificity) for 1 h. Heat inactivate ThermoScript at 85 °C for 5 min. Chill on ice.

3.4.2. Magnetic Bead Capture and Enrichment

1. Prepare Dynabeads using a magnetic particle concentrator (MPC). Swirl the Dynabead bottle to completely resuspend the beads and then take out enough for all of the samples. Usually, 20 µL/sample is adequate for 20 pmol of the biotinylated primer 1.

2. Transfer 200 µL (i.e., 20 µL per sample × 10 samples) of well-resuspended bead solution to a 1.5-mL tube; use the MPC to separate the beads from the supernatant and remove the supernatant. Wash beads twice with 400 µL 2X BW by pipeting up and down and then use the MPC to remove supernatant.
3. Resuspend beads in 200 µL 2X BW. After RT, add 20 µL of the bead solution to each sample in **Subheading 3.4.1.** (*Note: Do not* vortex while the Dynabeads are present; a quick spin at low speed is okay, e.g., a few seconds at <less than> 1000 rpm).
4. Immobilize DNA on the beads by rotating the mixture at room temperature for 15–60 min (see the instructions from the manufacturer). Do a low-speed spin to bring down material and remove supernatant using the MPC.
5. Wash once with 50 µL of 2X BW and remove supernatant using the MPC.
6. Remove the RNA from the RNA–DNA hybrid by addition of 50 µL 0.15 *M* NaOH (freshly diluted from a 5 *M* stock) to the beads, mix, and incubate at 37 °C for 5–10 min.
7. Remove supernatant using the MPC. The cDNA will remain on the beads.
8. Wash beads once with another 50 µL 0.15 *M* NaOH. Do a quick low-speed spin to bring down any residual NaOH. Remove supernatant using the MPC.
9. Wash beads twice with 100 µL of TE, pH 7.5. Remove supernatant using the MPC.
10. Resuspend beads in 10 µL 0.1X TE, pH 7.5.

3.4.3. TdT Step

1. Prepare TdT mix (10 µL each): Prepare enough for total number of samples plus two extra (*see* **Table 3**).
2. Add 10 µL TdT mix to each sample, mix well by pipeting, and incubate at 37 °C for 15 min.
3. Wash twice with 100 µL TE, 7.5. Remove supernatant using the MPC.
4. Resuspend in 15 µL of 0.1X TE, 7.5.

3.4.4. Ligation

1. Prepare the ligation mix (15 µL each): Prepare enough for total number of samples plus two extra (*see* **Table 4**).

Table 3.
Terminal Deoxynucleotidyl Transferase (TdT) Mix

Component	1x µL
ddH_2O	4.93
5X TdT buffer	4.0
100 m*M* GTP (Roche)	0.4
TdT (15 U/µL, BRL)[a]	0.67
Total	10

ddH_2O, double-distilled water; GTP, guanosine 5′-triphosphate.
[a]*See* **Notes 6** and **7**.

Table 4.
Ligation Mix

Component	1x µL
ddH_2O	7.95
1 *M* Tris-HCl, pH 7.5	1.5
1 *M* $MgCl_2$	0.3
1 *M* DTT	0.3
100 m*M* ATP	0.3
10 mg/mL BSA	0.15
20 µ*M* TDPCR linker	3.0
T4 DNA ligase[a]	1.5
Total	15

ATP, adenosine 5′-triphosphate; BSA, bovine serum albumin; ddH_2O, double-distilled water; DTT, dithiothreitol; TDPCR, terminal transferase-dependent polymerase chain reaction.
[a]*See* **Notes 6** and **7**.

2. Add 15 µL of ligation mix to each sample; mix well by pipeting.
3. Add a lid lock to each tube and incubate at 17 °C overnight.

3.4.5. PCR Amplification

1. Do a quick low-speed spin to bring down the ligation mix and remove supernatant using the MPC.
2. Wash the beads twice with 100 µL TE, pH 8.0, remove supernatant using the MPC, and resuspend each bead sample in 10–30 µL 0.1X TE, pH 8.0. If 20 or 30 µL of 0.1X TE, pH 8.0, were used to resuspend the beads, keep the leftover at 4 °C (*do not freeze when the magnetic beads are present*).
3. Do the following for **step 3**:
   a. Prepare PCR mix using AmpliTaq, 40 µL per sample: Prepare enough for total number of samples plus two extra (*see* **Table 5**). Add 40 µL of AmpliTaq PCR mix to each 10-µL resuspended sample in **step 2**. Mix well by pipeting; do not spin (total volume should be ~ 50 µL).
   b. Prepare PCR master mixes 1 (15 µL) and 2 (25 µL) using Expand Long Template PCR system (read the instructions from Roche carefully before deciding which 10X buffer to use). Prepare enough for the total number of samples plus two extra (*see* **Table 5**). First, add 15 µL of master mix 1 to each 10-µL resuspended sample in **step 2**; mix well by pipeting and do not spin. Then, add 25 µL of master mix 2 and mix well by pipeting; do not spin (total volume should be 50 µL).
4. Add 30 µL of mineral oil on top. The oil can be omitted if the thermocycler has a hot bonnet; in this case, remember to prestart the program and set at pause to allow the hot bonnet to equilibrate to temperature.
5. After an initial 3 min at 95 °C, perform 18–20 cycles (*see* **Note 13**) of 45 s at 95 °C, 2 min at 63 °C or the melting temperature of the second primer, and 3 min at 74 °C. Do 30–40 cycles if band isolation is desired ***(6)*** (*see* **Note 14**).

Table 5.
Polymerase Chain Reaction Mixes

AmpliTaq		
Component	Without betaine, 1X µL	With betaine, 1X µL
ddH_2O	23.5	8.5
5X Taq buffer[a]	10	10
25 m*M* $MgCl_2$[a]	4	4
25 m*M* dNTPs	0.5	0.5
20 µ*M* primer 2 (P2)	0.5	0.5
20 µ*M* LP25	0.5	0.5
5 *M* betaine	0.0	15
Mix thoroughly and do a quick spin	—	—
AmpliTaq[a]	1.0	1.0
Total	40	40

Expand Long Template

Master mix 1		Master mix 2	
Component	1X µL	Component	1X µL
ddH_2O	12.5	ddH_2O	19.5
25 m*M* dNTP mix	1.0	10 x buffer 3	5.0
20 µ*M* P2	0.75	Enzyme mix[b]	0.75
20 µ*M* LP25	0.75	—	—
Total	15	Total	25

ddH_2O, double-distilled water; dNTP, deoxynucleotide 5′-triphosphate.
[a]*See* **Notes 2–7**.
[b]*See* **Note 6**.

6. Keep the PCR products at 4 °C (*do not freeze*) for more labeling reactions. Although most of the thermostable DNA polymerases we have used remained active for months if PCR mixtures were stored properly at 4 °C, we add 0.1 µ**L** additional polymerase into the final labeling reaction.

3.4.6. Labeling Using ^{32}P

3.4.6.1. Labeling of Primer 3

1. Prepare kinase mix (*see* **Table 6**).
2. Mix well, place on ice, quick spin, place on ice, take to the designated radioactivity area, add 1 µL (1X) or 5 µL (5X) ^{32}P-γ-ATP, mix well, and quick spin.
3. Incubate at 37 °C for 1–1.5 h, then at 65 °C for 15 min. Purify using a G-25 spin column (suitable for 50–100 µL per column). Take 0.5–1.0 µL to count. Use immediately or store at –20 °C.

Table 6.
Primer 3 Kinasing Mix

Component	1X µL	5X µL
ddH_2O	5.9	29.5
10X kinase buffer	1.0	5.0
20 µ*M* P3	1.1	5.5
T4 DNA kinase[a]	1.0	5.0
Total	9.0	45.0

ddH_2O, double-distilled water.
[a]*See* **Notes 6** and **7**.

3.4.6.2. Primer Extension

1. Transfer 10 µL from each PCR reaction (use MPC to withheld beads) to a new tube for direct labeling. *Optional*: Take 9 µL from each PCR reaction (use MPC to withheld beads) to a new tube, add 1 µL *E. coli* Exo I (diluted to 1 U/µL), and incubate at 37 °C for 30 min to eliminate unincorporated primers (P1, P2, and LP25) left from the previous steps. Inactivate Exo I by incubation at 72 °C for 15 min (*see* **Note 5**).
2. Add 1 µL or more (depending on the counts [cpm]/µL, usually we use 1 µL of 1–5 × 10^6 cpm/µL kinased P3 and use up to 3 µL if counts are below 1 × 10^6 cpm/µL) of ^{32}P-γ-ATP-labeled primer 3, spin, and add 10 µL mineral oil unless the thermal cycler is equipped with a hot bonnet.
3. Perform primer extension using a thermal cycler. Pause at 95 °C and put the tubes in, then denature at 95 °C for 2 min, followed by three to nine cycles at 95 °C for 45 s, annealing for 2 min at the melting temperature of primer 3, 72 °C for 3 min, and another 10 min at 72 °C.
4. Preparation for gel loading: Add an equal volume of 2X FD to each sample, denature at 95 °C for 2 min, and then cool on ice before loading (usually 10 µL per lane) onto a 6% or 8% acrylamide/7.5 *M* urea sequencing gel. Labeled samples can be stored at –20 °C.

3.3.7. Labeling With LI-COR Dyes IRD-700 or IRD-800

1. Order primer 3 from LI-COR tagged at the 5′ end with either the fluorescent infrared dye IRD-700 or IRD-800 (*see* **Notes 8** and **9**). Resuspend the primer (which arrives as a dry pellet from LI-COR) according to LI-COR's instruction to obtain 1 µ*M* final concentration. Store aliquots at –20 °C and always handle LI-COR primers under dimmed light.
2. For **step 2**, do the following:
   a. After PCR, transfer 9 µL from each reaction to a new tube, add 1 µL of 1 µ*M* LI-COR primer 3, and perform primer extension as above for ^{32}P labeling.
   b. If Exo I treatment was done (*see* **Note 5**), add the LI-COR primer 3 directly to the Exo-treated sample and perform primer extension.

3. Add 3 µL of LI-COR gel loading dye (IR^2 Stop solution, LI-COR) to each sample, denature at 95 °C for 2 min, cool on ice, and load (usually 1.5–2 µL per lane) onto a 5% or 6% LI-COR sequencing gel (*see* **Notes 10** and **11**) along with 1 µL LI-COR molecular weight STR markers (50–700 bp either IRD-700- or IRD-800-labeled; LI-COR.).
4. Store labeled samples in the dark at –20 °C.

3.3.8. Gel Electrophoresis

1. ^{32}P-γ-ATP-labeled samples: Any DNA-sequencing electrophoresis apparatus and power supply (6000 V) is suitable. Usually, the gel is run at 75 W for 4–5 h until the xylene cyanol reaches the bottom of the gel. The gel is transferred to a 3 mm Whatman paper, covered with a piece of plastic wrap, and dried under vacuum at 80 °C. The gel is exposed overnight using a PhosphorImager cassette and scanned the next morning.
2. IRD-700- or IRD-800-labeled samples: To run LI-COR IRD-primer-labeled samples, we use an IR^2 Long Readir DNA Sequencer from LI-COR. The data are automatically saved as a TIFF file during the run, and the TIFF file can be opened directly by a PhotoShop program (**Fig. 3**) or analyzed quantitatively by programs that accept TIFF files.
3. Both PhosphorImager and LI-COR TIFF files can be read and quantitated using Gene ImagIR (an upgraded version of RFLP) from Scanalytics.

4. Notes

1. Choice of reverse transcriptase: All reverse transcriptases should be tested as described in **ref. *6*** to obtain reliable results. AMV and M-MLV reverse transcriptases, which have a temperature optimum of 42–45 °C, often work well but can pause at some secondary structures in RNA molecules. Reverse transcriptases such as ThermoScript and C. therm., which have higher temperature optimums, can transcribe through most secondary structures (*see* **Fig. 2**). ThermoScript and C. therm. have been used in our laboratory for RT-TDPCR for their better extension capability at higher temperature and their ability to overcome secondary structures present in RNA molecules. Currently, we are using only ThermoScript because of a formulation change by Roche from the single C. therm. enzyme to a double-enzyme cocktail of C. therm. plus Taq. The single enzyme used in **Fig. 2B** is no longer available; the cocktail enzyme does not adequately reproduce the results in **Fig. 2B**, so we have switched to using ThermoScript, whose patterns closely match those of C. therm. alone. For determining secondary structures of RNA, AMV RT or M-MLV RT in addition to ThermoScript are useful ***(6)***.
2. 10X Taq buffer from PerkinElmer is also suitable, but our homemade 5X Taq buffer with pH 8.9 works better in our hands. The final [Mg^{+2}] should be between 1.5 and 2.0 m*M* to optimize PCR, depending on primer and template.
3. Use of betaine to a final concentration of 1.5 *M* in the AmpliTaq PCR mix during the PCR amplification step (**Subheading 3.4.5.**) has proved to be advantageous for high G+C regions with apparent secondary structure problems (*see* **Fig. 2**).

4. Expand Long Template PCR system (Roche) is currently used in our lab for the PCR step, and it works well to allow for longer readouts when using the LI-COR Long Readir gel electrophoresis system (*see* **Fig. 3**). We follow the instructions from Roche to prepare PCR master mixtures.
5. (Optional) *Escherichia coli* exonuclease I treatment after PCR amplification: This step eliminates all primers left after the PCR reaction, allowing primer 3 to be the only primer present in the direct labeling step. For Exo I treatment:
   a. Use the MPC to attract the beads, transfer 9 μL of each PCR solution to a new tube, add 1 μL of Exo I (diluted to 1 U/μL with water or 1X PCR buffer from the Amersham stock, which is 10 U/μL).
   b. Incubate at 37 °C for 30 min to eliminate unincorporated primers (P1, P2, and LP25) left from the previous steps. To inactivate Exo I, incubate at 72 °C for 15 min.
6. RT-TDPCR is a procedure that involves many steps and numerous pipetings. Each step is relatively robust but take special care to make sure that all components of the reaction mixtures are at the correct concentrations and active, especially the enzymes. Always add together all the components except the enzyme; mix well by gentle pipeting up and down, do not vortex, do a quick spin, then add the enzyme, mix well by gentle pipeting up and down, do not vortex, and spin again.
7. To ensure the reproducibility of results, performing experiments in duplicate is highly recommended. In general, thin-wall tubes give better results.
8. The LI-COR primers IRD-800 and IRD-700 are light sensitive; handling under dimmed light or yellow light is recommended.
9. Both IRD-700 and IRD-800 give similar sensitivity. A useful feature is that both can be detected simultaneously, allowing two differently labeled samples to be run together in the same lane.
10. LI-COR gels often show evenly spaced bands to 1000 bp, although good-quality data usually do not go beyond 500 bp. In **Fig. 3**, good resolution stops at around 350 bp (6% was used); using a lower percent gel plus deionized urea solution [treated with AG^{R}501-X8 (D) Resin (20–50 mesh); Bio-Rad, Hercules, CA] would improve the resolution.
11. To avoid bubble formation when pouring the sequencing gels, the plates should be very clean and air-dried completely. According to LI-COR, there are two types of air bubbles: one can be seen during or after pouring the gels, and the second kind, caused by invisible tiny dirt particles, exerts its effect on the collected image during electrophoresis when the temperature is high. The white line (lane 7) presented in **Fig. 3** illustrates the pattern of the first kind, and the black lines (lanes 5 and 8) illustrate the pattern of the second type. Soaking the LI-COR plates overnight in 0.1 *M* NaOH solution before cleaning can help prevent the formation of these unwanted lines.
12. RT-TDPCR theoretically and experimentally requires little RNA. We have found that amounts as low as 50 ng of total mammalian RNA gives good signal in our hands. A pilot experiment using varied amounts of RNA, such as 0, 50, 100, 500, 1000, and 2000 ng, should be done first to optimize the amount of RNA.

13. If band isolation is desired, the PCR step can be done with 30–40 cycles. When quantitation is desired, one must stay in the linear, quantitative range of PCR, as is usually the case for the first 20 cycles. In general, if more than 20 cycles are needed, it is likely that at least one of the primers is poor or at least one step is inefficient. If this is the case, the patterns seen may not be reproducible.
14. TDPCR generally shows a weak band at every position. This is normal and is due to polymerase pausing and variability in tailing by TdT. When Taq polymerase is used for PCR, every LMPCR or TDPCR band has a shadow band. This is due to variability in the addition of an extra base by Taq and indicates that the enzyme is no longer adequately active in the last few cycles of PCR. Note also that size purity of the linker primer and labeling primers is important because the final labeled DNA fragments need to be separated with single-nucleotide resolution.
15. In vitro RNase T1-treated samples could be phenol/chloroform extracted first before ethanol precipitation. The footprints were similar either with or without phenol/chloroform extraction.

Acknowledgments

We thank Steven Bates for culturing HepG2 and X8 cells and Louise Shively for critical reading of the manuscript. This work was supported by National Institutes of Health grant GM50575 to A.D.R.

References

1. Komura, J., and Riggs, A. D. (1998). A sensitive and versatile method of genomic sequencing: ligation-mediated PCR with ribonucleotide tailing by terminal deoxynucleotidyl transferase. *Nucleic Acids Res.* **26**, 1807–1811.
2. Komura, J., Ikehata, H., Hosoi, Y., Riggs, A. D., and Ono, T. (2001) Psoralen cross-links at the nucleotide level in mammalian cells: suppression of cross-linking by transcription factor- or nucleosome-binding. *Biochemistry* **40**, 4096–4105.
3. Kontaraki, J., Chen, H.-H., Riggs, A. D., and Bonifer, C. (2000) Chromatin fine structure profiles for a developmentally regulated gene: reorganization of the lysozyme locus before trans-activator binding and gene expression. *Genes Dev.* **15**, 2106–2122.
4. Chen, H.-H., Kontaraki, J., Bonifer, C., and Riggs, A. D. Terminal transferase-dependent PCR (TDPCR) for in vivo UV photofootprinting of vertebrate cells. *Sci. STKE*, **77**, PL1, 2001.
5. Besaratinia, A., Bates, S. E., and Pfeifer, G. P. (2002) Mutational signature of the proximate bladder carcinogen *N*-hydroxy-4-acetylaminobiphenyl: inconsistency with the p53 mutational spectrum in bladder cancer. *Cancer Res.* **62**, 4331–4338.
6. Chen, H.-H., Castanotto, D., LeBon, J. M., Rossi, J. J., and Riggs, A. D. (2000) In vivo, high-resolution analysis of yeast and mammalian RNA–protein interactions, RNA structure, RNA splicing and ribozyme cleavage by use of terminal transferase-dependent PCR. *Nucleic Acids Res.* **28**, 1656–1664.

7. Buettner, V. L., LeBon, J. M., Gao, C., Riggs, A. D., and Singer-Sam, J. (2000) Use of terminal transferase dependent antisense RNA amplification to determine the transcription start site of the Snrpn gene in individual neurons. *Nucleic Acids Res.* **28**, E25.
8. Chen, H. H., Castanotto, D., Rossi, J. J., and Riggs, A. D. (1999) RNA analysis by terminal transferase-dependent PCR. In: *Intracellular Ribozyme Applications: Principles and Protocols* (Rossi, J. J., and Couture, L., eds.), Horizon Scientific Press, Norfolk, U.K., pp. 217–229.
9. Chen, H. H., Castanotto, D., Rossi, J. J., and Riggs, A. D. (2004) In vivo detection of ribozyme cleavage products and RNA structure by use of terminal transferase-dependent PCR. *Methods Mol. Biol.* **252**, 109–124.
10. Scherr, M., LeBon, J., Riggs, A. D., and Rossi, J. J. (1999) The use of cell extracts and antisense deoxyribo-oligo nucleotides for identifying ribozyme cleavage sites on messenger RNAs. In: *Interacellular Ribozyme Applications, Principles, and Protocols* (Rossi, J. J., and Couture, L. A., eds.), Horizon Scientific Press, Norfolk, U.K., pp. 47–56.
11. Scherr, M., LeBon, J., Castanotto, D., et al. (2001) Detection of antisense and ribozyme accessible sites on native mRNAs: application to NCOA3 mRNA. *Mol. Ther.* **4**, 454–460.
12. Zahringer, J., Baliga, R. S., and Munro, H. N. (1976) Novel mechanism for translational control in regulation of ferritin synthesis by iron. *Proc. Natl. Acad. Sci. U. S. A.* **73**, 857–61.
13. Hentze, M. W., Caughman, S. W., Rouault, T. A., et al. (1987) Identification of the iron-responsive element for the translational regulation of human ferritin mRNA. *Science* **238**, 1570–1573.
14. Ehresmann, C., Baudin, F., Mougel, M., Romby, P., Ebel, J. P., and Ehresmann, B. (1987) Probing the structure of RNAs in solution. *Nucleic Acids Res.* **15**, 9109–9128.
15. Bertrand, E., Fromont-Racine, M., Pictet, R., and Grange, T. (1993) Visualization of the interaction of a regulatory protein with RNA in vivo. *Proc. Natl. Acad. Sci. U. S. A.* **90**, 3496–3500.
16. Henke, W., Herdel, K., Jung, K., Schnorr, D., and Loening, S. A. (1997) Betaine improves the PCR amplification of GC-rich DNA sequences. *Nucleic Acids Res.* **25**, 3957–3958.
17. Lyon, M. F. (1961) Gene action in the X chromosome of the mouse (*Mus musculus* L.). *Nature* **190**, 372–373.
18. Lyon, M. F. (1996) Molecular genetics of X-chromosome inactivation. *Adv. Genome Biol.* **4**, 119–151.
19. Brown, C. J., Hendrich, B. D., Rupert, J. L., et al. (1992) The human XIST gene: analysis of a 17 kb inactive X-specific RNA that contains conserved repeats and is highly localized within the nucleus. *Cell* **71**, 527–542.
20. Brockdorff, N., Ashworth, A., Kay, G. F., et al. (1992) The product of the mouse *Xist* gene is a 15-kb inactive X-specific transcript containing no conserved ORF and located in the nucleus. *Cell* **71**, 515–526.
21. Adner, P., and Heard, E. (2001) X-chromosome inactivation: counting choice and initiation. *Nature Rev. Genet.* **2**, 59–67.

22. Huynh, K. D., and Lee, J. T. (2001) Imprinted X inactivation in eutherians: a model of gametic execution and zygotic relaxation. *Curr. Opin. Cell Biol.* **13**, 690–697.
23. Brockdorff, N. (2002) X-chromosome inactivation: closing in on proteins that bind Xist RNA. *Trends Genet.* **18**, 352–358.
24. Caparros, M.-L., Alexiou, M., Webster, Z., and Brockdorff, N. (2002) Functional analysis of the highly conserved exon IV of Xist RNA. *Cytogenet. Genome Res.* **99**, 99–105.
25. Gartler, S. M., and Riggs, A. D. (1983) Mammalian X-chromosome inactivation. *Annu. Rev. Genet.* **17**, 155–190.
26. Pfeifer, G. P., and Riggs, A. D. (1991) Chromatin differences between active and inactive X chromosomes revealed by genomic footprinting of permeabilized cells using DNase I and ligation-mediated PCR. *Genes Dev.* **5**, 1102–1113.

22

Duplex Unwinding and RNP Remodeling With RNA Helicases

Eckhard Jankowsky and Margaret E. Fairman

Summary

RNA helicases are essential for the adenosine 5′-triphosphate (ATP)-driven rearrangement of many RNAs and RNA–protein complexes (ribonucleoproteins, RNPs) throughout RNA metabolism. We describe assays to measure RNA and RNP remodeling by RNA helicases in vitro. We show how to prepare substrates for these reactions and how to monitor unwinding of RNA duplexes and displacement of proteins from RNA using standard molecular biology techniques.

Key Words: DEAD-box; DExH/D; PAGE; protein displacement; RNA; RNA helicase; RNP; RNPase; unwinding.

1. Introduction

Virtually all aspects of RNA metabolism involve controlled and coordinated conformational rearrangements of RNA or RNA–protein (ribonucleoprotein, RNP) assemblies. These essential processes are frequently catalyzed by proteins of the helicase superfamilies 1 or 2 (RNA helicases) or by multicomponent complexes that contain these enzymes *(1)*.

RNA helicases, which play a pivotal role throughout RNA metabolism, have been shown in vitro and in vivo to couple nucleotide 5′-triphosphate (NTP) hydrolysis to conformational changes of RNA *(2)*. It is often desirable to measure this NTP-driven conformational work on RNA in vitro, either to demonstrate the presence of such activities in an enzyme or complex or to gain insight into molecular mechanisms of RNA helicases.

Most frequently, the ability of a RNA helicase or of a helicase complex to perform adenosine 5′-triphosphate (ATP)-driven conformational work on RNA is monitored through analysis of NTP-dependent unwinding of purified RNA duplexes. Recently, NTP-driven remodeling of RNA–protein interactions has

From: *Methods in Molecular Biology, vol. 488: RNA-Protein Interaction Protocols*
Edited by: Ren-Jang Lin © Humana Press Inc, Totowa, NJ

also been measured in vitro *(3)*. This is significant because in a cellular context RNA is invariably bound to other proteins *(4)*.

We describe strategies and protocols to analyze ATP-driven conformational work on RNA/RNPs by RNA helicases in vitro. We focus on unwinding of purified RNA duplexes and remodeling of RNA–protein interactions using standard molecular biology methods.

1.1. Substrate Selection for Duplex Unwinding Reactions

Most RNA helicases function in a non-sequence-specific manner in vitro *(1)*. However, many helicases require specific structural elements to be effective, such as regions of unstructured RNA at certain locations within the substrate *(5, 6)*. In addition, most RNA helicases have only a limited ability to unwind very long and stable RNA duplexes in vitro *(7–10)*.

To obtain an optimal signal for RNA unwinding or RNP remodeling, it is therefore important to carefully design the substrates, even though the sequence of the RNA normally does not pose constraints. In the following section, we list useful considerations for designing substrates (1) to qualitatively demonstrate RNA helicase activity in vitro, (2) to elucidate basic mechanistic parameters of RNA helicases, and (3) to test the ability of an RNA helicase to remodel RNP complexes.

1.1.1. Qualitative Demonstration of RNA Helicase Activity

It is often of primary concern to demonstrate that a "putative" RNA helicase actually has the ability to perform ATP-driven conformational work on RNA in vitro. Without specific knowledge about substrate preferences of the helicase, this question is most easily addressed by conducting unwinding reactions utilizing a permissive substrate.

The efficiency of duplex unwinding for many RNA helicases decreases with length or stability of the duplexes *(8)*. Therefore, effective RNA unwinding (i.e., strand separation with a fast reaction rate constant or high reaction amplitude) is more likely to be observed with duplexes that are as short as possible. A practical lower limit for RNA duplexes is approx 10–12 bp for reactions at room temperature and 14 bp for reactions at 37 °C, although the minimal length depends clearly on the duplex stability *(8)*.

Most RNA helicases require unpaired nucleotides adjacent to the duplex regions for efficient strand separation. For some RNA helicases, both length and position of these single-stranded "overhangs" are critical for the efficiency of the unwinding reaction *(6)*; for other enzymes, only the presence of a single-stranded region is important, not the position *(5)*. Without knowledge about substrate requirements for a given RNA helicase, it is useful to start with a substrate containing at least two overhangs (one 5′ and one 3′ to the duplex) that are longer than approx 20 nt (*see* **Note 1**). This length appeared to be a critical limit for efficient unwinding activities for a number of RNA helicases, although some enzymes function relatively well with shorter overhangs *(7)*.

1.1.2. Probing Mechanistic Parameters of RNA Helicases

If elucidation of specific mechanistic parameters of a given RNA helicase is desired, it is pivotal to devise a minimal, yet completely functional, substrate. This is important because multiple points of initiation or loading of a given helicase reaction have to be prevented to circumvent complication of the kinetic analysis. In addition, nonproductive and possibly inhibitory binding of the helicase to segments of the substrate has to be avoided.

To satisfy these conditions, a substrate should be used that contains only one single-stranded overhang, either 3′ or 5′ to the duplex region, depending on the preference of the RNA helicase under study. The remainder of the duplex should be perfectly basepaired, with no additional overhangs, such that the point of helicase initiation or loading is well defined. It is usually more convenient to utilize chemically synthesized oligonucleotides for mechanistic studies because these RNAs have exactly the desired length. In vitro-transcribed RNAs can be employed as well; however, the potentially heterogeneous ends should be trimmed exactly to the appropriate length using either DNAzymes ***(11, 12)*** or an alternative method such as ribonuclease (RNase) H digestion or ribozyme cleavage.

1.2. Protein Displacement

Protein displacement monitors the dissociation of a protein (or protein complex) from an RNA substrate (**Fig. 1**). To date, protein displacement by RNA helicases has been demonstrated with five different systems in vitro ***(13–15)***. Basically, displacement of a protein from RNA can be monitored as long as a means is devised for separating the RNA–protein complex from the free RNA. Although we focus on direct, polyacrylamide gel electrophoresis (PAGE)-based separation of RNA and RNP in this chapter, many RNPs cannot be analyzed using PAGE. It is however possible to employ alternative RNA/RNP separation techniques such as nuclease digestion or immunoprecipitation ***(14, 15)***.

Analogous to the principle of monitoring RNA helicase activity, which probes whether the dissociation of two complimentary RNA strands is accelerated by the enzyme under study, protein displacement tests whether a given RNA helicase can hasten the dissociation of another protein from RNA in an NTP-driven fashion (**Fig. 1**). To unambiguously distinguish helicase-catalyzed, NTP-driven protein displacement from spontaneous dissociation of the protein, an RNA–protein complex with a considerable half-life should be used (a good orientation value is $t_{1/2} > 60$ min or longer, i.e., $\boldsymbol{k}_{\mathrm{off}} < 10^{-2}\ \mathrm{min}^{-1}$).

Finally, it is important to consider that it is experimentally less challenging to utilize a protein with a binding site that can be incorporated into a bipartite

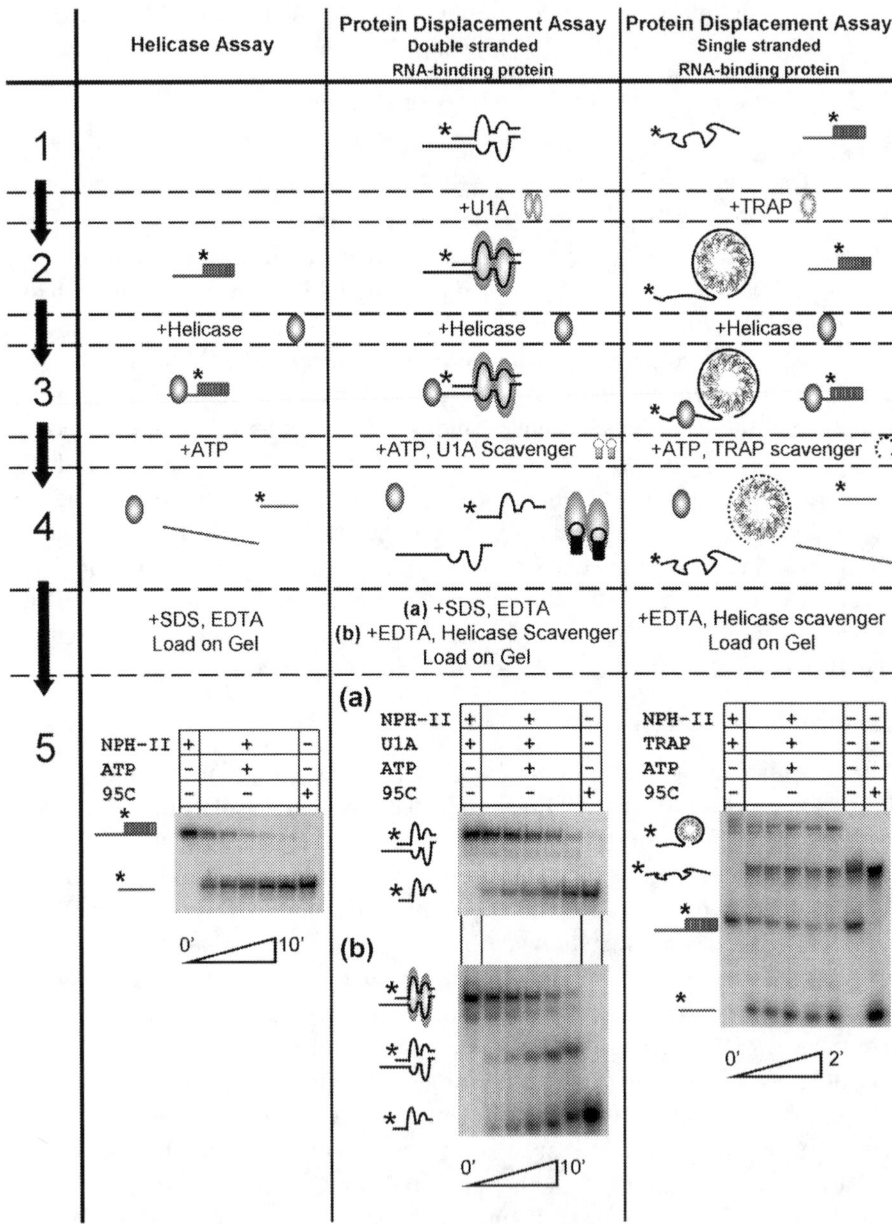

Fig. 1. Reaction scheme and representative polyacrylamide gel electrophoresis (PAGE) for duplex unwinding and protein displacement reactions. Left column: Duplex unwinding reaction (helicase assay). Middle column: Remodeling of a ribonucleoprotein (RNP) involving a bipartite RNA complex (e.g., RNP based on the RNA binding protein U1A) *(13)*. Right column: Protein displacement from single-stranded

RNA. This is because protein displacement can then be monitored "indirectly" by measuring the separation of the two RNA strands (**Fig. 1**). If a bipartite RNA is not an option, the reaction must be monitored by observing the separation of the RNP from the free RNA. Such reactions might require extensive optimization if PAGE-based techniques are used. For example, it is critical to establish conditions for quenching the reaction such that the helicase is removed from the RNA but not the other protein (**Fig. 1**). Proven techniques to accomplish selective removal of helicases from the RNA are described in this chapter.

2. Materials

An established sequence for unwinding reactions is based on the following duplex (13-bp duplex shown):

Substrates with 3′ single-stranded regions:

```
                     5′-AGCACCGUAAAGA-3′
                        |||||||||||||
3′-(A4C)4AAAAUUCGUGGCAUUUCU-5′
```

Substrates with 5′ single-stranded regions:

```
5′-(A4C)4AAAAUAGCACCGUAAAGA-3′
                    |||||||||||||
                 3′-UCGUGGCAUUUCU-5′
```

Fig.1 (contiued) RNA (e.g., displacement of the RNA binding protein TRAP from its cognate RNA) ***(14)***. Rows 1 and 2: Radiolabeled substrates are added to reaction mixture. For protein displacement assays, RNA–protein complex is formed. Row 3: Helicase is added and incubated to allow formation of helicase–substrate complex. Row 4: Adenosine 5′-triphosphate (ATP) is added, and the unwinding/remodeling reaction starts. Reactions are quenched with the regimes indicated, and samples are loaded on PAGE. Row 5: Representative gels of unwinding/protein displacement reactions. Left panel: A duplex unwinding reaction is stopped with sodium dodecyl sulfate (SDS); therefore, only duplex and single-stranded RNA, but no RNA helicase complexes, are observed (cartoons indicate the RNA species). Middle panel: (**a**) RNP remodeling reaction is stopped with SDS, and only duplex and single-stranded RNA, but no RNA–protein complexes, are observed (cartoons indicate the RNA species). (**b**) RNP remodeling reaction is stopped with ethylenediaminetetraacetic acid (EDTA) and helicase scavenger. The helicase scavenger removes the helicase from the RNA; therefore, no helicase–RNA complexes are observed. Other protein-bound RNA species (here U1A-RNA complexes) remain intact and are visible, in addition to free duplex and single-stranded RNA (cartoons indicate the RNA and RNA–protein species). Right panel: Protein displacement from single-stranded RNA is stopped with EDTA and helicase scavenger. No helicase–RNA complexes are thus observed, yet other protein-bound RNA species (here TRAP–RNA complexes) remain intact and are visible (cartoons indicate the RNA and RNA–protein species). The reaction includes a generic RNA duplex to monitor the effect of the reaction conditions on the duplex unwinding activity and to rule out inactivation of the helicase by a reaction component.

However, helicase activity has been shown with numerous different sequences.

2.1. Duplex Preparation

1. RNA oligonucleotides (deprotected and purified).
2. T4 polynucleotide kinase (10,000 U/μL) (New England Biolabs).
3. 10X T4 polynucleotide kinase (PNK) Buffer: 700 m*M* Tris-HCl (pH 7.6), 100 m*M* $MgCl_2$, 50 m*M* dithiothreitol.
4. $\gamma^{32}P$ γ-labeled ATP (5 mCi/mL, 7000 Ci/mmol = 23.8 μ*M*).
5. 50 μ*M* ATP.
6. 10X duplex annealing buffer: 100 m*M* 4-morpholinepropanesulfonic acid (MOPS), pH 6.5, 10 m*M* ethylenediaminetetraacetic acid (EDTA), 0.5 *M* KCl.
7. 10X TBE: 89 m*M* Tris base, 89 m*M* boric acid, 2 m*M* EDTA.
8. 1X TBE, 0.5 X TBE (dilutions of **item 7**).
9. Denaturing polyacrylamide gel: 20% acrylamide:bis 19:1, 7 *M* urea, 1X TBE.
10. Nondenaturing polyacrylamide gel: 15% acrylamide:bis 19:1, 0.5X TBE.
11. 5X denaturing loading buffer: 80% formamide, 0.1% bromophenol blue (BPB), 0.1% xylene cyanol (XC).
12. 5X nondenaturing gel loading buffer: 50% glycerol, 0.1% BPB, 0.1% XC.
13. Gel elution buffer: 300 m*M* NaOAc, 1 m*M* EDTA, 0.5% sodium dodecyl sulfate (SDS).
14. 100% EtOH.
15. 50% glycerol.

2.2. Duplex Unwinding/Protein Displacement Reaction

1. 400 m*M* Tris-HCl, pH 8.0.
2. 5 m*M* $MgCl_2$.
3. 0.1% Nonidet P-40 (diluted from 10%; Boehringer Mannheim).
4. 10 n*M* labeled RNA duplex.
5. Purified helicase (in excess, optimal concentration varies per protein).
6. 35 m*M* equimolar ATP/$MgCl_2$.
7. 2X helicase reaction SDS stop buffer (HRSB): 50 m*M* EDTA, 1% SDS, 0.1% BPB, 0.1% XC, 20% glycerol.
8. 2X helicase reaction nondenaturing stop buffer (HRNDSB): 50 m*M* EDTA, 30% glycerol, 10X helicase scavenger (*see* below under protocols).
9. Nondenaturing polyacrylamide gel: 15% acrylamide:bis 19:1, 0.5X TBE.
10. Whatman chromatography paper (Fisher).
11. Gel dryer (equipped with membrane or oil pump).
12. PhosphorImager screen/PhosphorImager.

3. Methods

3.1. RNA Labeling

1. Mix 1 μL 100 μ*M* top-strand RNA (short strand for duplexes with overhangs), 1 μL 10X T4 PNK buffer, 2 μL 50 μ*M* ATP, 2.1 μL $\gamma^{32}P$-ATP (23.8 μ*M*), and

2.4 μL water. This yields a fraction of 0.33 radioactive labels/total labels. This fraction can be varied depending on the reference date for the γ^{32}P-ATP to give a stronger or weaker signal.

2. Add 1.5 μL T4 PNK. The final volume is 10 μL.
3. Incubate at 37 °C for 60 min (labeling efficiency is usually higher than 80%; *see* **Note 2**).
4. Inactivate the kinase by adding 2 μL denaturing gel loading buffer and heating to 95 °C for 2 min.
5. Prerun 20% denaturing gel (0.8-mm thick) for roughly 30 min to reach separation temperature (~50 °C).
6. Load sample on gel and run at approx 30 V/cm for at least 2 h, depending on the length of the RNA.
7. Remove glass plates and expose gel to film (Kodak) to localize the labeled RNA. Cut out labeled strand, crush gel slice into a 1.5-mL tube using a syringe, and add 600 μL elution buffer. The buffer has to cover the gel. Elute overnight at 4 °C under gentle shaking.

3.2. Duplex Formation

1. Centrifuge to sediment residual acrylamide debris (~16,000 *g*, ~5 min).
2. Carefully remove elution buffer, split into two 1.5-mL tubes, and add 3X volume of 100% EtOH and 2 μL 50% glycerol (aids quantitative precipitation).
3. Place on dry ice for at least 1 h and centrifuge (16,000 *g*) for 30 min at 4 °C.
4. Remove supernatant and dry pellet in SpeedVac for at least 15 min.
5. Combine pellets by mixing one pellet in 10 μL water, resuspend, and transfer to other tube. Rinse tube with 6 μL water and transfer again. Completion of transfer of labeled material can be verified with a Geiger counter at all steps.
6. To the tube with combined, dissolved pellets, add 2 μL 100 μ*M* unlabeled complementary RNA and 2 μL 10X duplex annealing buffer.
7. Heat to 95 °C and gradually cool to room temperature over at least 30 min.
8. After room temperature is reached, add 4 μL 5X nondenaturing gel loading buffer and load on 15% nondenaturing PAGE (0.8-mm thick). Ensure that gel has been prerun for at least 30 min and load sample on running gel. Run gel at room temperature (~20 V/cm) for at least 1 h (longer for duplexes exceeding 30 bp). It is advisable to run labeled single-stranded RNA as a size marker on a parallel lane.
9. Remove glass plates, expose gel to X-ray film, localize and cut out labeled duplex. Crush gel slice into a 1.5-mL tube using a syringe and add 600 μL elution buffer. The buffer has to cover the gel. Elute overnight at 4 °C under gentle shaking.
10. Repeat **steps 1–5**.
11. Measure concentration of purified duplex by aliquoting 1 μL into a scintillation vial and count in a scintillation counter.

3.3. Duplex Unwinding Reaction

1. Mix 3 μL 400 m*M* Tris-HCl, 3 μL 5 m*M* $MgCl_2$, 3 μL 0.1% Nonidet P-40, 3 μL 10 n*M* labeled RNA duplex, 3 μL 10X helicase (in the desired protein concentration), and 15 μL water to a final volume of 30 μL.

2. Incubate for 5 min at reaction temperature.
3. Aliquot 3 μL reaction into 3 μL 2X HRSB for zero timepoint (**Fig. 1**, left column, gel panel, leftmost lane) and place aliquot on ice.
4. Add 3 μL 35 m*M* equimolar ATP/$MgCl_2$ mix to initiate unwinding reaction (*see* **Note 3**).
5. Aliquot 3 μL reaction into 3 μL HRSB at desired timepoints and place on ice.
6. Aliquot 3 μL reaction into 3 μL HRSB and heat to 95 °C for 2 min. This serves as the single-stranded (SS) size marker (**Fig. 1**, left column, gel panel, rightmost lane).
7. Load aliquots on prerunning (10 V/cm) 15% nondenaturing PAGE (0.8-mm thick).
8. Run for at least 30 min, depending on the size of the duplex.
9. Remove glass plates and place gel on Whatman chromatography paper. Cover gel with plastic wrap and dry on gel dryer. Quantify radioactivity in duplex (I^{Duplex}) and product ($I^{Product}$) bands using a PhosphorImager. The fraction product [Frac P] formed at each timepoint is calculated according to the following:

$$[\text{Frac P}] = I^{Product} / (I^{Product} + I^{Duplex}) \tag{1}$$

3.4. Protein Displacement

For protein displacement, materials are identical to those used in the unwinding reaction except that the protein to be displaced is prebound to the RNA (**Fig. 1**, middle and right columns). In addition, HRSB is replaced with HRNDSB that contains a helicase scavenger. The helicase scavenger (generic RNA oligonucleotide that binds the helicase) serves to remove the helicase from the RNA and thus prevents helicase–RNA complexes from interfering with the analysis of the reaction products (**Fig. 1**). If displacement of proteins from single-stranded RNA is measured, it is advisable to include an RNA duplex in the reaction to monitor the effect of the reaction conditions on the unwinding activity (**Fig. 1**, right column).

1. Mix 3 μL 400 m*M* Tris-HCl, 3 μL 5 m*M* $MgCl_2$, 3 μL 0.1% Nonidet P-40, 3 μL 10 n*M* labeled RNA substrate, 3 μL 10X RNA binding protein, 3 μL 10 n*M* labeled RNA duplex (optional), and 12 μL (9 μL) water.
2. Incubate RNA binding protein with RNA for 10 min at reaction temperature (verify full binding of the protein to the RNA; if necessary, incubate longer).
3. Add 3 μL 10X helicase and incubate with reaction for 5 min at room temperature. Final reaction volume is 30 μL.
4. Aliquot 3 μL reaction into 3 μL 2X HRNDSB for control timepoint. Load immediately on prerunning 8% nondenaturing gel. (It is usually advisable to run the gels at 4 °C. Therefore, it is practical to perform the reactions in a heat block set to 20 °C in a cold room.)
5. Add 3 μL 35 m*M* equimolar ATP/$MgCl_2$ to initiate unwinding reaction (*see* **Note 3**).
6. Aliquot 3 μL reaction into 3 μL HRNB at desired timepoints and load immediately on gel.

7. Aliquot 3 µL reaction into 3 µL HRSB and heat to 95 °C for 2 min. This serves as the SS size marker. An additional size marker lane, containing the undenatured duplex (if used) in the absence of protein, should also be included. Load size markers on gel.
8. Run gel for at least 1 h, depending on size of free RNA and protein complex.
9. Remove glass plates and place gel on Whatman chromatography paper. Cover gel with plastic wrap and dry on gel dryer. Quantify radioactivity (I) in RNP, duplex, and product bands using a PhosphorImager. The fraction product [Frac P] formed at each step is calculated as follows:

$$[\text{Frac P}] = \text{I}^{\text{Product}} / (\text{I}^{\text{Product}} + \text{I}^{\text{Duplex}} + \text{I}^{\text{RNP}}) \quad (2)$$

If a protein is utilized that binds a bipartite RNA (**Fig. 1**, middle panel) it is possible to monitor protein displacement by measuring strand separation, and it is not necessary to add an additional duplex to the reaction. In this case, reactions are stopped with SDS, which removes all protein from the RNA. It is then possible to follow the experimental procedure for measuring duplex unwinding (**Subheading 3.3., steps 5–9**).

4. Notes

1. The single-stranded region should be largely unstructured. That is, the sequences should be analyzed for potential formation of internal hairpins, for potential duplex formation, and if more than one single-stranded region is present in the substrate, for the potential of basepairing between the different single-stranded regions. In addition, extensive G tracts should be avoided.
2. If the concentration of duplex RNA is determined by measuring the radioactivity present in the labeled RNA, it is essential to maintain a reproducibly high efficiency of labeling with ^{32}P. This is because the concentration of labeled duplex in the reaction is critical (*see* **Note 4**). Variations of more than 30–50% in labeling efficiency can lead to significant fluctuations in the actual duplex concentration in the reaction and might critically affect the observed kinetics or the reaction amplitude. It is therefore important to verify labeling efficiency for each oligonucleotide used.
3. Reaction conditions such as concentrations of duplex and proteins (*see* **Note 4**), temperature (*see* **Note 5**), salt concentrations, the nature of the ions as well as the pH of the buffer (*see* **Note 6**), the ratio of ATP to Mg^{2+} (*see* **Note 7**), preincubation time of RNA with protein (*see* **Note 8**), and the order of addition (*see* **Note 9**) are critical for both duplex unwinding and protein displacement. These conditions will need to be optimized for each enzyme.
4. The concentration of RNA duplex in the reaction should not exceed approximately 1 n*M* for reactions at room temperature and 0.5 n*M* for reactions at 37 °C. This is because at higher concentrations of RNA, reannealing of the separated strands becomes significant ***(10)***. Reannealing affects the observed unwinding rates by decreasing the ratio of product to reactant, which results in an apparent amplitude of the reaction. If a distinct amplitude is observed that increases with decreasing duplex concentration, reannealing is most likely occurring. Spontaneous annealing

rate constants under usual reaction conditions (room temperature, 50–100 m*M* monovalents) are approximately k_{on} ~ 10^{-7} M−1 min^{-1} ($t_{1/2}$ ~ 20 min at 5 n*M* of the respective strands). In addition, RNA helicases appear to accelerate the spontaneous reannealing by a factor of roughly 10 ***(10)***. Therefore, at 5 n*M* duplex, $t_{1/2}$ for annealing becomes $t_{1/2}$ ~ 2 min in the presence of helicase. Most importantly, several RNA helicases specifically facilitate strand annealing ***(7, 10)***. For example, the DEAD-box proteins Ded1p and Mss116p are among the strongest strand annealers known to date ***(7, 10)***. These enzymes facilitate the bimolecular strand-annealing reaction by several orders of magnitude, up to the physical limit imposed by diffusion ***(7, 10)***. It is therefore advisable to explicitly determine the rate of strand annealing in the presence of the helicase under study (in the absence and in the presence of ATP). The maximal duplex concentration should then be selected to yield an observed annealing rate constant that is, if possible, at least 20-fold lower than the observed annealing rate constant for helicase-mediated duplex unwinding. If it becomes necessary to conduct the reaction with subnanomolar duplex concentrations, it is important to verify that the enzyme binds to the substrate with sufficient velocity (*see* **Note 8**).

5. The reaction temperature is critical not only because of the effect of temperature on enzyme function but also because of the effect of temperature on RNA annealing rates. The rate of spontaneous reannealing increases by a factor of approximately 5 to 10 from 20 °C to 37 °C (dependent on the ionic strength), which has to be taken into account when determining which duplex concentration is to be used in the reaction (*see also* **Note 4**). Moreover, observed unwinding rate constants for DEAD-box proteins show especially strong temperature dependence ***(8)***. For some enzymes, a temperature difference of 1 °C can alter the observed rate constants by a factor of four (own unpublished results). To ensure reproducibility of the measured rate constants, it is advisable to conduct such reactions in a temperature-controlled environment, such as a temperature-controlled aluminum block or a PCR machine.
6. The concentration of salt in the reaction significantly affects rate and efficiency of duplex unwinding by RNA helicases. For example, increasing monovalent salt concentrations from 20 m*M* to 150 m*M* decreased unwinding rates by approx 200-fold for the DEAD-box protein Ded1p (own unpublished results). Similar degrees of deceleration have been observed for the RNA helicase NPH-II (own unpublished results). In addition, helicase processivity is usually negatively affected by increased salt concentrations. We have not observed strong differences between distinct monovalent cations and anions for the helicase activities tested in this laboratory. However, significant effects of chloride on the activity of the DEAD-box protein eIF4A have been reported ***(16)***. If an unknown RNA helicase is being tested, it might therefore be advisable to utilize acetate or glutamate rather than chloride salts. Different divalent ions can have a much wider range of effects due to their interaction with ATP binding. For example, cobalt has been shown to significantly decrease the rate of unwinding for the helicase NPH-II as compared to magnesium ***(12)***.

7. Most RNA helicases that have been investigated in vitro have affinities for ATP with $K_{d(ATP)}$ in the range of 0.1–10 m*M*. ATP concentrations around 1 m*M* are thus a good initial point when optimizing reaction conditions. ATP has been found to fuel unwinding reactions for all DExH/D proteins with demonstrated helicase activity. Several DExH/D proteins can utilize other NTPs and dNTPs as well. DEAD-box proteins appear to be specific for ATP or dATP *(17)*. It is important to note that the ratio of ATP to Mg^{2+} significantly affects reaction rates and affinities of the enzyme for the RNA. To be hydrolyzed, ATP has to be bound to Mg^{2+}. If [ATP] > [Mg^{2+}], the uncomplexed ATP, which in most cases can still bind to the helicase, essentially acts as an inhibitor. Since excess ATP over Mg^{2+} can thus complicate quantitative interpretation of the data, excess of Mg^{2+} over ATP is clearly preferable for mechanistic studies. However, some helicases, such as HCV NS3, are extremely sensitive to excess Mg^{2+} (e.g., 0.5 m*M* free Mg^{2+} reduces the unwinding rate by a factor of approximately 10 for HCV NS3; *18*). Optimizing the ratio of ATP to Mg^{2+} for the respective set of experiments is therefore important.
8. Generally, unwinding reactions are conducted by preforming the helicase substrate complex and starting the reaction by addition of ATP. In most cases, preincubation of helicase and substrate for approximately 10 min is sufficient. However, there are some notable exceptions. For example, the helicase HCV NS3 requires a prolonged preincubation of more than 30 min to be fully active *(18)*. Preincubation time should therefore be optimized for unknown helicases. Some RNA helicases unwind duplexes faster when preincubated with ATP (rather than with RNA) and when the reaction is being started by addition of the RNA substrate (own unpublished results). This reaction scheme might be useful if only weak activity is detected with other reaction regimes. It is important to note that some RNA helicases have a high basal rate of ATP hydrolysis in the absence of nucleic acids *(19)*; therefore, this reaction scheme is not always an option. Most DEAD-box proteins, however, do not hydrolyze ATP without nucleic acids at appreciable rates *(17)*.
9. It is occasionally desirable to observe only a single cycle of the unwinding reaction. Although multiple unwinding cycles generally lead to a larger fraction of product, mechanistic information is more readily obtained from single-cycle conditions. To accomplish single-cycle conditions, the RNA/RNP substrate is saturated with enzyme (*see* **Note 8**), and the reaction is started with ATP and a helicase scavenger, which functions to immediately bind free helicase. Use of the helicase scavenger also prevents dissociated enzyme from rebinding the substrate *(12)*. The concentration and composition of the helicase scavenger are both critical to accomplish single-cycle conditions. Usually, scavengers that are similar to the substrate under investigation have been found to be effective, although concentrations of scavenger that exceed substrate concentrations by as much as a 1000-fold are often necessary. The scavenger should also be tested for eventual interactions with the RNA substrate at the concentrations used. Homopolymer RNAs such as poly-U or poly-C are frequently not suitable as scavengers without further purification because they often contain short nucleotide fragments that interfere with the helicase reaction. To test for effective helicase scavenging and to verify

whether a reaction is truly under a single-cycle regime with a given scavenger concentration, it is advisable to include a control for which the scavenger is added simultaneously with substrate and prior to the addition of ATP. In this control reaction, helicase activity should not be detectable.

References

1. Tanner, N. K., and Linder, P. (2001) DExD/H box RNA helicases. From generic motors to specific dissociation functions. *Mol, Cell* **8**, 251–261.
2. Jankowsky, E., and Fairman, M. (2007) RNA helicases—one fold for many functions. *Curr. Opin. Struct. Biol.* **17**, 316–324.
3. Jankowsky, E., and Bowers, H. (2006) Remodeling of ribonucleoprotein complexes with DExH/D RNA helicases. *Nucleic Acids Res.* **34**, 4181–4188.
4. Linder, P. (2004) The life of RNA with proteins. *Science* **304**, 694–695.
5. Yang, Q., and Jankowsky, E. (2006) The DEAD-box protein Ded1 unwinds RNA duplexes by a mode distinct from translocating helicases. *Nat. Struct. Mol. Biol.* **13**, 981–986.
6. Shuman, S. (1993) Vaccinia virus RNA helicase. Directionality and substrate specificity. *J. Biol. Chem.* **268**, 11798–11802.
7. Halls, C., Mohr, S., Del Campo, M., Yang, Q., Jankowsky, E., and Lambowitz, A. M. (2007) Involvement of DEAD-box proteins in group I and group II intron splicing. Biochemical characterization of Mss116p, ATP hydrolysis-dependent and -independent mechanisms, and general RNA chaperone activity. *J. Mol. Biol.* **365**, 835–855.
8. Rogers, G. W., Richter, N. J., and Merrick, W. C. (1999) Biochemical and kinetic characterization of the RNA helicase activity of eukaryotic initiation factor 4A. *J. Biol. Chem.* **274**, 12236–12244.
9. Rogers, G. W. J., Lima, W. F., and Merrick, W. C. (2001) Further characterization of the helicase activity of eIF4A. Substrate specificity. *J. Biol. Chem.* **276**, 12598–12608.
10. Yang, Q., and Jankowsky, E. (2005) ATP- and ADP-dependent modulation of RNA unwinding and strand annealing activities by the DEAD-box protein DED1. *Biochemistry* **44**, 13591–13601.
11. Pyle, A. M., Chu, V. T., Jankowsky, E., and Boudvillain, M. (2000) Using DNAzymes to cut, process, and map RNA molecules for structural studies or modification. *Meth. Enzymol.* **317**, 140–146.
12. Jankowsky, E., Gross, C. H., Shuman, S., and Pyle, A. M. (2000) The DExH protein NPH-II is a processive and directional motor for unwinding RNA. *Nature* **403**, 447–451.
13. Jankowsky, E., Gross, C. H., Shuman, S., and Pyle, A. M. (2001) Active disruption of an RNA–protein interaction by a DExH/D RNA helicase. *Science* **291**, 121–125.
14. Fairman, M., Maroney, P. A., Wang, W., et al. (2004) Protein displacement by DExH/D RNA helicases without duplex unwinding. *Science* **304**, 730–734.
15. Bowers, H. A., Maroney, P. A., Fairman, M. E., et al. (2006) Discriminatory RNP remodeling by the DEAD-box protein DED1. *RNA* **12**, 903–912.

16. Lorsch, J. R., and Herschlag, D. (1998) The DEAD box protein eIF4A. 1. A minimal kinetic and thermodynamic framework reveals coupled binding of RNA and nucleotide. *Biochemistry* **37**, 2180–2193.
17. Linder, P. (2006) Dead-box proteins: a family affair—active and passive players in RNP-remodeling. *Nucleic Acids Res.* **34**, 4168–4180.
18. Pang, P. S., Jankowsky, E., Planet, P., and Pyle, A. M. (2002) The hepatitis C viral NS3 protein is a processive DNA helicase with co-factor enhanced RNA unwinding. *EMBO J.* **21**, 1168–1176.
19. Shuman, S. (1992) Vaccinia virus RNA helicase: an essential enzyme related to the DE-H family of RNA-dependent NTPases. *Proc. Natl. Acad. Sci. U. S. A.* **89**, 10935–10939.

23

Preparation of Efficient Splicing Extracts From Whole Cells, Nuclei, and Cytoplasmic Fractions

Naoyuki Kataoka and Gideon Dreyfuss

Summary

Pre-mRNA (messenger RNA) splicing is an essential step for gene expression in higher eukaryotes. Splicing reactions have been well studied in vitro using extracts prepared from cultured cells. We describe protocols for the preparation of splicing-competent extracts from whole cells, nuclei, and cytoplasmic fractions. The nuclear and whole-cell extracts are fully active in splicing, while S100 extracts are able to support splicing only when SR (Serine/Arginine-rich) proteins are supplied. The simple method described here to prepare splicing active extracts from whole cells is particularly useful in studying pre-mRNA splicing in many different cell types.

Key Words: Nuclear extracts; S100 extracts; sonication; splicing; whole-cell extracts.

1. Introduction

In eukaryotic cells, the excision of introns from pre-mRNAs (messenger RNAs) by pre-mRNA splicing is an essential step for gene expression *(1,2)*. The pre-mRNA splicing reaction has been extensively studied in vitro using cell extracts. In vitro analyses revealed that splicing occurs via a two-step reaction. During the first step, cleavage at the 5′ splice site and the formation of the phosphodiester bond between the branch point and the 5′ end of the intron occurs. This results in the formation of two intermediates, the 5′ exon and the intron lariat-downstream exon. In the second step, the cleavage at the 3′ splice site occurs, resulting in ligation of 5′ and 3′ exons to form the mRNA and the release of the intron lariat.

Analyses of the splicing reaction also revealed that splicing takes place in a large RNA–protein complex, termed a *spliceosome*, which consists of five U-rich small nuclear ribonucleoproteins (U snRNPs) and numerous non-snRNP splicing factors *(1,2)*. Each U snRNP contains a small nuclear RNA (snRNA)

From: *Methods in Molecular Biology, vol. 488: RNA-Protein Interaction Protocols*
Edited by: Ren-Jang Lin © Humana Press Inc, Totowa, NJ

and several proteins, including the Sm proteins. It is estimated that at least 100 proteins are required for splicing ***(3)***.

Splicing-competent whole-cell extracts have been prepared from *Saccharomyces cerevisiae* and *Drosophila melanogaster* cells ***(4,5)***. To study the splicing reaction in mammalians, cultured HeLa cells have been used as the most common source of extracts. HeLa cells are easy to grow in tissue culture, contain relatively low amounts of ribonucleases, and procedures have been developed for preparation of extracts from them for transcription studies ***(6)***.

We describe a detailed protocol we developed for preparation of efficient splicing extracts from HeLa cells ***(7)***. This system is based on methods originally developed for preparation of extracts for in vitro transcription ***(6)*** with slight modifications. The protocol described is for preparation of both nuclear and cytoplasmic S100 extracts from HeLa cells in an analytical and small-to-medium preparative scale. The nuclear extracts are active in splicing of many different pre-mRNAs, while S100 extracts are not. The likely reason for the more limited efficiency of S100 extracts is that they have a limited amount of essential splicing factors, SR (Serine/Arginine-rich) proteins, although they contain many other splicing factors. Indeed, addition of one or more SR proteins to the S100 extract improves its pre-mRNA splicing efficiency ***(8–10)***. Both nuclear extracts and S100 extracts supplemented with SR protein have been used to study constitutive as well as alternative splicing of several pre-mRNAs, including the splicing of minor spliceosome pathway (AT-AC) introns ***(11–16)***.

A simplified procedure for a small-scale whole-cell extract preparation is also described, which is useful for rapidly testing and studying pre-mRNA splicing in many different cell types. It is also of use in producing extracts from cells expressing transfected tagged proteins ***(7,17–19)***. This whole-cell extract system can be used for in vitro splicing followed by immunoprecipitation ***(7,17–19)***.

2. Materials

All reagents should be prepared with high-quality water such as Milli-Q (Millipore, Bedford, MA) or equivalent. Sterilization of all the reagents used for extract preparation is carried out by autoclaving except for solutions 1 and 2 (*see* **Subheading 2.3.**).

2.1. HeLa Cell Suspension Culture

1. Medium: Joklik's modification of Eagle's minimal essential medium for suspension culture (ICN Pharmaceuticals, Costa Mesa, CA cat. no. 10-323-24) or equivalent. A 5X stock solution is prepared by dissolving this powdered medium in Milli-Q water and is sterilized by filtration through 0.22-µm filters (Millipore).

Prepare culture medium by diluting this stock solution with filter-sterilized water and adding sterile calf serum (Gibco-BRL, Grand Island, NY).

2. HeLa cells: Use a HeLa strain that is adapted to growth in suspension culture (e.g., the S-3 strain, which is available from ATCC, Manassas, VA).

2.2. HEK293T Cell Culture on Plates

As a culture medium, use Dulbecco's modified Eagle's medium with high glucose (Gibco-BRL, cat. no. 11965-084) or equivalent. Add sterile fetal bovine serum (Gibco-BRL) and penicillin-streptomycin solution (Gibco-BRL, cat. no. 15140-122) to 100 mL/L and 10 mL/L, respectively.

2.3. Preparation of Extracts

Reagents 1 and 2 are stock solutions that have to be added to reagents 4–8 immediately before use. These stock solutions should be stored at –20 °C.

1. 1 *M* Dithiothreitol (DTT).
2. 20 mg/mL phenylmethanesufonyl fluoride (PMSF) dissolved in ethanol.
3. Phosphate-buffered saline (PBS): 137 m*M* NaCl, 2.7 m*M* KCl, 8 m*M* Na_2HPO_4, 1.5 m*M* KH_2PO_4, 0.5 m*M* $MgCl_2$.
4. Buffer A: 10 m*M* HEPES KOH, pH 7.9, 10 m*M* KCl, 1.5 m*M* $MgCl_2$, 0.5 m*M* PMSF, 1 m*M* DTT.
5. Buffer B: 300 m*M* HEPES KOH, pH 7.9, 1.4 *M* KCl, 30 m*M* $MgCl_2$.
6. Buffer C: 20 m*M* HEPES KOH, pH 7.9, 600 m*M* KCl, 1.5 m*M* $MgCl_2$, 0.2 m*M* EDTA, 25% (v/v) glycerol, 0.5 m*M* PMSF, 1 m*M* DTT.
7. Buffer D: 20 m*M* HEPES-KOH, pH 7.9, 100 m*M* KCl, 0.2 m*M* EDTA, 20% (v/v) glycerol, 0.5 m*M* PMSF, 1 m*M* DTT.
8. Buffer E: 20 m*M* HEPES KOH, pH 7.9, 100 m*M* KCl, 0.2 m*M* EDTA, 10% (v/v) glycerol, 1 m*M* DTT.
9. 40-mL dounce glass homogenizer with loose-fitting pestle (Kontes, Vineland, NJ, 0.003–0.006 inch).
10. Dialysis tubes with 8000 nominal molecular weight cutoff (Spectrum, Houston, TX).
11. Ultrasonic processor model VC130PB (Sonics and Materials, Newtown, PA).

2.4. In Vitro Splicing Reaction

All reagents should be prepared with high-quality water (e.g., Milli-Q or equivalent).

1. m^7GpppG (New England Biolabs, Ipswich, MA, cat. no. S1404).
2. 10X SP mixture: 5 m*M* adenosine 5′-triphosphate (ATP), 200 m*M* creatine phosphate, 16 m*M* $MgCl_2$. Make small aliquots (0.2 mL) in microcentifuge tubes and store at –20 °C from the following stocks:

   a. 100 m*M* ATP, pH 7.5 (GE Healthcare, Bucks, U.K., cat. no. 27-2056-01).
   b. Creatine phosphate, disodium salt (Calbiochem, San Diego, CA, cat. no. 2380).
   c. 1 *M* $MgCl_2$ (Ambion, Austin, TX, cat. no. AM9530G).

3. 2X proteinase K (PK) buffer: 0.2 m*M* Tris-HCl, pH 7.5, 25 m*M* ethylenediamine-tetraacetic acid (EDTA), 0.3 *M* NaCl, 2% sodium dodecyl sulfate (SDS). Do not autoclave. Sterilize by filtration through a 0.2-μm filter and store at room temperature.
4. 20 mg/mL PK solution: Dissolve proteinase K in water. Sterilization is not required if high-quality water is used. Aliquot in 0.2-mL microcentrifuge tubes and store at –20 °C.

3. Methods

3.1. HeLa Cell Suspension Culture

HeLa cells are grown in spinner flasks at 37 °C. The cell density needs to be maintained between 1 and 5 × 10^5 cells/mL (*see* **Note 1**). The cell density can be monitored using a hematocytometer. It is better to avoid direct contact of the bottom of the flask to the stirring plate so the heat from the magnetic stirrer is not transferred to the suspension culture.

3.2. HEK293T Cell Culture

HEK293T cells are grown on 10-cm plates for cell culture (Falcon) at 37 °C up to 80–100% confluency. The cells should not be overgrown.

3.3. Preparation of Nuclear Extract From HeLa Cells

1. Harvest HeLa cells at the logarithmic growth stage (4–6 × 10^5 cells/mL) by low-speed centrifugation at approx 700 *g*, 4 °C (e.g., Beckman JA-10 rotor, 2000 rpm, 10 min). Calculate the total cell number by checking the cell density on a hematocytometer.
2. Resuspend the cells with ice-cold PBS by gently pipeting, centrifuge again, and measure the packed cell volume (PCV). All the subsequent steps should be done on ice, and solutions used in each step must be ice cold.
3. Gently resuspend the cell pellet in 5X PCVs of buffer A and keep on ice for 10 min to swell the cells before disruption.
4. Pellet the cells by low-speed centrifugation at approx 500 *g* (e.g., Beckman JA-25.50 rotor, 2000 rpm) for 10 min at 4 °C and resuspend the pellet in 2X PCVs of buffer A.
5. Homogenize cells in a dounce homogenizer with 10–15 strokes. The number of strokes may vary depending on the homogenizer. The cell lysis can be checked on a hematocytometer (*see* **Note 2**).
6. Transfer the lysate to tubes and centrifuge at approx 1900 *g* (e.g., Beckman JA25.50 rotor, 4000 rpm) for 10 min at 4 °C to precipitate nuclei. Save the supernatant for further preparation of the cytoplasmic S100 extract (*see* **Subheading 3.4.**).
7. Centrifuge the nuclei pellet again in the same tubes at approx 39,000 *g* (e.g., Beckman JA25.50 rotor, 18,000 rpm) for 20 min at 4 °C to pack the pellet tightly. Remove the supernatant carefully and discard it.

8. Transfer the pellet into a clean dounce homogenizer without adding buffer C. Add buffer C to the tubes to recover the remainder of the pellet. The volume of buffer C to add is determined as 1.4 mL/10^9 cells.
9. Suspend the pellet by several strokes with a loose pestle (*see* **Note 3**). Normally, 10–15 strokes are sufficient to obtain a homogeneous suspension.
10. Transfer the suspension into a screw-capped centrifuge tube and mix gently by rocking 30 min at 4 °C.
11. Pellet the nuclei by high-speed centrifugation at approx 39,000 *g* (e.g., Beckman JA25.50 rotor, 18,000 rpm) for 30 min and carefully save the supernatant into a clean disposable tube.
12. Dialyze the supernatant against buffer D (usually 0.5–2 L) 5–8 h at 4 °C.
13. After dialysis, the precipitate is always formed. Remove the precipitate by high-speed centrifugation at approx 33,000 *g* (e.g., Beckman JA25.50 rotor, 16,500 rpm) for 20 min and transfer the supernatant carefully to a clean disposable tube. This is the nuclear extract.
14. Make aliquots of the extract into microcentrifuge tubes, in 0.1- to 1-mL portions, as desired. Freeze them quickly in liquid nitrogen and store at –80 °C or lower. The extract remains active for at least 1 yr if stored appropriately.

3.4. Preparation of HeLa Cell Cytosolic S100 Extract

HeLa cell cytosolic S100 extract can be obtained from the supernatant of the centrifugation in **Subheading 3.3.6.** S100 extract contains cytosolic components as well as most, but not all, nuclear components.

1. Measure the volume of supernatant derived from **Subheading 3.3.6.** Add 0.11 volume of buffer B and mix gently.
2. Centrifuge the mixture at approx 100,000 *g* (e.g., 38,000 rpm, Beckman 60Ti rotor) for 1 h at 4 °C.
3. Save the supernatant in a disposable tube and dialyze it against 2 L of buffer D twice: once overnight and once at least 4 h in the cold room.
4. Remove precipitated insoluble materials by centrifugation at high speed at approx 33,000 *g* (e.g., 16,500 rpm, Beckman JA25.50 rotor) for 20 min at 4 °C.
5. Aliquot the extract into 0.1- to 1-mL portions in microcentrifuge tubes as desired. Quickly freeze them in liquid nitrogen and store at –80 °C or lower. The extract is active for at least 1 yr if stored properly (*see* **Note 4**).

3.5. Preparation of HEK293T Whole-Cell Extract

1. The cells were washed three times with 10 mL of ice-cold PBS each time on the plates. Since HEK293T cells can be easily detached from the plates, add PBS gently to cells.
2. The cells are detached from the plates by pipeting and resuspended in 10 mL PBS (*see* **Note 5**). Harvest cells by low-speed centrifugation at approx 500 *g* (e.g., Beckman JA25.50 rotor, 2000 rpm) for 5 min at 4 °C and remove PBS completely.

3. The resulting pellet is resuspended in ice-cold buffer E. Typically, 200 μL of buffer E is used per 10-cm plate.
4. Disrupt cells by sonication with an ultrasonic processor model VC130PB (Sonics and Materials) or equivalent. Sonication is carried out at 30% continuous for 5 s three times with a 30-s incubation on ice between bursts (*see* **Note 6**).
5. Transfer the sonicated lysate into a microcentrifuge tube and centrifuge at approx 15,000 *g* (e.g., TOMY, Tokyo, Japan, MX-300, 13,000 rpm), 20 min at 4 °C. Carefully save the supernatant, which is the whole-cell extract, into a microfuge tube placed on ice.
6. Aliquot the extract in 20- to 50-μL portions into microcentrifuge tubes as desired. Freeze quickly by dropping the tubes into liquid nitrogen and store at −80 °C or lower. The extract remains active for at least 1 yr if stored at −80 °C.

3.6. In Vitro Splicing Assays Using Nuclear Extracts and Whole-Cell Extracts

The in vitro splicing reaction is typically carried out with nuclear extracts, S100 extracts supplemented with SR proteins, or whole-cell extracts. The preparation procedures for those extracts are described in the previous subheadings. The pre-mRNA substrate for in vitro splicing is usually prepared by in vitro runoff transcription with a bacteriophage RNA polymerase by using a linearized plasmid. The pre-mRNAs are capped by priming the transcription with a dinucleotide primer m^7GpppG and labeled internally with ^{32}P-UTP (uridine 5′-triphosphate). After the in vitro splicing reaction, RNAs are purified and analyzed by denaturing polyacrylamide gel electrophoresis (PAGE) and autoradiography.

1. Prepare in vitro splicing mixture by mixing pre-mRNA and reagents. A recipe for a typical splicing reaction of 20 μL is as follows (*see* **Note 7**):
   a. 1 μL pre-mRNA (20 fmol).
   b. 2 μL 10X SP mixture.
   c. 12 μL cell extracts.
   d. 5 μL Milli-Q water.
2. Incubate at 30 °C for 30 min to 4 h. The kinetics and efficiency of splicing depends on which pre-mRNA is used for the reaction.
3. Stop the splicing reaction by adding 100 μL of 2X PK buffer and 80 μL of Milli-Q water.
4. Add 4 μL of 20 mg/mL PK solution and incubate at 37 °C for 30 min.
5. Extract with Tris buffer-saturated phenol and precipitate RNA with ethanol.
6. Analyze recovered RNAs on denaturing PAGE.

RNA recovered from in vitro splicing reactions using either HeLa cell nuclear extract or HEK293T whole-cell extract is shown in **Fig. 1**.

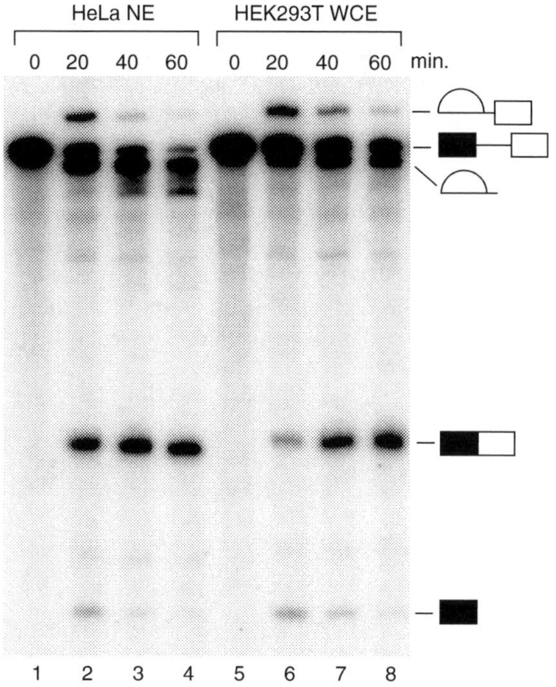

Fig. 1. In vitro splicing reactions in HeLa cell nuclear extracts and HEK293T whole-cell extracts. ^{32}P-labeled chicken δ-crystallin pre-mRNA (messenger RNA) ***(20)*** was incubated with either HeLa cell nuclear extracts (lanes 1–4) or HEK293T whole-cell extracts under splicing conditions. Recovered RNAs were re-solved by electrophoresis on 6% denaturing polyacrylamide gel. The identity of the RNA is shown at the right side of the panel. Boxes and lines represent exons and introns, respectively.

4. Notes

1. Using healthy, growing cells under optimal growth conditions is essential. Slowly growing or overgrown cells should be avoided as they produce much less active extracts.
2. The disruption of cells is a critical process for extract preparation. Homogenization with a loose pestle can disrupt cells but keeps the nuclei intact, which prevents excess leakage of nuclear splicing factors. Alternatively, passing through a 25-gauge needle four to six times can be used for cell disruption.
3. This homogenization is not intended to destroy nuclei but to mix buffer rapidly with nuclei to prevent transient exposure to high-salt concentration. The salt concentration should be between 0.20 *M* and 0.25 *M*. Measuring the conductivity can be used to check the final salt concentration.
4. The activity of S100 cytoplasmic extracts is not so reproducible as that of nuclear extracts. Some batches may fail to be complemented by SR proteins, while some

may have a splicing activity even without addition of SR proteins. It is better to prepare and check more than one batch of S100 extract.

5. For other cells that attach well to tissue culture plates, like HeLa cells, use trypsin-EDTA (1X, Gibco-BRL, cat. no. 25200-056) to detach cells from the plates. Briefly, add 0.5 mL trypsin-EDTA with 2 mL PBS to the cells on a 10-cm plate after PBS wash. Incubate cells at 37 °C for 3 min and add 7.5 mL of culture medium to stop the trypsinization. Resuspend the cells and transfer to a clean disposable tube to harvest cells.
6. The setting for sonication may vary between different sonicators. It is necessary to determine the optimal settings for cell disruption for each type of sonicator. Avoid foaming the cell suspension during sonication since it causes inactivation of the extract.
7. The in vitro splicing condition described here is applicable for many kinds of pre-mRNAs ***(20)***. However, optimal concentrations of magnesium and salt vary with different pre-mRNAs, which shall be determined empirically.

References

1. Reed, R. (2000) Mechanisms of fidelity in pre-mRNA splicing. *Curr. Opin. Cell Biol.* **12**, 340–345.
2. Hastings, M. L., and Krainer, A. R. (2001) Pre-mRNA splicing in the new millennium. *Curr. Opin. Cell Biol.* **13**, 302–309.
3. Jurica, M. S., and Moore, M. J. (2003) Pre-mRNA splicing: awash in a sea of proteins. *Mol. Cell.* **12**, 5–14.
4. Rio, D. C. (1988) Accurate and efficient pre-mRNA splicing in *Drosophila* cell-free extracts. *Proc. Natl. Acad. Sci. U. S. A.* **85**, 2904–2908.
5. Lin, R. J., Newman, A. J., Cheng, S. C., and Abelson, J. (1985) Yeast mRNA splicing in vitro. *J. Biol. Chem.* **260**, 14780–14792.
6. Dignam, J. D., Lebovitz, R. M., and Roeder, R. G. (1983) Accurate transcription initiation by RNA polymerase II in a soluble extract from isolated mammalian nuclei. *Nucleic Acids Res.* **11**, 1475–1489.
7. Kataoka, N., and Dreyfuss, G. (2004) A simple whole-cell lysate system for in vitro splicing reveals a stepwise-assembly of the exon–exon junction complex. *J. Biol. Chem.* **279**, 7009–7013.
8. Krainer, A. R., Conway, G. C., and Kozak, D. (1990) The essential pre-mRNA splicing factor SF2 influences 5′ splice site selection by activating proximal sites. *Cell* **62**, 35–42.
9. Fu, X. D. (1995) The superfamily of arginine/serine-rich splicing factors. *RNA* **1**, 663–680.
10. Graveley, B. R. (2000) Sorting out the complexity of SR protein functions. *RNA* **6**, 1197–1211.
11. Mayeda, A., and Krainer, A. R. (1992) Regulation of alternative pre-mRNA splicing by hnRNP A1 and splicing factor SF2. *Cell* **68**, 365–375.
12. Mayeda, A., Screaton, G. R., Chandler, S. D., Fu, X. D., and Krainer, A. R. (1999) Substrate specificities of SR proteins in constitutive splicing are determined by

their RNA recognition motifs and composite pre-mRNA exonic elements. *Mol. Cell. Biol.* **19**, 1853–1863.

13. Tarn, W. Y., and Steitz, J. A. (1996) A novel spliceosome containing U11, U12, and U5 snRNPs excises a minor class (AT-AC) intron in vitro. *Cell* **84**, 801–811.
14. Wu, Q., and Krainer, A. R. (1996) U1-mediated exon definition interactions between AT-AC and GT-AG introns. *Science* **274**, 1005–1008.
15. Pellizzoni, L., Kataoka, N., Charroux, B., and Dreyfuss, G. (1998) A novel function for SMN, the spinal muscular atrophy disease gene product, in pre-mRNA splicing. *Cell* **95**, 615–624.
16. Kataoka, N., Yong, J., Kim, V. N., et al. (2000) Pre-mRNA splicing imprints mRNA in the nucleus with a novel RNA-binding protein that persists in the cytoplasm. *Mol. Cell.* **6**, 673–682.
17. Kataoka, N., Diem, M. D., Kim, V. N., Yong, J., and Dreyfuss, G. (2001) Magoh, a human homolog of *Drosophila mago nashi* protein, is a component of splicing-dependent exon-exon junction complex. *EMBO J.* **20**, 6424–6433.
18. Kim, V. N., Kataoka, N., and Dreyfuss, G. (2001) Role of the nonsense-mediated decay factor hUpf3 in the splicing-dependent exon-exon junction complex. *Science* **293**, 1832–1836.
19. Chan, C. C., Dostie, J., Diem, M. D., et al. (2004) eIF4A3 is a novel component of the exon junction complex. *RNA* **10**, 200–209.
20. Sakamoto, H., Ohno, M., Yasuda, K., Mizumoto, K., and Shimura, Y. (1987) In vitro splicing of a chicken delta-crystallin pre-mRNA in a mammalian nuclear extract. *J. Biochem. (Tokyo)* **102**, 1289–1301.

24

Designing and Utilization of siRNAs Targeting RNA Binding Proteins

Dong-Ho Kim, Mark Behlke, and John J. Rossi

Summary

Small interfering RNA (siRNA)-mediated RNA interference (RNAi) is a very powerful tool for triggering posttranscriptional gene silencing in several organisms. We discuss the improvement of two different sources of siRNAs synthesized either chemically or by an enzymatic method. When the siRNAs are synthesized by in vitro transcription using a phage polymerase, the initiating triphosphates trigger a potent interferon induction that can lead to misinterpretation of the data. A novel method is presented to minimize the nonspecific effect of enzymatic siRNAs while maintaining the advantages of lower cost and less turnaround time. When chemical siRNAs are used, the expense and long turnaround time can be a problem, especially if the selected siRNAs are not highly functional in triggering RNAi. The new format for making double-stranded RNAs (dsRNAs) is described to achieve more efficient suppression. The format has been tested by creating siRNAs targeting two RNA binding proteins, La and hnRNP (heterogeneous nuclear ribonucleoprotein) H, and has shown better potency at lower concentrations than the conventional 21-mer siRNA.

Key Words: Dicer; Dicer substrates; in vitro transcription; RNA; RNA interference; siRNA; triphosphate.

1. Introduction

RNA interference (RNAi) is a powerful mechanism for eliciting posttranscriptional gene silencing in mammalian cells *(1)*. Small interfering RNA (siRNA) duplexes 21 to 25 nt in length trigger RNAi, resulting in posttranscriptional RNA degradation *(2)*. The siRNAs are potent reagents for directed posttranscriptional gene silencing and a major new genetic tool for mammalian cells. Since longer double-stranded RNA (dsRNA) may elicit nonspecific interferon responses mediated by PKR activation, successful posttranscriptional suppression of gene expression utilizing siRNAs 19–21 nt long with 2-nt

From: *Methods in Molecular Biology, vol. 488: RNA-Protein Interaction Protocols*
Edited by: Ren-Jang Lin © Humana Press Inc, Totowa, NJ

3′overhangs *(2)* has become increasingly popular. In evaluating siRNA targets, there is often profound variation of RNAi efficacy depending on the siRNA and the targeted region of the RNA in question *(3)*. Because of this, the total cost of this approach can be significant. Although there are several algorithms for siRNA design *(4,5)*, there is so far no absolute way to reduce the cost of testing multiple siRNAs against refractory targets. We describe approaches for siRNA synthesis that are more cost-effective as well as a new format for siRNA targeting that greatly improves efficacy.

Currently, three different methods to generate siRNAs are available. The most cost-effective and convenient is the phage polymerase-mediated in vitro transcription (IVT) using published methods *(6,7)* or commercially available kits (Silencer siRNA construction kit, Ambion, Austin, TX). Each strand of RNA can be readily transcribed in vitro, hybridized to form siRNAs, and purified prior to use. The DNA templates for IVT are prepared using DNA oligos, including the phage promoter. The complement to the siRNA sequence is on one oligo that also contains 8–10 bases of complementary sequence to the bacteriophage promoter. A duplex is made using Klenow DNA polymerase and deoxynucleotide 5′-triphosphates (dNTPs). As described in detail next, the siRNAs synthesized by this method can induce interferon-α and -β *(8)* if not properly designed and treated posttranscriptionally.

Another approach is to in vitro transcribe longer dsRNAs that can be processed in vitro into 21- to 23-mer duplexes by the ribonuclease (RNase) III family member Dicer *(9)*. The siRNAs produced from IVT and sequential dicing reactions can also be used for RNAi-mediated target degradation. Although this approach is cost-effective and relatively facile, there are some potential problems with it. First, long dsRNA that may contaminate the preps due to incomplete dicing can activate cellular PKR and lead to cellular toxicity *(10)*. Since the kinetics of Dicer are slow *(9)*, the percentage of unprocessed RNA can be high, thereby making it necessary to purify the 21- to 23-mer siRNAs to eliminate the longer uncut precursors. Second, siRNA-mediated off-target effects *(11)* are more of a potential problem when a mixture of siRNAs of unknown sequence are applied to cells. However, this possibility needs to be experimentally confirmed.

The third method for obtaining siRNAs is chemical synthesis. This can be expensive and may have a longer turnaround time, but the quantities and purity of the chemically synthesized siRNAs usually far exceed anything that can be obtained via IVT.

2. Materials

1. HEK293 (human embryonic kidney 293) and HeLa cells with the medium (Dulbecco's modified Eagle's medium [DMEM] containing 10% serum, 2 m*M* L-glutamine, and a penicillin-streptomycin mixture).

2. Opti medium (Gibco) and lipofectamine 2000 (Invitrogen).
3. Vector expressing green fluorescent protein (GFP) and red fluorescent protein (RFP) for the cotransfection assay.
4. Fluorimeter employing two filters (D490 for excitation, and D520 for emission).
5. For enzymatic synthesis of siRNA without interferon induction, the reagents described in **Subheading 3.2.2.** are required. Alternatively, the Silencer siRNA construction kit (Ambion) may be used. In this case, the supplied T7 siRNA should be replaced by the new oligo (5′-TAATAGACTCATATA-3′). To remove the interferon inducing 5′ppp, dephosphorylation reaction use of CIP (Promega) is essential.
6. RNA STAT-60 (Tel-Test B, Friendswood, TX) for total RNA preparation for Northern blot.
7. ^{32}P end-labeled oligos (La, 5′-CCAAAGGTACCCAGCCTTCATCCAGTT-3′; β-actin, 5′-GTGAGGATGCCTCTCTTGCTCTGGGCCTCG-3′) for detection of RNA level of La and β-actin by a Northern blot. For Northern blot, hybridization solution 1 mL 50X Denhardt's solution (5 g Ficoll,5 g polyvinylpyrrolidone, 5 g bovine serum albumin [BSA] in 500 mL H_2O), 3 mL 20X SSPE (174 g NaCl, 27.6 g NaH_2PO_4, 7.4 g ethylenediaminetetraacetic acid [EDTA] in 800 mL H_2O, pH 7.4), and 0.5 mL 10% sodium dodecyl sulfate (SDS) are required.
8. Anti-hnRNP (heterogeneous nuclear ribonucleoprotein) H rabbit antibody and antirabbit antibody conjugated with alkaline phosphatase (Sigma).
9. An antiactin antibody (Sigma) and antimouse antibody conjugated with alkaline phosphatase (Sigma).
10. Klenow buffer: 1X Klenow buffer (50 m*M* NaCl, 10 m*M* Tris-HCl, pH.7.5).
11. Klenow enzyme (New England Biolabs).
12. 1X RNA polymerase reaction buffer: 40 m*M* Tris-HCl, pH. 7.9, 6 m*M* $MgCl_2$, 2 m*M* spermidine, 10 m*M* dithiothreitol (DTT), and 0.5 m*M* of each nucleotide 5′-triphosphate (NTP).
13. 10X T1/DNase (deoxyribonuclease) I reaction buffer: 100 m*M* NaCl, 50 m*M* Tris-HCl, pH 7.9, 10 m*M* $MgCl_2$, and 1 m*M* DTT.

3. Methods

3.1. Transfection Assays for RNAi

Efficient delivery of siRNAs is essential for studying the downregulation of a targeted transcript. Transfection efficiencies can vary depending on the cells used. As a model system, we utilize HEK293 or HeLa cells. Typically, up to 80% transfection efficiencies can be obtained using the following protocol. For transfection assays, a reporter vector and an siRNA directed against a sequence in the target are cotransfected with a third vector containing an irrelevant reporter gene as a transfection control.

1. HEK293 or HeLa cells are plated at 40% confluency in an appropriate size tissue culture plate such as 6 or 24 wells in DMEM medium containing 10% serum,

2 m*M* L-glutamine, and a penicillin-streptomycin mixture. The assay described here is for 24-well plates; for 6-well plates, every reagent should be scaled up fivefold.

2. After 16- to 24-h incubation, transfections are carried out in Eppendorf tubes using the designated amounts of siRNAs. The total amount of siRNAs should not exceed 30 n*M* to minimize the nonspecific effects.
3. In a separate tube, the reporter vector and transfection control vector are mixed with Opti-MEM I medium. For an enhanced GFP (EGFP) target using RNAi in a 24-well plate, 100–200 ng of EGFP plasmid and 20 ng of an RFP control vector are premixed with 50 μL of Opti medium and aliquoted to new sterile plastic tubes.
4. 50 μL of Opti are mixed with 1.5 μL of Lipofectamine 2000 and dispensed into each tube. We have observed concentration-dependent cytotoxic effects when greater quantities of Lipofectamine 2000 are used. The mixture should be incubated for 15 min at room temperature to form a complex of the nucleic acid and transfection reagent.
5. Old media are removed and replaced with 0.4 mL (for 24-well plates) or 2 mL (for 6-well plates) of fresh DMEM. The nucleic acid/lipofectin complexed mixtures can be added to the media directly.
6. RNAi can be assayed either in 24 h (for experiments with a cotransfected reporter system such as EGFP) or 24–48 h for endogenous targets. As a transfection control, RFP or EGFP reporter assays can be carried out using a Fluorimeter employing two filters (D490 for excitation and D520 for emission). Extract measurements of EGFP are carried out as follows: 1×10^5 cells are suspended in 300 μL of PBS and sonicated for 10 s followed by a 2-min microcentrifugation. The supernatants are used for fluorescence measurements. Percentages of EGFP expression are determined relative to extracts prepared from untreated cells.

3.2. Enzymatic synthesis of siRNAs

The enzymatic synthesis of siRNAs from synthetic DNA templates is perhaps the most cost-effective method for siRNA production. Unfortunately, the in vitro-transcribed siRNAs can be potent inducers of interferons *(8)*. We have identified the initiating 5′ppp of in vitro-transcribed siRNAs as the inducer (**Fig. 1**). siRNAs synthesized via three different enzymatic methods all were inducers of interferon *(8)*. To circumvent this problem, we have developed a modified protocol for producing siRNAs via IVT.

3.2.1. Effective Elimination of Initiating 5′ ppp

When the initiating triphosphate is present on the in vitro-transcribed RNAs, the level of interferon induction is proportional to the amounts of 5′ pppG (*see* below). To make homogeneous transcripts, we utilize the T7 promoter oligo (5′-TAATAGACTCATATA-3′) and unique template oligos containing the complement to either the sense or antisense strand along with a common sequence at the 3′ end (5′-AA—N(19–21)—CCCTATAGTGAGTCGTA-3′). Since siRNAs have

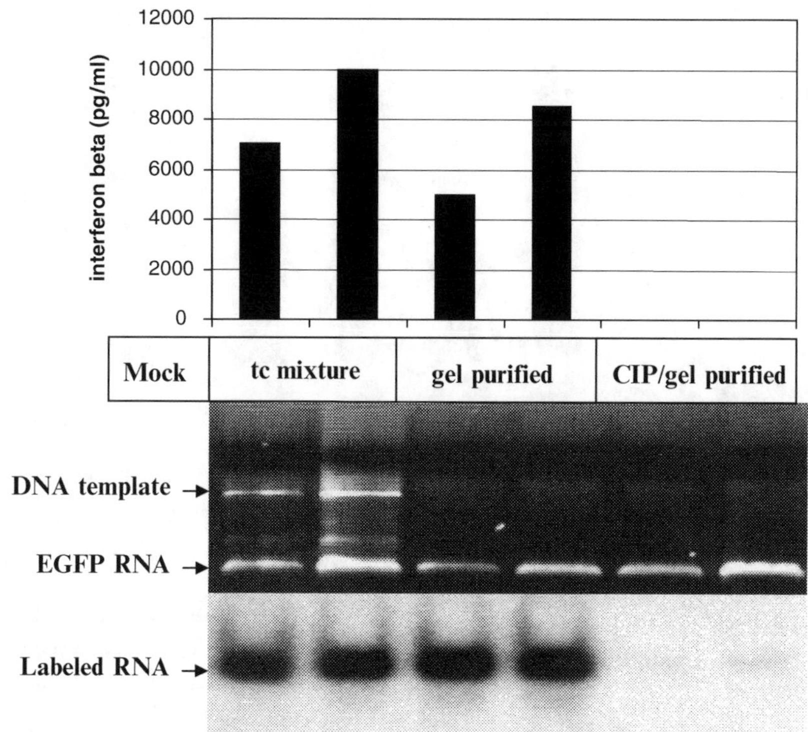

Fig. 1. The 5′ triphosphates on transcribed single-stranded (ssRNA) are sufficient for induction of interferon-β. The enhanced green fluorescent protein (EGFP) RNA was transcribed in the presence of γ-^{32}P-GTP (guanosine 5′-triphosphate) and transfected into cells without further modification (columns 2 and 3), after gel purification (columns 4 and 5), and following gel purification and CIP treatment (columns 6 and 7). The induced levels of interferon-β were determined by an enzyme-linked immunosorbent assay (ELISA; top panel). Transcribed RNAs used for transfection reactions were analyzed in a nondenaturing agarose gel (middle panel). Removal of the triphosphate by CIP was monitored in the bottom gel. The ELISA determinations represent the average of two independent experiments.

two-base 3′ overhangs, we synthesized primers containing the following classes of two-base 3′ overhang (19-UU, 19-AA, and 21-AA) as diagrammed in **Fig. 2**. The standard treatment protocol for in vitro-transcribed siRNAs is to anneal the upper and lower strands, followed by treatment with the single-stranded endonuclease T1, which is used to digest the initiating pppGGG from the siRNAs. When the 19-UU siRNA strands are hybridized, the 3′ Us can potentially form wobble basepairs with the initiating pppGGG. This wobble pairing could protect the pppGGG from digestion by T1 (**Fig. 3**, lane 2). Thus, when one is making siRNAs using the Tuschl

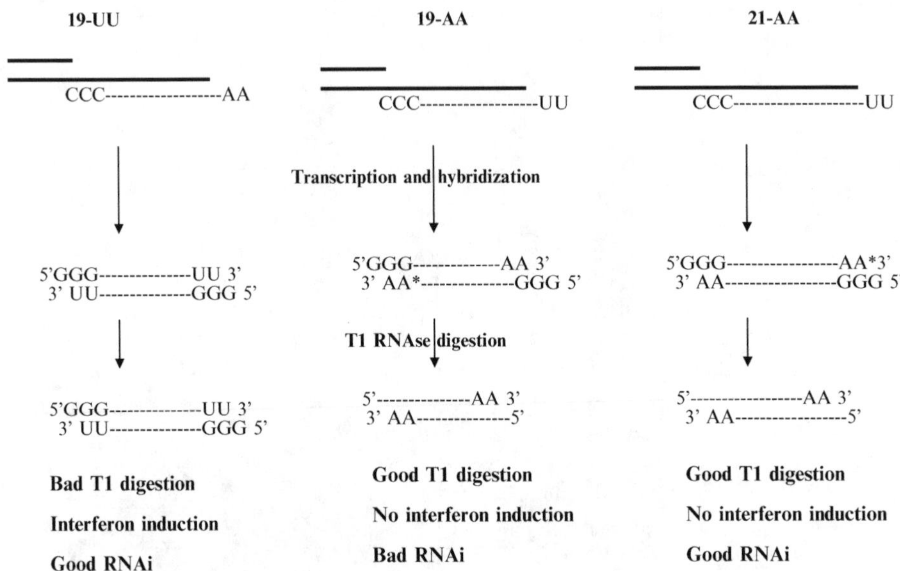

Fig. 2. Diagrams of three formats of enzymatic small interfering RNAs (siRNAs) in these studies. The antisense RNA of 19-UU has all 21 nt complementary to the target site of enhanced green fluorescent protein (EGFP). The antisense of 19-AA has 19 nt complementary to target and 2 nt of mismatched AA residues at the 3′ overhang. 21-AA is to have 21 nt complementary to the target and maintain two AA 3′ overhangs. RNAi, RNA interference; RNase, ribonuclease.

rules, which stipulate identifying an AA in the target followed by 17 or 19 bases for the siRNA, the 3′ UU are necessary components at the 3′ end of the antisense strand of the siRNA.

To test whether the wobble pairing of the Gs with UU was the problem, siRNAs with 3′AA (19-AA) were in vitro transcribed. When the transcripts contained a 3′ AA, the 5′ pppGGG was more susceptible to ribonuclease T1 cleavage, as expected (**Fig. 3**, lanes 4 to 6) since the 5′ Gs do not form basepairs with the adenosines. When a 19-mer siRNA with 3′ AA residues was tested in a cotransfection assay (19-AA), RNAi efficacy was poor (**Fig. 4**). Therefore, a new siRNA with 21 nt of sequence complementary to the target and the 3′ AAs was designed and tested (21-AA) (**Fig. 2**). This siRNA was processed by T1 and did not induce interferons (**Table 1**) and had good RNAi efficacy (**Fig. 4**).

3.2.2. Enzymatic Synthesis of siRNAs Without Interferon Induction

Synthesis can be carried out using the two described primers and either using the Silencing siRNA kit or serial enzymatic reactions. If the kit is used, two modifications should be made. First, the two unique primers should be used

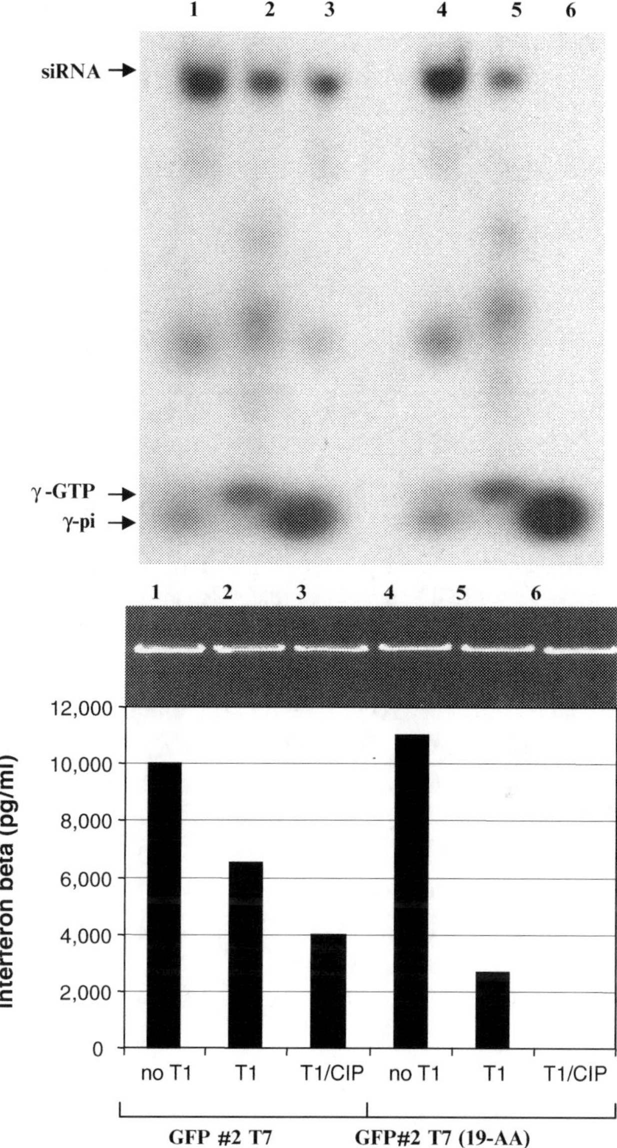

Fig. 3. Triphosphate-mediated interferon induction. γ-GTP (guanosine 5′-triphosphate)-labeled small interfering RNAs (siRNAs) were treated using each of the conditions described here and electrophoresed in a native gel (top panel). RNAs were electrophoresed in a 15% polyacrylamide gel and stained with ethidium bromide (middle panel). Of the siRNAs, 20 n*M* were transfected into human embryonic kidney 293 (HEK293) cells and assayed for interferon-β (bottom panel). Column 1, the enhanced green fluorescent protein (EGFP) 2 (19+UU) T7 siRNA without T1 endonuclease treatment; column 2, with T1 endonuclease treatment; column 3, with T1 endonuclease and CIP treatment. Column 4, EGFP 2 T7 (19+AA) siRNA without T1 endonuclease treatment; column 5, with T1 endonuclease digestion; column 6, with T1 endonuclease and CIP treatment.

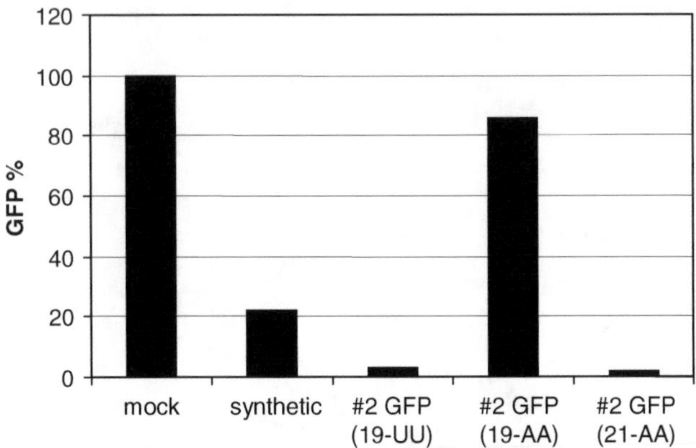

Fig. 4. T7 small interfering RNAs (siRNAs) with N 21-AA are effective in RNA interference (RNAi). The human embryonic kidney 293 (HEK293) cells were cotransfected with the enhanced green fluorescent protein (EGFP) reporter plasmid and each of the siRNAs. The percentages of EGFP expression relative to the non-siRNA-treated controls were used as the assay for RNAi. Each value is the average of two independent assays.

(T7 primer; 5′-TAATAGACTCATATA-3′, and a unique template primer for the 21-AA RNA; 5′ TT-N19) - CCCTATAGTGAGTCGTA-3′. Second, additional alkaline phosphatase (CIP) treatment steps should be included after the ribonuclease T1 and DNase I treatments. The CIP reaction can be carried out without changing the buffer. The following lists step-by-step procedures for the enzymatic synthesis of siRNAs.

1. Mix 2 μL of 100 n*M* of T7 oligo and each template oligo in a final volume of 10 μL in 1X Klenow buffer.
2. The mixture is heated to 70 °C for 5 min and then left at room temperature for 10 min.

Table 1.
Induction of Interferon by T7 Small Interfering RNAs (siRNAs)

siRNA (10 n*M*)	Amount of interferon-α (pg/mL)	Amount of interferon-β (pg/mL)
Mock	0.2 ± 0.3	3 ± 2
Synthetic siRNA	2 ± 0.5	5 ± 5
T7 siRNA 1 (anti-La 2)	300 ± 85	4,000 ± 300
T7 siRNA 2 (anti-Ro 1)	250 ± 35	3,500 ± 300
T7 siRNA 1[a] (21-AA/CIP)	25 ± 5	50 ± 25
T7 siRNA 2[a] (21-AA/CIP)	30 ± 10	120 ± 30

[a]New siRNAs synthesized in 21-AA format and CIP treated to remove the 5′ ppp.

3. Add the hybridized oligonucleotides to 20 μL 1X Klenow buffer containing 50 μ*M* of dNTP and 5 U of Klenow fragment and incubate for 30 min.
4. Take 2 μL of the reaction mixture and perform IVT reactions in 20 μL of 1X RNA polymerase reaction buffer containing 50 U of T7 RNA polymerase. Incubate the reaction mixture for 1 h at 37 °C.
5. Combine the two transcription mixtures (one sense strand and one antisense strand) and heat to 70 °C for 5 min, then leave at room temperature for 30 min.
6. Mix the reaction with 8 μL of 10X T1/DNase I reaction buffer with 10 U of RNase T1 and 10 U of DNase I in 100 μL final volume and incubate for 2 h at 37 °C.
7. Add 10 U of alkaline phosphatase (CIP) and incubate for an additional 1 h without changing the buffer.
8. siRNAs are purified by using Microspin G-25 columns from Amersham (Piscataway, NJ) according to the manufacturer's protocol.

3.2.3. Enzyme-Linked Immunosorbent Assays for Interferon Induction

Interferon induction can be tested using commercially available interferon-α or -β enzyme-linked immunosorbent assay (ELISA) kits. When HEK293 or HeLa cells were used for these assays, the media was taken from the transfected cells after 24 h and further tested by the kit as described by the manufacturer's protocol.

3.3. Chemically Synthesized Dicer-Substrate RNAs

In attempts to identify RNAi triggers that effectively function at lower concentrations, it was found that synthetic RNA duplexes 25 to 30 nt in length are more potent than corresponding conventional 21-mer siRNAs ***(12)***. Some sites that are refractory to silencing by 21-mer siRNAs can be effectively targeted by 27-mer blunt-ended duplexes, and silencing lasts up to 10 d. Importantly, the 27-mers do not induce interferon or activate PKR. The enhanced potency of the longer duplexes is attributed to the fact that they are Dicer substrates, directly linking the production of siRNAs to incorporation in the RNA-induced silencing complex (RISC). This group of RNAs is called *disRNAs*, or *Dicer-substrate RNAs* ***(12)***. The synthetic siRNAs can be purchased from any vendor that provides chemically synthesized RNAs. The RNAs are maintained in water treated with DEPC (diethylpyrocarbonate) and kept frozen at –20 °C until ready for use. Most vendors provide the RNAs already in duplex form.

3.3.1. Longer siRNAs Are Processed Into 21-mer siRNAs.

During investigation of cellular interferon induction caused by dsRNAs made with bacteriophage T7 IVT products ***(8)***, it was observed that some of the IVT dsRNAs appeared to have greater potency than synthetic 21-mer siRNAs directed to the same target site, and that this property seemed to correlate with length. To test that the longer disRNAs are processed into 21-mer siRNAs by

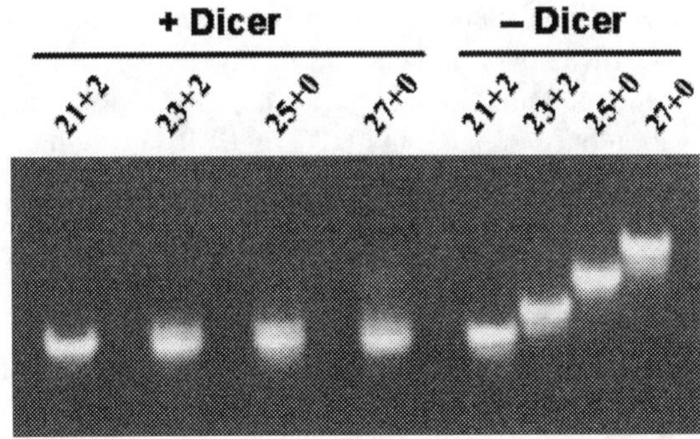

Fig. 5. Dicer cleavage of various double-stranded RNAs (dsRNAs). Each RNA was incubated in the presence or absence of recombinant Dicer for 24 h. The treated RNAs were electrophoresed in a 7.5% nondenaturing polyacrylamide gel and visualized by ethidium bromide staining. +, 3′ overhang; –, 5′ overhang.

Dicer, various lengths of duplexes were tested via in vitro dicing reactions and analyzed in nondenaturing gels. The 19 + 2 duplex was not cleaved, whereas 21 + 2, 23 + 2, and 27-mer blunt-ended duplexes were all substrates for Dicer cleavage (**Fig. 5**). For duplexes 25 bp and longer, RNAs with a 5′ overhang, 3′ overhang, or blunt ends were digested equally well (data not presented).

3.3.2. Longer siRNAs Have Better Efficacy

Since only dsRNAs with duplexes of 21 bp or longer were cleaved by Dicer, a series of dsRNAs of varying lengths was tested for RNAi. Synthetic RNA duplexes of varying lengths containing 3′ overhangs, 5′ overhangs, or blunt ends were tested for their relative potency in a model system targeting "site 2" in EGFP ***(13)***. HEK293 cells were cotransfected with the EGFP expression plasmid EGFP-C1 plus RNA duplexes at varying concentrations. Using duplex RNA at several concentrations, we observed that potency increased with length even with 5′ overhangs or a blunt-ended format (**Fig. 6**).

3.3.3. Designing of disRNA

Currently, the rule of thumb in designing disRNAs is to make asymmetric molecules in which the upper strand is 25 nt in length with two deoxynucleotides at the 3′ end and a bottom strand of 27 nt with a two-base 3′ overhang and a ribose 5′ end. We designate the bottom strand as the preguide strand since it will be selectively incorporated in the RISC. Interestingly, maximal inhibitory activity plateaued around a duplex length of 27, and no increase in potency was

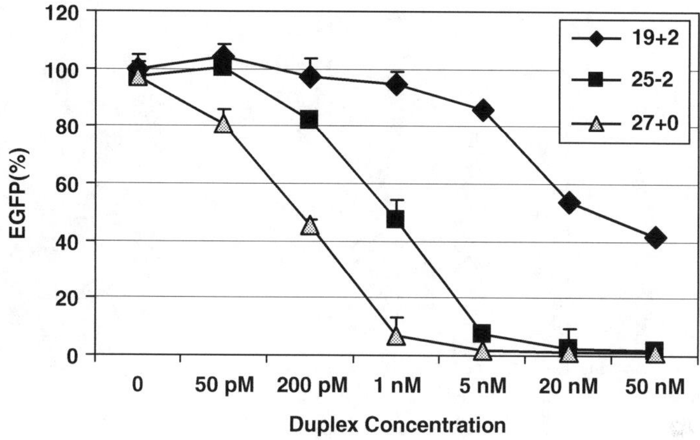

Fig. 6. Longer double-stranded RNAs (dsRNAs) are more potent effectors of RNA interference (RNAi) than a 21-mer small interfering RNA (siRNA). Dose-response curve of cotransfected dsRNAs and enhanced green fluorescent protein (EGFP) expression plasmid. Each graph or point represents the average (with standard deviation) of three independent measurements. +2, two-base 3′ overhang; −2, two-base 5′ overhang.

observed with 30- to 45-base duplexes (**Fig. 7**). Based on in vitro dicing reactions, we conclude that the low potency of longer dsRNAs is attributed to the fact that they are inefficiently processed by Dicer.

3.4. Downregulation of RNA Binding Proteins Using siRNAs

RNAi has been successfully adopted in several target suppression studies ***(1, 14)***. Since some RNA binding proteins, such as La (SS-B) and hnRNP H, are highly expressed, RNAi-mediated gene suppression can be challenging. Here, we tested the new format of disRNAs to downregulate transcripts from these two genes.

3.4.1. RNAi Suppression of La (SS-B)

One of the representative RNA binding proteins is La (SS-B). La was first found in humans as an autoantigen in the patients of lupus erythematosus and Sjogren's syndrome. The major form of La binds nascent RNA polymerase III transcripts that contain the UUU_{OH} at the 3′ end of all newly transcribed RNAs ***(15, 16)***. La is one of the most abundant proteins in human cells. RNAi-mediated suppression of La was tested using a 21-mer siRNA or 27-mer disRNA (**Fig. 8**). It can be seen that the disRNA was more potent than the 21-mer.

1. Transfect 1×10^6 HEK293 cells in 10-cm dishes as described in **Heading 2**. Mock or indicated amounts of siRNAs or disRNAs (**Table 2**) are transfected (**Fig. 8**).

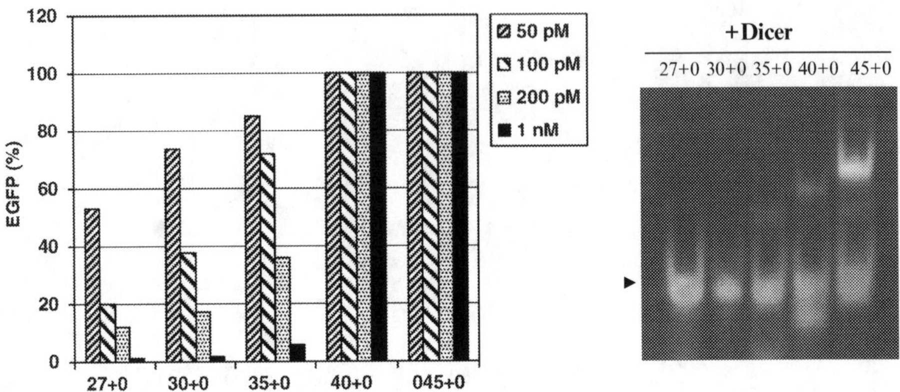

Fig. 7. Double-stranded RNAs (dsRNAs) longer than 27-mer show less efficacy than the 27-mer. Dose-response testing of longer dsRNAs. Transfections were performed with the indicated concentrations of dsRNA. Right panel depicts in vitro dicing reactions with the same longer RNAs. Arrowhead indicates the position of 21-mer siRNA. EGFP, enhanced green fluorescent protein.

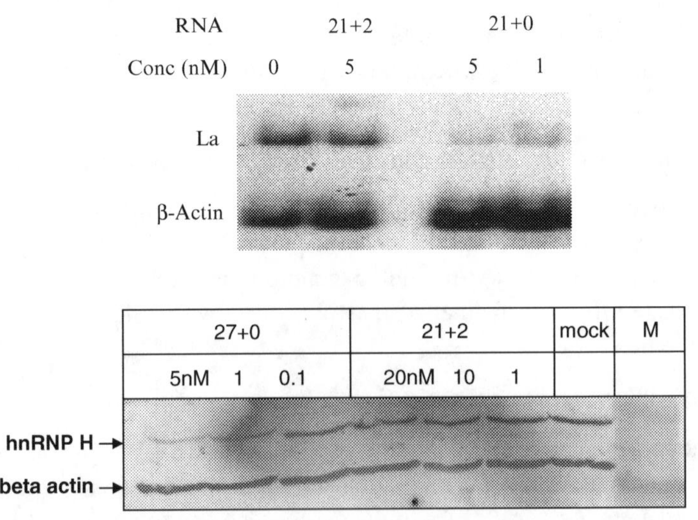

Fig. 8. Comparison of 21+2 small interfering RNA (siRNA) and 27-mer Dicer-substrate RNAs (disRNAs) in downregulation of RNA binding proteins. RNAs for a 21+2 siRNA and 27-mer disRNA were designed to target randomly selected sites in La mRNA (**A**) or heterogeneous nuclear ribonucleoprotein (hnRNP) H messenger RNA (**B**). HnRNP H knockdown was monitored by Western blot and La knockdown by Northern blot. The double-stranded RNAs (dsRNAs) were used at the indicated concentrations. β-Actin was used as an internal loading standard in both experiments.

Table 2.
List of Oligonucleotide Reagents

Sequence	Name
5′-GTTGAACTTGAATCAGAAGATGAAGTCAAATTGGC-3′	**hnRNP H site 1**
5′-CUUGAAUCAGAAGAUGAAGUC 3′-UUGAACUUAGUCUUCUACUUC	hnRNP H 21-mer
5′-AACUUGAAUCAGAAGAUGAAGUCAAAU 3′-UUGAACUUAGUCUUCUACUUCAGUUUA	hnRNP H 27+0
5′-ATAAAACTGGATGAAGGCTGGGTACCTTTGGAGAT-3′	**La site 1**
5′-CUGGAUGAAGGCUGGGUACUU 3′-UUGACCUACUUCCGACCCAUG	La 19+2
5′-AACUGGAUGAAGGCUGGGUACCUUUUU 3′-UUGACCUACUUCCGACCCAUGGAAACC	La 27+0UU

hnRNP, heterogeneous nuclear ribonucleoprotein.

2. At 48 h posttransfection, cells are harvested and mixed with 2 mL RNA STAT-60 (Tel-Test B), and approx 200 μg total RNA were extracted according to the manufacturer's protocol.
3. Electrophorese 20 μg of RNA in a 6% denaturing polyacrylamide gel electrophoresis (PAGE) gel, transfer to a nylon membrane, and probe with 1 pmol of ^{32}P-end-labeled oligos (La, 5′-CCAAAGGTACCCAGCCTTCATCCAGTT-3′; β-actin, 5′-GTGAGGATGCCTCTCTTGCTCTGGGCCTCG-3′). Hybridizations are carried out in 10 mL of hybridization solution for 3 h at 37 °C. The blot is washed in 2X SSPE at 37 °C for 4 min. The washing solution is changed every 4 min and monitored for radioactivity.

3.4.2. Suppression of hnRNP H

It has been shown that hnRNP H is involved in alternative splicing regulation as an RNA binding protein ***(17, 18)***. hnRNP H is a component of splicing enhancer complexes that activate a c-src alternative exon in neuronal cells ***(19)***. It is highly expressed in HEK293 and HeLa cells. Different amounts of 21-mer siRNA or 27-mer disRNA were transfected into HEK293 cells and the level of protein expression was monitored by Western blotting (**Fig. 8**).

1. Plate 5 × 10^5 HEK293 cells in 6-well plates. On the following day, the cells at 30% confluency are transfected as described in **Subheading 3.1.1.** The media is changed the next day.
2. Cells from each well are harvested in 300 μL of PBS after 72 h. Extracts are prepared from one 6-well plate of cells by suspending them in 300 μL of PBS followed by sonication for 10 s and a 2-min microcentrifugation at top speed in an Eppendorf microfuge.

3. For Western blotting, 2µL of each cell extract are loaded in a 10% SDS-PAGE gel. If too much extract is loaded, the sensitivity of the assay is compromised. The hnRNP H is detected using an anti-hnRNP H rabbit antibody *(17)* and antirabbit antibody conjugated with alkaline phosphatase (Sigma). β-Actin is detected by an antiactin antibody (Sigma) and antimouse antibody conjugated with alkaline phosphatase (Sigma).

4. Notes

1. As an alternative to the chemical synthesis of siRNAs, siRNAs transcribed using bacteriophage T7 RNA polymerase can be used. Since the T7 transcribed siRNAs trigger a potent induction of interferon-α and -β in a variety of cell lines, we need to modify the method to minimize nonspecific effect of siRNAs. Analyses of the potential mediators of this response revealed that the initiating 5′ triphosphate is required for interferon induction. A novel method for T7 siRNA synthesis is described in this chapter. siRNAs synthesized by this method alleviate the interferon response while maintaining full efficacy of the siRNAs in minimum of expanse and turnaround time.
2. In attempts to identify RNAi triggers that effectively function at lower concentrations, we found that synthetic RNA duplexes 25 to 30nt in length can be more potent than corresponding conventional 21-mer siRNAs. We have shown that the long dsRNAs can be processed into 21-mer siRNAs by Dicer. Some sites that are refractory to silencing by 21-mer siRNAs can be effectively targeted by 27-mer blunt-ended duplexes and silencing lasts up to 10d *(12)*. Importantly, the 27-mers do not induce interferon or activate PKR *(12)*. These results provide a new strategy for eliciting RNAi-mediated target cleavage using very low RNA concentrations of synthetic RNAs that are substrates for cellular Dicer-mediated cleavage. Based on efficacy tests, disRNAs show variation that depends on the target sequence, so empirical testing of several disRNAs may be required to achieve maximal levels of target inhibition.

References

1. Hannon, G. J. (2002) RNA interference. *Nature* **418**, 244–251.
2. Elbashir, S.M., Harborth, J., Lendeckel, W., Yalcin, A., Weber, K., and Tuschl, T. (2001) Duplexes of 21-nucleotide RNAs mediate RNA interference in cultured mammalian cells. *Nature* **411**, 494–498.
3. Scherer, L. J., and Rossi, J. J. (2003) Approaches for the sequence-specific knockdown of mRNA. *Nat. Biotechnol.* **21**, 1457–1465.
4. Schwarz, D. S., Hutvágner, G., Du, T., Xu, Z., Aronin, N., and Zamore, P. D. (2003) Asymmetry in the assembly of the RNAi enzyme complex. *Cell* **115**, 199–208.
5. Reynolds, A., Leake, D., Boese, Q., Scaringe, S., Marshall, W. S., and Khvorova, A. (2004) Rational siRNA design for RNA interference. *Nat. Biotechnol.* **22**, 326–330.

6. Sohail, M., Doran, G., Riedemann, J., Macaulay, V., and Southern, E. M. (2003) A simple and cost-effective method for producing small interfering RNAs with high efficacy. *Nucleic Acids Res.* **31**, e38.
7. Donze, O., and Picard, D. (2002) RNA interference in mammalian cells using siRNAs synthesized with T7 RNA polymerase. *Nucleic Acids Res.* **30**, e46.
8. Kim, D. H., Longo, M., Han, Y., Lundberg, P., Cantin, E., and Rossi, J. J. (2004) Interferon induction by siRNAs and ssRNAs synthesized by phage polymerase. *Nat. Biotechnol.* **22**, 321–325.
9. Zhang, H., Kolb, F. A., Brondani, V., Billy, E., and Filipowicz, W. (2002) Human Dicer preferentially cleaves dsRNAs at their termini without a requirement for ATP. *EMBO J.* **21**, 5875–5885.
10. Manche, L., Green, S. R., Schmedt, C., and Mathews, M. B. (1992) Interactions between double-stranded RNA regulators and the protein kinase DAI. *Mol. Cell Biol.* **12**, 5238–5248.
11. Jackson, A. L., Bartz, S. R., Schelter, J., et al. (2003) Expression profiling reveals off-target gene regulation by RNAi. *Nat. Biotechnol.* **21**, 635–637.
12. Kim, D. H., Behlke, M. A., Rose, S. D., Chang, M. S., Choi, S., and Rossi, J. J. (2005) Synthetic dsRNA Dicer substrates enhance RNAi potency and efficacy. *Nat. Biotechnol.* **23**, 222–226.
13. Kim, D. H., and Rossi, J. J. (2003) Coupling of RNAi-mediated target downregulation with gene replacement. *Antisense Nucleic Acid Drug Dev.* **13**, 151–155.
14. Scherer, L., and Rossi, J. J. (2004) RNAi applications in mammalian cells. *Biotechniques* **36**, 557–561.
15. Wolin, S. L., and Cedervall, T. (2002) The La protein. *Annu. Rev. Biochem.* **71**, 375–403.
16. Stefano, J. E. (1984) Purified lupus antigen La recognizes an oligouridylate stretch common to the 3′ termini of RNA polymerase III transcripts. *Cell* **36**, 145–154.
17. Markovtsov, V., Nikolic, J. M., Goldman, J. A., Turck, C. W., Chou, M. Y., and Black, D. L. (2000) Cooperative assembly of an hnRNP complex induced by a tissue-specific homolog of polypyrimidine tract binding protein. *Mol. Cell Biol.* **20**, 7463–7479.
18. Chen, C. D., Kobayashi, R., and Helfman, D. M. (1999) Binding of hnRNP H to an exonic splicing silencer is involved in the regulation of alternative splicing of the rat beta-tropomyosin gene. *Genes Dev.* **13**, 593–606.
19. Chou, M. Y., Rooke, N., Turck, C. W., and Black, D. L. (1999) hnRNP H is a component of a splicing enhancer complex that activates a c-src alternative exon in neuronal cells. *Mol. Cell Biol.* **19**, 69–77.

25

The Use of *Saccharomyces cerevisiae* Proteomic Libraries to Identify RNA-Modifying Proteins

Jane E. Jackman, Elizabeth J. Grayhack, and Eric M. Phizicky

Summary

Biochemical assay of proteomic libraries derived from the *Saccharomyces cerevisiae* genome provides a powerful new tool for the assignment of activities to proteins. Particular advantages of this approach include the speed with which a protein can be identified and the generality for any biological activity for which an assay can be developed. We discuss the utility of this approach for the identification of RNA-modifying enzymes using a yeast proteomic library derived from a genomic set of strains expressing GST-ORF fusion proteins. This technique is also broadly applicable to other classes of RNA–protein interactions, including RNA binding and RNA degradation, and can be used with any of the proteomic libraries that are available.

Key Words: Biochemical genomics approach; proteomics; RNA modification; RNA processing; *Saccharomyces cerevisiae*.

1. Introduction

The recent explosion in genomic sequencing projects has resulted in the creation of the new field of proteomics, one goal of which is to classify the biochemical activities of proteins on a genome-wide scale. Thus, the entire set of proteins encoded by the genome of an organism can be queried in a single experiment to identify a protein associated with a given activity. The advantages of a proteomic approach for assignment of protein function include its speed, generality, comprehensiveness, and sensitivity. Once a particular activity has been shown to exist in an organism or extract and an assay has been established, the protein responsible can often be identified in a matter of days. Moreover, this approach inherently broadens the scope of proteins that can be identified since all proteins are in principle examined without bias. It also does not depend on any prior knowledge of the protein responsible for an activity or of its catalytic

From: *Methods in Molecular Biology, vol. 488: RNA-Protein Interaction Protocols*
Edited by: Ren-Jang Lin

motifs or domains, as determined by bioinformatic analysis. Therefore, this approach is well suited to the identification of novel proteins. Finally, the method has very high sensitivity, particularly for enzymatic activities; thus, one can obtain numerous proteins with particular biochemical activities in a single experiment.

A large number of proteins interact with RNA, including those that participate in transcription, translation, RNA processing, assembly or maintenance of ribonucleoproteins, and RNA trafficking. Proteins that interact with RNA can function as chaperones or helicases; as degradation, processing, editing, or modifying enzymes; or as various structural components of ribonucleoprotein complexes.

The biochemical genomics approach involves the preparation of purified proteins from pools and subpools of strains derived from proteomic expression libraries, and the subsequent biochemical assay of the resulting pools of purified proteins, to identify previously unknown proteins ***(1,2)***. This approach has been applied to several proteomic expression libraries derived from the yeast *Saccharomyces cerevisiae* ***(3,4)*** and has resulted in the identification of a number of proteins that act on RNA, several of which have potentially novel catalytic or binding motifs. Some of these include a two-component methyltransferase required for formation of m^7G_{46} of transfer RNA (tRNA) ***(5)***; a tRNA m^1G_9 methyltransferase that is phylogenetically distinct from other classes of methyltransferases ***(6)***; a family of four dihydrouridine synthases with specificity for different sites ***(7,8)***; a highly unusual guanylyltransferase responsible for addition of a GMP (guanosine monophosphate) residue to the 5′ end of $tRNA^{His}$ ***(9)***; a pseudouridylase responsible for modification of U2 RNA ***(10)***; a cyclic phosphodiesterase involved in the metabolism of the nucleotide byproduct of tRNA splicing ***(4)***; and a 2′-*O*-methyltransferase that acts at position 4 of a diverse set of tRNA species ***(11)***. The biochemical genomics approach is generally applicable to any catalytic activity acting on RNA and can be easily extended to RNA binding reactions by using the appropriate assay. We illustrate the utility of this technique by describing the example of Trm5, a tRNA m^1G_{37} methyltransferase previously identified by other methods ***(12)***.

2. Materials

1. Radiolabeled tRNA substrate (produced by site-specific labeling techniques ***(13)*** or by standard in vitro transcription with T7 RNA polymerase).
2. Enzymes for production of site-specifically labeled RNA: ribonuclease (RNase) H, alkaline phosphatase, T4 polynucleotide kinase, and their respective reaction buffers (Roche).
3. [γ-^{32}P] adenosine 5′-triphosphate (ATP): Labeling grade, 7000 Ci/mmol (MPX Biomedicals).
4. Urea-polyacrylamide gel equipment for purification of RNA substrates.

5. Methyltransferase reaction buffer: 100 m*M* Tris-HCl, pH 8.0, 5 m*M* $MgCl_2$, 2 m*M* dithiothreitol (DTT), 100 m*M* ammonium acetate, 0.1 m*M* ethylenediaminetet raacetic acid (EDTA), and 1 m*M* *S*-adenosylmethionine.
6. 1.0 *M* Tris-HCl, pH 8.0, for termination of methyltransferase reactions.
7. Ribonuclease T2 (RNase T2) digestion buffer: 20 m*M* sodium acetate, pH 5.2, 1 m*M* EDTA.
8. *S*-Adenosylmethionine (Sigma): 20 m*M* stock solution made in 1 m*M* HCl, stored at −80 °C, and added as final component to methyltransferase reaction buffer.
9. Bulk yeast RNA: 10 mg/mL stock solution in water.
10. Phenol saturated with 0.5 *M* Tris-HCl, pH 7.5: The preparation of buffer-saturated phenol can take several days and is as follows: A fresh bottle of phenol is thawed at room temperature and approximately 0.5 volumes of 0.5 *M* Tris-HCl, pH 7.5, is added to it and mixed thoroughly by inversion. The phases are allowed to separate, the aqueous phase is removed, and a fresh layer of 0.5 *M* Tris-HCl, pH 7.5, is added. This is repeated until the pH of the aqueous phase is around 7.5 (usually two or three times), and then a small amount of 8-hydroxyquinoline is added as a preservative. The resulting phenol is stored at 4 °C protected from light for up to 6 mo.
11. Phenol/chloroform/isoamyl alcohol (PCA) solution: 25:24:1 v/v solution made using buffer-saturated phenol (*see* **step 10**) and commercially available chloroform and isoamyl alcohol with no further purification. Then 0.5 *M* Tris-HCl, pH 7.5 (1/10 of the final volume) is added to the final solution to maintain proper pH, and the resulting solution is stored at 4 °C protected from light for up to 6 mo.
12. RNase T2 (Invitrogen): 1 U/μL stock solution made in 50 m*M* sodium acetate, pH 5.2, and stored at −80 °C. The solution is stable for several years; however, small aliquots are used to avoid repeated freeze-thawing.
13. Thin-layer chromatographic (TLC) tank and plastic-backed polyethyleneimine (PEI)-F cellulose and cellulose TLC plates (EM Science).
14. Chromatography solvent: Isobutyric acid/water/ammonium hydroxide (66:33:1 v/v/v).
15. PhosphorImager and phosphor screens (Molecular Dynamics).
16. *Saccharomyces cerevisiae* GST-ORF library strain collection or equivalent genomic library of strains ***(3)*** (*see* also **Note 5**).
17. Affinity-purified preparations of pools (64 each) of 96 GST-ORF strains. Detailed methods for the preparation of the protein pools are provided elsewhere ***(4,14)***. Briefly, the library strains are pooled according to their location on the 96-well plates such that each pool of cells contains all of the library strains from a given plate. The pool strains are grown, the expression of proteins is induced (in the case of the GST-ORF library, by the addition of copper sulfate), the cells are harvested after induction, and the resulting proteins from each pool of strains are purified. Purification of the proteins from the GST-ORF library is performed on glutathione-agarose affinity resin.

3. Methods

The procedures that follow outline (1) the development of an assay to detect the interaction with the targeted RNA, (2) the biochemical assay of the proteomic expression library and subsequent identification of a positive signal, and (3) assignment of that signal to a single open reading frame (ORF) by deconvolution and confirmation of the identity of the ORF by standard biochemical techniques.

3.1. Development of m^1G_{37} Methyltransferase Activity Assay

3.1.1. Preparation of Radiolabeled $tRNA^{Leu}$ Substrate

The main concern when designing an assay to use with a proteomic expression library is the sensitivity of the detection system; the signal produced from the desired activity should be strong enough that it can be detected when the desired protein is present as only one component of a mixture of purified proteins. Fortunately, for analysis of protein interactions with nucleic acids, this is fairly straightforward given the many readily accessible and well-known techniques of labeling the RNA at the position of interest with ^{32}P. Thus, when looking for an activity that modifies a G residue in tRNA, tRNA could be simply transcribed with T7 RNA polymerase in the presence of [α-^{32}P]-GTP, resulting in the incorporation of ^{32}P throughout the backbone of the tRNA substrate. Likewise, another labeled nucleotide could be used in transcription for detection of modifications of other residues. An alternative, and sometimes advantageous, approach (*see* **Note 1**) is to use a site-specifically labeled substrate, such as the substrate used for the assay described here, which consists of $tRNA^{Leu}$ specifically labeled with ^{32}P at the phosphate 3′ of the G residue at position 37 (the position known to be methylated to m^1G_{37} in vivo; ***15***). This substrate is referred to as $G37^*Leu$ (**Fig. 1**). Site-specific labeling of RNA is accomplished by (1) cleavage of the RNA substrate with RNase H directed at the appropriate position by annealing with a chimeric 2′-*O*-methylated/deoxyoligonucleotide, (2) postcleavage labeling of the resulting 3′ half-molecule with [γ-^{32}P] ATP, and (3) religation of the RNA pieces to yield $G37^*Leu$ tRNA (for specific details, *see* **refs. *6*** and ***13***).

3.1.2. m^1G_{37} Methyltransferase Activity Assay

Methyltransferase reactions are first performed with yeast crude extract as a source of Trm5 methyltransferase activity to determine the limits of detection of activity using this assay. After incubation of $G37^*Leu$ tRNA with the crude extract in the presence of SAM, the RNA is digested to 3′-phosphorylated nucleotides with ribonuclease T2. The resulting Gp (from unreacted substrate) and m^1Gp (from methylated product) are resolved by TLC (**Fig. 1**; *see* **Note 2**).

1. Decreasing concentrations of yeast crude extract (typically approximately 1 mg/mL to 10 μg/mL total protein, by factors of 4–5) are incubated with 10,000 cpm of $G37^*Leu$ tRNA in methyltransferase reaction buffer at 30 °C for 4–16 h. Reaction

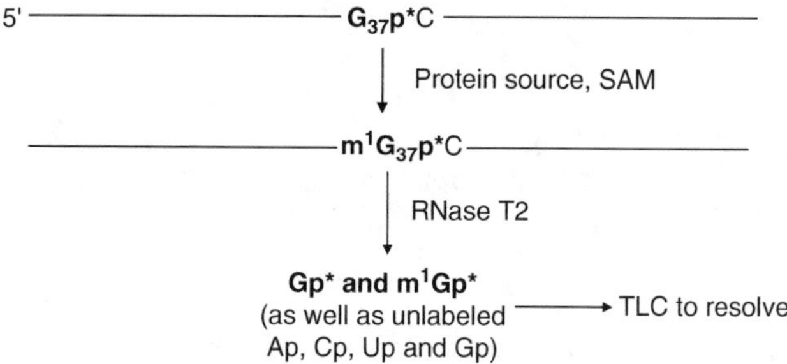

Fig. 1. Schematic representation of m^1G_{37} methyltransferase assay with G37*Leu transfer RNA (tRNA) substrate. The m^1G_{37} methyltransferase activity can be detected by using a substrate tRNA with a radiolabeled phosphate 3´ of G37 (*see* **Subheading 3.1.2.**). The ribonuclease used for this assay is ribonuclease (RNase) T2, which yields 3´-phosphorylated nucleotides, but different RNases may be used for other labeled RNA molecules, depending on the position of the radiolabeled phosphate (*see* **Note 2**). The chromatography system used to resolve m^1Gp from Gp is an isobutyric acid/water/ammonium hydroxide system (66:33:1 v/v/v).

mixtures containing all components except crude extract are assembled on ice in a volume of 8 µL, and reactions are started by the addition of 2 µL of each crude extract dilution.

2. Add to each reaction: 50 µL of 1.0 *M* Tris-HCl, pH 8.0, 38 µL of water, and 2 µL of 10 mg/mL bulk yeast RNA (*see* **Note 3**).
3. Add 100 µL of 0.5 *M* Tris-HCl-saturated (pH 7.5) phenol/chloroform/isoamyl alcohol (25:24:1) and vortex. Spin in microcentrifuge at 13,000 rpm for 5 min at room temperature and remove aqueous layer to new microfuge tube.
4. Add 200 µL of 100% ethanol to aqueous layer, mix, and incubate at –20 °C for at least 30 min. Spin in microcentrifuge at 13,000 rpm for 20 min at 4 °C and remove supernatant, being careful not to disturb the RNA pellet.
5. Add 100 µL 70% ethanol, mix, and spin in microcentrifuge at 13,000 rpm for 5 min at 4 °C. Remove supernatant, again being careful not to disturb the RNA pellet, and air dry.
6. Resuspend dried pellet in 4 µL of RNase T2 digestion buffer and 2 U of RNase T2. Incubate at 50 °C for 30 min to 1 h.
7. Spot 2 µL of the digested reaction to flexible plastic-backed cellulose TLC plates and allow to air dry.
8. Thin-layer plates are developed (for at least 12 h) by chromatography in a solvent system of isobutyric acid/water/ammonium hydroxide (66:33:1 v/v/v).
9. After chromatography, the TLC plates are air dried, exposed to phosphor screens, and visualized by PhosphorImager (**Fig. 2**).

Trm5 methyltransferase activity in this assay is visible in yeast crude extract even at a total protein concentration of approximately 15 µg/mL (**Fig. 2**, lane 5).

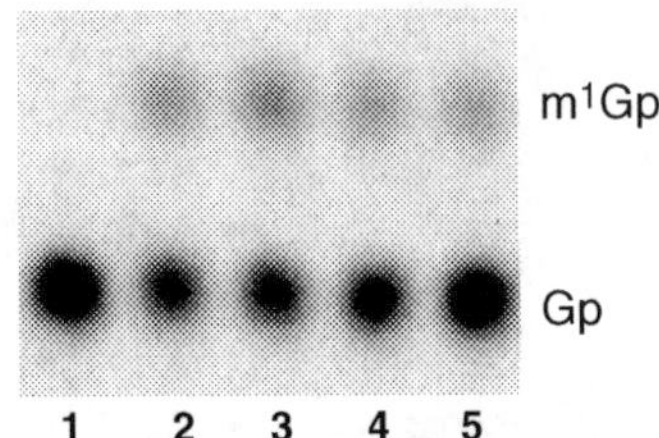

Fig. 2. Assay for m^1G_{37} methyltransferase activity with G37*Leu transfer RNA (tRNA). The m^1G_{37} methyltransferase activity was performed as described (*see* **Subheading 3.1.2.**) using decreasing concentrations of yeast crude extract as the source of m^1G_{37} methyltransferase activity. Lane 1, no crude extract; lane 2, 1 mg/mL total protein; lane 3, 0.25 mg/mL total protein; lane 4, 60 µg/mL total protein; lane 5, 15 µg/mL total protein.

This is typical of the level of activity visible in crude extract for several of the RNA-modifying enzymes that have been found using proteomic expression libraries and is sufficient for detection of Trm5 as well, as demonstrated below (*see* **Note 4**).

3.2. Assay of the Yeast Proteomic Expression Library

To identify an activity using the biochemical genomics approach, individual yeast strains are combined into multiple pools of 96 proteins each (derived from 96-well microtiter plates), which are grown and purified together ***(3,4)***. The resulting pools contain 96 different proteins with a final concentration of about 1–2 µg/mL per individual protein (**Fig. 3**). The GST-ORF library used for the test case described here consists of a set of approximately 6400 yeast strains each expressing a single N-terminally GST-tagged yeast open reading frame and therefore is assayed in a set of 64 pools (*see* **Note 5**). To determine which pool contains the protein of interest, in this case the tRNA m^1G_{37} methyltransferase Trm5, G37*Leu tRNA is assayed exactly as described in **Subheading 3.1.2., steps 1–9**), except with 2 µL of each pool of the GST-ORF library replacing the yeast crude extract as the source of enzymatic activity, resulting in an initial assay consisting of 64 total tubes (**Fig. 4A**). The known tRNA m^1G_{37} methyltransferase, Trm5 (YHR070w), is located on plate 14 of the GST-ORF library. As expected, this assay yields a positive response for m^1G_{37} methyltransferase activity in pool 14 (**Fig. 4A**); thus, one experiment narrows down the possible candidates for the m^1G_{37} methyltransferase, or a protein that copurifies with the methyltransferase, to 96 proteins of the initial 6400. Similar approaches to identify novel protein activities have been used with other proteomic libraries derived from *S. cerevisiae* ***(11,16)*** (*see* **Note 5**), and the biochemical genomics

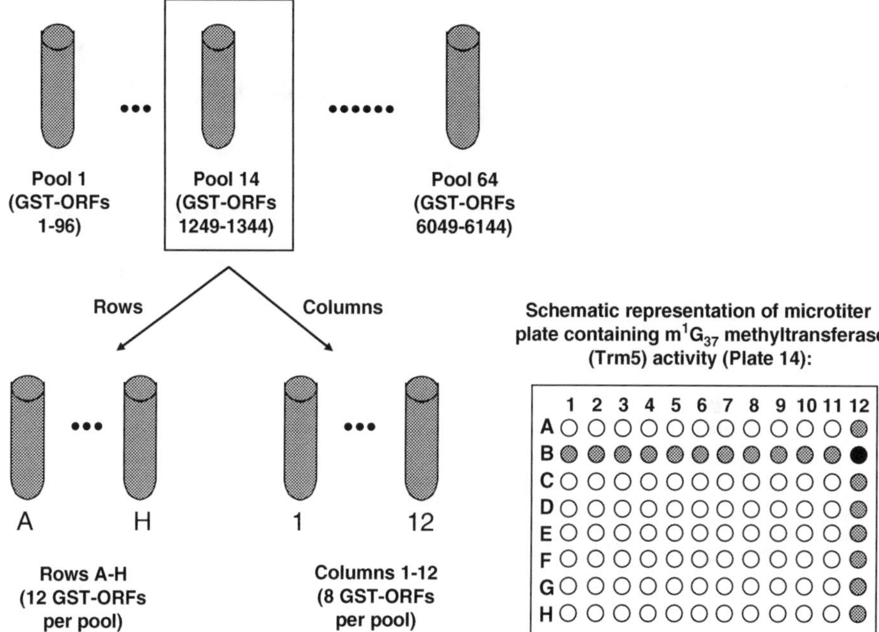

Fig. 3. Overview of the biochemical genomics approach. The GST-ORF fusion protein library consists of 6144 strains of *Saccharomyces cerevisiae*, each of which bears a plasmid that can be induced to express the protein product of one open reading frame (ORF). These strains are stored on 64 microtiter plates and grown in pools. The overexpressed proteins can be purified as 64 separate pools of 96 proteins each and each pool assayed for a desired activity. Once that activity is found, subpools of the rows and columns from the original plate can be constructed and assayed to pinpoint the ORF responsible for the activity, as shown here for Trm5 on plate 14.

approach can also be easily applied to proteomic libraries from other organisms *(17–19)*.

3.3. Assignment of the Positive Signal to Trm5 and Verification of the Identity of the ORF

Once the positive signal is observed in a given protein pool, the individual yeast strain containing the ORF with the desired activity can be readily identified. In this case, the 96-well plate of strains comprising pool 14 was regrouped into intersecting sets of strains representing the rows (A–H) or columns (1–12) of the plate, and proteins from these subpools were purified in the same way as the original library (**Fig. 3**). In this experiment, the number of proteins in an individual pool has been decreased to 8 (columns) or 12 (rows).

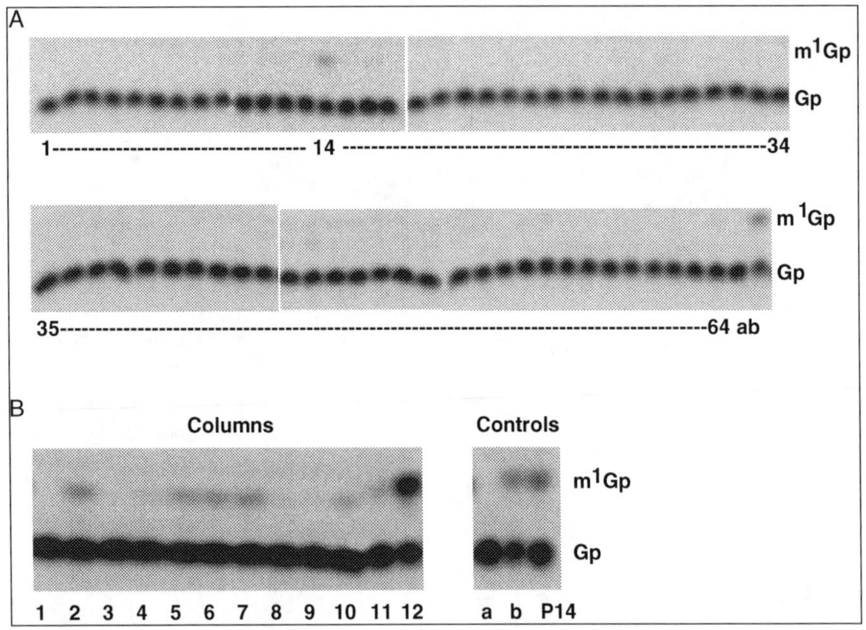

Fig. 4. Assay of the GST-ORF library to identify Trm5. (**A**) Assay of the 64 initial pools with G37*Leu transfer RNA (tRNA). Lanes 1–64, each of the 64 pools of proteins (approximately 96 proteins per pool) assayed as described in **Subheading 3.1.2.**, with the positive m^1Gp signal observed in pool 14. Lane a, buffer-only control; lane b, yeast extract positive control (1 mg/mL total protein). (**B**) Deconvolution of the positive signal to identify the individual open reading frame (ORF) by assay of the individual columns (lanes 1–12) of plate 14 with G37*Leu tRNA. Lane a, buffer-only control; lane b, yeast crude extract control (1 mg/mL total protein); lane P14, pool 14 control.

Again, G37*Leu tRNA is assayed using the standard methyltransferase protocol. The strongest signals are found in row B (data not shown) and column 12 (**Fig. 4B**). The intersection of these two pools should be the position on the 96-well plate of the individual ORF expressing m^1G_{37} methyltransferase activity or of an ORF that copurifies with the protein that expresses this activity (**Fig. 3**). In this case, the ORF located in plate 14 at position B12 is Trm5, the yeast protein that has been previously demonstrated by other methods to catalyze the methylation of G_{37} to m^1G in tRNA *(12)*.

Once the identity of an individual protein is known, standard biochemical techniques can be used to prove that this is the ORF responsible for the desired activity and to rule out a requirement for additional yeast proteins for the desired activity. These include purification of the individual protein to prove that it is associated with activity; cloning of the identified ORF into a vector for purification from *Escherichia coli* to determine if the protein alone is sufficient for catalytic activity; and analysis of the corresponding yeast deletion strain to determine if the protein is responsible for the activity in vivo.

Notably, neither Trm5 nor another tRNA methyltransferase, Trm10, ***(6)*** contains any previously known SAM-binding motifs; in fact, neither of these proteins was identified in a search of *S. cerevisiae* proteins for putative methyltransferases ***(20)***. In the case of Trm5, the protein was only identified subsequent to the identification of another distantly related homolog of the *E. coli* protein known to catalyze m^1G_{37} modification ***(12)***. Furthermore, in the case of Trm10, the protein is a member of a previously undescribed family of proteins that also bears no homology to any other known methyltransferases, and therefore it is highly unlikely that this protein could have been found by any method that relies on sequence comparison.

4. Notes

1. Use of a uniformly labeled RNA, such as would result from in vitro transcription in the presence of $\alpha[^{32}P]$-NTPs, is simple and can be a good alternative to creating more complicated site-specifically labeled substrates. However, use of a uniquely labeled RNA molecule as the substrate for assay of the proteomic expression library improves the sensitivity of the assay 15- to 25-fold when looking for an enzyme that modifies a known position. This increase in sensitivity can be crucial when looking for a single protein within a pool of purified proteins, such as in the first step of the biochemical genomics approach by which the entire expression library is assayed (*see* **Subheading 3.2.**). Nonetheless, there may be cases when observation of a specific RNA residue is not the goal of the assay, particularly when looking for general RNA binding activities; in these cases, uniformly labeled RNA substrates may be preferred.
2. The assay procedure described here could, in principle, be used for any nucleotide modification of any RNA substrate. For uniformly labeled RNA, digestion of the substrate after reaction with library proteins would be performed with P1 nuclease since the radiolabeled phosphate would be 5 of the nucleotide of interest. Also, the TLC system to be used in the assay may be different for different modified nucleotides to be able to resolve the original unmodified nucleotide from its modified form. Other TLC systems that have been used successfully to visualize modified nucleotides include cellulose TLC systems resolved in one or two dimensions ***(21)*** or systems using PEI-cellulose plates resolved in formate buffer or lithium chloride ***(22,23)***.
3. The inclusion of bulk yeast RNA makes the pellet from the ethanol precipitation step readily visible and has the additional advantage of permitting visualization of the digestion products by fluorescence quenching after chromatography when the TLC plates are fluorescently labeled.
4. If the activity observed in the titration of crude extract is significantly weaker than that shown here, it may still be possible to observe the activity in the proteomic

expression library. Purification of the library may remove contaminating proteins that compete with the desired protein in a crude extract; this may be particularly important when looking for RNA binding activities for which nonspecific interactions with other proteins in a crude extract may mask the desired activity. However, if on assay of the proteomic expression library a signal is not detected, it may be necessary to optimize the assay by varying experimental conditions such as time, temperature, or buffer conditions.

5. The proteomic expression library used for the example shown here is the yeast GST-ORF library developed by Martzen et al. ***(4)***, and the proteins represented in it are derived from the yeast genome annotation at about 1996. In recent years, the annotation of the genome has been updated to add some new ORFs as well as remove some sequences that had been incorrectly described. Therefore, although this earlier yeast expression library has been successfully used to identify at least 10 previously unknown RNA interacting proteins, this collection does not contain all known yeast proteins as of 2007. A more recently described *S. cerevisiae* proteomic expression library, the MORF (Moveable Open Reading Frame) library, has several advantages over the earlier GST-ORF library in that it is derived from the most recently annotated version of the yeast genome, it allows for tight control of protein expression under a galactose-inducible promoter, and it is extensively sequence verified. The ORFs in the MORF library contain a tripartite tag at the C-terminus that allows for purification by either immunoglobulin G (IgG)-Sepharose affinity columns or immobilized metal-ion affinity chromatography, as well as for antibody detection through an HA epitope and cleavage of part of the tag at a 3C protease site ***(3)***.

Acknowledgment

This work was supported by National Institutes of Health grant GM52347 to E.M.P.

References

1. Grayhack, E. J., and Phizicky, E. M. (2001) Genomic analysis of biochemical function. *Curr. Opin. Chem. Biol.* **5**, 34–9.
2. Phizicky, E. M., and Grayhack, E. J. (2006) Proteome-scale analysis of biochemical activity. *Crit. Rev. Biochem. Mol. Biol.* **41**, 315–327.
3. Gelperin, D. M., White, M. A., Wilkinson, M. L., et al. (2005) Biochemical and genetic analysis of the yeast proteome with a movable ORF collection. *Genes Dev.* **19**, 2816–2826.
4. Martzen, M. R., McCraith, S. M., Spinelli, S. L., et al. (1999) A biochemical genomics approach for identifying genes by the activity of their products. *Science* **286**, 1153–1155.
5. Alexandrov, A., Martzen, M. R., and Phizicky, E. M. (2002) Two proteins that form a complex are required for 7-methylguanosine modification of yeast tRNA. *RNA* **8**, 1253–1266.

6. Jackman, J. E., Montange, R. K., Malik, H. S., and Phizicky, E. M. (2003) Identification of the yeast gene encoding the tRNA m1G methyltransferase responsible for modification at position 9. *RNA* **9**, 574–585.
7. Xing, F., Martzen, M. R., and Phizicky, E. M. (2002) A conserved family of Saccharomyces cerevisiae synthases effects dihydrouridine modification of tRNA. *RNA* **8**, 370–381.
8. Xing, F., Hiley, S. L., Hughes, T. R., and Phizicky, E. M. (2004) The specificities of four yeast dihydrouridine synthases for cytoplasmic tRNAs. *J. Biol. Chem.* **279**, 17850–17860.
9. Gu, W., Jackman, J. E., Lohan, A. J., Gray, M. W., and Phizicky, E. M. (2003) tRNAHis maturation: an essential yeast protein catalyzes addition of a guanine nucleotide to the 5′ end of tRNAHis. *Genes Dev.* **17**, 2889–2901.
10. Ma, X., Zhao, X., and Yu, Y. T. (2003) Pseudouridylation (Psi) of U2 snRNA in S. cerevisiae is catalyzed by an RNA-independent mechanism. *EMBO J.* **22**, 1889–1897.
11. Wilkinson, M., Crary, S., Jackman, J. E., Grayhack, E., and Phizicky, E. (2007) The 2′-*O*-methyltransferase responsible for modification of yeast tRNA at position 4. *RNA* **13**, 404–413.
12. Bjork, G. R., Jacobsson, K., Nilsson, K., Johansson, M. J., Bystrom, A. S., and Persson, O. P. (2001) A primordial tRNA modification required for the evolution of life? *EMBO J.* **20**, 231–239.
13. Yu, Y. T. (1999) Construction of 4-thiouridine site-specifically substituted RNAs for cross-linking studies. *Methods* **18**, 13–21.
14. Phizicky, E. M., Martzen, M. R., McCraith, S. M., et al. (2002) Biochemical genomics approach to map activities to genes. *Methods Enzymol.* **350**, 546–559.
15. Sprinzl, M., Horn, C., Brown, M., Ioudovitch, A., and Steinberg, S. (1998) Compilation of tRNA sequences and sequences of tRNA genes. *Nucleic Acids Res.* **26**, 148–153.
16. Zhu, H., Bilgin, M., Bangham, R., et al. (2001) Global analysis of protein activities using proteome chips. *Science* **293**, 2101–2105.
17. Dricot, A., Rual, J. F., Lamesch, P., et al. (2004) Generation of the Brucella melitensis ORFeome version 1.1. *Genome Res.* **14**, 2201–2206.
18. Wei, C., Lamesch, P., Arumugam, M., et al. (2005) Closing in on the C. elegans ORFeome by cloning TWINSCAN predictions. *Genome Res.* **15**, 577–582.
19. Lamesch, P., Li, N., Milstein, S., et al. (2007) hORFeome v3.1: a resource of human open reading frames representing over 10,000 human genes. *Genomics* **89**, 307–315.
20. Niewmierzycka, A., and Clarke, S. (1999) S-Adenosylmethionine-dependent methylation in Saccharomyces cerevisiae. Identification of a novel protein arginine methyltransferase. *J. Biol. Chem.* **274**, 814–824.
21. Nishimura, S., and Kuchino, Y. (1983) *Methods of DNA and RNA Sequencing*, Praeger, New York.
22. Bochner, B. R., and Ames, B. N. (1982) Complete analysis of cellular nucleotides by two-dimensional thin layer chromotography. *J. Biol. Chem.* **257**, 9759–9769.
23. Randerath, K., Reddy, M. V., and Gupta, R. C. (1981) 32P-labelling test for DNA damage. *Proc. Natl. Acad. Sci. U. S. A.* **78**, 6126–6129.

Index

L

M

N

P

S

Y